National Fuel Gas Code Handbook

FOURTH EDITION

Edited by

Theodore C. Lemoff, P.E.
Principal Gases Engineer

With the complete text of the 1999 edition of NFPA 54,
National Fuel Gas Code

National Fire Protection Association
Quincy, Massachusetts

Product Manager: Pam Powell
Project Editor: Robin MacFarlane
Copy Editor: Marilyn Morrison
Text Processing: Beth Turner
Composition: Argosy
Art Coordinator: Cheryl Langway and Nancy Maria
Illustrations: Todd K. Bowman and George Nichols
Cover Design: Groppi Advertising Design
Manufacturing Buyer: Ellen Glisker
Printer: R.R. Donnelley

Notice Concerning Liability: Publication of this handbook is for the purpose of circulating information and opinion among those concerned for fire and electrical safety and related subjects. While every effort has been made to achieve a work of high quality, neither the NFPA nor the contributors to this handbook guarantee the accuracy or completeness of or assume any liability in connection with the information and opinions contained in this handbook. The NFPA and the contributors shall in no event be liable for any personal injury, property, or other damages of any nature whatsoever, whether special, indirect, consequential, or compensatory, directly or indirectly resulting from the publication, use of, or reliance upon this handbook.

This handbook is published with the understanding that the NFPA and the contributors to this handbook are supplying information and opinion but are not attempting to render engineering or other professional services. If such services are required, the assistance of an appropriate professional should be sought.

Notice Concerning Code Interpretations: This fourth edition of the *National Fuel Gas Code Handbook* is based on the 1999 edition of NFPA 54, *National Fuel Gas Code*. All NFPA codes, standards, and recommended practices, and guides are developed in accordance with the published procedures of the NFPA technical committees comprised of volunteers drawn from a broad array of relevant interests. The handbook contains the complete text of NFPA 54 and any applicable Formal Interpretations issued by the Association. These documents are accompanied by explanatory commentary and other supplementary materials.

The commentary and supplementary materials in this handbook are not a part of the Code and do not constitute Formal Interpretations of the NFPA (which can be obtained only through requests processed through the responsible technical committee in accordance with the published procedures of the NFPA). The commentary and supplementary materials, therefore, solely reflect the personal opinions of the editor or other contributors and do not necessarily represent the official position of the NFPA or its technical committees.

NFPA No.: F9-54HB99
ISBN: 0-87765-446-8
Library of Congress Card Catalog No.: 99-075206

Printed in the United States of America
04   03   02   01   00   5   4   3   2

This handbook is dedicated to the memory of my parents,
Jacob and Anne Lemoff. Their guidance led me to the path of
education and the field of engineering and eventually to NFPA.

Contents

Acknowledgments   x

History of the *National Fuel Gas Code*   xi

PART ONE   NFPA 54 *NATIONAL FUEL GAS CODE*, 1999 EDITION, WITH COMMENTARY   1

1   GENERAL   3

1.1   Scope   3
1.2   Alternate Materials, Equipment, and Procedures   11
1.3   Retroactivity   11
1.4   Qualified Agency   11
1.5   Interruption of Service   12
1.6   Prevention of Accidental Ignition   13
1.7   Definitions   14

2   GAS PIPING SYSTEM DESIGN, MATERIALS, AND COMPONENTS   39

2.1   Piping Plan   40
2.2   Provision for Location of Point of Delivery   40
2.3   Interconnections Between Gas Piping Systems   41
2.4   Sizing of Gas Piping Systems   41
2.5   Piping System Operating Pressure Limitations   44
2.6   Acceptable Piping Materials and Joining Methods   47
2.7   Gas Meters   56
2.8   Gas Pressure Regulators   58
2.9   Overpressure Protection Devices   61
2.10   Back Pressure Protection   65
2.11   Low-Pressure Protection   66
2.12   Shutoff Valves   66
2.13   Expansion and Flexibility   67

3   GAS PIPING INSTALLATION   69

3.1   Piping Underground   70
3.2   Aboveground Piping Outside   78

3.3   Piping in Buildings   78
3.4   Concealed Piping in Buildings   81
3.5   Piping in Vertical Chases   83
3.6   Gas Pipe Turns   84
3.7   Drips and Sediment Traps   85
3.8   Outlets   87
3.9   Branch Pipe Connection   89
3.10   Manual Gas Shutoff Valves   89
3.11   Prohibited Devices   90
3.12   Systems Containing Gas–Air Mixtures Outside the
Flammable Range   90
3.13   Systems Containing Flammable Gas–Air Mixtures   91
3.14   Electrical Bonding and Grounding   93
3.15   Electrical Circuits   94
3.16   Electrical Connections   94

4   INSPECTION, TESTING, AND PURGING   97
4.1   Pressure Testing and Inspection   97
4.2   System and Equipment Leakage Test   101
4.3   Purging   103

5   EQUIPMENT INSTALLATION   107
5.1   General   108
5.2   Accessibility and Clearance   123
5.3   Air for Combustion and Ventilation   124
5.4   Equipment on Roofs   144
5.5   Equipment Connections to Building Piping   147
5.6   Electrical   152
5.7   Room Temperature Thermostats   153

6   INSTALLATION OF SPECIFIC EQUIPMENT   157
6.1   General   158
6.2   Air-Conditioning Equipment
(Gas-Fired Air Conditioners and Heat Pumps)   160
6.3   Central Heating Boilers and Furnaces   167
6.4   Clothes Dryers   183
6.5   Conversion Burners   187
6.6   Decorative Appliances for Installation in Vented Fireplaces   187
6.7   Gas Fireplaces, Vented   190
6.8   Direct Make-up Air Heaters   192
6.9   Direct Gas-Fired Industrial Air Heaters   194
6.10   Duct Furnaces   196
6.11   Floor Furnaces   198
6.12   Food Service Equipment, Floor-Mounted   201

6.13 Food Service Equipment Counter Appliances 207
6.14 Hot Plates and Laundry Stoves 207
6.15 Household Cooking Appliances 208
6.16 Illuminating Appliances 211
6.17 Incinerators, Commercial-Industrial 213
6.18 Incinerators, Domestic 213
6.19 Infrared Heaters 215
6.20 Open-Top Broiler Units 217
6.21 Outdoor Cooking Appliances 218
6.22 Pool Heaters 219
6.23 Refrigerators 220
6.24 Room Heaters 221
6.25 Stationary Gas Engines 224
6.26 Gas-Fired Toilets 224
6.27 Unit Heaters 225
6.28 Wall Furnaces 228
6.29 Water Heaters 232
6.30 Compressed Natural Gas (CNG) Vehicular Fuel Systems 237
6.31 Appliances for Installation in Manufactured Housing 238

7 VENTING OF EQUIPMENT 241
7.1 General 242
7.2 Specification for Venting 244
7.3 Design and Construction 248
7.4 Type of Venting System to Be Used 253
7.5 Masonry, Metal, and Factory-Built Chimneys 256
7.6 Gas Vents 267
7.7 Single-Wall Metal Pipe 278
7.8 Through the Wall Vent Termination 282
7.9 Condensation Drain 285
7.10 Vent Connectors for Category I Gas Utilization Equipment 285
7.11 Vent Connectors for Category II, Category III, and
      Category IV Gas Utilization Equipment 292
7.12 Draft Hoods and Draft Controls 292
7.13 Manually Operated Dampers 294
7.14 Automatically Operated Vent Dampers 294
7.15 Obstructions 295

**8 PROCEDURES TO BE FOLLOWED TO PLACE EQUIPMENT
   IN OPERATION 297**
8.1 Adjusting the Burner Input 297
8.2 Primary Air Adjustment 301
8.3 Safety Shutoff Devices 302
8.4 Automatic Ignition 302

8.5   Protective Devices   303
8.6   Checking the Draft   303
8.7   Operating Instructions   303

9   SIZING TABLES   305
9.1   Tables for Sizing Gas Piping Systems   305

10   SIZING OF CATEGORY I VENTING SYSTEMS   339
10.1   Additional Requirements to Single Appliance Vent Tables
10.1 Through 10.5   341
10.2   Additional Requirements to Multiple Appliance Vent Tables
10.6 Through 10.13(a) and (b)   347

11   REFERENCED PUBLICATIONS   385

APPENDIXES   389

A   Explanatory Material   389

B   Coordination of Gas Utilization Equipment Design, Construction, and Maintenance   391

C   Sizing and Capacities of Gas Piping   397

D   Suggested Method for Checking for Leakage   411

E   Suggested Emergency Procedure for Gas Leaks   415

F   Flow of Gas Through Fixed Orifices   417

G   Sizing of Venting Systems Serving Appliances Equipped with Draft Hoods, Category I Appliances, and Appliances Listed for Use with Type B Vents   429

H   Recommended Procedure for Safety Inspection of an Existing Appliance Installation   457

I   Recommended Procedure for Installing Electrically Operated Automatic Vent Damper Devices in Existing Vents   459

J   Recommended Procedure for Installing Mechanically Actuated Automatic Vent Damper Devices in Existing Vents   463

K   Recommended Procedure for Installing Thermally Actuated Automatic Vent Damper Devices in Existing Vents   467

L   Example of Air Opening Design for Combustion and Ventilation   469

M   Referenced Publications   473

PART TWO   SUPPLEMENTS   481

1   Development of Revised Venting Guidelines   483

2   Corrugated Stainless Steel Tubing Gas Piping Systems   515

3   Technical Background for Residential Carbon Monoxide Responders   525

4   Revision to American National Standard for Gas Water Heaters — Volume 1 — Storage Water Heaters with Input Ratings of 75,000 Btu per Hour or Less   547

5   Procedure to Estimate Infiltration Rate for Residential Structures   559

Index   567

Acknowledgments

The contributions of all those who participated in the development of this fourth edition of the handbook are very much appreciated. First and foremost, I must express my gratitude to the National Fuel Gas Code Committee, the authors of the code. The committee's attention to technical detail and hard work within the consensus code-making process make NFPA 54 a document that commands respect.

Two contributing authors, Robert A. Borgeson and Bradford Wong, assisted the editor in revising commentary and artwork for this edition.

Bob Borgeson is a member of the gas research community. For 11 years, he has conducted research for the Research Department of the American Gas Association (now known as AGA Research). He specializes in venting of gas appliances, application of vent-free appliances, appliance efficiency, flammable vapor issues, and combustion. Now an independent consultant to the natural gas industry, Bob is also executive director of the American Society of Gas Engineers.

Brad Wong has extensive experience as a code user. He is service coordinator for Commonwealth Gas Company, which serves 240,000 natural gas customers in Massachusetts. He chairs the customer service group and instructs for the Gas Operation School of the New England Gas Association. Brad chairs the Massachusetts Code Update and Review Committee, is secretary of the Central Massachusetts Plumbing and Gas Inspectors Association, and is a member of the New England Plumbing, Gas, and Mechanical Inspectors Association.

Bob's technical expertise and Brad's practical experience made them particularly insightful contributing authors, and I am grateful for their efforts.

The *National Fuel Gas Code* is a joint project of NFPA and the American Gas Association. The editor gratefully recognizes the contributions of his AGA counterpart, Paul Cabot, in the development of the code and for technical review of draft handbook materials.

Of course, any edition of a book builds on the editions that precede it. The inaugural edition of the *National Fuel Gas Code Handbook* in 1988 was the work of contributors Richard White, Orrin Burwell, J. Herbert Witte, and Forrest G. Hammaker, Jr., editor. Joseph Drechsler and Allen Callahan, editor, contributed significantly to the second edition, published in 1992. Committee members Ross Burnside and Mike Gorham were contributing authors for the third edition.

Finally, I thank my wife, Sharon, for her continued understanding of the time away from home and the time at home spent on this book and other NFPA projects.

Theodore C. Lemoff

History of the *National Fuel Gas Code*

The *National Fuel Gas Code* covers the installation of gas piping and gas appliances in buildings. The history of the code dates back approximately 75 years and involves a number of different organizations, all with the common goal of providing for the safe and satisfactory use of gas-burning equipment.

The following chronology highlights the significant steps in the development of the *National Fuel Gas Code*.

1912. Following joint preparation of specifications for gas ranges, committees of the American Gas Institute (AGI) and the National Commercial Gas Association (NCGA) recognized the need for a uniform gas installation code for the guidance of architects and builders. At that time, many municipalities, states, and federal bodies promulgated their own codes.

1913. The National Fire Protection Association's Committee on Explosives and Combustibles, whose scope covered flammable gases, appointed a subcommittee to prepare a fire protection code for city gas installation.

1915. The first, definite step toward unifying the requirements for installing gas-burning equipment on a national scale was taken when the AGI and the NCGA, in cooperation with the U.S. Bureau of Standards, began gathering the information needed to prepare a uniform gas safety code.

1919. World War I interrupted the program of the AGI and the NCGA. These organizations combined to form the American Gas Association, Inc. (A.G.A.), which continued their standardization activities.

1920. After several years of cooperative work with the U.S. Bureau of Standards and other interested agencies, the NFPA Committee on Explosives and Combustibles presented the NFPA with a preliminary code, which was adopted in 1920. In the same year, the National Board of Fire Underwriters adopted the preliminary code and printed it for general distribution.

The A.G.A. asked the American Engineering Standards Committee to draft a *Gas Safety Code* on installation practices under its customary procedure for standards development. The American Engineering Standards Committee became the American Standards Association in

1928, the United States of America Standards Institute in 1964, and the American National Standards Institute (ANSI) in 1969.

1925. In 1925, the A.G.A. Laboratories were established, and the A.G.A.'s Approval Requirements Committee was appointed to supervise the standardization of appliance and installation specifications. One of the first subcommittees appointed was the Subcommittee on Requirements for House Piping and Appliance Installation, whose assignment was to prepare installation standards to supplement the basic gas piping and appliance requirements of the *Gas Safety Code* then being prepared.

1927–1928. On March 10, 1927, the American Standards Association approved the *American Standard Gas Safety Code for Installation and Work in Buildings*, K2-1927. This document was the forerunner of a number of internationally known standards for gas equipment and installation requirements.

A tentative set of specifications prepared by the Subcommittee on Requirements for House Piping and Appliance Installation was distributed for industry comment in August 1927 and, following substantive revision, was redistributed for comment as a second edition in December 1927. This text, as finally revised, was approved by the A.G.A. Approval Requirements Committee on April 9, 1928. This action was ratified by the A.G.A.'s executive board, and the first edition of the A.G.A. *Requirements for House Piping and Appliance Installation* was published in June 1928.

1930. In September, the A.G.A. Approval Requirements Committee became the Sectional Committee on Approval and Installation Requirements for Gas Burning Appliances, Project Z21, of the American Standards Association, with the A.G.A. as its sponsor. The Subcommittee on Requirements for House Piping and Appliance Installation automatically became a subcommittee of the Z21 Committee.

1933. The NFPA's code for city gas installation, as revised in 1932 with the active cooperation of the A.G.A. and the U.S. Bureau of Standards, was approved by the American Standards Association on March 6, 1933, as the *American Recommended Practice for the Installation, Maintenance and Use of Piping and Fittings for City Gas*, Z27-1933. The National Board of Fire Underwriters also published this code as NBFU Pamphlet No. 54. Ultimately, it was printed by the NFPA in 1938 in the *National Fire Codes*, Vol. I, "Flammable Liquids, Gases, Chemicals and Explosives." (It was not until 1950 that the NFPA established a general policy of identifying its standards by a number, as well as a title.)

1936–1940. By late 1936, increased gas use, improvements in domestic gas equipment, and new gas appliance standards indicated a need to revise and update the existing A.G.A. *Requirements for House Piping and Appliance Installation*. Since these specifications were widely used as a basis for various ordinances enacted by municipalities, states, and federal agencies, it was believed they also should be correlated with the latest practices of industry and the applicable NFPA regulations.

The Z21 Subcommittee made comprehensive revisions, which were distributed for industry comment and subsequently were approved by the Z21 Committee at its December

1940 meeting. The revised standard was submitted to the American Standard Association, but it was not approved because two existing publications, K2-1927 and Z27-1933, already covered such installations. Consequently, this revised text was published in 1940 as the second edition of the A.G.A. *Requirements and Recommended Practice for House Piping and Appliance Installation.*

1943. The NFPA adopted revisions to Z27 and published the resulting standard in the *National Fire Codes*, Vol. I, in 1943 and 1945. The National Board of Fire Underwriters also published the standard in a revised Pamphlet No. 54. However, this revised standard was not submitted to the American Standards Association for approval.

1948–1950. Following World War II, it became evident that municipalities, state and federal bodies, architects, and builders needed an up-to-date American Standard covering the installation of gas piping and gas appliances in buildings. To provide one, the Z21 Committee and its house piping and appliance installation subcommittee were expanded to include a representative of the National Board of Fire Underwriters, who would also serve as liaison with the NFPA Committee on Gases. Meetings during 1948 produced a proposed standard, which was distributed for comment in February 1949 to gas companies, gas equipment manufacturers, fire insurance underwriters, safety and accident prevention organizations, interested bodies of utility commissioners, building officials, manufacturers' associations, labor unions, and others. In the meantime, the American Standards Association withdrew K2-1927 and Z27-1933 in 1949 by mutual agreement with their sponsors.

Following review of voluminous industry comments by its subcommittee, the Z21 Committee approved the revised standard at its March 1950 meeting. Subsequent consideration by the NFPA Committee on Gases resulted in a few additional changes, which were approved by the Z21 Committee in June 1950.

The resulting text was adopted by the NFPA at its May 1950 meeting and was approved by the American Standards Association on December 5, 1950. This standard, *Installation of Gas Appliances and Gas Piping*, was the first of several having a common text to be published under two separate covers as ASA Z21.30-1950 and NFPA 54-1950.

1954. The 1954 edition of the standard contained revisions that effected considerable correlation with those sections of the *Southern Standard Building Code* that covered the installation of gas piping and gas appliances.

1959. The Liquefied Petroleum Gas Association, Inc. (now the National Propane Gas Association) proposed expanding the standard to cover undiluted liquefied petroleum gas so there would be a single standard for the installation of gas appliances. The 1959 edition of the standard incorporated the essential features of NFPA 52, *Liquefied Petroleum Gas Piping and Appliance Installation in Buildings*. Other revisions extended the coverage of gas piping to include all piping downstream of the outlet of the meter set assembly.

1960. The 1960 edition of ASA Z21.30 and NFPA 54 reflected the increasing use of gas and new types of domestic and commercial gas appliances. The venting section was standardized and expanded to include more specific data on gas appliance venting systems.

In a separate but related activity, the Industrial Gas Practices Committee of the Industrial and Commercial Gas Section of the A.G.A., in 1952, had begun to develop a standard for industrial gas piping and industrial gas utilization equipment similar to ASA Z21.30 and NFPA 54. This task was undertaken because ASA Z21.30 and NFPA 54 was limited to domestic and certain commercial installations and specifically excluded industrial installations.

The first document released by this A.G.A. committee was Information Letter No. 44, *Nomenclature on Industrial Gas Equipment.* Soon after its release, Information Letter Nos. 70, *A.G.A. Recommended Good Practice Requirements for the Installation of Consumer-Owned Gas Piping on Industrial and Commercial Premises,* and 70A, *A.G.A. Recommended Good Practice Requirements for Installation of Gas Equipment on Industrial and Commercial Premises*, were released. In 1958, Information Letters Nos. 70 and 70A were revised, combined, and issued as Information Letter No. 90, *Proposed American Standard for Installation of Consumer-Owned Gas Piping and Gas Equipment on Industrial and Commercial Premises.*

The committee initiated action to have these recommendations approved as an American Standard. Following a general conference, in March 1960 the American Standards Association approved the formation of a Sectional Committee on Industrial Gas Equipment Installation and Utilization, Z83, with the A.G.A. as its sponsor.

1968. A draft standard for industrial gas installations was developed by the Z83 Committee in cooperation with the NFPA Committee on Gases. It was based on the above noted Information Letter No. 90 and coordinated with the latest revisions of ASA Z21.30 and NFPA 54 and with Sections 2 and 8 of the American Society of Mechanical Engineers (ASME) *Standard Code for Pressure Piping*, USASI B31.1. This first edition of *Installation of Gas Piping and Gas Equipment on Industrial Premises* was tentatively adopted by the National Fire Protection Association as NFPA 54A-T in May 1966, and was approved by the United States of America Standards Institute in March 1968 as USASI Z83.1-1968.

1969. Further revisions to Z21.30 and NFPA 54 included coverage for appliances installed on roofs and recognition of plastic piping materials for use only outside and underground. Due to unresolved objections, Z21.30 was not approved by the American National Standards Institute. However, it was adopted as NFPA 54-1969 in May 1969.

1971. There was an expressed need within the gas industry and among public safety authorities, insurance groups, architects, designers, and builders for one code that would cover all facets of fuel gas piping and appliance installations downstream of meter set assemblies and other facilities that made up the gas service entrance to consumers' premises. In October 1967, a Conference Group on Piping and Installation Standards, comprised of representatives of the A.G.A., the ASME, and the NFPA, was held to consider the development of a national fuel gas code.

A working group of this conference group developed the objectives and scope of a proposed National Standards Committee, which envisioned combining American National Standards Z21.30 and NFPA 54, Z83.1 and NFPA 54A, and B31.2, *Fuel Gas Piping*, into a national fuel gas code. The proposed scope limited the coverage of piping systems to a maximum operating gas pressure of 60 psig. National Standards Committees Z21, Z83, and B31 agreed to relinquish Z21.30, Z83.1, and that portion of B31.2 covering piping systems at pressures up to and including 60 psig. The NFPA agreed to relinquish NFPA 54 and NFPA 54A.

On August 13, 1971, ANSI approved the formation and the scope of activities of an American National Standards Committee on *National Fuel Gas Code*, Z223, cosponsored by the A.G.A., the ASME, and the NFPA.

1972. In November 1972, ANSI approved a second edition of Z83.1, which included added coverage for piping system operating pressure limitations.

To establish a national fuel gas code that would satisfy industry's immediate needs for a single installation code, the Z223 Committee combined NFPA 54-1969 and Z83.1-1972 at its organizational meeting in December 1972, making only those editorial revisions necessary to accomplish the combination and to reflect the new scope of the code. The Committee recognized that further revisions would be necessary to incorporate coverage of fuel gas piping from B31.2-1968.

1974. ANSI and the NFPA approved the first edition of the *National Fuel Gas Code* in 1974. This single publication identified itself on the front cover as both ANSI Z223.1 and NFPA 54 for the first time.

Following the publication of the first edition of the *National Fuel Gas Code*, the ASME relinquished its role as the code's cosponsor. Shortly thereafter, the National Fuel Gas Code Committee, Z223, also became the NFPA Technical Committee on the *National Fuel Gas Code*.

The energy crisis of the mid-1970s prompted an early revision of the 1974 edition of the code to cover modifications to existing appliance installations for the purpose of energy conservation. This was issued as Z223.1a-1978/TIA-54-74-1 to NFPA 54.

1980. The second edition of the code consolidated and reorganized Parts 1 and 2 of the previous edition and added pertinent portions of B31.2. It eliminated the need to differentiate between residential, commercial, and industrial installations and updated the code in line with current installation practices.

1984. The third edition of the code was updated in line with industry developments and current installation practices. The energy crisis also motivated the development of more efficient gas appliance designs that required vent systems capable of handling low flue gas temperatures, condensing flue gases, and positive vent pressures. To recognize such systems, revisions to "Venting of Equipment," Part 7 of the code, were issued as Z223.1a-1987/TIA-54-84-1 to NFPA 54 in 1987.

1988. In the fourth edition of the code, the maximum pressure for gas piping was increased from 60 psig to 125 psig, and the vent system coverage was updated for more efficient appliances. In addition, flexible corrugated stainless steel conduit for gas piping was recognized, and a method for sizing multistory gas vents was added.

1992. The fifth edition of the code included several revisions. New coverage, including new vent sizing tables, had been added for venting Category I, fan-assisted appliances. Also, the code addressed the installation of corrugated stainless steel tubing house piping systems that were required to comply with the *Standard for Corrugated Stainless Steel Tubing*, ANSI/A.G.A. LC 1-1991. In addition, the code recognized the use of listed gas convenience outlets and addressed the installation of decorative gas appliances in vented fireplaces in bedrooms and bathrooms. Coverage was added for direct gas-fired industrial air heaters, and the code referenced NFPA 52-1992, *Standard for Compressed Natural Gas (CNG) Vehicular Fuel Systems*, when it addressed the installation of natural gas vehicle fueling stations. The code was revised extensively and editorially to make it more usable, adoptable, and enforceable.

1996. The sixth edition of the code includes a change to the scope to relocate the point of delivery for propane piping systems to the discharge of the final stage pressure regulator (excluding line pressure regulators) from the discharge of the first stage regulator. This edition also includes a new option for one opening to the outdoors for air for combustion and ventilation for confined spaces. Several new tables were added for sizing on polyethylene tubing and corrugated stainless steel tubing and for sizing of masonry chimneys exposed to the outdoors below the roofline.

1999. The seventh edition of the *National Fuel Gas Code* makes the code applicable to the installation of appliances in manufactured housing, after the initial sale. This addition is especially important because the Federal Requirements for Manufactured Housing are only applicable to the construction of Manufactured Housing.

Other changes include revised Table 6.2.3(a), formerly Table V, which is now applicable only to unlisted gas-fired furnaces, boilers, and air conditioners, and a revised definition of *unusually tight construction* to reflect current construction practices.

NFPA 54
National Fuel Gas Code
1999 Edition, with Commentary

Part One of this handbook includes the complete text and figures of the 1999 edition of NFPA 54, *National Fuel Gas Code*. The text and figures from the code are printed in black and are the official requirements of NFPA 54. Line drawings and photographs from the code are labeled Figures.

Paragraphs that begin with the letter A are extracted from Appendix A of the code. Although printed in black ink, this nonmandatory material is purely explanatory in nature. For ease of use, this handbook places Appendix A material immediately after the code paragraph to which it refers.

Part One also includes Formal Interpretations of NFPA 54, which apply to previous and subsequent editions for which the requirements remain substantially unchanged. Formal Interpretations are not part of the code and are printed in shaded blue boxes.

In addition to code text, appendixes, and Formal Interpretations, Part One includes commentary that provides the history and other background information for specific paragraphs in the code. This insightful commentary takes the reader behind the scenes, into the reasons underlying the requirements.

Commentary text, captions, and tables are printed in blue, to clarify identification of commentary material. So the reader can easily distinguish between line drawings and photographs from the code and those in the commentary, line drawings, graphs, and photographs in the commentary are labeled Exhibits, and each is enclosed in a blue box. The distinction between figures in the code and exhibits in the commentary is new for this edition of the *National Fuel Gas Code Handbook*.

This edition of the handbook also includes supplements. Part Two contains five supplements that explore the background of selected topics related to NFPA 54 in more detail than the commentary.

General

Chapter 1 provides the scope and general requirements of the code. The scope defines the application (and nonapplication) of NFPA 54, which is important because users of this, or any, code must know whether the right document is being used.

Chapter 1 covers the following:

- Scope and application of the code, including installations covered and not covered (in Section 1.1)
- Alternative materials, equipment, and procedures not specifically addressed in the code, including the basis for the authority having jurisdiction to permit alternatives (in Section 1.2)
- Retroactivity of the code. NFPA 54 is retroactive only when it is specifically stated. In the 1999 edition, there are no retroactive requirements (in Section 1.3).
- Limitation of installations, modifications, and repairs performed under the code to a qualified agency (in Section 1.4).
- What must be done prior to and following an interruption of gas service (in Section 1.5).
- Prevention of accidental ignition while repairs are made to gas piping (in Section 1.6).
- Definitions, glossary of terms, and abbreviations used throughout this code (in Section 1.7).

1.1 Scope

1.1.1 Applicability.

(a) This code is a safety code that shall apply to the installation of fuel gas piping systems, fuel gas utilization equipment, and related accessories as follows:

The *National Fuel Gas Code* is the American National Standard that applies to the installation of fuel gas piping systems and fuel gas utilization equipment that are supplied with natural gas, manufactured gas, liquefied petroleum (LP) gas in the vapor phase only, liquefied petroleum gas–air mixtures, mixtures of these gases, and gas–air mixtures in the flammable range.

Natural gas, as its name implies, is a naturally occurring product found in various parts of the world. It is recovered by drilling wells into underground pockets of natural gas. The recovered gas is piped via collection, cleanup, transmission, and utility distribution piping to homes and businesses. Another source of natural gas is synthetic natural gas, which is produced by cracking naphtha or other chemical feedstocks to supplement natural gas supplies. Under suitable conditions of elevated pressure and low temperature, natural gas can be liquefied to reduce its volume for purposes of storage or transportation.

There are no standards or conventions that specify a composition of natural gas. Natural gas consists principally of methane, but it also contains ethane and small amounts of propane, butane, and higher hydrocarbons. Also, it can contain small amounts of nitrogen, carbon dioxide, hydrogen sulfide, and helium. Normally, natural gas contracts specify a heating value [usually 940 Btu/ft^3–1080 Btu/ft^3 (35.0 kJ/m^3 –40.2 kJ/m^3)] and a maximum amount of hydrogen sulfide [typically 0.3 grains per 100 ft^3 (6.8 mg/m^3)]. Hydrogen sulfide is a natural gas contaminant that is corrosive to copper and brass. Some contractual arrangements also limit the carbon dioxide and water content of natural gas.

Manufactured gas was the only fuel gas used in most communities in North America until the early 1950s, when the natural gas transmission system was significantly expanded in the United States. Manufactured gas is a low Btu fuel gas produced by one of several processes, and it was used primarily for lighting and cooking.

Liquefied petroleum gases include propane and butane, which are used as fuel gases. Pure propane, due to its low boiling point of approximately −44°F (−42°C), vaporizes rapidly and is used extensively as a fuel gas. Pure normal butane, due to its higher boiling point of approximately 31°F (−0.6°C), requires a vaporizer for most applications, and it is not widely used in the United States as a fuel gas. It is used in warmer countries as a fuel gas and for other applications not covered by this code.

Propane–air mixtures are used by some gas utility companies to supplement natural gas supplies during peak demand periods, such as in extremely cold weather, and by some large users of natural gas as a standby fuel during periods of curtailment of natural gas supplies.

Propane has a higher heating value per unit of vapor volume than natural gas [approximately 2500 Btu/ft^3 (93 kJ/m^3) versus approximately 1000 Btu/ft^3 (37 kJ/m^3) for natural gas], and mixing it with air provides a fuel with burning characteristics similar to natural gas. Propane–air mixtures are always well above the upper flammable limit of approximately 10 percent for propane in air.

Table 1.1 gives typical properties of several fuel gases. Their compositions can vary widely, and they should not be used for design purposes. Contact the gas supplier for data on the fuel gas that will be used.

Not included under the scope of the code is liquid LP-Gas. This category is covered by NFPA 58, *Liquefied Petroleum Gas Code.* LP-Gas has two sources: It is a by-product of natural gas production that is removed during the cleanup of natural gas prior to its entering the gas transmission system, and it is a product of petroleum refineries.

Table 1.1 Typical Properties of Fuel Gases

Typical Fuel Gases % Constituents and Properties		Nat. Gas	Com'l Propane	Propane HD 5	Com'l Butane	Water Gas	Producer Gas	Coke Oven Gas
Butane(s)		0.6			95 min. combined			
Butylene								
Carbon dioxide		0.4				5.1	4.3	2.0
Carbon monoxide						40.2	26.7	6.2
Ethane		6.0						
Hydrogen						50.0	13.9	53.2
Methane		81.1				0.7	3.1	26.7
Nitrogen		9.2				4.0	51.5	7.0
Pentane(s)		0.2						
Propane		2.1	95 min. combined	95				
Propylene								
Others, unknown, miscellaneous		0.4	5 max.	5 max.	5 max.	Some	0.5	4.9
Heat of combustion Btu per ft^3	Gross	1000	2500	2500	3200	260	165	570
	Net	900	2300	2300	2950	240	155	510
Specific gravity (air = 1.00)		0.65	1.52	1.45	1.95	0.70	0.86	0.40
Flammability limits % volume in air	Lower	3.9	2.4	2.1	1.9	6.9	16.8	5.0
	Upper	15.0	9.6	10.1	8.6	69.5	73.7	28.0
Combustion	Per ft^3 gas	10	25	25	32	2.6	1.7	5.7
Air ft^3 (approx.)	Per 100 Btu	1	1	1	1	1	1	1

1 Btu/ft^3 = 3788 J/m^3; 1 ft/sec = 0.305 m/sec; 1 ft^3 = 0.028 m^3; 1 Btu = 1056 J.

(1) Coverage of piping systems shall extend from the point of delivery to the connections with each gas utilization device. For other than undiluted liquefied petroleum gas systems, the point of delivery shall be considered the outlet of the service meter assembly or the outlet of the service regulator or service shutoff valve where no meter is provided. For undiluted liquefied petroleum gas systems, the point of delivery shall be considered the outlet of the final pressure regulator, exclusive of line gas regulators, in the system.

For natural gas, the point of delivery is the outlet of the gas meter, or it is the outlet of the service regulator or service shutoff valve when no meter is provided. The installation of transmission and distribution piping and the gas meter in the United States is covered by the *Code of Federal Regulations, 49 CFR*, Part 192, and ANSI Z380, *Guide for Gas Transmission and Distribution Piping.*

For LP-Gas systems, the point of delivery is the outlet of the final pressure regulator. Prior to the 1996 edition of this code, the point of delivery was the outlet of the first-stage pressure regulator. This modification recognized a change in NFPA 58, *Liquefied Petroleum Gas Code*, which mandated two-stage pressure regulation for most LP-Gas systems covered by the *National Fuel Gas Code.*

(2) The maximum operating pressure shall be 125 psi (862 kPa).

The maximum operating gas pressure covered by the *National Fuel Gas Code* is 125 psi (862 kPa). Prior to the 1988 edition, gas piping for pressures of 60 psi to 125 psi (410 kPa to 862 kPa) was covered by the American Society of Mechanical Engineers (ASME) *Code for Pressure Piping Fuel Gas Piping*, ASME B31.2. Originally, ASME had planned to incorporate the ASME B31.2 standard into a new standard, ASME B31.9, which would cover all building service piping systems, including gas piping above 60 psi (410 kPa). Instead, the ASME B31.9 Committee chose to delete all coverage for fuel gas piping systems from proposed ASME B31.9, *Building Services Piping Code*, and to withdraw ASME B31.2. Since there was no longer any standard covering fuel gas piping systems operating at pressures in excess of 60 psi (410 kPa) downstream of the point of delivery, the *National Fuel Gas Code* was revised to address this area.

For piping systems' supplies with pressures in excess of 125 psi (862 kPa), piping installations should be designed and constructed for specific equipment that requires these inlet pressures. The piping installation should be constructed in accordance with comprehensive specifications that are consistent with piping design and installation practice for pressures being used. Additional information regarding the design of gas piping systems above 125 psi (862 kPa) can be found in ASME B31.1, *Power Piping*, or ASME B31.3, *Chemical Plant and Petroleum Refinery Piping.* (See Exhibit 1.1.)

Exception No. 1: Piping systems for gas–air mixtures within the flammable range are limited to a maximum pressure of 10 psi (69 kPa).

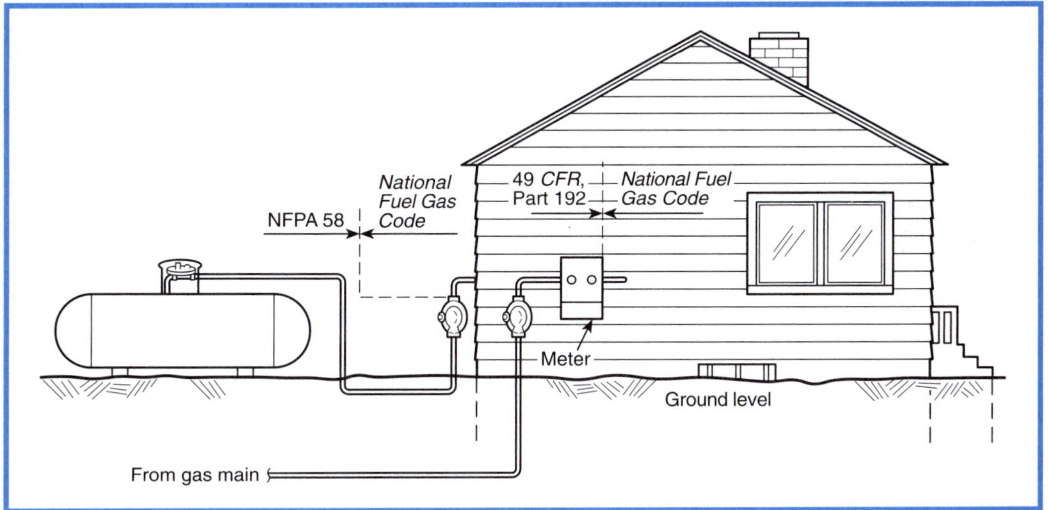

Exhibit 1.1 *A typical propane supply (showing the split between* National Fuel Gas Code *and NFPA 58) and a typical natural gas supply (showing the split between National Fuel Gas Code and Federal Pipeline Regulations, 49 CFR, Part 192). Normally, only one gas supply will serve a building. (Courtesy of T. Lemoff.)*

The pressure of flammable gas–air mixtures (5 percent to 15 percent natural gas in air and 2.15 percent to 9.6 percent propane in air) is limited to 10 psi (69 kPa) to minimize any hazard.

Exception No. 2: LP-Gas piping systems are limited to 20 psi (140 kPa), except as provided in 2.5.2.

The pressure of LP-Gas piping systems is limited to 20 psi (140 kPa) to ensure that propane remains in the vapor phase. Below this pressure, propane will liquefy at temperatures below approximately −5°F (−21°C). Systems designed to operate below −5°F (−21°C) and butane systems are permitted in 2.5.2, but they must be designed either to accommodate liquid LP-Gas or to prevent LP-Gas vapor from condensing.

(3) Piping systems requirements shall include design, materials, components, fabrication, assembly, installation, testing, inspection, operation, and maintenance.

(4) Requirements for gas utilization equipment and related accessories shall include installation, combustion, and ventilation air and venting.

(b) This code shall not apply to the following (reference standards for some of which appear in Appendix M):

(1) Portable LP-Gas equipment of all types that are not connected to a fixed fuel piping system

This paragraph clarifies that LP-Gas equipment of any type that is not connected to fixed fuel piping systems, such as outdoor gas grills, handheld soldering torches, and butane lighters, is not covered by this code. Some of this equipment is covered by other codes. An example is a portable outdoor gas grill for connection to a self-contained LP-Gas supply system, for which the propane cylinder is covered by NFPA 58, *Liquefied Petroleum Gas Code,* and the construction of the grill is covered by ANSI Z21.58, *Outdoor Cooking Gas Appliances.* Handheld soldering torches are also covered by NFPA 58. The use of butane lighters is not covered by any national standard.

(2) Installation of farm equipment such as brooders, dehydrators, dryers, and irrigation equipment

The vast majority of farms are not served by natural gas distribution mains. Most gas-fired farm equipment operates on LP-Gas directly from a container. This equipment is covered by NFPA 58, *Liquefied Petroleum Gas Code.*

(3) Raw material (feedstock) applications except for piping to special atmosphere generators

Since this code limits its scope to fuel gases, exclusion of other applications is appropriate because the committee has no expertise in certain specialized areas. Natural gas or LP-Gas, when used as a raw material or feedstock for a chemical, petrochemical, or other application, is outside of the scope of this code and, therefore, is not covered. Such uses fall under the scope of ASME B31.3, *Chemical Plant and Petroleum Refinery Piping.*

Piping to special atmosphere generators and special use ovens that employ a reducing atmosphere, however, are covered under this code. These appliances include gas-burning devices in greenhouses. They tend to be smaller units and are somewhat similar to other appliances covered under this code.

(4) Oxygen–fuel gas cutting and welding systems

Oxygen–fuel gas cutting and welding systems are covered by NFPA 51, *Standard for the Design and Installation of Oxygen-Fuel Gas Systems for Welding, Cutting, and Allied Processes.*

(5) Industrial gas applications using such gases as acetylene and acetylenic compounds, hydrogen, ammonia, carbon monoxide, oxygen, and nitrogen

These gases and other industrial gases are excluded specifically from coverage in the *National Fuel Gas Code* because of their inherent differences in properties from methane, propane, and similar hydrocarbons. This code covers only fuel gases that are commercially distributed.

(6) Petroleum refineries, pipeline compressor or pumping stations, loading terminals, compounding plants, refinery tank farms, and natural gas processing plants

(7) Large integrated chemical plants or portions of such plants where flammable or combustible liquids or gases are produced by chemical reactions or used in chemical reactions

(8) LP-Gas installations at utility gas plants

LP-Gas installations at utility gas plants are covered by NFPA 59, Liquefied Petroleum Gas Code at Utility Gas Plants. These plants mix propane and air to make a heating fuel used by gas utility companies to supplement natural gas supplies during periods of peak demand, such as in extreme cold weather. The heating value of the propane–air mixture is adjusted to provide an equivalent natural gas input rate to gas appliances.

(9) Liquefied natural gas (LNG) installations

Liquefied natural gas installations are covered by NFPA 59A, Standard for the Production, Storage, and Handling of Liquefied Natural Gas (LNG). The use of liquefied natural gas as a vehicle fuel is covered in NFPA 57, Liquefied Natural Gas (LNG) Vehicular Fuel Systems Code.

(10) Fuel gas piping in power and atomic energy plants

Fuel gas piping in these power plants is covered by ASME B31.1, Power Piping, which is written by a committee of the American Society of Mechanical Engineers.

(11) Proprietary items of equipment, apparatus, or instruments such as gas generating sets, compressors, and calorimeters

These items represent specialized equipment that is not of a general nature, and, therefore, they are excluded from the code. Gas generator sets, however, are covered by NFPA 37, Standard for the Installation and Use of Stationary Combustion Engines and Gas Turbines.

(12) LP-Gas equipment for vaporization, gas mixing, and gas manufacturing

This equipment is covered by NFPA 58, Liquefied Petroleum Gas Code.

(13) LP-Gas piping for buildings under construction or renovations that is not to become part of the permanent building piping system—that is, temporary fixed piping for building heat

Temporary LP-Gas piping is covered by NFPA 58, Liquefied Petroleum Gas Code.

(14) Installation of LP-Gas systems for railroad switch heating

Railroad switch heaters are covered by NFPA 58, Liquefied Petroleum Gas Code.

(15) Installation of LP-Gas and compressed natural gas systems on vehicles

LP-Gas vehicle fuel systems are covered by NFPA 58, Liquefied Petroleum Gas Code. Compressed natural gas fuel systems for vehicles are covered by NFPA 52, Standard for Compressed Natural Gas (CNG) Vehicular Fuel Systems.

(16) Gas piping, meters, gas pressure regulators, and other appurtenances used by the serving gas supplier in distribution of gas, other than undiluted LP-Gas

These items are provided and installed by the serving gas supplier. They are covered by the regulations of the U.S. Department of Transportation. In LP-Gas installations, the second-stage regulator, or the single-stage regulator in single-stage regulation systems, is covered by NFPA 58, Liquified Petroleum Gas Code.

(17) Building design and construction, except as specified herein

This paragraph clarifies that the National Fuel Gas Code is not intended to cover building design and construction, including means for limiting the spread of heat and smoke, which are the responsibility of NFPA 101®, Life Safety Code®, and local building codes. Installers of fuel gas system components should be aware of the possible impact of the gas system on the spread of heat and smoke, especially in multistory buildings, and coordinate their activities with building safety authorities.

(18) Fuel gas systems on recreational vehicles manufactured in accordance with NFPA 1192, *Standard on Recreational Vehicles*

1.1.2 Other Standards.

In applying this code, reference shall also be made to the manufacturers' instructions and the serving gas supplier regulations.

For a list of standards specifically referenced in applying this code and other standards that can be used, see Chapter 11 and Appendix M.

The code lists the rules that govern the fuel gas industry. The code recognizes that appliances vary considerably and that it is impractical for the committee that writes the code to include all variations. Therefore, compliance with the manufacturers' installation instructions is required for appliances, equipment, and some piping and venting materials. Manufacturers' instructions contain the specific instructions and installation rules for the proper installation of a manufacturer's appliance. The installation instructions for listed products have been reviewed by the listing laboratory. They include items specific to the product and reflect the testing of that product.

The gas supplier may also add rules for installations, based on the conditions in the area. Such issues as climate, geographical conditions, and the condition of the fuel supply can affect how appliances and piping should be installed to ensure a safe installation. When a state, city, or other political entity adopts the code, it can also be amended based on local experience.

1.2 Alternate Materials, Equipment, and Procedures

The provisions of this code are not intended to prevent the use of any material, method of construction, or installation procedure not specifically prescribed by this code, provided any such alternate is acceptable to the authority having jurisdiction *(see Section 1.7)*. The authority having jurisdiction shall require that sufficient evidence be submitted to substantiate any claims made regarding the safety of such alternates.

The intent of the National Fuel Gas Code Committee is not to prohibit safe practices in the installation of gas piping and equipment that have not been developed yet, nor to prohibit new technology. This paragraph allows the authority having jurisdiction to require evidence to substantiate any claims and, with that evidence, to permit an installation using an alternate or new method or procedure. (See A.1.7, Authority Having Jurisdiction.)

1.3 Retroactivity

Unless otherwise stated, the provisions of this code shall not be applied retroactively to existing systems that were in compliance with the provisions of the code in effect at the time of installation.

The matter of retroactivity in a standard that is widely adopted by public safety regulatory agencies and used in litigation, as is the *National Fuel Gas Code,* is of considerable importance. Although the National Fuel Gas Code Committee has the authority to make a provision retroactive, it has never elected to do so. If a matter proposed a hazard to life and property, it could induce the committee to take such an action.

The code, therefore, is not intended to be retroactive. Any equipment or system complying with the code at the time of installation can be maintained in use as long as the equipment is not changed significantly. Retroactivity could be required by a state, city, or other government body adopting the code. If a question exists on this subject, the authority having jurisdiction should be queried or local legislation reviewed.

1.4 Qualified Agency

Installation, testing, and replacement of gas piping, gas utilization equipment, or accessories, and repair and servicing of equipment, shall be performed only by a qualified agency.

This paragraph reflects the committee's strong belief that fuel gases are potentially dangerous if they are not handled properly. For this reason, the code specifically states (and defines) that only qualified persons can install and maintain gas utilization equipment. Many states, counties, cities, and municipalities take this statement further and permit only licensed plumbers or gas fitters to install and maintain gas distribution systems and equipment.

1.5 Interruption of Service

Section 1.5 provides a comprehensive set of safety procedures to use when turning off the main supply of gas in nonemergency situations. They clearly define the responsibility of the agency doing the work. For example, if a gas utility or LP-Gas supplier must shut down a valve that supplies gas to several buildings, it is the supplier's responsibility to notify all affected users and to take the necessary steps to shut down and reactivate service properly, including re-lighting pilot lights. If interruptions in work occur, all open-ended piping must be capped or plugged. This section provides specific steps for reducing pressure in purging sections prior to work and it also provides the following:

- A reference to Section 4.3 for a purging procedure
- A reference to Appendix D for a leak-testing procedure
- Clarification of who can perform hot taps

1.5.1 Notification of Interrupted Service.

When the gas supply is to be turned off, it shall be the duty of the qualified agency to notify all affected users.

Where two or more users are served from the same supply system, precautions shall be exercised to ensure that service only to the proper user is turned off.

Exception: In cases of emergency, affected users shall be notified as soon as possible of the actions taken by the qualified agency.

1.5.2* Before Turning Gas Off.

Before the gas is turned off to the premises, or section of piping to be serviced, for the purpose of installation, repair, replacement, or maintenance of gas piping or gas utilization equipment, all equipment shutoff valves shall be turned off.

A leakage test shall be performed to determine that all equipment is turned off in the piping section affected.

A.1.5.2 See Appendix D for a method of leakage testing.

Exception: In cases of emergency, these paragraphs shall not apply.

1.5.3 Turn Gas Off.

All gas piping installations, equipment installations, and modifications to existing systems shall be performed with the gas turned off and the piping purged in accordance with Section 4.3.

Exception: Hot taps shall be permitted if they are installed by trained and experienced crews utilizing equipment specifically designed for such a purpose.

1.5.4 Work Interruptions.

When interruptions in work occur while repairs or alterations are being made to an existing piping system, the system shall be left in a safe condition.

1.6 Prevention of Accidental Ignition

1.6.1 Potential Ignition Sources.

Where work is being performed on piping that contains or has contained gas, the following shall apply:

(a) Provisions for electrical continuity shall be made before alterations are made in a metallic piping system.

(b) Smoking, open flames, lanterns, welding, or other sources of ignition shall not be permitted.

(c) A metallic electrical bond shall be installed around the location of cuts in metallic gas pipes made by other than cutting torches. If cutting torches, welding, or other sources of ignition are unavoidable, it shall be determined that all sources of gas or gas–air mixtures have been secured and that all flammable gas or liquids have been cleared from the area. Piping shall be purged as required in Section 4.3 before welding or cutting with a torch is attempted.

(d) Artificial illumination shall be restricted to listed safety-type flashlights and safety lamps. Electric switches shall not be operated, on or off.

Section 1.6.1 provides work rules for preventing accidental ignition when working on piping that contains, or has contained, gas. Electrical continuity must be maintained to prevent sparks, which are a source of ignition. Smoking, open flames, lanterns, welding, and other sources of ignition are not permitted. Artificial illumination is restricted to electrical devices listed for Class I, Division I, Group D locations as defined in NFPA 70, *National Electrical Code®,* and to locations where operation of electric switches is prohibited.

For additional information on safety in cutting and welding, refer to NFPA 51B, *Standard for Fire Prevention During Welding, Cutting, and Other Hot Work.*

1.6.2 Handling of Flammable Liquids.

(a) *Drip Liquids.* Liquid that is removed from a drip in existing gas piping shall be handled to avoid spillage or ignition. The gas supplier shall be notified when drip liquids are removed.

(b) *Other Flammable Liquids.* Flammable liquids used by the installer shall be handled with precautions and shall not be left within the premises from the end of one working day to the beginning of the next.

Flammable liquids are ignited easily and are of great concern in the workplace. Extreme caution must be used when storing and transferring flammable liquids. For additional information, refer to NFPA 30, *Flammable and Combustible Liquids Code.*

Gasoline is an extremely flammable liquid that usually is present as a fuel for portable welding machines, generators, or other internal combustion engines. Extreme caution must be observed when handling or storing this fuel.

Drip liquids, which are liquids that condense from natural gas or LP-Gas in piping systems, can be highly flammable and can cause operational problems for valves and other piping components. Extreme caution must be observed when drip liquids are found in piping systems that are open for work.

Although natural gas is mostly "dry," the lubrication used at compressor stations on the pipeline facilities results in oil in the system. This oil may contain PCBs, which are highly toxic. PCBs have not been intentionally used for at least 25 years. Traces, however, can still be present in pipeline and distribution systems. When oil is found in piping systems, care must be taken because the type and source of the liquid cannot be determined by sight. If it is suspected that the oil is from the transmission line, the local utility should be contacted as soon as possible and the oil should be treated as potentially hazardous until it is identified.

1.7 Definitions

The terms used in this code are defined within the parameters of the scope of the code.

Accessible. Having access to but which first may require the removal of a panel, door, or similar covering of the item described.

Accessible, Readily. Having direct access without the need of removing or moving any panel, door, or similar covering of the item described.

Agency, Qualified. *See* Qualified Agency.

Air, Circulating. Air for cooling, heating, or ventilation distributed to habitable spaces.

Air Conditioner, Gas-Fired. A gas-burning, automatically operated appliance for supplying cooled and/or dehumidified air or chilled liquid.

Air Conditioning. The treatment of air so as to control simultaneously its temperature, humidity, cleanness, and distribution to meet the requirements of a conditioned space.

Air Mix. That portion of an injection- (Bunsen) type burner into which the primary air is introduced.

Air Shutter. An adjustable device for varying the size of the primary air inlet(s).

Ambient Temperature. The temperature of the surrounding medium; usually used to refer to the temperature of the air in which a structure is situated or a device operates.

Anodeless Riser. A transition assembly where plastic piping is permitted to be installed and terminated above ground outside of a building. The plastic piping is piped from below grade to an aboveground location inside a protective steel casing and terminates in either a factory-assembled transition fitting or a field-assembled service head adapter-type transition fitting.

Appliance (Equipment). Any device that utilizes gas as a fuel or raw material to produce light, heat, power, refrigeration, or air conditioning.

Appliance, Automatically Controlled. Appliance equipped with an automatic burner ignition and safety shutoff device and other automatic devices that (1) accomplish complete turn-on and shutoff of the gas to the main burner or burners and (2) graduate the gas supply to the burner or burners, but do not effect complete shutoff of the gas.

Appliance Categorized Vent Diameter/Area. The minimum vent area/diameter permissible for Category I appliances to maintain a nonpositive vent static pressure when tested in accordance with nationally recognized standards.

Appliance, Fan-Assisted Combustion. An appliance equipped with an integral mechanical means to either draw or force products of combustion through the combustion chamber or heat exchanger.

Appliance, Low-Heat. An appliance such as a food service range, pressing machine boiler operating at any pressure, bake oven, candy furnace, stereotype furnace, drying and curing appliance, and other process appliances in which materials are heated or melted at temperatures (excluding flue-gas temperatures) not exceeding 600°F (315°C).

Approved.* Acceptable to the authority having jurisdiction.

A-1.7 Approved. The American Gas Association and the National Fire Protection Association do not approve, inspect, or certify any installations, procedures, equipment, or materials; nor do they approve or evaluate testing laboratories. In determining the acceptability of installations, procedures, equipment, or materials, the authority having jurisdiction may base acceptance on compliance with NFPA or other appropriate standards. In the absence of such standards, said authority may require evidence of proper installation, procedure, or use. The authority having jurisdiction may also refer to the listings or labeling practices of an organization that is concerned with product evaluations and is thus in a position to determine compliance with appropriate standards for the current production of listed items.

Atmospheric Pressure. The pressure of the weight of air and water vapor on the surface of the earth, approximately 14.7 pounds per square inch (psia) (101 kPa absolute) at sea level.

In the largest sense, atmospheric pressure is the weight of air (including water vapor) at any place being discussed, whether it is at the surface of the earth or not.

Authority Having Jurisdiction.* The organization, office, or individual responsible for approving equipment, materials, an installation, or a procedure.

A-1.7 Authority Having Jurisdiction. The phrase "authority having jurisdiction" is used in this Code in a broad manner since jurisdictions and "approval" agencies vary as do their responsibilities. Where public safety is primary, the "authority having jurisdiction" may be a federal, state, local, or other regional department or individual such as a fire chief, fire marshal, chief of a fire prevention bureau, labor department, health department, building official, electrical inspector, or others having statutory authority. For insurance purposes, an insurance inspection department, rating bureau, or other insurance company representative may be the "authority having jurisdiction." In many circumstances the property owner or his delegated agent assumes the role of the "authority having jurisdiction"; at government installations, the commanding officer or departmental official may be the "authority having jurisdiction."

Automatic Damper Regulator. A mechanically or electrically actuated device designed to maintain a constant draft on combustion equipment.

Automatic Firecheck. A device for stopping the progress of a flame front in burner mixture lines (flashback) and for automatically shutting off the fuel–air mixture. Present units are customarily equipped with spring- or weight loaded valves released for closure by a fusible link or by movement of bimetallic elements; they are also equipped with metallic screens for stopping the progress of a flame front.

Automatic Gas Shutoff Device. A device constructed so that the attainment of a water temperature in a hot water supply system in excess of some predetermined limit acts in such a way as to cause the gas to the system to be shut off.

Automatic Ignition. Ignition of gas at the burner(s) when the gas controlling device is turned on, including reignition if the flames on the burner(s) have been extinguished by means other than by the closing of the gas controlling device.

Back Pressure. Pressure against which a fluid is flowing, resulting from friction in lines, restrictions in pipes or valves, pressure in vessel to which fluid is flowing, hydrostatic head, or other impediment that causes resistance to fluid flow.

Backfire Preventer. *See* Safety Blowout.

Baffle. An object placed in an appliance to change the direction of or retard the flow of air, air–gas mixtures, or flue gases.

Barometric Draft Regulator. A balanced damper device attached to a chimney, vent connector, breeching, or flue gas manifold to protect combustion equipment by controlling chimney draft. A double-acting barometric draft regulator is one whose balancing damper is free to move in either direction to protect combustion equipment from both excessive draft and backdraft.

Boiler, Hot Water Heating. A boiler in which no steam is generated, from which hot water is circulated for heating purposes and then returned to the boiler, and that operates at water pressures not exceeding 160 psi (1100 kPa) and at water temperatures not exceeding 250°F (121°C) at or near the boiler outlet.

Boiler, Hot Water Supply. A boiler, completely filled with water, that furnishes hot water to be used externally to itself and that operates at water pressures not exceeding

160 psi (1100 kPa) and at water temperatures not exceeding 250°F (121°C) at or near the boiler outlet.

Boiler, Low-Pressure. A self-contained gas-burning appliance for supplying steam or hot water.

Low-pressure boilers provide hot water for heating, but they do not meet the requirements for providing water for cooking, dishwashing, and bathing. Hot water for these purposes is provided by water heaters.

Boiler, Steam Heating. A boiler in which steam is generated and that operates at a steam pressure not exceeding 15 psi (100 kPa).

Branch Line. Gas piping that conveys gas from a supply line to the appliance.

Breeching. *See* Vent Connector.

Broiler. A general term including broilers, salamanders, barbecues, and other devices cooking primarily by radiated heat, excepting toasters.

Btu. Abbreviation for British thermal unit, which is the quantity of heat required to raise the temperature of 1 pound of water 1 degree Fahrenheit (equivalent to 1055 joules).

Burner. A device for the final conveyance of gas, or a mixture of gas and air, to the combustion zone.

Burner, Forced-Draft. *See* Burner, Power.

Burner, Induced-Draft. A burner that depends on draft induced by a fan that is an integral part of the appliance and is located downstream from the burner.

Burner, Injection (Atmospheric). A burner in which the air at atmospheric pressure is injected into the burner by a jet of gas.

Burner, Injection (Bunsen) Type. A burner employing the energy of a jet of gas to inject air for combustion into the burner and mix it with the gas.

Burner, Power. A burner in which either gas or air, or both, are supplied at a pressure exceeding, for gas, the line pressure, and for air, atmospheric pressure; this added pressure being applied at the burner. A burner for which air for combustion is supplied by a fan ahead of the appliance is commonly designated as a forced-draft burner.

Burner, Power, Premixing. A power burner in which all or nearly all of the air for combustion is mixed with the gas as primary air.

Burner, Power, Fan-Assisted. A burner that uses either induced or forced draft.

Carbon Steel. By common custom, steel that is considered to be carbon steel when no minimum content is specified or required for aluminum, boron, chromium, cobalt, columbium, molybdenum, nickel, titanium, tungsten, vanadium, zirconium, or any other element added to obtain a desired alloying effect; when the specified minimum for copper does not exceed 0.40 percent; or when the maximum content specified for any of the following elements does not exceed the percentages noted: manganese, 1.65; silicon, 0.60; copper, 0.60.

In effect, carbon steel has little or no requirement for any of the alloying elements discussed. It is a general-purpose product, suitable for many applications.

Central Premix System. A system that distributes flammable gas–air mixtures to two or more remote stations and is employed to provide one or more of the following: (1) wide in-plant distribution of a centrally controlled gas–air mixture; (2) a wider range of mixture pressures (frequently in the 1 to 5 psi range) (7 to 34 kPa) than is available from other gas–air mixing equipment; (3) close control of gas–air ratios over a wide turndown range (often 20 to 1 or more); or (4) ability to change total connected burner port area without installing new mixing devices or inserts. Central premix systems either may proportion the flows of pressurized air and pressurized gas for subsequent mixing in a downstream tee or comparable fitting, or they may draw room air at essentially atmospheric pressure through a proportioning mixing valve and then through a blower or compressor downstream.

Chimney. *(See also Gas Vent, and Venting System.)* One or more passageways, vertical or nearly so, for conveying flue or vent gases to the outside atmosphere.

Chimney, Factory-Built. A chimney composed of listed factory-built components assembled in accordance with the terms of listing to form the completed chimney.

Chimney, Masonry. A field-constructed chimney of solid masonry units, bricks, stones, listed masonry chimney units, or reinforced portland cement concrete, lined with suitable chimney flue liners.

Chimney, Metal. A field-constructed chimney of metal.

Clothes Dryer. A device used to dry wet laundry by means of heat derived from the combustion of fuel gases.

Clothes Dryer, Type 1. Factory-built package, multiply produced. Primarily used in family living environment. May or may not be coin-operated for public use. Usually the smallest unit physically and in function output.

Clothes Dryer, Type 2. Factory-built package, multiply produced. Used in business with direct intercourse of the function with the public. May or may not be operated by public or hired attendant. May or may not be coin-operated. Not designed for use in individual family living environment. May be small, medium, or large in relative size.

Combustible Material. As pertaining to materials adjacent to or in contact with heat-producing appliances, vent connectors, gas vents, chimneys, steam and hot water pipes, and warm air ducts, shall mean materials made of or surfaced with wood, compressed paper, plant fibers, or other materials that are capable of being ignited and burned. Such material shall be considered combustible even though flame-proofed, fire-retardant treated, or plastered.

Combustion. As used herein, the rapid oxidation of fuel gases accompanied by the production of heat or heat and light. Complete combustion of a fuel is possible only in the presence of an adequate supply of oxygen.

Combustion Chamber. The portion of an appliance within which combustion occurs.

Combustion Products. Constituents resulting from the combustion of a fuel with the oxygen of the air, including the inert but excluding excess air.

Note that excess air is not a "combustion product," but it is air that passes through the combustion chamber and the appliance flue in excess of that which theoretically is required for complete combustion.

Concealed Gas Piping. Gas piping that, when in place in a finished building, would require removal of permanent construction to gain access to the piping.

Condensate (Condensation). The liquid that separates from a gas (including flue gas) due to a reduction in temperature or an increase in pressure.

Consumption. The maximum amount of gas per unit of time, usually expressed in cubic feet per hour, or Btu per hour, required for the operation of the appliance or appliances supplied.

Control Piping. All piping, valves, and fittings used to interconnect air, gas, or hydraulically operated control apparatus or instrument transmitters and receivers.

Controls. Devices designed to regulate the gas, air, water, or electrical supply to a gas appliance. These may be manual or automatic.

Convenience Outlet, Gas. A permanently mounted, hand-operated device providing a means for connecting and disconnecting an appliance or an appliance connector to the gas supply piping. The device includes an integral, manually operated gas valve with a nondisplaceable valve member so that disconnection can be accomplished only when the manually operated gas valve is in the closed position.

Conversion Burner, Gas. A unit consisting of a burner and its controls utilizing gaseous fuel for installation in an appliance originally utilizing another fuel.

Conversion Burner, Gas, Firing Door Type. A conversion burner specifically for boiler or furnace firing door installation.

Conversion Burner, Gas, Inshot Type. A conversion burner normally for boiler or furnace ash pit installation and fired in a horizontal position.

Conversion Burner, Gas, Upshot Type. A conversion burner normally for boiler or furnace ash pit installation and fired in a vertical position at approximately grate level.

Counter Appliances, Gas. *See* Food Service Equipment, Gas Counter Appliance.

Cubic Foot (ft^3) of Gas. The amount of gas that would occupy 1 ft^3 (0.3 m) when at a temperature of 60°F (16°C), saturated with water vapor and under a pressure equivalent to that of 30.0 in. (7.5 kPa) mercury column.

Decorative Appliance for Installation in a Vented Fireplace. A self-contained, free-standing, fuel-gas burning appliance designed for installation only in a vented fireplace and whose primary function lies in the aesthetic effect of the flame.

Decorative Appliance for Installation in a Vented Fireplace, Coal Basket. An open-flame-type appliance consisting of a metal basket that is filled with simulated coals and gives the appearance of a coal fire when in operation.

Decorative Appliance for Installation in a Vented Fireplace, Fireplace Insert. Consists of an open-flame, radiant-type appliance mounted in a decorative metal panel to cover the fireplace or mantel opening and having provisions for venting into the fireplace chimney.

Decorative Appliance for Installation in a Vented Fireplace, Gas Log. An open-flame-type appliance consisting of a metal frame or base supporting simulated logs.

Decorative Appliance for Installation in a Vented Fireplace, Radiant Appliance. An open-front appliance designed primarily to convert the energy in fuel gas to radiant heat by means of refractory radiants or similar radiating materials. A radiant heater has no external jacket. A radiant appliance is designed for installation in a vented fireplace.

Deep Fat Fryer. *See* Food Service Equipment, Gas Deep Fat Fryer.

Design Certification. The process by which a product is evaluated and tested by an independent laboratory to affirm that the product design complies with specific requirements.

Design Pressure. The maximum operating pressure permitted by this code, as determined by the design procedures applicable to the materials involved.

This term defines the maximum gas pressure allowed for a given installation, taking into account the nature of the equipment, type of occupancy, piping location, and so forth.

Dilution Air. Air that enters a draft hood or draft regulator and mixes with the flue gases.

Direct Gas-Fired Industrial Air Heater. A heater in which all of the products of combustion generated by the gas-burning device are released into the airstream being heated; whose purpose is to offset the building heat loss by heating incoming outside air, inside air, or a combination of both.

Direct Gas-Fired Makeup Air Heater. A heater in which all the products of combustion generated by the fuel-gas burning device are released into the outside airstream being heated.

Direct Vent Appliances. Appliances that are constructed and installed so that all air for combustion is derived directly from the outside atmosphere and all flue gases are discharged to the outside atmosphere.

Diversity Factor. Ratio of the maximum probable demand to the maximum possible demand.

Draft. The flow of gases or air through chimney, flue, or equipment, caused by pressure differences.

Draft, Mechanical or Induced. The draft developed by fan or air or steam jet or other mechanical means.

Draft, Natural. The draft developed by the difference in temperature of hot gases and outside atmosphere.

Even though this definition refers to "difference in temperature," the difference in densities of hot gases and the outside atmosphere is the important point.

Draft Hood. A nonadjustable device built into an appliance, or made a part of the vent connector from an appliance, that is designed to (1) provide for the ready escape of the flue gases from the appliance in the event of no draft, backdraft, or stoppage beyond the draft hood, (2) prevent a backdraft from entering the appliance, and (3) neutralize the effect of stack action of the chimney or gas vent upon the operation of the appliance.

Draft Regulator. A device that functions to maintain a desired draft in the appliance by automatically reducing the draft to the desired value.

Drip. The container placed at a low point in a system of piping to collect condensate and from which it may be removed.

Dry Gas. A gas having a moisture and hydrocarbon dew point below any normal temperature to which the gas piping is exposed.

Duct Furnace. A furnace normally installed in distribution ducts of air conditioning systems to supply warm air for heating. This definition applies only to an appliance that depends for air circulation on a blower not furnished as part of the furnace.

Excess Air. Air that passes through the combustion chamber and the appliance flues in excess of that which is theoretically required for complete combustion.

Explosion Heads (Soft Heads or Rupture Discs). A protective device for relieving excessive pressure in a premix system by bursting of a rupturable disc.

Exterior Masonry Chimneys. Masonry chimneys exposed to the outdoors on one or more sides below the roof line.

Fan-Assisted Combustion System. An appliance equipped with an integral mechanical means to either draw or force products of combustion through the combustion chamber or heat exchanger.

FAN Max. The maximum input rating of a Category I, fan-assisted appliance attached to a vent or connector.

FAN Min. The minimum input rating of a Category I, fan-assisted appliance attached to a vent or connector.

FAN+FAN. The maximum combined appliance input rating of two or more Category I, fan-assisted appliances attached to the common vent.

FAN+NAT. The maximum combined appliance input rating of one or more Category I, fan-assisted appliances and one or more Category I, draft hood-equipped appliances attached to the common vent.

Fireplace. A fire chamber and hearth constructed of noncombustible material for use with solid fuels and provided with a chimney.

Fireplace, Factory-Built. A fireplace composed of listed factory-built components assembled in accordance with the terms of listing to form the completed fireplace.

Fireplace, Masonry. A hearth and fire chamber of solid masonry units such as bricks, stones, listed masonry units, or reinforced concrete, provided with a suitable chimney.

Flame Arrester. A nonvalve device for use in a gas–air mixture line containing a means for temporarily stopping the progress of a flame front (flashback).

Flames:
   Bunsen. The flame produced by premixing some of the air required for combustion with the gas before it reaches the burner ports or point of ignition.
   Yellow, Luminous, or Nonbunsen. The flame produced by burning gas without any premixing of air with the gas.

Floor Furnace. A completely self-contained unit furnace suspended from the floor of the space being heated, taking air for combustion from outside this space.

Floor Furnace, Fan-Type. A floor furnace equipped with a fan that provides the primary means for circulation of air.

Floor Furnace, Gravity-Type. A floor furnace depending primarily on circulation of air by gravity. This classification also includes floor furnaces equipped with booster-type fans that do not materially restrict free circulation of air by gravity flow when such fans are not in operation.

Flue, Appliance. The passage(s) within an appliance through which combustion products pass from the combustion chamber of the appliance to the draft hood inlet opening on an appliance equipped with a draft hood or to the outlet of the appliance on an appliance not equipped with a draft hood.

Flue, Chimney. The passage(s) in a chimney for conveying the flue or vent gases to the outside atmosphere.

Flue Collar. That portion of an appliance designed for the attachment of a draft hood, vent connector, or venting system.

Flue Gases. Products of combustion plus excess air in appliance flues or heat exchangers.

Note that flue gases include combustion products and excess air. (See commentary concerning Combustion Products.) Vent gases include dilution air, too. (See definitions of Vent Gases and Dilution Air.)

Food Service Equipment, Gas Counter Appliance. An appliance such as a gas coffee brewer and coffee urn and any appurtenant water heating equipment, food and dish warmer, hot plate, and griddle.

Food Service Equipment, Gas Deep Fat Fryer. An appliance, including a cooking vessel in which oils or fats are placed to such a depth that the cooking food is essentially supported by displacement of the cooking fluid or a perforated container immersed in the cooking fluid rather than by the bottom of the vessel, designed primarily for use in hotels, restaurants, clubs, and similar institutions.

Food Service Equipment, Gas-Fired Kettle. An appliance with a cooking chamber that is heated either by a steam jacket in which steam is generated by gas heat or by direct gas heat applied to the cooking chamber.

Food Service Equipment, Gas Oven, Baking and Roasting. An oven primarily intended for volume food preparation that may be composed of one or more sections or units of the following types: (1) *cabinet oven,* an oven having one or more cavities heated by a single burner or group of burners; (2) *reel-type oven,* an oven employing trays that are moved by mechanical means; or (3) *sectional oven,* an oven composed of one or more independently heated cavities.

Food Service Equipment, Gas Range. A self-contained gas range providing for cooking, roasting, baking, or broiling, or any combination of these functions, and not designed specifically for domestic use.

Food Service Equipment, Gas Steam Cooker. A gas appliance that cooks, defrosts, or reconstitutes food by direct contact with steam.

Food Service Equipment, Gas Steam Generator. A separate appliance primarily intended to supply steam for use with food service equipment.

Furnace, Central. A self-contained, gas-burning appliance for heating air by transfer of heat of combustion through metal to the air and designed to supply heated air through ducts to spaces remote from or adjacent to the appliance location.

Furnace, Direct Vent Central. A system consisting of (1) a central furnace for indoor installation, (2) combustion air connections between the central furnace and the outdoor atmosphere, (3) flue-gas connections between the central furnace and the vent cap, and (4) a vent cap for installation outdoors, supplied by the manufacturer and constructed so that all air for combustion is obtained from the outdoor atmosphere and all flue gases are discharged to the outdoor atmosphere.

Furnace, Downflow. A furnace designed with airflow discharge vertically downward at or near the bottom of the furnace.

Furnace, Enclosed. A specific heating, or heating and ventilating, furnace incorporating an integral total enclosure and using only outside air for combustion.

Furnace, Forced-Air. A furnace equipped with a fan or blower that provides the primary means for circulation of air.

Furnace, Forced Air, with Cooling Unit. A single-package unit, consisting of a gas-fired, forced-air furnace of the downflow, horizontal, or upflow type combined with an electrically or gas-operated summer air-conditioning system, contained in a common casing.

Furnace, Gravity. A furnace depending primarily on circulation of air by gravity.

Furnace, Gravity, with Booster Fan. A furnace equipped with a booster fan that does not materially restrict free circulation of air by gravity flow when the fan is not in operation.

Furnace, Gravity, with Integral Fan. A furnace equipped with a fan or blower as an integral part of its construction and operable on gravity systems only. The fan or blower is used only to overcome the internal furnace resistance to airflow.

Furnace, Horizontal. A furnace designed for low headroom installation with airflow across the heating element essentially in a horizontal path.

Furnace, Upflow. A furnace designed with airflow discharge vertically upward at or near the top of the furnace. This classification includes "highboy" furnaces with the blower mounted below the heating element and "lowboy" furnaces with the blower mounted beside the heating element.

Garage, Repair. A building, structure, or portions thereof wherein major repair or painting or body and fender work is performed on motorized vehicles or automobiles, and includes associated floor space used for offices, parking, and showrooms.

Garage, Residential. A building or room in which not more than three self-propelled passenger vehicles are or may be stored and that will not normally be used for other than minor service or repair operations on such stored vehicles.

Gas Fireplace, Direct Vent. A system consisting of (1) an appliance for indoor installation that allows the view of flames and provides the simulation of a solid fuel fireplace, (2) combustion air connections between the appliance and the vent-air intake terminal, (3) flue-gas connections between the appliance and the vent-air intake terminal, (4) a vent-air intake terminal for installation outdoors, constructed such that all air for combustion is obtained from the outdoor atmosphere and all flue gases are discharged to the outdoor atmosphere.

Gas Fireplace, Vented. A vented appliance that allows the view of flames and provides the simulation of a solid fuel fireplace.

Gas Main or Distribution Main. A pipe installed in a community to convey gas to individual services or other mains.

Gas-Mixing Machine. Any combination of automatic proportioning control devices, blowers, or compressors that supply mixtures of gas and air to multiple burner installations where control devices or other accessories are installed between the mixing device and burner.

Gas Utilization Equipment. Any device that utilizes gas as a fuel or raw material or both.

Gas Vent. A passageway composed of listed factory-built components assembled in accordance with the terms of listing for conveying vent gases from gas appliances or their vent connectors to the outside atmosphere.

For an item to be called a gas vent, it must be factory made and listed for conveying vent gases, which are defined as products of combustion plus excess air plus dilution air.

Gas Vent, Special Type. Gas vents for venting listed Category II, III, and IV gas appliances.

Gas Vent, Type B. A vent for venting listed gas appliances with draft hoods and other Category I gas appliances listed for use with Type B gas vents.

The Type B gas vent is a double-wall type with a relatively low permissible temperature of vent gases.

Gas Vent, Type B-W. A vent for venting listed gas-fired vented wall furnaces.

The Type B-W gas vent is usually an oval-shaped, double-wall type, designed to fit within a conventional stud wall.

Gas Vent, Type L. A vent for venting appliances listed for use with Type L vents and appliances listed for use with Type B gas vents.

The Type L vent is rated for higher temperature vent gases than the Type B vent. Type L vents can be used with appliances that are listed for Type B, but not vice versa.

Gases. Include natural gas, manufactured gas, liquefied petroleum (LP) gas in the vapor phase only, liquefied petroleum gas–air mixtures, and mixtures of these gases, plus gas–air mixtures within the flammable range, with the fuel gas or the flammable component of a mixture being a commercially distributed product.

Governor, Zero. A regulating device that is normally adjusted to deliver gas at atmospheric pressure within its flow rating.

Gravity. *See* Specific Gravity.

Heat Pump, Gas-Fired. A gas-burning, automatically operated appliance utilizing a refrigeration system for supplying either heated air or liquid or heated and/or cooled air or liquid.

Heating Value (Total). The number of British thermal units produced by the combustion, at constant pressure, of 1 ft^3 (0.03 m^3) of gas when the products of combustion are cooled to the initial temperature of the gas and air, when the water vapor formed during combustion is condensed, and when all the necessary corrections have been applied.

Hoop Stress. The stress in a pipe wall, acting circumferentially in a plane perpendicular to the longitudinal axis of the pipe and produced by the pressure of the fluid in the pipe.

Hot Plate. *See* Food Service Equipment, Gas Counter Appliance.

Hot Plate, Domestic. A fuel-gas burning appliance consisting of one or more open-top-type burners mounted on short legs or a base.

Hot Taps. Piping connections made to operating pipelines or mains or other facilities while they are in operation. The connection of the branch piping to the operating line and the tapping of the operating line are done while it is under gas pressure.

Household Cooking Gas Appliance. A gas appliance for domestic food preparation, providing at least one function of (1) top or surface cooking, (2) oven cooking, or (3) broiling.

Household Cooking Gas Appliance, Built-In Unit. A unit designed to be recessed into, placed upon, or attached to the construction of a building, but not for installation on the floor.

Household Cooking Gas Appliance, Broiler. A unit that cooks primarily by radiated heat.

Household Cooking Gas Appliance, Floor-Supported Unit. A self-contained cooking appliance for installation directly on the floor. It has a top section and an oven section. It may have additional sections.

Incinerator, Domestic. A domestic, fuel-gas burning appliance, used to reduce refuse material to ashes, that is manufactured, sold, and installed as a complete unit.

Indirect Oven. An oven in which the flue gases do not flow through the oven compartment.

Infrared Heater. A heater that directs a substantial amount of its energy output in the form of infrared energy into the area to be heated. Such heaters may be of either the vented or unvented type.

Insulating Millboard. A factory fabricated board formed with noncombustible materials, normally fibers, and having a thermal conductivity in the range of 1 Btu/in./ft^2/°F/hr (0.14 W/m/K).

Joint. A connection between two lengths of pipe or between a length of pipe and a fitting.

Joint, Adhesive. A joint made in plastic piping by the use of an adhesive substance that forms a continuous bond between the mating surfaces without dissolving either one of them.

Joint, Solvent Cement. A joint made in thermoplastic piping by the use of a solvent or solvent cement that forms a continuous bond between the mating surfaces.

Kettle, Gas-Fired. *See* Food Service Equipment, Gas-Fired Kettle.

Labeled. Equipment or materials to which has been attached a label, symbol, or other identifying mark of an organization acceptable to the authority having jurisdiction and concerned with product evaluation, that maintains periodic inspection of production of labeled equipment or materials, and by whose labeling the manufacturer indicates compliance with appropriate standards or performance in a specified manner.

Laundry Stove, Domestic. A fuel-gas burning appliance consisting of one or more open-top-type burners mounted on high legs or having a cabinet base.

Leak Check. An operation performed on a complete gas piping system and connected equipment prior to placing it into operation following initial installation and pressure testing or interruption of gas supply to verify that the system does not leak.

Leak Detector. An instrument for determining concentration of gas in air.

Limit Control. A device responsive to changes in pressure, temperature, or liquid level for turning on, shutting off, or throttling the gas supply to an appliance.

Listed.* Equipment, materials, or services included in a list published by an organization that is acceptable to the authority having jurisdiction and concerned with evaluation of products or services, that maintains periodic inspection of production of listed equipment or materials or periodic evaluation of services, and whose listing states that either the equipment, material, or service meets appropriate designated standards or has been tested and found suitable for a specified purpose.

A-1.7 Listed. The means for identifying listed equipment may vary for each organization concerned with product evaluation; some organizations do not recognize equipment as listed unless it is also labeled. The authority having jurisdiction should utilize the system employed by the listing organization to identify a listed product.

Loads, Connected. Sum of the rated Btu input to individual gas utilization equipment connected to a piping system. May also be expressed in cubic feet per hour.

Main Burner. A device or group of devices essentially forming an integral unit for the final conveyance of gas or a mixture of gas and air to the combustion zone and on which combustion takes place to accomplish the function for which the appliance is designed.

Manifold, Gas. The conduit of an appliance that supplies gas to the individual burners.

Manufactured Home. A structure, transportable in one or more sections, that is 8 body-ft (24.4 cm) or more in width or 40 body-ft (1219 cm) or more in length in the traveling mode or, when erected on site, is 320 ft^2 (28 m^2) or more; which is built on a chassis and designed to be used as a dwelling, with or without a permanent foundation, when connected to the required utilities, including the plumbing, heating, air conditioning, and electrical systems contained therein. Calculations used to determine the number of square feet in a structure will be based on a structure's exterior dimensions, measured at the largest horizontial projections when erected on site. These dimensions include all expandable rooms, cabinets, and other projections containing interior space, but do not include bay windows.

This definition was originally published in NFPA 501, *Standard on Manufactured Housing.*

Maximum Working Pressure. The maximum pressure at which a piping system may be operated in accordance with the provisions of this code. It is the pressure used in determining the setting of pressure-relieving or pressure-limiting devices installed to protect the system from accidental overpressuring.

Maximum working pressure can be determined from allowable stress considerations for piping material, or from the nature of the installation itself and the code limitations thereon. Compare this term with Design Pressure, which limits the pressure based on system considerations or appliance limitations. As defined in this section, the two terms are nearly interchangeable.

Mechanical Exhaust System. Equipment installed in and made a part of the vent, which will provide a positive induced draft.

Meter. An instrument installed to measure the volume of gas delivered through it.

Mixing Blower. A motor-driven blower to produce gas–air mixtures for combustion through one or more gas burners or nozzles on a single-zone industrial heating appliance or on each control zone of a multizone industrial appliance or on each control zone of a multizone installation. The blower shall be equipped with a gas-control valve at its air entrance so arranged that gas is admitted to the airstream, entering the blower in proper proportions for correct combustion by the type of burners employed; the said gas-control valve being of either the zero governor or mechanical ratio valve type that controls the gas and air adjustment simultaneously. No valves or other obstructions shall be installed between the blower discharge and the burner or burners.

NA. Vent configuration is not permitted due to potential for condensate formation or pressurization of the venting system, or not applicable due to physical or geometric restraints.

NAT Max. The maximum input rating of a Category I, draft hood-equipped appliance attached to a vent or connector.

NAT+NAT. The maximum combined appliance input rating of two or more Category I draft hood-equipped appliances attached to the common vent.

Noncombustible Material. For the purpose of this code, noncombustible material shall mean material that is not capable of being ignited and burned, such as material consisting entirely of, or of a combination of, steel, iron, brick, tile, concrete, slate, asbestos, glass, and plaster.

Nondisplaceable Valve Member. A nondisplaceable valve member that cannot be moved from its seat by a force applied to the handle or by a force applied by a plane surface to any exterior portion of the valve.

Orifice. The opening in a cap, spud, or other device whereby the flow of gas is limited and through which the gas is discharged to the burner.

Orifice Cap (Hood). A movable fitting having an orifice that permits adjustment of the flow of gas by the changing of its position with respect to a fixed needle or other device.

Orifice Spud. A removable plug or cap containing an orifice that permits adjustment of the flow of gas either by substitution of a spud with a different sized orifice or by the motion of a needle with respect to it.

Outdoor Cooking Gas Appliance. As used in this code, a post-mounted, fuel-gas burning outdoor cooking appliance for installation directly on and attachment to a post provided as a part of the appliance by the manufacturer.

Oven, Gas Baking and Roasting. *See* Food Service Equipment, Gas Oven.

Parking Structure. A building, structure, or portion thereof used for the parking of motor vehicles.

Parking Structure, Basement or Underground. A parking structure or portion thereof located below grade.

Parking Structure, Enclosed. Having exterior enclosing walls that have less than 25 percent of the total wall area open to atmosphere at each level using at least two sides of the structure.

Pilot. A small flame that is utilized to ignite the gas at the main burner or burners.

Pipe. Rigid conduit of iron, steel, copper, brass, aluminum, or plastic.

Pipe, Equivalent Length. The resistance of valves, controls, and fittings to gas flow expressed as equivalent length of straight pipe for convenience in calculating pipe sizes.

Piping. As used in this code, either pipe, tubing, or both. *See* Pipe, Tubing.

Piping System. All piping, valves, and fittings from the outlet of the point of delivery from the supplier to the outlets of the equipment shutoff valves.

Plenum. A compartment or chamber to which one or more ducts are connected and that forms part of the air distribution system.

Pool Heater. An appliance designed for heating nonpotable water stored at atmospheric pressure, such as water in swimming pools, therapeutic pools, and similar applications.

Pressure. Unless otherwise stated, is expressed in pounds per square inch above atmospheric pressure.

Pressure Control. Manual or automatic maintenance of pressure, in all or part of a system, at a predetermined level, or within a selected range.

Pressure Drop. The loss in pressure due to friction or obstruction in pipes, valves, fittings, regulators, and burners.

Pressure Limiting Device. Equipment that under abnormal conditions will act to reduce, restrict, or shut off the supply of gas flowing into a system in order to prevent the gas pressure in that system from exceeding a predetermined value.

Pressure Test. An operation performed to verify the gastight integrity of gas piping following its installation or modification.

Primary Air. The air introduced into a burner that mixes with the gas before it reaches the port or ports.

Purge. To free a gas conduit of air or gas, or a mixture of gas and air.

Qualified Agency. Any individual, firm, corporation, or company that either in person or through a representative is engaged in and is responsible for (a) the installation, testing, or replacement of gas piping or (b) the connection, installation, testing, repair, or servicing of equipment; that is experienced in such work; that is familiar with all precautions required; and that has complied with all the requirements of the authority having jurisdiction.

Quick-Disconnect Device. A hand-operated device that provides a means for connecting and disconnecting an appliance or an appliance connector to a gas supply and that is equipped with an automatic means to shut off the gas supply when the device is disconnected.

Range. *See* Food Service Equipment, Gas Range.

Refrigerator (Using Gas Fuel). A fuel-gas-burning appliance that is designed to extract heat from a suitable chamber.

Regulator, Gas Appliance. A pressure regulator for controlling pressure to the manifold of gas equipment.

Regulator, Gas Appliance, Adjustable. (1) *Spring type, limited adjustment:* a regulator in which the regulating force acting upon the diaphragm is derived principally from a spring, the loading of which is adjustable over a range of not more than ±15 percent of the outlet pressure at the midpoint of the adjustment range; (2) *spring type, standard adjustment:* a regulator in which the regulating force acting on the diaphragm is derived principally from a spring, the loading of which is adjustable; the adjustment means shall be concealed.

Regulator, Gas Appliance, Multistage. A regulator for use with a single gas whose adjustment means can be positioned manually or automatically to two or more predetermined outlet pressure settings. Each of these settings may be either adjustable or nonadjustable. The regulator may modulate outlet pressures automatically between its maximum and minimum predetermined outlet pressure settings.

Regulator, Gas Appliance, Nonadjustable. (1) *Spring type, nonadjustable:* a regulator in which the regulating force acting on the diaphragm is derived principally from a spring, the loading of which is not field adjustable; (2) *weight type:* a regulator in which the regulating force acting upon the diaphragm is derived from a weight or combination of weights.

Regulator, Line Gas. A pressure regulator placed in a gas line between the service regulator and the gas appliance regulator.

Regulator, Monitoring. A pressure regulator set in series with another pressure regulator for the purpose of automatically taking over in an emergency the control of the pressure downstream of the regulator in case that pressure tends to exceed a set maximum.

Regulator, Pressure. A device placed in a gas line for reducing, controlling, and maintaining the pressure in that portion of the piping system downstream of the device.

Regulator, Series. A pressure regulator in series with one or more other pressure regulators. The regulator nearest to the gas supply source is set to continuously limit the pressure on the inlet to the regulator downstream to some predetermined value (between the pressure of the gas supply source and the pressure of the system being controlled) that can be tolerated in the downstream system.

Regulator, Service. A pressure regulator installed by the serving gas supplier to reduce and limit the service line gas pressure to delivery pressure.

Regulator Vent. The opening in the atmospheric side of the regulator housing permitting the in and out movement of air to compensate for the movement of the regulator diaphragm.

Relief Opening. The opening provided in a draft hood to permit the ready escape to the atmosphere of the flue products from the draft hood in the event of no draft, backdraft, or stoppage beyond the draft hood and to permit inspiration of air into the draft hood in the event of a strong chimney updraft.

Room Heater, Unvented. An unvented, self-contained, freestanding, nonrecessed [except as noted under (b) of the following classifications], fuel-gas-burning appliance for furnishing warm air by gravity or fan circulation to the space in which installed, directly from the heater without duct connection. Unvented room heaters do not normally have input ratings in excess of 40,000 Btu/hr (11,720 W).

Room Heater, Unvented Circulator. A room heater designed to convert the energy in fuel gas to convected and radiant heat by direct mixing of air to be heated with the combustion products and excess air inside the jacket. Unvented circulators have an external jacket surrounding the burner and may be equipped with radiants with the jacket open in front of the radiants.

Room Heater, Vented. A vented, self-contained, freestanding, nonrecessed, fuel-gas-burning appliance for furnishing warm air to the space in which installed, directly from the heater without duct connections.

Room Heater, Vented Circulator. A room heater designed to convert the energy in fuel gas to convected and radiant heat, by transfer of heat from flue gases to a heat exchanger surface, without mixing of flue gases with circulating heated air. Vented circulators may be equipped with transparent panels and radiating surfaces to increase radiant heat transfer as long as separation of flue gases from circulating air is maintained. Vented circulators may also be equipped with an optional circulating air fan but should perform satisfactorily with or without the fan in operation.

Room Heater, Vented Circulator, Fan Type. A vented circulator equipped with an integral circulating air fan, the operation of which is necessary for satisfactory appliance performance.

Room Heater, Vented Overhead Heater. A room heater designed for suspension from or attachment to or adjacent to the ceiling of the room being heated and transferring the energy of the fuel gas to the space being heated primarily by radiation downward from a hot surface, and in which there is no mixing of flue gases with the air of the space being heated.

Room Heater, Wall Heater, Unvented Closed Front. An unvented circulator having a closed front, for insertion in or attachment to a wall or partition. These heaters are marked "UNVENTED HEATER" and do not have normal input ratings in excess of 25,000 Btu/hr (7325 W).

Room Large in Comparison with Size of Equipment. Rooms having a volume equal to at least 12 times the total volume of a furnace or air-conditioning appliance and at least 16 times the total volume of a boiler. Total volume of the appliance is determined from exterior dimensions and is to include fan compartments and burner vestibules, when used. When the actual ceiling height of a room is greater than 8 ft (2.4 m), the volume of the room is figured on the basis of a ceiling height of 8 ft (2.4 m).

This term is used in the requirements covering the installation of central heating boilers, furnaces, and air-conditioning equipment, in Chapter 6, with regard to the size of the room in which a central heating boiler or furnace is installed. It must not be confused with the term *confined space* (or unconfined space), which is concerned with air for combustion and ventilation.

Safety Blowout (Backfire Preventer). A protective device located in the discharge piping of large mixing machines, incorporating a bursting disc for excessive pressure release, means for stopping a flame front, and an electric switch or other release mechanism for actuating a built-in or separate safety shutoff. A check valve, signaling means, or both may also be incorporated.

Safety Shutoff Device. A device that will shut off the gas supply to the controlled burner(s) in the event the source of ignition fails. This device can interrupt the flow of gas to main burner(s) only or to pilot(s) and main burner(s) under its supervision.

Secondary Air. The air externally supplied to the flame at the point of combustion.

Service Head Adapter. A transition fitting for use with plastic piping (which is encased in non-pressure-carrying metal pipe) that connects the metal pipe casing and plastic pipe and tubing to the remainder of the piping system.

Service Meter Assembly. The piping and fittings installed by the serving gas supplier to connect the inlet side of the meter to the gas service and to connect the outlet side of the meter to the customer's house or yard piping.

Service Pipe. The pipe that brings the gas from the gas main to the meter.

Service Regulator. *See* Regulator, Pressure and Regulator, Service.

Shall. Indicates a mandatory requirement.

Shutoff. *See* Valve.

Sources of Ignition. Devices or equipment that, because of their intended modes of use or operation, are capable of providing sufficient thermal energy to ignite flammable gas–air mixtures.

Space, Confined. For the purposes of this code, a space whose volume is less than 50 ft^3 per 1000 Btu/hr (4.8 m^3/kW) of the aggregate input rating of all appliances installed in that space.

Space, Unconfined. For purposes of this code, a space whose volume is not less than 50 ft^3 per 1000 Btu/hr (4.8 m^3/kW) of the aggregate input rating of all appliances installed in that space. Rooms communicating directly with the space in which the appliances are installed, through openings not furnished with doors, are considered a part of the unconfined space.

All the space that connects to the region that contains the appliance(s) can be combined to calculate the volume, provided there are no doors intervening.

Specific Gravity. As applied to gas, the ratio of the weight of a given volume to that of the same volume of air, both measured under the same conditions.

Steam Cooker. *See* Food Service Equipment, Gas Steam Cooker.

Steam Generator. *See* Food Service Equipment, Gas Steam Generator.

Stress. The resultant internal force that resists change in the size or shape of a body acted on by external forces. In this code, *stress* is often used as being synonymous with unit stress, which is the stress per unit area (psi).

Tensile Strength. The highest unit tensile stress (referred to the original cross section) a material can sustain before failure (psi).

Thermostat, Electric Switch Type. A device that senses changes in temperature and controls electrically, by means of separate components, the flow of gas to the burner(s) to maintain selected temperatures.

Thermostat, Integral Gas Valve Type. An automatic device, actuated by temperature changes, designed to control the gas supply to the burner(s) in order to maintain temperatures between predetermined limits and in which the thermal actuating element is an integral part of the device: (1) *graduating thermostat,* a thermostat in which the motion of the valve is approximately in direct proportion to the effective motion of the thermal element induced by temperature change; (2) *snap-acting thermostat,* a thermostat in which the thermostatic valve travels instantly from the closed to the open position, and vice versa.

Thread Joint Compounds. Nonhardening materials used on pipe threads to ensure a seal.

Tubing. Semirigid conduit of copper, steel, aluminum, or plastic.

Type B Gas Vent. *See* Gas Vent, Type B.

Type B-W Gas Vent. *See* Gas Vent, Type B-W.

Type L Vent. *See* Gas Vent, Type L.

Unit Broiler. A broiler constructed as a separate appliance.

Unit Heater, High-Static Pressure Type. A self-contained, automatically controlled, vented, fuel-gas-burning appliance having integral means for circulation of air against 0.2 in. (15 mm) H_2O or greater static pressure. It is equipped with provisions for attaching an outlet air duct and, when the appliance is for indoor installation remote from the space to be heated, is also equipped with provisions for attaching an inlet air duct.

Unit Heater, Low-Static Pressure Type. A self-contained, automatically controlled, vented, fuel-gas burning appliance, intended for installation in the space to be heated without the use of ducts, having integral means for circulation of air, normally by a propeller fan(s), and may be equipped with louvers or face extensions made in accordance with the manufacturers' specifications.

Unusually Tight Construction. Construction where (1) walls and ceilings exposed to the outside atmosphere have a continuous water vapor retarder with a rating of 1 perm (6×10^{-11} kg/pa-sec-m^2) or less with openings gasketed or sealed; and (2) weatherstripping has been added on openable windows and doors; and (3) caulking or sealants are applied to areas such as joints around window and door frames, between sole plates and floors, between wall–ceiling joints, between wall panels, at penetrations for plumbing, electrical, and gas lines, and at other openings; and (4) the building has an average air infiltration rate of less than 0.35 air changes per hour.

This definition was revised in the 1999 edition with the addition of (4). The new (4) adds a fourth criterion, air volume, to the list of specific construction details in determining whether a building will have sufficient combustion air supplied solely through air infiltration. The new criterion (4) gives the builder and the appliance installer the opportunity to demonstrate to the authority that a building's air volume and its air infiltration rate are sufficient.

The definition is important because of the requirements for air for combustion and ventilation in Section 5.3. In 5.3.2, additional openings to the outdoors must be provided when the building in which gas utilization equipment is located is of unusually tight construction.

There are several ways to estimate the air exchange rate for a proposed or existing building. Calculations can be performed based on a number of ASHRAE models as described in ASHRAE *Fundamentals Handbook,* Chapter 25. For residential buildings, the ASHRAE handbook describes residential calculation examples utilizing effective air-leakage areas. The air exchange of an existing building can be estimated by performing calculations or measured by a "blower door" test.

It is not the definition's intent that all new construction have calculations conducted or be tested for air exchange. If any of the first three construction details are not used, then the building is not considered unusually tight and (4) need not be considered. If, however, all three details have been incorporated into a building and the builder/installer still wishes to demonstrate that the building is not unusually tight (due to large air volume, design, etc.), then (4) could be used for that purpose.

Utility Gases. Natural gas, manufactured gas, liquefied petroleum gas–air mixtures, or mixtures of any of these gases.

Valve. A device used in piping to control the gas supply to any section of a system of piping or to an appliance.

Valve, Automatic. An automatic or semiautomatic device consisting essentially of a valve and operator that control the gas supply to the burner(s) during operation of an appliance. The operator may be actuated by application of gas pressure on a flexible diaphragm, by electrical means, by mechanical means, or by other means.

Valve, Automatic Gas Shutoff. A valve used in conjunction with an automatic gas shutoff device to shut off the gas supply to a fuel-gas-burning water heating system. It may be constructed integrally with the gas shutoff device or be a separate assembly.

Valve, Equipment Shutoff. A valve located in the piping system, used to shut off individual equipment.

Valve, Individual Main Burner. A valve that controls the gas supply to an individual main burner.

Valve, Main Burner Control. A valve that controls the gas supply to the main burner manifold.

Valve, Manual Main Gas Control. A manually operated valve in the gas line for the purpose of completely turning on or shutting off the gas supply to the appliance, except to a pilot or pilots that are provided with independent shutoff.

Valve, Manual Reset. An automatic shutoff valve installed in the gas supply piping and set to shut off when unsafe conditions occur. The device remains closed until manually reopened.

Valve, Pressure Relief. A valve that automatically opens and closes a relief vent, depending on whether the pressure is above or below a predetermined value.

Valve, Relief. A safety valve designed to forestall the development of a dangerous condition by relieving either pressure, temperature, or vacuum in a hot water supply system.

Valve, Service Shutoff. A valve, installed by the serving gas supplier between the service meter or source of supply and the customer piping system, to shut off the entire piping system.

Valve, Temperature Relief. A valve that automatically opens and automatically closes a relief vent, depending on whether the temperature is above or below a predetermined value.

Valve, Vacuum Relief. A valve that automatically opens and closes a vent for relieving a vacuum within the hot water supply system, depending on whether the vacuum is above or below a predetermined value.

Valve Member. That part of a gas valve rotating within or in respect to the valve body that, by its position with respect to the valve body, controls the flow of gas.

Vent. A passageway used to convey flue gases from gas utilization equipment or their vent connectors to the outside atmosphere.

Vent Connector. The pipe or duct that connects a fuel-gas-burning appliance to a vent or chimney.

Vent Damper Device, Automatic. A device that is intended for installation in the venting system, in the outlet of or downstream of the appliance draft hood, of an individual automatically operated fuel-gas-burning appliance and that is designed to automatically open the venting system when the appliance is in operation and to automatically close off the venting system when the appliance is in a standby or shutdown condition.

Vent Damper Device, Automatic, Electrically Operated. An automatic vent damper device that employs electrical energy to control the device.

Vent Damper Device, Automatic, Mechanically Actuated. An automatic vent damper device dependent for operation on the direct application or transmission of mechanical energy without employing any type of energy conversion.

Vent Damper Device, Automatic, Thermally Actuated. An automatic vent damper device dependent for operation exclusively on the direct conversion of the thermal energy of the vent gases into mechanical energy.

Vent Gases. Products of combustion from fuel-gas-burning appliances plus excess air, plus dilution air in the venting system above the draft hood or draft regulator.

Vented Appliance, Category I.* An appliance that operates with a nonpositive vent static pressure and with a vent gas temperature that avoids excessive condensate production in the vent.

Vented Appliance, Category II.* An appliance that operates with a nonpositive vent static pressure and with a vent gas temperature that may cause excessive condensate production in the vent.

Vented Appliance, Category III.* An appliance that operates with a positive vent static pressure and with a vent gas temperature that avoids excessive condensate production in the vent.

Vented Appliance, Category IV.* An appliance that operates with a positive vent static pressure and with a vent gas temperature that may cause excessive condensate production in the vent.

A-1.7 Vented Appliance Categories I–IV.   For additional information on appliance categorization, see the appropriate Z21 and Z83 American National Standards.

Vented Wall Furnace.   A self-contained, vented, fuel-gas-burning appliance complete with grilles or equivalent, designed for incorporation in or permanent attachment to the structure of a building and furnishing heated air, circulated by gravity or by a fan, directly into the space to be heated through openings in the casing. Such appliances should not be provided with duct extensions beyond the vertical and horizontal limits of the casing proper, except that boots not to exceed 10 in. (250 mm) beyond the horizontal limits of the casing for extension through walls of nominal thickness are permitted. When such boots are provided, they shall be supplied by the manufacturer as an integral part of the appliance. This definition excludes floor furnaces, unit heaters, direct vent wall furnaces, and central furnaces.

Vented Wall Furnace, Fan-Type.   A wall furnace that is equipped with a fan.

Vented Wall Furnace, Gravity-Type.   A wall furnace that depends on circulation of air by gravity.

Venting.   Removal of combustion products as well as process fumes to the outer air.

Venting System.*   A continuous open passageway from the flue collar or draft hood of a gas-burning appliance to the outside atmosphere for the purpose of removing flue or vent gases.

A-1.7 Venting System.   A venting system is usually composed of a vent or a chimney and vent connector(s), if used, assembled to form the open passageway.

Venting System, Mechanical Draft.   A venting system designed to remove flue or vent gases by mechanical means, which may consist of an induced draft portion under nonpositive static pressure or a forced draft portion under positive static pressure.

Venting System, Mechanical Draft, Forced.   A portion of a venting system using a fan or other mechanical means to cause the removal of flue or vent gases under positive static vent pressure.

Venting System, Mechanical Draft, Induced.   A portion of a venting system using a fan or other mechanical means to cause the removal of flue or vent gases under nonpositive static vent pressure.

Venting System, Mechanical Draft, Power.   *See* Venting System, Mechanical Draft, Forced.

Venting System, Natural Draft.   *See* Draft.

Wall Furnace, Direct Vent.   A system consisting of an appliance, combustion air, and flue gas connections between the appliance and the outdoor atmosphere, and a vent cap supplied by the manufacturer and constructed so that all air for combustion is obtained from the outdoor atmosphere and all flue gases are discharged to the outdoor atmosphere. The appliance shall be complete with grilles or equivalent, designed for incorporation in or permanent attachment to the structure of a building, mobile home, or recreational vehicle, and shall furnish heated air circulated by gravity or by a fan directly into the space to be heated through openings in the casing. Such appliances shall not be provided with duct extensions beyond the vertical and horizontal limits of the appliance casing, except that boots not to exceed 10 in. (250 mm) beyond the horizontal limits of the casing for extension through walls or nominal thickness may be permitted. This definition excludes floor furnaces, unit heaters, vented wall furnaces, and central furnaces as defined in appropriate American National Standards.

Wall Head Adapter. A transition fitting for terminating plastic pipe inside of buildings at the building wall.

Water Heater. An appliance for supplying hot water for domestic or commercial purposes.

Water Heater, Automatic Circulating Tank. A water heater that furnishes hot water to be stored in a separate vessel. Storage tank temperatures are controlled by means of a thermostat installed on the water heater. Circulation may be either gravity or forced.

Water Heater, Automatic Instantaneous. A water heater that has a rated input of at least 4000 Btu/hr/gal (5 kW/L) of self-stored water. Automatic control is obtained by water-actuated control, thermostatic control, or a combination of water-actuated control and thermostatic control. This classification includes faucet-type water heaters designed to deliver water through a single faucet integral with or directly adjacent to the appliance.

Water Heater, Coil Circulation. A water heater whose heat transfer surface is composed primarily of water tubes less than $1^1/_2$ in. (38 mm) in internal diameter and that requires circulation.

Water Heater, Commercial Storage. A water heater that heats and stores water at a thermostatically controlled temperature for delivery on demand. Input rating: 75,000 Btu/hr (21,980 W) or more.

Water Heater, Countertop Domestic Storage. (1) *Concealed type:* a vented automatic storage heater that is designed for flush installation beneath a countertop 36 in. (910 mm) high, wherein the entire heater is concealed; (2) *flush type:* a vented automatic storage water heater that has flat sides, top, front, and back and is designed primarily for flush installation in conjunction with or adjacent to a counter 36 in. (910 mm) high, wherein the front and top of the heater casing are exposed; and (3) *recessed type:* a vented automatic storage water heater that has flat sides, top, front, and back and is designed for flush installation beneath a counter 36 in. (910 mm) high, wherein the front of the heater casing is exposed.

Water Heater, Domestic Storage. A water heater that heats and stores water at a thermostatically controlled temperature for delivery on demand. Input rating may not exceed 75,000 Btu/hr (21,980 W).

Water Heater, Nonautomatic Circulating Tank. A water heater that furnishes hot water to be stored in a separate vessel. Storage tank temperatures are controlled by means of a thermostat installed in the storage vessel.

References Cited in Commentary

The following publications are available from the American Society of Mechanical Engineers, United Engineering Center, 345 East 47th Street, New York, NY 10017.

ASME B31.1, *Power Piping,* 1995.
ASME B31.3, *Chemical Plant and Petroleum Refinery Piping,* 1990.

The following publications are available from CSA International, 8501 East Pleasant Valley Road, Cleveland, OH 44131.

ANSI Z21.58, *Outdoor Cooking Gas Appliances,* 1995.
ANSI Z380, *Guide for Gas Transmission and Distribution Piping,* 1998.

The following publications are available from the National Fire Protection Association, 1 Batterymarch Park, P.O. Box 9101, Quincy, MA 02269-9101.

NFPA 30, *Flammable and Combustible Liquids Code,* 1996 edition.
NFPA 37, *Standard for the Installation and Use of Stationary Combustion Engines and Gas Turbines,* 1998 edition.
NFPA 51, *Standard for the Design and Installation of Oxygen-Fuel Gas Systems for Welding, Cutting, and Allied Processes,* 1997 edition.
NFPA 51B, *Standard for Fire Prevention During Welding, Cutting, and Other Hot Work,* 1999 edition.
NFPA 52, *Standard for Compressed Natural Gas (CNG) Vehicular Fuel Systems,* 1995 edition.
NFPA 57, *Liquefied Natural Gas (LNG) Vehicular Fuel Systems Code,* 1999 edition.
NFPA 58, *Liquefied Petroleum Gas Code,* 1998 edition.
NFPA 59, *Liquefied Petroleum Gas Code at Utility Gas Plants,* 1998 edition.
NFPA 59A, *Standard for the Production, Storage, and Handling of Liquefied Natural Gas (LNG),* 1996 edition.
NFPA 70, *National Electrical Code®,* 1999 edition.
NFPA *101®, Life Safety Code®,* 2000 edition.

The following publication is available from the U.S. Government Printing Office, Washington, DC 20401.

Code of Federal Regulations, 49 *CFR,* Part 192.

The following publication is available from American Society of Heating, Refrigerating and Air Conditioning Engineers, Inc., 1791 Tullie Circle, N.E., Atlanta, GA 30329-2305.

ASHRAE *Fundamentals Handbook,* 1997.

Gas Piping System Design, Materials, and Components

Chapter 2 covers the requirements for piping design, materials, and components. It includes detailed requirements for the following:

- A piping plan, which can be required by the authority having jurisdiction for new and modified gas piping systems (in Section 2.1)
- Provisions for location of the point of delivery (in Section 2.2)
- Interconnection between gas piping systems to prevent contamination and inaccurate metering of gas (in Section 2.3)
- Sizing of gas piping systems to ensure that sufficient gas can be provided to each appliance (tables are included in Chapter 9 to size piping systems, and other engineering methods can be used) (in Section 2.4)
- Piping system operating pressure limitations, which limit gas pressure in piping systems in residential and most commercial buildings (in Section 2.5)
- Acceptable piping materials and joining methods. This section contains comprehensive requirements for all aspects of pipe joining, threads, flanges, and gaskets. This section also contains requirements for both metallic and plastic piping (in Section 2.6).
- Gas meters and pressure regulators (in Sections 2.7 and 2.8)
- Gas pressure regulators, including requirements for their location, protection, ventilation, and identification (in Section 2.8)
- Overpressure protection devices to prevent the high pressures of gas distribution systems and propane tanks from entering building piping systems (high pressure can result in gas control failure and leakage of gas into a building) (in Section 2.9)
- Back-pressure protection to prevent contamination of piped air, oxygen, or alternative fuel gas (in Section 2.10)
- Low-pressure protection to prevent air from entering pipes (in Section 2.11)
- Shutoff valves (in Section 2.12)
- Expansion and flexibility of gas piping systems, including thermal and seismic stresses (in Section 2.13)

2.1 Piping Plan

2.1.1 Installation of Piping System.

Where required by the authority having jurisdiction, a piping sketch or plan shall be prepared before proceeding with the installation. This plan shall show the proposed location of piping, the size of different branches, the various load demands, and the location of the point of delivery.

Examples of piping plans are shown in Exhibit 2.1 and in Appendix C.

2.1.2 Addition to Existing System.

When additional gas utilization equipment is being connected to a gas piping system, the existing piping shall be checked to determine whether it has adequate capacity *(see 2.4.3)*. If inadequate, the existing system shall be enlarged as required, or separate gas piping of adequate capacity shall be provided.

These provisions recognize that preparing a plan or sketch of the proposed piping system is necessary, whether it is new or an extension of an existing system. (See Exhibit 2.1.) By showing the loads and lengths of runs, the sizing can be determined by methods described in this chapter (see Section 2.4), and having any added gas utilization equipment overload the gas piping system can be avoided.

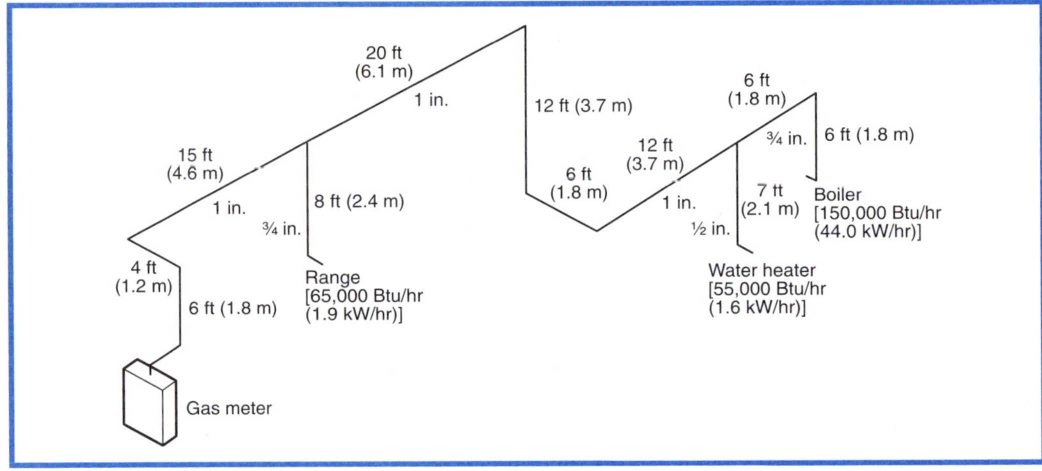

Exhibit 2.1 *Typical piping layout. (Courtesy of Richard E. White.)*

2.2 Provision for Location of Point of Delivery

The location of the point of delivery shall be acceptable to the serving gas supplier.

To avoid unnecessary changes and reworking, the gas supplier should be consulted to determine the best point of delivery.

2.3 Interconnections Between Gas Piping Systems

2.3.1 Interconnections Supplying Separate Users.

Where two or more meters, or two or more service regulators where meters are not provided, are located on the same premises and supply separate users, the gas piping systems shall not be interconnected on the outlet side of the meters or service regulators.

Interconnected gas systems can cause a potential safety problem. Maintenance, emergency, and utility personnel have no way of knowing that all services must be turned off to take the system out of service safely.

2.3.2 Interconnections for Standby Fuels.

Where a supplementary gas for standby use is connected downstream from a meter or a service regulator where a meter is not provided, a device to prevent backflow shall be installed. A three-way valve installed to admit the standby supply and at the same time shut off the regular supply shall be permitted to be used for this purpose.

This paragraph recognizes that while proper design of the gas service system prevents damage to its components (e.g., meter, regulator, pressure relief valve, etc.), connection to a standby gas system can subject these components to unanticipated conditions, possibly causing their failure. While most three-way valves (motorized or manual) allow backflow while in motion, this limited exposure is not a hazard in normal operations. The use of such valves is permitted expressly by the code. Installation of a backflow check valve downstream of the meter or service regulator also satisfies this requirement and has the advantage of automatically opening in the event of standby system failure. The requirement of this paragraph also can be met by installing a combination manual valve and backflow check valve device that is manufactured for this purpose.

2.4 Sizing of Gas Piping Systems

2.4.1* General Considerations.

Gas piping systems shall be of such size and so installed as to provide a supply of gas sufficient to meet the maximum demand without undue loss of pressure between the point of delivery and the gas utilization equipment.

This provision lists the general considerations for any fluid distribution system, and the paragraphs that follow (2.4.2 through 2.4.4) describe the topics in greater detail.

The code makes reference to the tables in Chapter 9, "Sizing Tables," and the additional material provided in Appendix C.

While the code requires that gas piping be sized to provide sufficient gas to meet the maximum load, it does not specify the method to be used to determine the sizing. The tables in Chapter 9 are provided for this reason. Appendix C gives an example based on the tables in Chapter 9 and mentions the Polyflo Computer (Dallas, TX), which can also be used for sizing. A third, graphical sizing method is given in the commentary at the end of Appendix C. Gas pipe sizing computer software also is available.

The tables in Chapter 9 are the most widely used method of sizing gas piping. They contain gas-carrying capacities for different sizes and lengths of iron pipe (or equivalently dimensioned rigid pipe), semi-rigid tubing, and corrugated stainless steel tubing (CSST). The tables are in order by type of gas (natural or LP), type of piping (pipe, tube, or CSST), pressure, and pressure drop allowed.

This method is straightforward and can be used by the gas piping installer. The tables for pipe and tubing do not attempt to optimize pipe size to minimize material costs, because the potential savings to be had on a house or a small multifamily dwelling or commercial building will be much less than the cost of the time it would take an engineer to calculate the system precisely. In larger installations, an architect or engineer is usually involved, and it can be economical to use another sizing method because significant savings can result.

Gas piping systems must supply the volume of gas required by each piece of gas utilization equipment at a pressure within the design range established by the manufacturer. Also, ensuring that gas pressure to the connected equipment will remain within the manufacturer's design limits when the equipment cycles "off" is important. A gas piping system's operating pressure will have an effect on capacity, with greater pressures producing greater capacities for a given pipe or tube size. Gas piping systems that operate at pressures exceeding the connected gas utilization equipment's rated inlet pressure or that deliver gas at a pressure that varies outside the equipment manufacturer's design pressure limits must incorporate pressure regulators so that gas delivered to the equipment is within the design pressure range. See Section 2.8 for line gas pressure regulator requirements.

A.2.4.1 The size of gas piping depends on the following factors:

(1) Allowable loss in pressure *(see 2.4.4)* from point of delivery to equipment
(2) Maximum gas demand
(3) Length of piping and number of fittings
(4) Specific gravity of the gas
(5) Diversity factor
(6) Foreseeable future demand

2.4.2* Maximum Gas Demand.

The volume of gas to be provided (in cubic feet per hour) shall be determined directly from the manufacturers' input ratings of the gas utilization equipment served. Where the input rating is not indicated, the gas supplier, equipment manufacturer, or a qualified agency shall be contacted for estimating the volume of gas to be supplied.

The total connected hourly load shall be used as the basis for piping sizing, assuming all equipment is operating at full capacity simultaneously.

Natural gas has a nominal heating value of 1000 Btu/ft^3. Propane gas has a nominal heating value of about 2500 Btu/ft^3. The undiluted LP-Gas tables in Chapter 9 are calculated for propane gas and present the flow data in terms of thousands of Btu per hour, which negates any need for conversion from cubic feet to Btu.

In some commercial and industrial applications, equipment requires pressure greater than the normal 7-in. (1.7-kPa) water column for natural gas or 11-in. (2.7-kPa) water column for propane. In this case, the serving supplier should be contacted to determine the available pressure; the piping system should be designed according to that supply pressure.

Exception: Smaller sized piping shall be permitted where a diversity of load is established.

Diversity of load becomes important in buildings such as apartment buildings, where there can be many gas ranges, water heaters, or furnaces on a single gas piping system. Logic dictates that we will never see all of these appliances operating at full input (all range burners plus the ovens, water heaters, and furnaces) simultaneously. Load diversity factors applicable to your area can be obtained from the gas supplier.

A.2.4.2 To obtain the cubic feet per hour of gas required, divide the Btu per hour rating by the Btu per cubic foot heating value of the gas supplied. The heating value of the gas can be obtained from the local gas supplier.

Where the ratings of the equipment to be installed are not known, Table C.2 in Appendix C shows the approximate demand of typical appliances by types.

2.4.3* Gas piping shall be sized in accordance with one of the following:

(1) The tables in Chapter 9
(2) Other approved engineering methods acceptable to the authority having jurisdiction

Although the code does not mandate the use of the tables in Chapter 9, most code users follow them because they are easy to use and because they provide adequate pipe size with an ample margin for future expansion. Other engineering methods, such as the graphical method provided in Appendix C, are also permitted, but only when approved by the authority having jurisdiction.

A.2.4.3 *Gas Piping Size.* The gas-carrying capacities for different sizes and lengths of iron pipe, or equivalent rigid pipe, and semirigid tubing are shown in the capacity tables in Chapter 9.

Tables 9.1 through 9.12 indicate approximate capacities for single runs of piping. If the specific gravity of the gas is other than 0.60, correction factors should be applied. Correction factors for use with these tables are given in Table 9.13.

For any gas piping system, for special gas utilization equipment, or for conditions other than those covered by the capacity tables in Chapter 9, such as longer runs, greater gas demands, or greater pressure drops, the size of each gas piping system should be determined by standard engineering methods acceptable to the authority having jurisdiction.

A suggested procedure with an example of using tables to size a gas piping system is presented in Appendix C.

2.4.4 Allowable Pressure Drop.

The design pressure loss in any piping system under maximum probable flow conditions, from the point of delivery to the inlet connection of the gas utilization equipment, shall be such that the supply pressure at the equipment is greater than the minimum pressure required for proper equipment operation.

The tables in Chapter 9 for piping systems that are without line gas pressure regulators or are downstream from them limit pressure drops to relatively low values. This requirement is set to provide pressure stability to the equipment whether it is "off" or at full load. If other engineering methods are used for pipe sizing, consideration should be given to what happens to the delivered pressure when the equipment cycles "off." This consideration limits the application of large pressure drops without the use of line gas pressure regulators.

2.5 Piping System Operating Pressure Limitations

2.5.1 Maximum Design Operating Pressure.

The maximum design operating pressure for piping systems located inside buildings shall not exceed 5 psi (34 kPa) unless one or more of the following conditions are met:

(1)* The piping system is welded.

A.2.5.1(1) For welding specifications and procedures that can be used, see the API 1104, *Standard for Welding Pipelines and Related Facilities;* AWS B2.1,*Standard for Welding Procedure and Performance Qualification*; or ASME *Boiler and Pressure Vessel Code,* Section IX.

(2) The piping is located in a ventilated chase or otherwise enclosed for protection against accidental gas accumulation.

(3) The piping is located inside buildings or separate areas of buildings used exclusively for one of the following:

  a. Industrial processing or heating
  b. Research
  c. Warehousing
  d. Boiler or mechanical equipment rooms

The requirement of obtaining the approval of the authority having jurisdiction to use pressure higher than 5 psi (34 kPa) in buildings has been removed in the 1999 edition. This requirement was believed to be unnecessary because the criteria for granting approval are presented, making specific approval unnecessary for each installation.

If a leak occurs in a gas piping system, the flow of gas will be proportional to the pressure. Most gas utilization equipment is designed to require substantially less than 5 psi (34 kPa) for proper operation. The 5 psi (34 kPa) general pressure limit for threaded gas piping systems in buildings recognizes these facts. For piping above 5 psi (34 kPa), welding is required because welded pipe joints have a lower possibility of leakage compared to threaded pipe joints.

The user is permitted to exceed this limit if special techniques in 2.5.1(1) and 2.5.1(2) are used or if the piping system is limited to special buildings or areas of buildings [2.5.1(3)]. The conditions described in 2.5.1(3) recognize that gas utilization equipment in these settings can require high-pressure gas and that it is unlikely that any leakage will be quickly diluted by the higher ventilation in these areas. This paragraph is applicable to both natural gas and LP-Gas systems.

(4) The piping is a temporary installation for buildings under construction.

2.5.2 Liquefied Petroleum Gas Systems.

The operating pressure for undiluted LP-Gas systems shall not exceed 20 psi (140 kPa). Buildings having systems designed to operate below −5°F (−21°C) or with butane or a propane–butane mix shall be designed to either accommodate liquid LP-Gas or prevent LP-Gas vapor from condensing back into a liquid.

The 20-psi (140-kPa) limit is a practical one because, at higher pressures, propane droplets will begin to form at temperatures below −5°F (−21°C), a temperature that can be expected in most parts of the country at some time during the year. LP-Gas in the liquid form, entering a burner designed for vapor, will result in a large increase in flame size, which can be a significant fire hazard. When vaporized, LP-Gas expands by a factor of 270 to 1 and, though a lesser volume of liquid will pass through an orifice than vapor, a much greater quantity of Btu will be available to be burned.

The second sentence recognizes that condensation of LP-Gas can occur at low temperatures, especially when butane is the fuel gas. (While butane is not used as a residential fuel in the United States, its cost can be sufficiently low at times to be used in industrial applications.) When this condensation cannot be prevented, the piping system must be designed to retain the liquid so that only vapor is delivered to the appliances. This configuration can be made by the use of pressure regulators that are rated for such service through creating liquid retention areas in the piping system or by other means. Typical larger residential installations using ASME tanks have a run of 10-psi (280-kPa) piping from the first-stage regulator, located in the tank dome, to the second-stage regulator located on an exterior building wall. If the piping is run underground, the portion from the underground section to the regulator is exposed to ambient temperature. In winter, if the temperature falls below $-5°F$ $(-21°C)$, liquid will condense and drain to the underground section. The liquid will stay in the underground section until the heat in the soil vaporizes it. This is an example of a piping system that accommodates liquid.

Table 2.1 shows the pressure at which liquefaction will occur for the minimum expected system temperature for propane systems. Table 2.2 presents similar data for butane systems.

Table 2.1 Liquefaction of Propane

Minimum Anticipated Temperature (°F)	Liquefaction Pressure (psi)
−5	20
−10	16.7
−15	13.6
−20	10.7
−25	8.0
−30	5.6
−35	3.4
−40	1.5

For SI units, °C = 5/9 (°F − 32); 1 psi = 6.894 kPa.

Table 2.2 Liquefaction of Butane

Minimum Anticipated Temperature (°F)	Liquefaction Pressure (psi)
80	22.9
75	19.8
70	16.9
65	14.2
60	11.6
55	9.1
50	6.9
45	4.9
40	3.0
35	1.3

For SI units: °C = 5/9 (°F − 32); 1 psi = 6.894 kPa.

Exception: Buildings or separate areas of buildings constructed in accordance with Chapter 7 of NFPA 58, Liquefied Petroleum Gas Code, and used exclusively to house industrial processes, research and experimental laboratories, or equipment or processing having similar hazards.

Chapter 7 of NFPA 58, *Liquefied Petroleum Gas Code,* sets requirements for buildings, attached buildings, and rooms within buildings where LP-Gas operations are housed, and it also provides requirements for construction, heating, and venting of deflagrations.

2.6 Acceptable Piping Materials and Joining Methods

2.6.1 General.

(a) *Material Application.* Materials and components conforming to standards or specifications listed herein or acceptable to the authority having jurisdiction shall be permitted to be used for appropriate applications, as prescribed and limited by this code.

(b) *Used Materials.* Pipe, fittings, valves, or other materials shall not be used again unless they are free of foreign materials and have been ascertained to be adequate for the service intended.

(c) *Other Materials.* Material not covered by the standards specifications listed herein shall be investigated and tested to determine that it is safe and suitable for the proposed service and, in addition, shall be recommended for that service by the manufacturer and shall be acceptable to the authority having jurisdiction.

As indicated in the commentary for Chapter 1 following Section 1.2, Alternate Materials, Equipment, and Procedures, the code's intent is not to prohibit the use of new or different materials or technology. Similarly, the intent of 2.6.1(c) is to provide the authority having jurisdiction a basis on which to judge the acceptability of piping materials not recognized specifically in the code (i.e., requiring evidence to substantiate the safety and suitability of the new piping materials or practices).

2.6.2 Metallic Pipe.

(a) Cast-iron pipe shall not be used.

(b) Steel and wrought-iron pipe shall be at least of standard weight (Schedule 40) and shall comply with one of the following standards:

(1) ANSI/ASME B36.10, *Standard for Welded and Seamless Wrought-Steel Pipe*
(2) ASTM A 53, *Standard Specification for Pipe, Steel, Black and Hot-Dipped, Zinc-Coated Welded and Seamless*
(3) ASTM A 106, *Standard Specification for Seamless Carbon Steel Pipe for High-Temperature Service*

(c)* Copper and brass pipe shall not be used if the gas contains more than an average of 0.3 grains of hydrogen sulfide per 100 scf of gas (0.7 mg/100 L).

Threaded copper, brass, or aluminum alloy pipe in iron pipe sizes shall be permitted to be used with gases not corrosive to such material.

This hydrogen sulfide concentration is equivalent to a trace and should not have any significant, corrosive effect on copper or brass pipe.

A.2.6.2(c) An average of 0.3 grains of hydrogen sulfide per 100 scf (0.7 mg/100 L) is equivalent to a trace as determined by ANSI/ASTM D 2385, *Method of Test for Hydrogen Sulfide and Mercaptan Sulfur in Natural Gas (Cadmium Sulfate — Iodometric Titration Method)*, or ANSI/ASTM D 2420, *Method of Test for Hydrogen Sulfide in Liquefied Petroleum (LP) Gases (Lead Acetate Method)*.

(d) Aluminum alloy pipe shall comply with ASTM B 241, *Specification for Aluminum-Alloy Seamless Pipe and Seamless Extruded Tube* (except that the use of alloy 5456 is prohibited) and shall be marked at each end of each length indicating compliance. Aluminum alloy pipe shall be coated to protect against external corrosion where it is in contact with masonry, plaster, or insulation or is subject to repeated wettings by such liquids as water, detergents, or sewage.

Aluminum alloy pipe shall not be used in exterior locations or underground.

The marking requirement for aluminum tubing is made so that technicians and inspectors can see readily that installed tubing meets the proper specification. The coating requirement and prohibition of underground and exterior applications are due to corrosion concerns.

Strong acids and strong bases readily corrode aluminum and its alloys. The assistance of a metallurgist or qualified corrosion specialist is strongly recommended in selecting an aluminum alloy, because the corrosion behavior of aluminum alloys varies significantly.

2.6.3 Metallic Tubing.

Seamless copper, aluminum alloy, or steel tubing shall be permitted to be used with gases not corrosive to such material.

Copper, aluminum alloy, or steel tubing is permitted for the types and grades listed, provided that the gas is not corrosive to the tubing material. Note that aluminum alloy tubing is subject to the same restrictions as aluminum alloy pipe.

(a) Steel tubing shall comply with ASTM A 539, *Standard Specification for Electric Resistance-Welded Coiled Steel Tubing for Gas and Fuel Oil Lines*, or ASTM A 254, *Standard Specification for Copper Brazed Steel Tubing*.

(b)* Copper and brass tubing shall not be used if the gas contains more than an average of 0.3 grains of hydrogen sulfide per 100 scf of gas (0.7 mg/100 L). Copper tubing shall comply with standard Type K or L of ASTM B 88, *Specification for Seamless Copper Water Tube,* or ASTM B 280, *Specification for Seamless Copper Tube for Air Conditioning and Refrigeration Field Service.*

The prohibition of the use of copper and brass tubing when gas contains more than a trace of hydrogen sulfide was added in the 1999 edition to repeat this prohibition that previously was listed only in 2.6.2(d) for copper and brass pipe. All natural gas that is distributed in pipelines, and all LP-Gas, in the United States today is treated to remove hydrogen sulfide. If natural gas is obtained directly from a gas well, this prohibition is important.

A.2.6.3(b) See A.2.6.2(c).

Copper and brass tubing and fittings (except tin-lined copper tubing) should not be used if the gas contains more than an average of 0.3 grains of hydrogen sulfide per 100 scf of gas (0.7 mg/100 L).

(c) Aluminum alloy tubing shall comply with ASTM B 210, *Specification for Aluminum-Alloy Drawn Seamless Tubes,* or ASTM B 241, *Specification for Aluminum Alloy Seamless Pipe and Seamless Extruded Tube.* Aluminum alloy tubing shall be coated to protect against external corrosion where it is in contact with masonry, plaster, or insulation or is subject to repeated wettings by such liquids as water, detergent, or sewage.

Aluminum alloy tubing shall not be used in exterior locations or underground.

(d) Corrugated stainless steel tubing shall be tested and listed in compliance with the construction, installation, and performance requirements of ANSI/AGA LC-1, *Standard for Fuel Gas Piping Systems Using Corrugated Stainless Steel Tubing.*

Corrugated stainless steel tubing (CSST) was introduced to the North American market in the mid-1980s as a new gas piping material. It has been accepted and is being used throughout most areas. Supplement 2, "Corrugated Stainless Steel Gas Piping Systems," provides more detail. See Exhibit 2.2.

In 1993, ANSI/AGA LC 1, *Standard for Interior Fuel Gas Piping Systems Using Corrugated Stainless Steel Tubing,* was published. The *National Fuel Gas Code* permits the installation of systems using CSST listed in accordance with the requirements of ANSI/AGA LC 1.

2.6.4 Plastic Pipe, Tubing, and Fittings.

Plastic pipe, tubing, and fittings shall be used outside underground only and shall conform with ASTM D 2513, *Standard Specification for Thermoplastic Gas Pressure Pipe, Tubing, and Fittings.* Pipe to be used shall be marked "gas" and "ASTM D 2513."

Anodeless risers shall comply with the following:

(a) Factory-assembled anodeless risers shall be recommended by the manufacturer for the gas used and shall be leak tested by the manufacturer in accordance with written procedures.

Exhibit 2.2 *Corrugated stainless steel tubing. (Courtesy of Foster-Miller, Inc.)*

(b) Service head adapters and field-assembled anodeless risers incorporating service head adapters shall be recommended by the manufacturer for the gas used by the manufacturer and shall be design-certified to meet the requirements of Category I of ASTM D 2513, *Standard Specification for Thermoplastic Gas Pressure Pipe, Tubing, and Fittings,* and the *Code of Federal Regulations,* Title 49, Part 192.281(e). The manufacturer shall provide the user qualified installation instructions as prescribed by the *Code of Federal Regulations,* Title 49, Part 192.283(b).

This section was revised in the 1996 edition by the addition of requirements for risers. Risers have been in use for gas piping that is upstream of the utility meter.

A riser is an assembly that makes the transition from underground plastic pipe or tubing to aboveground metal pipe or tubing. Risers are available as one-piece, factory-assembled units and as field-assembled units. Factory-assembled risers are coated in the factory with a baked, protective coating that makes the use of a sacrificial anode for corrosion protection unnecessary; hence they are called "anodeless."

Risers are used when polyethylene pipe or tubing is run underground to make the transition to metal piping aboveground. (See Exhibit 2.3.) Examples of such use include serving a gas light or providing gas to a detached garage for heating.

(c) The use of plastic pipe, tubing, and fittings in undiluted liquefied petroleum gas piping systems shall be in accordance with NFPA 58, *Liquefied Petroleum Gas Code.*

2.6.5 Workmanship and Defects.

Gas pipe or tubing and fittings shall be clear and free from cutting burrs and defects in structure or threading and shall be thoroughly brushed, and chip and scale blown.

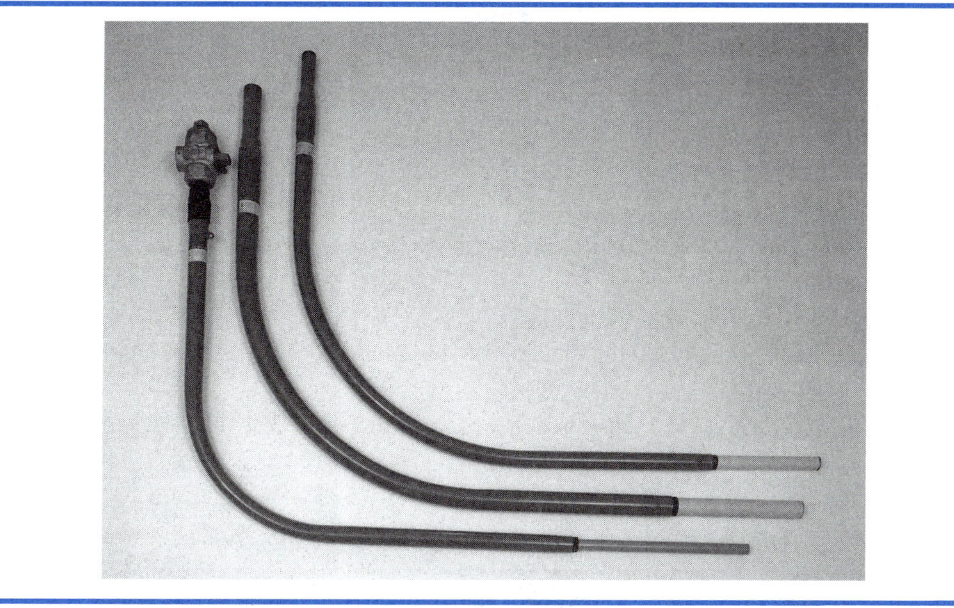

Exhibit 2.3 *Risers, used to connect underground polyethylene to aboveground metallic piping. (Courtesy of R. W. Lyall & Co.)*

Defects in pipe or tubing or fittings shall not be repaired. When defective pipe, tubing, or fittings are located in a system, the defective material shall be replaced. *[See 4.1.1(c).]*

This paragraph requires the installer to inspect and clean piping materials before assembly to ensure compliance with the code. The defects specified could cause problems in the small clearances between moving parts and in orifices. Defects in pipe, tubing, or fittings may not be repaired. When defective pipe, tubing, or fittings are located in a system, the defective material must be replaced. [See 4.1.1(c).]

Repairing defective components is not an appropriate remedy. This paragraph requires each installation to be made with defect-free materials. Paragraph 4.1.1(c) covers testing of any repaired systems.

2.6.6 Protective Coating.

Where in contact with material or atmosphere exerting a corrosive action, metallic piping and fittings coated with a corrosion-resistant material shall be used.

External or internal coatings or linings used on piping or components shall not be considered as adding strength.

Commonly used materials include paint, polyethylene jacketing over a mastic primer, and fusion-bonded epoxy. Generally, paint is used on abovegrade applications and the others on underground pipe. Using a corrosion-resistant material for these applications is often more satisfactory than providing corrosion protection for a material that is less resistant. When using coated steel pipe underground, consideration generally is given to providing it with cathodic protection.

2.6.7 Metallic Pipe Threads.

(a) *Specifications for Pipe Threads.* Metallic pipe and fitting threads shall be taper pipe threads and shall comply with ANSI/ASME B1.20.1, *Standard for Pipe Threads, General Purpose (Inch).*

(b) *Damaged Threads.* Pipe with threads that are stripped, chipped, corroded, or otherwise damaged shall not be used. If a weld opens during the operation of cutting or threading, that portion of the pipe shall not be used.

(c) *Number of Threads.* Field threading of metallic pipe shall be in accordance with Table 2.6.7.

Table 2.6.7 Specifications for Threading Metallic Pipe

Iron Pipe Size (in.)	Approximate Length of Threaded Portion (in.)	Approximate No. of Threads to Be Cut
$^1/_2$	$^3/_4$	10
$^3/_4$	$^3/_4$	10
1	$^7/_8$	10
$1^1/_4$	1	11
$1^1/_2$	1	11
2	1	11
$2^1/_2$	$1^1/_2$	12
3	$1^1/_2$	12
4	$1^5/_8$	13

For SI units, 1 in. = 25.4 mm.

Properly cut pipe threads are tapered (e.g., the threads nearest the end of the pipe have a smaller diameter than the threads farthest away). If common threading dies (e.g., fixed cutting heads) are run too far onto the pipe during the threading process, the resulting threads will be straight for a portion of the threads, then they will taper. These straight threads are known as running threads. Since pipe fittings depend on a tapered thread on the pipe to seal properly, an improperly threaded pipe can easily result in a leak at the joint. Proper use of a fixed die requires the threading operation to stop as soon as the end of the pipe exits the die. Table 2.6.7 shows the approximate number of threads necessary to properly thread a pipe.

(d) *Thread Compounds.* Thread (joint) compounds (pipe dope) shall be resistant to the action of liquefied petroleum gas or to any other chemical constituents of the gases to be conducted through the piping.

> Dry pipe taper threads will not seal joints. Thread compounds that are not resistant to the conducted gas can break down and cause leaks. Thread compounds should be applied to only the male pipe connection to avoid having the compound enter the piping system during assembly. Teflon® tape is also a thread compound. Application of Teflon tape must be done properly to prevent pieces of Teflon from entering the gas piping system and causing problems downstream. Component manufacturer's instructions (for example, for equipment such as gas valves) should be consulted regarding restrictions on Teflon tape.

2.6.8 Metallic Piping Joints and Fittings.

The type of piping joint used shall be suitable for the pressure-temperature conditions and shall be selected giving consideration to joint tightness and mechanical strength under the service conditions. The joint shall be able to sustain the maximum end force due to the internal pressure and any additional forces due to temperature expansion or contraction, vibration, fatigue, or the weight of the pipe and its contents.

> This section lists considerations that influence the selection of a joining method. A threaded joint is not desirable for a system that is exposed to significant vibration or external mechanical loads.

(a)* *Pipe Joints.* Pipe joints shall be threaded, flanged, or welded, and nonferrous pipe shall be permitted to also be brazed with materials having a melting point in excess of 1000°F (538°C). Brazing alloys shall not contain more than 0.05 percent phosphorus.

> Since some fuel gases can contain trace amounts of sulfur, brazing alloys containing more than 0.05 percent phosphorus are not permitted; sulfur and phosphorous will combine to destroy the joint. This prohibition applies even if a current gas supply contains no sulfur, because the gas supply can shift to a source containing a trace of sulfur. Most brazing alloys contain less than this amount of phosphorus.

A.2.6.8(a) For welding and brazing specifications and procedures that can be used, see API 1104, *Standard for Welding Pipelines and Related Facilities;* AWS B2.1, *Standard for Welding Procedure and Performance Qualification;* AWS B2.2, *Standard for Brazing Procedure and Performance Qualification;* or *ASME Boiler and Pressure Vessel Code,* Section IX.

(b) *Tubing Joints.* Tubing joints shall either be made with approved gas tubing fittings or be brazed with a material having a melting point in excess of 1000°F (538°C). Brazing alloys shall not contain more than 0.05 percent phosphorus.

The phrase *approved gas tubing fittings* means gas tubing fittings that have been approved by the authority having jurisdiction. This category generally includes flare and compression fittings.

(c) *Flared Joints.* Flared joints shall be used only in systems constructed from nonferrous pipe and tubing where experience or tests have demonstrated that the joint is suitable for the conditions and where provisions are made in the design to prevent separation of the joints.

(d) *Metallic Fittings (Including Valves, Strainers, Filters).*

(1) Threaded fittings in sizes larger than 4 in. (100 mm) shall not be used unless acceptable to the authority having jurisdiction.

Metallic threaded joints larger than 4 in. are limited to applications approved by the authority having jurisdiction because of the difficulty in developing and maintaining tight threaded joints in these sizes. Alternate fittings include welded and flanged fittings.

(2) Fittings used with steel or wrought-iron pipe shall be steel, brass, bronze, malleable iron, or cast iron.
(3) Fittings used with copper or brass pipe shall be copper, brass, or bronze.
(4) Fittings used with aluminum alloy pipe shall be of aluminum alloy.
(5) *Cast-Iron Fittings.*

   a. Flanges shall be permitted to be used.
   b. Bushings shall not be used.

These prohibitions are based on the brittle nature of cast iron. This brittleness makes bushings less able to withstand overtightening during assembly or external mechanical abuse in service. Flammable gas–air mixtures in piping systems require more shock resistance than cast iron fittings can provide. Large cast iron fittings [flanged, since 2.6.8(d)1 prohibits large threaded fittings] cannot resist external loads adequately and are only allowed on approval of the authority having jurisdiction.

   c. Fittings shall not be used in systems containing flammable gas–air mixtures.
   d. Fittings in sizes 4 in. (100 mm) and larger shall not be used indoors unless approved by the authority having jurisdiction.
   e. Fittings in sizes 6 in. (150 mm) and larger shall not be used unless approved by the authority having jurisdiction.

(6) *Brass, Bronze, or Copper Fittings.* Fittings, if exposed to soil, shall have a minimum 80 percent copper content.
(7) *Aluminum Alloy Fittings.* Threads shall not form the joint seal.

Flanges, welded fittings, flared fittings, and metallic ball sleeve compression fittings are acceptable under this requirement.

(8) *Zinc–Aluminum Alloy Fittings.* Fittings shall not be used in systems containing flammable gas–air mixtures.

(9) *Special Fittings.* Fittings such as couplings, proprietary-type joints, saddle tees, gland-type compression fittings, and flared, flareless, or compression-type tubing fittings shall be permitted to be used provided they are (1) used within the fitting manufacturers' pressure-temperature recommendations; (2) used within the service conditions anticipated with respect to vibration, fatigue, thermal expansion, or contraction; (3) installed or braced to prevent separation of the joint by gas pressure or external physical damage; and (4) acceptable to the authority having jurisdiction.

2.6.9 Plastic Piping, Joints, and Fittings.

Plastic pipe, tubing, and fittings shall be joined in accordance with the manufacturers' instructions. The following shall be observed when making such joints:

(a) The joint shall be designed and installed so that the longitudinal pullout resistance of the joint will be at least equal to the tensile strength of the plastic piping material.

(b) Heat-fusion joints shall be made in accordance with qualified procedures that have been established and proven by test to produce gastight joints at least as strong as the pipe or tubing being joined. Joints shall be made with the joining method recommended by the pipe manufacturer. Heat fusion fittings shall be marked "ASTM D 2513."

(c) Where compression-type mechanical joints are used, the gasket material in the fitting shall be compatible with the plastic piping and with the gas distributed by the system. An internal tubular rigid stiffener shall be used in conjunction with the fitting. The stiffener shall be flush with the end of the pipe or tubing and shall extend at least to the outside end of the pipe or tubing and shall extend at least to the outside end of the compression fitting when installed. The stiffener shall be free of rough or sharp edges and shall not be a force fit in the plastic. Split tubular stiffeners shall not be used.

(d) Plastic piping joints and fittings for use in liquefied petroleum gas piping systems shall be in accordance with *Liquefied Petroleum Gas Code,* NFPA 58.

Since the manufacturers' instructions must be followed for any joining procedures, their recommendations become the basis of all installation procedures. Systems that are in common use and are recommended by most manufacturers for the joining of plastic gas pipe include butt fusion, saddle fusion, socket fusion, electrofusion, and mechanical joints. The code places additional requirements in 2.6.9(a), 2.6.9(b), and 2.6.9(c) to ensure a safe system. Note that as with everything else involving the use of plastic pipe in LP-Gas systems, NFPA 58, *Liquefied Petroleum Gas Code,* is the applicable standard.

2.6.10 Flanges.

All flanges shall comply with ANSI/ASME B16.1, *Standard for Cast Iron Pipe Flanges and Flanged Fittings;* ANSI/ASME B16.20, *Standard for Ring-Joint Gaskets and Grooves for*

Steel Pipe Flanges; or MSS SP-6, *Standard Finishes for Contact Faces of Pipe Flanges and Connecting-End Flanges of Valves and Fittings.* The pressure-temperature ratings shall equal or exceed that required by the application.

(a) *Flange Facings.* Standard facings shall be permitted for use under this code. Where 150-psi (1090 kPa) steel flanges are bolted to Class 125 cast-iron flanges, the raised face on the steel flange shall be removed.

This requirement recognizes that if the face were left on the steel flange, the resulting structural forces on the cast iron flange could cause its breakage.

(b) *Lapped Flanges.* Lapped flanges shall be permitted to be used only above ground or in exposed locations accessible for inspection.

2.6.11 Flange Gaskets.

The material for gaskets shall be capable of withstanding the design temperature and pressure of the piping system and the chemical constituents of the gas being conducted without change to its chemical and physical properties. The effects of fire exposure to the joint shall be considered in choosing the material.

(1) Acceptable materials include the following:

    a. Metal or metal-jacketed asbestos (plain or corrugated)
    b. Asbestos
    c. Aluminum "O" rings and spiral-wound metal gaskets

"O" rings, spiral-wound metal gaskets, and some other types of flange gaskets either do not extend to the outside edge of the flange face or are not of full thickness all the way to the edge of the flange face. Either condition could place unacceptable mechanical strain on bronze or cast iron flanges.

(2) When a flanged joint is opened, the gasket shall be replaced.
(3) Full-face gaskets shall be used with all bronze and cast-iron flanges.

2.7* Gas Meters

Premises-owned meters are typically submeters that are downstream of the gas supplier's meter (master meter) in applications such as mobile home courts, apartment buildings, or industrial properties where measurement of the gas flow downstream of the gas supplier's meter is necessary. These provisions are intended to ensure safe, trouble-free service from the meters and to protect them from physical damage. (See Exhibit 2.4.) Meters installed and owned by the gas supplier are covered by the

Exhibit 2.4 *Protection of a gas meter from physical harm. (Courtesy of Richard E. White.)*

regulations of the U.S. Department of Transportation, *Code of Federal Regulations, 49 CFR*, Part 192.

A.2.7 This section applies to premises-owned meters *[see 1.1.1(b)16].*

2.7.1 Capacity.

Gas meters shall be selected for the maximum expected pressure and permissible pressure drop.

Small, residential-size meters measure their rated flow with a $^1/_2$-in. (0.1 kPa) water column pressure drop. Larger meters are dual-rated, and the designer can choose between the rated flow at a $^1/_2$-in. (0.1 kPa) water column pressure drop or a larger flow at a 2-in. (0.5-kPa) water column pressure drop.

2.7.2 Location.

(a) Gas meters shall be located in ventilated spaces readily accessible for examination, reading, replacement, or necessary maintenance.

The ventilated space requirement anticipates the release of a small amount of gas during meter change-out.

(b) Gas meters shall not be placed where they will be subjected to damage, such as adjacent to a driveway, under a fire escape, in public passages, halls, or coal bins, or where they will be subject to excessive corrosion or vibration.

(c) Gas meters shall be located at least 3 ft (0.9 m) from sources of ignition.

Most electric meters are not a source of ignition, because they use induction motors that do not arc or spark during operation. They can become a source of ignition during installation or removal, but it is assumed that a gas leak at a gas meter is not a normal occurrence and that the person installing or removing the electric meter would note the smell of any escaping gas and take appropriate action.

Electric meters that incorporate switching relays (e.g., time-of-day metering or interruptible meters) can be a source of ignition and should be treated accordingly. When there is a doubt, the premises owner, agent, or the electric utility company should be consulted for further information. The question that should be asked is whether the meter contains arcing or sparking devices.

(d) Gas meters shall not be located where they will be subjected to extreme temperatures or sudden extreme changes in temperature. Meters shall not be located in areas where they are subjected to temperatures beyond those recommended by the manufacturer.

2.7.3 Supports.

Gas meters shall be supported or connected to rigid piping so as not to exert a strain on the meters. Where flexible connectors are used to connect a gas meter to downstream piping at mobile homes in mobile home parks, the meter shall be supported by a post or bracket placed in a firm footing or by other means providing equivalent support.

2.7.4 Meter Protection.

Meters shall be protected against overpressure, back pressure, and vacuum, where such conditions are anticipated.

Gas meters have a maximum allowable operating pressure (MAOP) that must be observed. Small residential meters typically have MAOPs of 5 psi (34 kPa). Larger or special-application meters can carry MAOPs of 10 psi (70 kPa), 25 psi (170 kPa), or higher.

2.7.5 Identification.

Gas piping at multiple meter installations shall be marked by a metal tag or other permanent means attached by the installing agency, designating the building or the part of the building being supplied.

2.8* Gas Pressure Regulators

With increasing uses for gas utilization equipment and with the economy of using higher pressures for smaller-diameter gas piping systems, a need frequently exists for line gas pressure regulators to deliver proper gas pressures to equipment or groups of equipment. This system parallels using higher electrical voltage (pressure) and smaller wires (pipes) with transformers (regulators) to supply the loads properly.

A.2.8 This section applies to premises-owned regulators *[see 1.1.1(b)16].*

2.8.1 Where Required.

A line gas pressure regulator or gas equipment pressure regulator, as applicable, shall be installed where the gas supply pressure is higher than that at which the branch supply line or gas utilization equipment is designed to operate or varies beyond design pressure limits.

2.8.2 Location.

The gas pressure regulator shall be accessible for servicing.

In the 1999 edition, this requirement was revised. It previously required that the regulator be "readily accessible," meaning accessible without the use of tools. This was changed because access to regulators is not required in emergencies.

2.8.3 Regulator Protection.

Pressure regulators shall be protected against physical damage.

2.8.4 Venting.

(a) *Line Gas Pressure Regulators.*

(1) An independent vent to the outside of the building, sized in accordance with the regulator manufacturer's instructions, shall be provided where the location of a regulator is such that a ruptured diaphragm will cause a hazard. Where there is more than one regulator at a location, each regulator shall have a separate vent to the outside, or if approved by the authority having jurisdiction, the vent lines shall be permitted to be manifolded in accordance with accepted engineering practices to minimize back pressure in the event of diaphragm failure. (See 2.9.7 for information on properly locating the vent.) Materials for vent piping shall be in accordance with Section 2.6.

Large line gas pressure regulators are not able to be equipped with vent-limiting means, so they must be vented to the outside atmosphere. Since the manifolding of such vents presents two potential problems, the code specifies that such manifolding have the approval of the authority having jurisdiction. The first problem can occur when a diaphragm on one of the connected regulators fails. If the manifolded vent is not large enough, the flow from the failed diaphragm can build a back pressure in the manifold. This back pressure biases the remaining regulators to a higher discharge pressure, possibly creating a hazardous condition. Second, if the manifolded vent termination becomes plugged and a diaphragm subsequently fails, the resulting high pressure gas in the vent line will drive all of the remaining regulators wide open, completely negating their installation. Any approval of vent line manifolding should address these potential problems.

Note that vent-piping materials must be of materials specified in Section 2.6. This requirement limits vent piping to metallic piping aboveground.

Exception: A regulator and vent limiting means combination listed as complying with ANSI Z21.80, Standard for Line Pressure Regulators, shall be permitted to be used without a vent to the outdoors.

ANSI Z21.18, *Standard for Gas Appliance Pressure Regulators,* and Z21.78, *Standard for Combination Gas Controls for Gas Appliances,* limit vented gas to an amount of 2.5 ft^3/hr (0.07 m^3/hr) in the unlikely event of a diaphragm rupture. The requirement that the installation be made in a ventilated location is not intended to require mechanical ventilation, but rather to prohibit installation in spaces where flammable gas–air mixtures can accumulate. Line gas pressure regulators installed in places where gas appliances are installed under this code will, for instance, meet the requirement of this paragraph. See Exhibit 2.5 for examples of a line gas pressure regulator with vent-limiting means; note the vent hole. Common vent-limiting means can be installation sensitive. Manufacturer's instructions should be consulted to ensure proper installation and functioning. For example, vent-limiting means intended for upright horizontal installation cannot function properly in any other position.

The exception has been revised in the 1999 edition to allow a regulator incorporating vent-limiting means to be used indoors without a vent to the outdoors. In practice, they have been safely used in this manner for years.

(2) The vent shall be designed to prevent the entry of water, insects, or other foreign materials that could cause blockage.

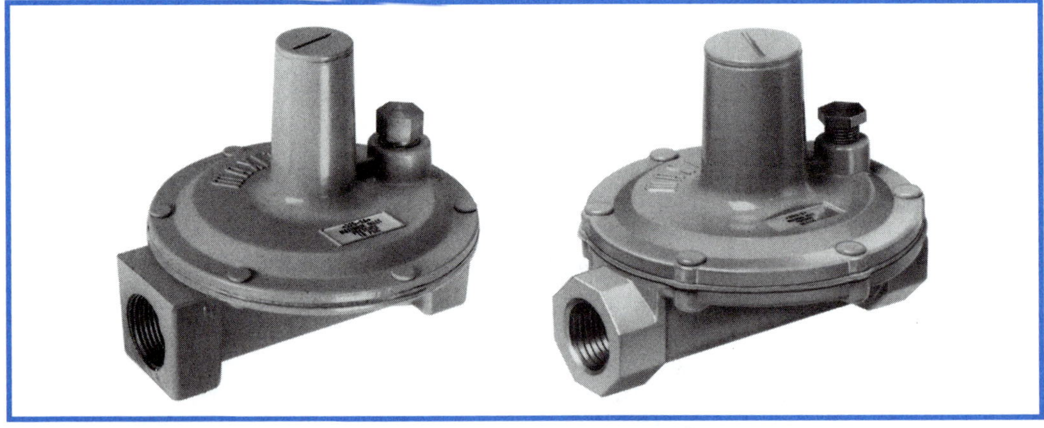

Exhibit 2.5 *Two styles of gas line regulators with vent-limiting means. (Courtesy of Maxitrol.)*

(3) At locations where regulators might be submerged during floods, a special antiflood-type breather vent fitting shall be installed, or the vent line shall be extended above the height of the expected flood waters.

(4) A regulator shall not be vented to the gas equipment flue or exhaust system.

> This requirement recognizes that the flue or exhaust system cannot safely handle the amount of gas that is likely to be discharged through a failed diaphragm and that sources of ignition can be present in such locations.

(b) *Gas Appliance Pressure Regulators.* For venting of gas appliance pressure regulators, see 5.1.18.

2.8.5 Bypass Piping.

Valved and regulated bypasses shall be permitted to be placed around gas line pressure regulators where continuity of service is imperative.

> If a bypass is necessary to ensure continued system operation in the event of a line gas pressure regulator failure, this paragraph ensures that the bypass includes a means of regulating the bypass flow.

2.8.6 Identification.

Line pressure regulators at multiple regulator installations shall be marked by a metal tag or other permanent means designating the building or the part of the building being supplied.

2.9 Overpressure Protection Devices

> This section spells out requirements for overpressure protection of gas piping systems, even though some of the components that can provide such protection are not within the scope of this code. For instance, natural gas service regulators are covered by U.S. DOT regulations, and LP-Gas second-stage regulators are covered by NFPA 58, *Liquefied Petroleum Gas Code.* Therefore, consultation with the gas supplier is necessary to ensure compliance with this section.

2.9.1 General.

Overpressure protection devices shall be provided to prevent the pressure in the piping system from exceeding that value that would cause unsafe operation of any connected and properly adjusted gas utilization equipment. *(See 2.9.5.)*

> Although the work is ongoing, most new residential and commercial gas utilization equipment rated for a maximum operating inlet pressure of 14-in. (3.5-kPa) water column will withstand an inlet pressure of 2 psi–2.5 psi (14 kPa –17 kPa) without

failing, although pressure control and leakage rates can be compromised. In addition, these controls will function normally (no permanent effects) once the overpressure is removed. Accordingly, limiting any possible overpressure to 2 psi–2.5 psi (14 kPa –17 kPa) in the case of these controls generally is accepted to be adequate. Maximum overpressure allowance determination on other equipment should be made by consulting the manufacturer's data.

(a) The requirements of this section shall be met and a piping system deemed to have overpressure protection where all of the following are included in the piping system:

(1) Two devices (a service or line pressure regulator plus one other device) are installed.
(2) Each device limits the pressure to a value that does not exceed the maximum working pressure of the downstream system.
(3) The failure of both devices occurs simultaneously in order to overpressure the downstream system.

Note that all three of these requirements in 2.9.1(a) must be met and also that the requirements ensure that a single-point failure will not result in a dangerous system overpressure. Several common means exist for providing the protection means specified.

For systems operating at inches-of-water-column pressures, a regulator [usually 7-in. (1.7-kPa) water column for natural gas, 11-in. (2.7-kPa) water column for LP-Gas] is equipped with either an integral or separate relief valve that will open upon an overpressure situation and flow enough gas to keep the outlet pressure of the failed regulator below 2 psi (14 kPa). If either the regulator or the relief valve fails, the connected equipment will not be damaged. Both must fail simultaneously for a dangerous overpressure to occur.

For piping systems operating at 2 psi (14 kPa) using line gas pressure regulators to reduce pressure to the appliances, a similar relief is used at the service regulator. This relief valve has a capacity sufficient to maintain the outlet pressure of the failed service regulator below the maximum inlet pressure of the line gas pressure regulator [generally, 10 psi (69 kPa) for common models]. Therefore, if the service regulator fails, the line gas pressure regulator is still able to handle the increased pressure. Additionally, if the line gas pressure regulator fails completely, the equipment will be able to handle the resulting 2-psi (14-kPa) pressure without failure.

Similar systems in which the gas pressure is further elevated [for example, to 5 psi (34 kPa)] must then have an additional element of overpressure protection, because a failure of the line gas pressure regulator would subject the downstream equipment to a pressure of 5 psi (34 kPa). ANSI Z21.80, *Standard for Line Pressure Regulators*, requires that line gas pressure regulators installed in such systems have an overpressure shutoff device that automatically shuts off the gas if a high downstream pressure is detected.

The systems just summarized give ample notification (e.g., discharge from relief valves, high flames, etc.) of problems from failed component failures.

The pressure regulating, limiting, and relieving devices shall be properly maintained, and inspection procedures shall be devised or suitable instrumentation installed to detect failures or malfunctions of such devices, and replacements or repairs shall be promptly made.

(b) A pressure relieving or limiting device shall not be required where (1) the gas does not contain materials that could seriously interfere with the operation of the service or line pressure regulator; (2) the operating pressure of the gas source is 60 psi (414 kPa) or less; and (3) the service or line pressure regulator has all of the following design features or characteristics.

(1) Pipe connections to the service or line regulator do not exceed 2-in. nominal diameter.
(2) It is self-contained with no external static or control piping.
(3) It has a single port valve with an orifice diameter no greater than that recommended by the manufacturer for the maximum gas pressure at the regulator inlet.
(4) The valve seat is made of resilient material designed to withstand abrasion of the gas, impurities in the gas, and cutting by the valve, and to resist permanent deformation where it is pressed against the valve port.
(5) It is capable, under normal operating conditions, of regulating the downstream pressure within the necessary limits of accuracy and of limiting the discharge pressure under no-flow conditions to not more than 150 percent of the discharge pressure maintained under flow conditions.

The requirements of 2.9.1(b) recognize that some systems are not likely to fail completely, and, therefore, the code allows systems meeting all of the restrictions to operate without overpressure protection devices. Special attention to maintaining unobstructed (from insects, ice, etc.) regulator vents should be given on systems of this type, because regulator vent blockage can cause an overpressure situation.

2.9.2 Devices.

Any of the following pressure-relieving or pressure-limiting devices shall be permitted to be used.

(1) Spring-loaded relief device
(2) Pilot-loaded back pressure regulator used as a relief valve so designed that failure of the pilot system or external control piping will cause the regulator relief valve to open
(3) A monitoring regulator installed in series with the service or line pressure regulator
(4) A series regulator installed upstream from the service or line regulator and set to continuously limit the pressure on the inlet of the service or line regulator to the maximum working pressure of the downstream piping system
(5) An automatic shutoff device installed in series with the service or line pressure regulator and set to shut off when the pressure on the downstream piping system reaches the maximum working pressure or some other predetermined pressure less than the maximum working pressure (This device shall be designed so that it will remain closed until manually reset.)

The manual reset requirement is made because if a system shuts down with open manual valves (e.g., on a range) and then comes back on some time later automatically, a dangerous condition could exist.

(6) A liquid seal relief device that can be set to open accurately and consistently at the desired pressure

The preceding devices shall be installed as an integral part of the service or line pressure regulator or as separate units. If separate pressure relieving or pressure limiting devices are installed, they shall comply with 2.9.3 through 2.9.8.

2.9.3 Construction and Installation.

All pressure relieving or pressure limiting devices shall meet the following requirements:

(1) Be constructed of materials so that the operation of the device will not be impaired by corrosion of external parts by the atmosphere or of internal parts by the gas.
(2) Be designed and installed so they can be operated to determine whether the valve is free. The devices shall also be designed and installed so they can be tested to determine the pressure at which they will operate and be examined for leakage when in the closed position.

2.9.4 External Control Piping.

External control piping shall be protected from falling objects, excavations, or other causes of damage and shall be designed and installed so that damage to any control piping shall not render both the regulator and the overpressure protective device inoperative.

Falling objects can include ice and snow shedding from roofs.

2.9.5 Setting.

Each pressure limiting or pressure relieving device shall be set so that the pressure shall not exceed a safe level beyond the maximum allowable working pressure for the piping and appliances connected.

See commentary following 2.9.1 for a discussion of overpressure parameters.

2.9.6 Unauthorized Operation.

Precautions shall be taken to prevent unauthorized operation of any shutoff valve that will make a pressure relieving valve or pressure limiting device inoperative. The following are acceptable methods for complying with this provision:

(1) Lock the valve in the open position. Instruct authorized personnel in the importance of leaving the shutoff valve open and of being present while the shutoff valve is closed so that it can be locked in the open position before leaving the premises.
(2) Install duplicate relief valves, each having adequate capacity to protect the system, and arrange the isolating valves or three-way valve so that only one safety device can be rendered inoperative at a time.

2.9.7 Vents.

The discharge stacks, vents, or outlet parts of all pressure relieving and pressure limiting devices shall be located so that gas is safely discharged into the outside atmosphere. Discharge stacks or vents shall be designed to prevent the entry of water, insects, or other foreign material that could cause blockage. The discharge stack or vent line shall be at least the same size as the outlet of the pressure relieving device.

> Manufacturer's data should be consulted to properly size any vent stacks except for very short ones, because a stack larger than the device outlet can be required.

2.9.8 Size of Fittings, Pipe, and Openings.

The openings, pipe, and fittings located between the system to be protected and the pressure-relieving device shall be sized to prevent hammering of the valve and to prevent impairment of relief capacity.

2.10 Back Pressure Protection

2.10.1 Where to Install.

Protective devices shall be installed as close to the utilization equipment as practical where the design of utilization equipment connected is such that air, oxygen, or standby gases could be forced into the gas supply system. Gas and air combustion mixers incorporating double diaphragm "zero" or "atmosphere" governors or regulators shall require no further protection unless connected directly to compressed air or oxygen at pressures of 5 psi (34 kPa) or more.

> Gas systems connected to zero-governor-equipped mixing equipment require no further back pressure prevention devices, because the standby gas will not flow until the blower is operating to "pull" the gas through the zero governor. This setup is very resistant to failure. If the system uses compressed air, back pressure protection is required because of the possibility of injecting air into the gas piping system.
>
> Venturi-type standby fuel systems require that the served gas piping system be provided with backflow protection, because a failure of the venturi gas valves could cause an overpressure of the gas piping system.

2.10.2 Protective Devices.

Protective devices shall include but not be limited to the following:

(1) Check valves
(2) Three-way valves (of the type that completely closes one side before starting to open the other side)
(3) Reverse flow indicators controlling positive shutoff valves
(4) Normally closed air-actuated positive shutoff pressure regulators

Perhaps the most common backflow prevention device is the utility service regulator. The standby fuel in this case is set up to operate at a pressure that "locks up" the service regulator. As long as the standby system has adequate means to ensure that the service regulator is not overpressured during operation of the standby system, this scheme is safe and effective. Most installations using this means also incorporate a secondary backflow check valve or a combination manual valve and back check valve ("plugaroo") in the utility gas piping immediately downstream of the meter.

2.11 Low-Pressure Protection

A protective device shall be installed between the meter and the gas utilization equipment if the operation of the equipment is such (i.e., gas compressors) that it could produce a vacuum or a dangerous reduction in gas pressure at the meter. Such devices include, but are not limited to, mechanical, diaphragm-operated, or electrically operated low-pressure shutoff valves.

If a low-pressure shutoff valve is installed, good practice would be to use a manual-reset device. This device ensures that the gas will not come back on unexpectedly. It is very important to protect the gas supply system from low pressure or vacuum. If the equipment should cause low pressure or vacuum in the supply system, it could adversely affect the supply pressure of other customers on the line. The low-pressure situation could create a dangerous condition for an unsuspecting user, and it would be difficult to determine the source if caused by another customer.

2.12 Shutoff Valves

Shutoff valves shall be approved and shall be selected giving consideration to pressure drop, service involved, emergency use, and reliability of operation. Shutoff valves of size 1 in. National Pipe Thread and smaller shall be listed.

Once again, "approved" means acceptable to the authority having jurisdiction. Pressure drop across a valve is a measure of its flow capacity. Although "full-ported" valves are available, they generally are not necessary for satisfactory results; although the flow port of a valve is smaller than that of the connecting pipe, the restriction is short and behaves more like a very large orifice at normal flows. Given the intended installation site, attention should be paid to mechanical strength of the valve and its operating means.

The requirement that shutoff valves smaller than 1 in. NPT be listed comes from adverse experience with some unlisted valves of these small sizes and recognizes that larger valves usually are not listed.

2.13 Expansion and Flexibility

Expansion and contraction problems usually are not encountered in fuel gas piping. If a branch run is taken from a very long main or a main that is subject to great changes in temperature, the branch should be made long enough to distribute the movement of the main over the longest practical distance of branch line. Swing joints that are used to provide flexibility necessitated by expansion and contraction will eventually leak. For this reason, such high-cycle activity can be better handled by flex connectors. Consideration also should be given to anchoring the piping in a manner so that expansion takes place where it is planned to take place.

2.13.1 Design.

Piping systems shall be designed to have sufficient flexibility to prevent thermal expansion or contraction from causing excessive stresses in the piping material, excessive bending or loads at joints, or undesirable forces or moments at points of connections to equipment and at anchorage or guide points. Formal calculations or model tests shall be required only where reasonable doubt exists as to the adequate flexibility of the system.

Flexibility shall be provided by the use of bends, loops, offsets, or couplings of the slip type. Provision shall be made to absorb thermal changes by the use of expansion joints of the bellows type or by the use of "ball" or "swivel" joints. Expansion joints of the slip type shall not be used inside buildings or for thermal expansion. If expansion joints are used, anchors or ties of sufficient strength and rigidity shall be installed to provide for end forces due to fluid pressure and other causes.

Pipe alignment guides shall be used with expansion joints according to the recommended practice of the joint manufacturer.

2.13.2 Special Local Conditions.

Where local conditions include earthquake, tornado, unstable ground, or flood hazards, special consideration shall be given to increased strength and flexibility of piping supports and connections.

References Cited in Commentary

The following publications are available from CSA International, 8501 East Pleasant Valley Road, Cleveland, OH 44131.

ANSI/AGA LC 1, *Standard for Interior Fuel Gas Piping Systems Using Corrugated Stainless Steel Tubing*, 1993.
ANSI Z21.18/CGA 6.3, *Standard for Gas Appliance Pressure Regulators*, 1995.
ANSI Z21.78, *Standard for Combination Gas Controls for Gas Appliances*, 1994.
ANSI Z21.80, *Standard for Line Pressure Regulators*, 1997.

The following publication is available from the National Fire Protection Association, 1 Batterymarch Park, P.O. Box 9101, Quincy, MA 02269-9101.

NFPA 58, *Liquefied Petroleum Gas Code,* 1998 edition.

The following publication is available from the U.S. Government Printing Office, Washington, DC 20401.

Code of Federal Regulations, 49 CFR, Part 192.

Gas Piping Installation

Chapter 3 includes requirements for the installation of piping systems. It complements Chapter 2, which specifies piping materials. Testing and purging of gas piping systems are covered in Chapter 4.

Chapter 3 covers the following:

- Piping installed underground, including protection against mechanical damage, corrosion, freezing, piping through foundation walls and beneath buildings, and the use of plastic pipe (in Section 3.1)
- Piping installed above ground outside (in Section 3.2)
- Piping in buildings, including restrictions on valves in concealed spaces, prohibited locations for gas piping, and pipe supports (in Section 3.3)
- Concealed piping in buildings, including fittings prohibited in these locations and repair methods for concealed spaces (in Section 3.4)
- Piping in vertical chases at pressures exceeding 5 psi (34 kPa), including location of pressure regulators (in Section 3.5)
- Gas pipe turns, including equipment and methods to make turns in piping (in Section 3.6)
- Drips, which are needed when the gas supplied contains liquids and sediment traps to prevent debris from clogging equipment and controls (in Section 3.7)
- Outlets, including detailed requirements for the location and capping of outlets (in Section 3.8)
- Branch pipe connection, which allows for future expansion of gas piping systems (in Section 3.9)
- Manual gas shutoff valves, which are required at gas pressure regulators and at buildings served by gas, for each individual unit supplied by gas. Also included are requirements for the exterior location and marking of emergency shutoff valves (in Section 3.10)

- Prohibited devices that could restrict the flow of gas (in Section 3.11)
- Systems containing gas–air mixtures outside the flammable range for which requirements are provided to prevent the mixture from entering the flammable range (in Section 3.12)
- Systems containing flammable gas–air mixtures, including minimum requirements for piping and equipment (in Section 3.13)
- Electrical bonding and grounding, electrical circuits, and electrical connections, which are consistent with NFPA 70, *National Electrical Code* (in Sections 3.14, 3.15, and 3.16)

3.1 Piping Underground

The intent of all of the coverage for underground piping is to provide protection for the installed piping from corrosion or physical damage.

3.1.1 Clearances.

Underground gas piping shall be installed with sufficient clearance from any other underground structure to avoid contact therewith, to allow maintenance, and to protect against damage from proximity to other structures. In addition, underground plastic piping shall be installed with sufficient clearance or shall be insulated from any source of heat so as to prevent the heat from impairing the serviceability of the pipe.

This requirement recognizes the possible hazard from the pipe's being in contact with, or in close proximity to, other underground structures. Such contact can be a hazard and cause harm to the pipe during backfilling or, later, due to settlement of the structure(s).

3.1.2 Protection Against Damage.

Means shall be provided to prevent excessive stressing of the piping where there is heavy vehicular traffic or soil conditions are unstable and settling of piping or foundation walls could occur. Piping shall be buried or covered in a manner so as to protect the piping from physical damage.

Piping shall be protected from physical damage where it passes through flower beds, shrub beds, and other such cultivated areas where such damage is reasonably expected.

(a) *Cover Requirements.* Underground piping systems shall be installed with at least 18 in. (460 mm) of cover. The cover shall be permitted to be reduced to 12 in. (300 mm) if external damage to the pipe is not likely to result. If a minimum of 12 in. (300 mm) of cover cannot be maintained, the pipe shall be installed in conduit or bridged (shielded).

(b) *Trenches.* The trench shall be graded so that the pipe has a firm, substantially continuous bearing on the bottom of the trench.

(c) *Backfilling.* Where flooding of the trench is done to consolidate the backfill, care shall be exercised to see that the pipe is not floated from its firm bearing on the trench bottom.

Protection from heavy vehicle loads must be ensured. In general, deeper burial [at a depth of 18 in. (460 mm) or more] or casing of the pipe carrying the gas is the means used to comply with this requirement. Deeper burial applies to cultivated areas, with depth of cover requirements depending on the anticipated risk, which can vary from hand garden tools to tillage equipment pulled by large tractors. Gas companies with distribution pipes in the area are a good resource to consult when encountering problems of this type.

Damage to pipes is not limited to plastic pipe. Copper tubing is severed easily with a shovel, and digging tools can damage the coating on steel pipe, leading to corrosion.

Frequently, installers ask what they can use to protect the pipe from physical damage. The intent of the section is twofold. One reason is to protect the pipe from damage by digging, so the protection must be of sufficient strength to protect the pipe. In most cases, the pipe is buried at a sufficient depth so that protective devices are not needed.

The other reason is to provide excavators with an indicator that a pipe is buried below. Many gas utilities use a terra tape (a yellow plastic tape printed with a warning that gas pipe is buried below) placed a few inches above the pipe to warn future diggers that a gas pipe is buried below. The tape is an indicator and not a protective device. See Exhibit 3.1.

If protective devices are needed, they must be of sufficient strength to protect the pipe and must be made of a material that will not corrode over time. Plastic pipe or PVC is often used as a protective material. Split along its length, it provides the necessary strength and does not corrode underground.

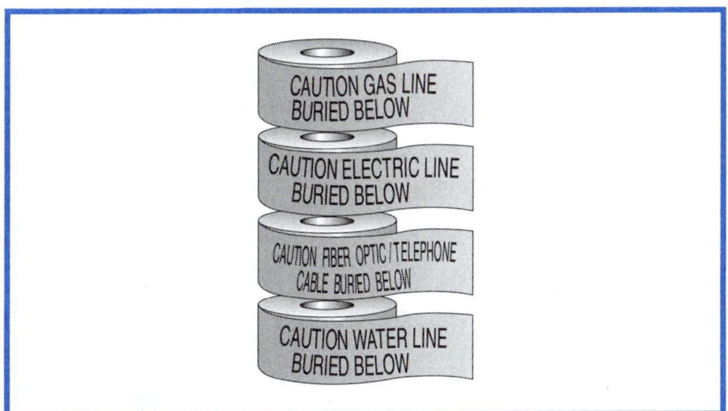

Exhibit 3.1 *Terra tape. (Courtesy of Empire Level Manufacturing Corp.)*

3.1.3* Protection Against Corrosion.

Gas piping in contact with earth or other material that could corrode the piping shall be protected against corrosion in an approved manner. When dissimilar metals are joined underground, an insulating coupling or fitting shall be used. Piping shall not be laid in contact with cinders.

Uncoated threaded or socket welded joints shall not be used in piping in contact with soil or where internal or external crevice corrosion is known to occur.

With the availability of polyethylene and other corrosion-resistant piping systems (e.g., plastic-coated steel, copper tube in most soils), the need to provide corrosion protection for steel often makes this material the last choice for underground installation.

Corrosion in steel pipe that is installed underground or in contact with water is through galvanic corrosion, which is an electrochemical process. With the availability of polyethylene and other corrosion-resistant piping systems (e.g., copper tube in most soils), providing corrosion protection for steel often makes this material a second choice for underground installation.

Corrosion of underground steel pipe or steel pipe in contact with water is through galvanic corrosion, an electrochemical process. This oxidation process can be prevented by stopping the flow of electricity between the pipe and the soil. Stopping electricity flow is accomplished by first coating the pipe with an electrically insulating material. Factory-applied coatings include polyethylene over a mastic primer and fusion-bonded epoxy. Field-applied coatings usually consist of either hot- or cold-applied tapes over a complementary primer.

The second step in properly protecting a steel pipe underground is to maintain an electrical charge on the pipe relative to the surrounding soil and adequate to ensure that the corrosion-causing transfer of electrons will not occur. This is called cathodic protection and is accomplished by the use of an impressed voltage supplied by either a passive sacrificial anode system or by an active rectifier system. Sizing and spacing of either the anodes or the rectifiers are important. Manufacturers of these components are a good source of guidance.

Partial steel pipe corrosion solutions, such as coating the pipe but not using cathodic protection, usually fare worse than unprotected pipe. Uncoated pipe will corrode over its entire surface area, while corrosion of coated pipe lacking cathodic protection will be concentrated at any pinhole or other imperfection (these imperfections are called "holidays") in the coating. Since obtaining and maintaining a perfect coating application is virtually impossible, protection of underground steel pipe should include cathodic protection. Any cathodically protected pipe should be electrically isolated from upstream and downstream components to prevent the loss of the required electrical charge. This isolation is accomplished with the use of dielectric unions or flange insulating kits.

This requirement also recognizes that threaded and socket-welded steel pipe joints are especially prone to the effects of corrosion and should not be used where corrosion is anticipated to occur. Similarly, the use of cinders as backfill is prohibited because they accelerate corrosion.

A.3.1.3 For information on corrosion protection of underground pipe, see NACE RP 0169, Control of External Corrosion on Underground or Submerged Metallic Piping Systems. Information on installation, maintenance, and corrosion protection may be available from the gas supplier.

3.1.4* Protection Against Freezing.

Where the formation of hydrates or ice is known to occur, piping shall be protected against freezing.

Ice problems are all but nonexistent with pipeline natural gas and propane, but they can still exist in gas from alternative sources such as local wells, landfills, or waste treatment plants. This requirement reminds the user to inquire from the gas supplier about the possibility of ice.

A.3.1.4 The gas supplier can be consulted for recommendations.

3.1.5 Piping Through Foundation Wall.

Underground piping, where installed through the outer foundation or basement wall of a building, shall be encased in a protective pipe. The space between the gas piping and the building shall be sealed to prevent entry of gas or water.

This requirement is included in the code because gas leaks in outdoor, underground piping can migrate along the trench and into the basement instead of venting to the atmosphere above grade. This condition can be extremely hazardous. The requirement to seal the casing against water entry is in keeping with the desire to exercise good workmanship. See Exhibit 3.2 for an illustration of a below-grade foundation wall penetration.

The use of an inert wrapping material provides an additional method of protection. The material must be of an approved type and suitable for the application. It must also be able to seal around the pipe and the wall to prevent the entry of gas or water.

3.1.6 Piping Underground Beneath Buildings.

Where the installation of gas piping underground beneath buildings is unavoidable, the piping shall be encased in an approved conduit designed to withstand the superimposed loads. The conduit shall extend into a normally usable and accessible portion of the building and, at the point where the conduit terminates in the building, the space between the conduit and the gas piping shall be sealed to prevent the possible entrance of any gas leakage. If the end sealing is

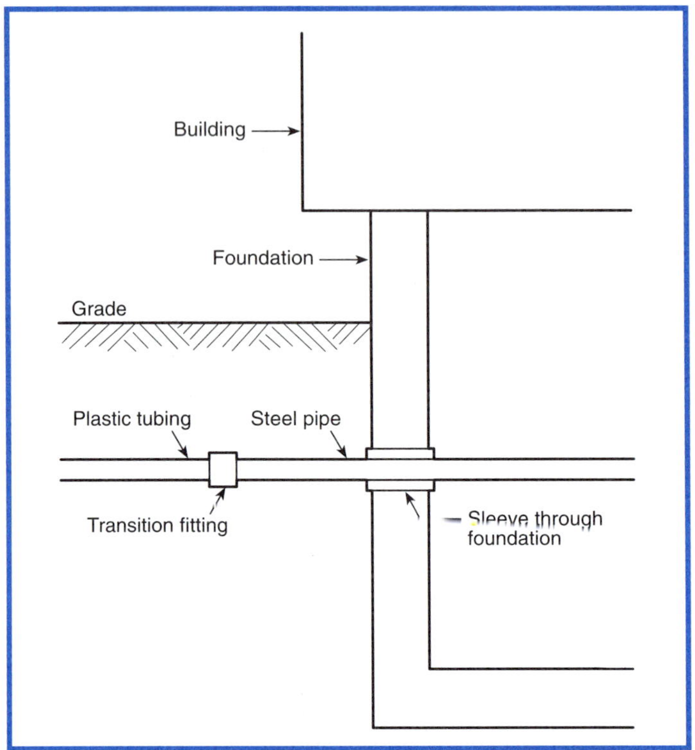

Exhibit 3.2 *Piping through a foundation wall. (Courtesy of J. J. Drechsler.)*

of a type that will retain the full pressure of the pipe, the conduit shall be designed for the same pressure as the pipe. The conduit shall extend at least 4 in. (100 mm) outside the building, be vented above grade to the outside, and be installed so as to prevent the entrance of water and insects.

Gas leaking from a pipe installed underneath a building can migrate into the building, which can be extremely hazardous. Therefore, the code requires that any such pipes be cased and that the annular space between the pipe and the case be vented to the outdoors. While the resulting installation is perfectly safe, the installation is difficult and the repairs are more so. For these reasons, practicality demands that most pipes run through buildings rather than under them.

3.1.7 Plastic Pipe.

Section 3.1.7 includes coverage of risers and wall head adapters (service head adapters). These fittings are used to install polyethylene pipe and tubing beyond the gas meter. For more than thirty years, polyethylene has been used successfully in gas distribution service upstream of the gas meter.

Plastic pipe and tubing is more susceptible to inadvertent damage during installation than most metallic pipe. For this reason special attention should be given to proper compaction below the pipe, to the elimination of shear points on connections during backfilling, and to the materials in the backfill, making certain that angular and large materials are not used near the pipe.

(a) *Connection of Plastic Piping.* Plastic pipe shall be installed outside, under ground only.

The use of anode-type risers and transition fittings with anodes (see Exhibits 3.3 and 3.4) has been allowed by prior editions of this code. In both of these applications, the plastic-to-steel transition is made below grade and meets this requirement.

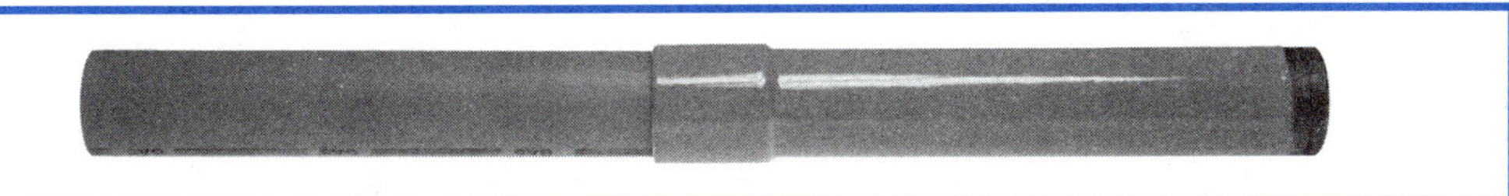

Exhibit 3.3 *Transition fitting. (Courtesy of Ray Murray, Inc.)*

Exception No. 1: Plastic pipe shall be permitted to terminate above ground where an anode-less riser is used.

Anodeless risers are constructed with the plastic pipe running inside steel all the way to the top end, where the pressure seal and the transition from plastic to steel are located. These risers are accepted by the code because their design addresses the concerns regarding plastic pipe terminating aboveground.

Exception No. 2: Plastic pipe shall be permitted to terminate with a wall head adapter above ground in buildings, including basements, where the plastic pipe is inserted in a piping material permitted for use in buildings.

This exception provides the code language that requires plastic gas piping to be terminated above ground and outside of buildings. This exception permits the use of fittings, known as wall head adapters, to terminate buried plastic pipe inside of buildings in basement areas. Wall head adapters often are used by utilities for insertion renewals of steel service lines. Within the scope of this code, wall head adapters can be used for renewals of steel pipes from meters at the property line or in renewals of existing underground steel pipes between buildings, as well as for making a clean-sleeved below-grade foundation wall penetration.

See Exhibit 3.5 for a typical steel line renewal using a wall head adapter. The line is renewed by inserting plastic pipe through the old steel pipe all the way from the outer end into the basement (i.e., insertion renewal). The inserted plastic in the

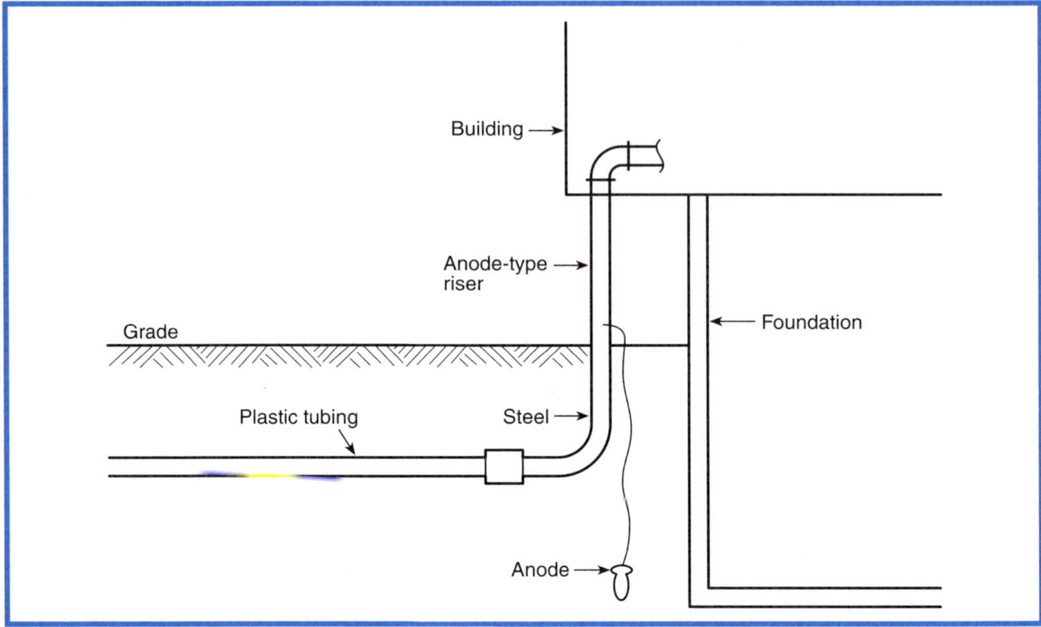

Exhibit 3.4 *Anode-type riser installation.*

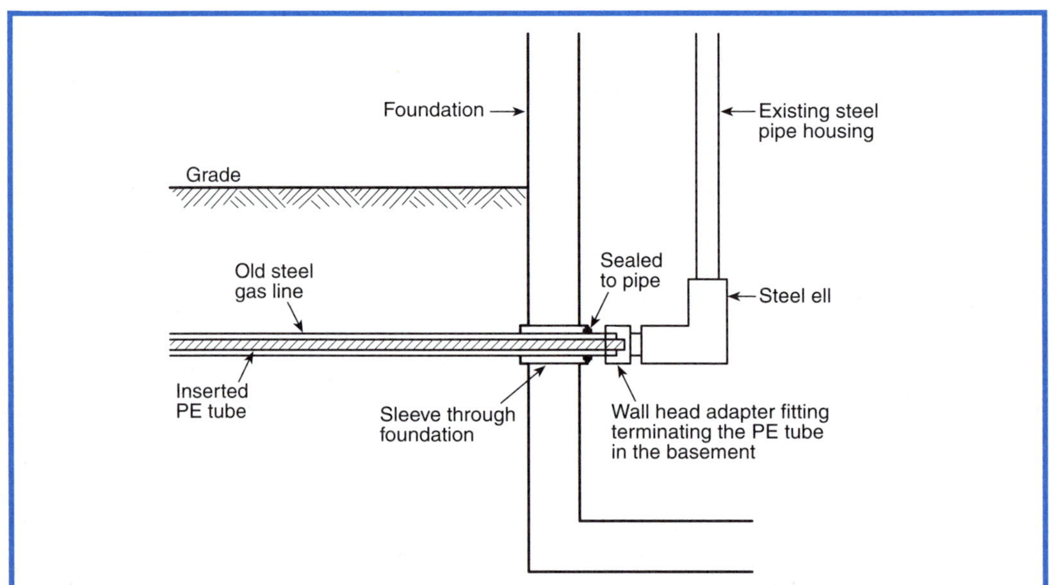

Exhibit 3.5 *Basement termination of plastic pipe. (Courtesy of Maxitrol.)*

basement is terminated by installing an appropriately sized wall head adapter to the end of the plastic pipe or tube in accordance with the manufacturer's qualified installation instructions. The wall head adapter/plastic pipe assembly is then pushed back toward and threaded onto the old steel service pipe that extends through the basement wall. Next, the house piping is fitted to the male threads of the wall head adapter. The plastic pipe and house piping are pressure tested. At this point the plastic is hooked up at the other end and purged, completing the renewal installation. See Exhibit 3.6 for typical wall head adapters.

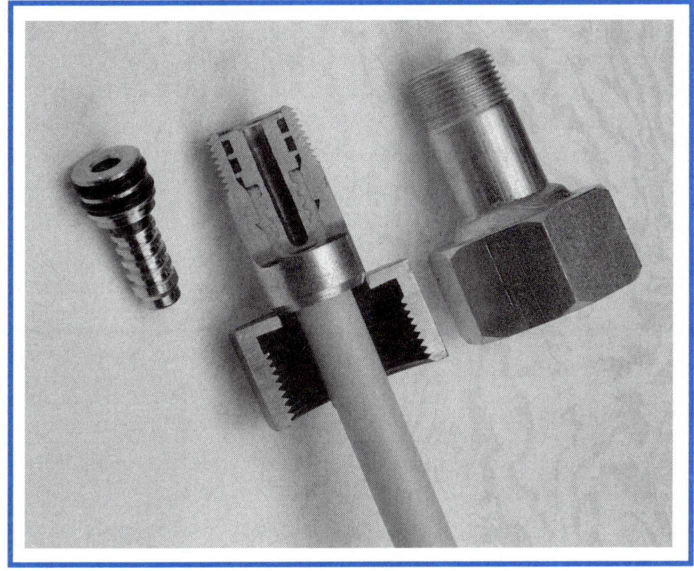

Exhibit 3.6 *Service/wall head adapters (cutaway). (Courtesy of R. W. Lyall & Co.)*

(b) Connections made outside and under ground between metallic and plastic piping shall be made only with ASTM D 2513, *Standard Specification for Thermoplastic Gas Pressure Pipe, Tubing, and Fittings,* Category I transition fittings.

(c) An electrically continuous corrosion-resistant tracer wire (minimum AWG 14) or tape shall be buried with the plastic pipe to facilitate locating. One end shall be brought above ground at a building wall or riser.

Since pipe locators cannot locate plastic pipe or tube, the requirement for tracer wire or tape is included. Tracer tape is a plastic tape with a continuous foil backing that can be detected by a metal detector. Tracer wire is used with special equipment that sends a current through the wire, creating a magnetic field that can then be located using a special detector.

3.2 Aboveground Piping Outside

Piping installed aboveground shall be securely supported and located where it will be protected from physical damage *(also see 3.1.4)*. Where passing through an outside wall, the piping shall also be protected against corrosion by coating or wrapping with an inert material approved for such applications. Where piping is encased in a protective pipe sleeve, the annular space between the gas piping and the sleeve shall be sealed at the wall to prevent the entry of water, insects, or rodents.

Piping that passes through an outside wall is required to be protected from corrosion, because this area can collect water. In addition, pipe in contact with building materials can corrode due to chemical reactions with those materials or due to galvanic corrosion from contact with metallic siding. The inert wrapping material must be approved for these applications.

The provision that the pipe be sealed at the wall is often overlooked. Usually the pipe is protected only when it passes through a concrete wall. However, when it passes through a noncorrosive wall such as wood, it must be sealed to prevent the entry of insects and water.

3.3 Piping in Buildings

3.3.1 Building Structure.

(1) The installation of gas piping shall not cause structural stresses within building components to exceed allowable design limits.
(2) Before any beams or joists are cut or notched, permission shall be obtained from the authority having jurisdiction.

The intent of this section is to ensure that the installation of gas piping does not cause damage to the building. Before granting approval to cut or notch beams or joists, the authority having jurisdiction should consult with the building architect or structural engineer. In general, penetrations of beams or joists should be made near the top or bottom edges (or notches cut) when the beam is carrying shear loads (i.e., near the ends of a beam). Notches generally should not be made anywhere near midspan of a beam, and any holes drilled into a beam at midspan should be made near the center. The preceding instructions assume that the beams are in simple bending (i.e., single span with ends free to rotate).

3.3.2 Other than Dry Gas.

Drips, sloping, protection from freezing, and branch pipe connections, as provided for in 3.1.4, 3.3.3, 3.7.1, and Section 3.9, shall be provided when other than dry gas is distributed and climatic conditions make such provisions necessary.

3.3.3 Gas Piping to Be Sloped.

Piping for other than dry gas conditions shall be sloped not less than $^1/_4$ in. in 15 ft (7 mm in 4.6 m) to prevent traps.

The section was revised in the 1999 edition by deleting instructions on how to slope the pipe, because it was confusing. The piping must to be sloped to prevent low spots or traps. See Exhibit 3.7.

This change allows the installer to determine the best way to install the pipe to prevent traps. In general, the horizontal piping should be sloped so that any accumulation of liquids will be trapped by the drips or sediment traps.

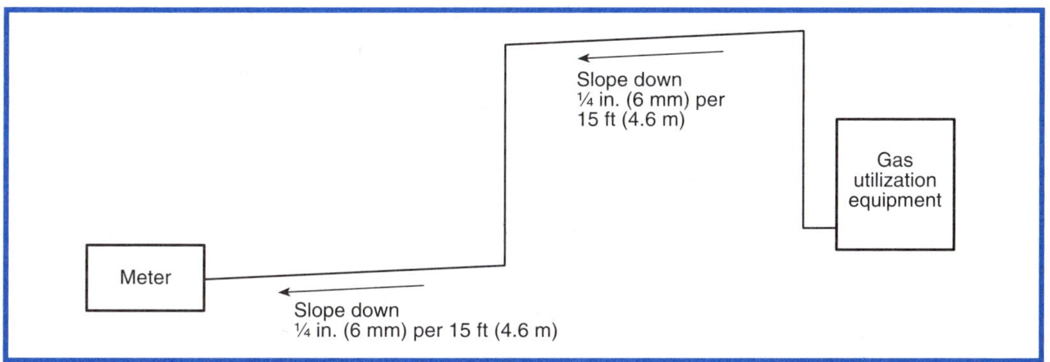

Exhibit 3.7 *Sloping of gas piping for other than dry gas. (Courtesy of J. J. Drechsler.)*

3.3.4 Ceiling Locations.

Gas piping shall be permitted to be installed in accessible spaces between a fixed ceiling and a dropped ceiling, whether or not such spaces are used as a plenum. Valves shall not be located in such spaces.

Exception: Equipment shutoff valves required by this code shall be permitted to be installed in accessible spaces containing vented gas utilization equipment.

The section was revised in the 1999 edition to clarify that the intent of the section is to allow the installation of valves in the space between the fixed ceiling and the dropped ceiling, where the space is accessible.

Although the code never prohibited the installation of gas piping in ceiling plenums, some provisions of the code have been interpreted to prohibit such installations. This provision remedies the problem.

The prohibition of valves in such locations recognizes that valves have a greater chance than gas piping to leak and that a leak would result in little dissipation due to the stagnant conditions in above-ceiling spaces. The exception was added in the

1996 edition of the code to recognize that if vented gas utilization equipment, such as a duct furnace, is installed in an above-ceiling space, ventilation is adequate to permit the installation of valves.

Above-ceiling spaces are spaces below a fixed ceiling that are enclosed by a second ceiling installed in a room. This description includes "drop ceilings" constructed of tiles that are held in place by metal hangers suspended from the fixed ceiling. They are accessible because the tiles can be removed easily, and they are removed periodically for maintenance of lighting fixtures installed as part of the drop ceiling. Attics are not above-ceiling spaces.

3.3.5 Prohibited Locations.

Gas piping inside any building shall not be installed in or through a circulating air duct, clothes chute, chimney or gas vent, ventilating duct, dumbwaiter, or elevator shaft. This provision shall not apply to ducts used to provide combustion and ventilation air in accordance with Section 5.3 or to above-ceiling spaces as covered in 3.3.4.

The intent of this provision is to prevent the mechanical transport of gas leakage throughout the building. Piping can be installed in combustion and ventilation ducts because they communicate directly to the area containing the utilization equipment. If there is any leakage, it will not be distributed throughout the building.

3.3.6 Hangers, Supports, and Anchors.

(a) Piping shall be supported with pipe hooks, metal pipe straps, bands, brackets, or hangers suitable for the size of piping, of adequate strength and quality, and located at intervals so as to prevent or damp out excessive vibration. Piping shall be anchored to prevent undue strains on connected equipment and shall not be supported by other piping. Pipe hangers and supports shall conform to the requirements of ANSI/MSS SP-58, *Pipe Hangers and Supports — Materials, Design and Manufacture.*

(b) Spacings of supports in gas piping installations shall not be greater than shown in Table 3.3.6.

(c) Supports, hangers, and anchors shall be installed so as not to interfere with the free expansion and contraction of the piping between anchors. All parts of the supporting equipment shall be designed and installed so they will not be disengaged by movement of the supported piping.

3.3.7 Removal of Pipe.

If piping containing gas is to be removed, the line shall be first disconnected from all sources of gas and then thoroughly purged with air, water, or inert gas before any cutting or welding is done. *(See Section 4.3.)*

Although the code permits the use of air for purging, an inert gas or water provides maximum safety, because a flammable gas–air mixture is never reached. Cutting is

Table 3.3.6 Support of Piping

Steel Pipe, Nominal Size of Pipe (in.)	Spacing of Supports (ft)	Nominal Size of Tubing (in. O.D.)	Spacing of Supports (ft)
$^1/_2$	6	$^1/_2$	4
$^3/_4$ or 1	8	$^5/_8$ or $^3/_4$	6
$1^1/_4$ or larger (horizontal)	10	$^7/_8$ or 1	8
$1^1/_4$ or larger (vertical)	every floor level		

For SI units, 1 ft = 0.305 m.

not limited to flame cutting because mechanical methods, such as cutting with a chop saw, can produce sparks that can ignite a gas–air mixture. Additional safety information can be found in NFPA 51B, *Standard for Fire Prevention During Welding, Cutting, and Other Hot Work.*

3.4 Concealed Piping in Buildings

3.4.1 General.

Gas piping shall be permitted to be installed in concealed locations in accordance with this section.

This paragraph calls attention to the intent of the National Fuel Gas Code Committee to allow concealed gas piping that is subject to the restrictions of this section.

3.4.2 Connections.

When gas piping that is to be concealed is being installed, unions, tubing fittings, right and left couplings, bushings, swing joints, and compression couplings made by combinations of fittings shall not be used. Pipe fittings such as elbows, tees, and couplings shall be permitted to be used.

Exception No. 1: Joining tubing by brazing [see 2.6.8(b)] shall be permitted.

Exception No. 2: Fittings listed for use in concealed spaces that have been demonstrated to sustain, without leakage, any forces due to temperature expansion or contraction, vibration, or fatigue based on their geographic location, application, or operation shall be permitted to be used.

Exception No. 3: Where necessary to insert fittings in gas pipe that has been installed in a concealed location, the pipe shall be permitted to be reconnected by welding, flanges, or the use of a ground joint union with the nut center-punched to prevent loosening by vibration.

The section is retitled "Connections" in the 1999 edition to meet the intent of the section. The reference to "running threads" — continuous, nontapered threads — was deleted because they are prohibited by 2.6.7. The prohibited list of fittings consists of those that are more likely to leak or are unnecessarily complicated for concealed installation. Brazed tubing joints are as strong as the tubing itself, and are, therefore, acceptable.

Exception 2 deals with the connections that must be concealed, such as those necessary for installation of listed gas convenience outlets. These fittings are a part of a listed system designed to meet the concerns, which prohibit other tubing fittings in concealed locations.

Exception 3 contains text relocated from 3.4.6 in previous editions. A prohibition of reconnection of tubing in concealed locations was deleted because it does not fall under the new scope of 3.4.2, Connections.

3.4.3 Piping in Partitions.

Concealed gas piping shall not be located in solid partitions.

Although gas piping cannot be installed in solid partitions, the partition may be modified to include a ventilated chase (see 3.5.3) into which gas piping can be installed, provided that the chase is vented to a safe location.

3.4.4 Tubing in Partitions.

This provision shall not apply to tubing that pierces walls, floors, or partitions.

Tubing shall be permitted to be installed vertically and horizontally inside hollow walls or partitions without protection along its entire concealed length where both of the following requirements are met:

(1) A steel striker barrier not less than 0.0508 in. (1.3 mm) thick, or equivalent, is installed between the tubing and the finished wall and extends at least 4 in. (100 mm) beyond concealed penetrations of plates, fire stops, wall studs, and so on.
(2) The tubing is installed in single runs and is not rigidly secured.

This provision applies to tubing that lies within a wall or partition. The protection specified is not as strict as that required by ANSI/AGA LC 1, *Standard for Interior Fuel Gas Piping Systems Using Corrugated Stainless Steel Tubing,* for CSST, which requires that the barrier extend 5 in. (130 mm) beyond the penetration. Material of 0.0508-in. (1.3-mm) thickness is approximately 16 gauge. Plates are required near penetrations. At this point, nails and screws are more apt to be able to pin down and penetrate a tube, because it is restrained by the penetration. This hazard is also the basis for the requirement that tubing away from penetrations must not be secured. If the tube is not secured, it can move. Thus, nails and screws will have difficulty penetrating the tubing because the tubing can be moved to the side by a nail or screw.

3.4.5 Piping in Floors.

Gas piping in solid floors such as concrete shall be laid in channels in the floor and covered to permit access to the piping with a minimum of damage to the building. Where piping in floor channels could be exposed to excessive moisture or corrosive substances, the piping shall be protected in an approved manner.

Exception: In other than industrial occupancies and where approved by the authority having jurisdiction, gas piping shall be permitted to be embedded in concrete floor slabs constructed with portland cement. Piping shall be surrounded with a minimum of $1^1/_2$ in. (38 mm) of concrete and shall not be in physical contact with other metallic structures such as reinforcing rods or electrically neutral conductors. All piping, fittings, and risers shall be protected against corrosion in accordance with 2.6.6. Piping shall not be embedded in concrete slabs containing quickset additives or cinder aggregate.

> Concrete containing cinder aggregate or quickset additives, usually calcium chloride, is not permitted, because such products attack the metals from which gas piping is made.

3.5 Piping in Vertical Chases

Where gas piping exceeding 5 psi (34 kPa) is located within vertical chases in accordance with 2.5.1(2), the requirements of 3.5.1 through 3.5.3 shall apply.

3.5.1 Pressure Reduction.

If pressure reduction is required in branch connections for compliance with 2.5.1, such reduction shall take place either inside the chase or immediately adjacent to the outside wall of the chase. Regulator venting and downstream overpressure protection shall comply with 2.8.4 and Section 2.9. The regulator shall be accessible for service and repair.

(1) Regulators equipped with a vent-limiting means shall be permitted to be vented into the chase.
(2) Regulators not equipped with a vent-limiting means shall be permitted to be vented either directly to the outdoors or to a point within the top 1 ft (0.3 m) of the chase.

Exception: If the fuel gas is heavier than air, the vent shall be vented only directly to the outdoors.

3.5.2 Construction.

Chase construction shall comply with local building codes with respect to fire resistance and protection of horizontal and vertical openings.

3.5.3* Ventilation.

A chase shall be ventilated to the outdoors and only at the top. The opening(s) shall have a minimum free area (in square inches) equal to the product of one-half of the maximum pressure in the piping (in psi) times the largest nominal diameter of that piping (in inches), or the cross-sectional area of the chase, whichever is smaller. Where more than one fuel gas piping system is present, the free area for each system shall be calculated and the largest area used.

This section contains specific provisions that are necessary to make safe installations in vertical chases. Note that because a chase is ventilated at the top, the inside of the chase is essentially outdoors. If the top of the chase terminates within the building — that is, equipment room on the top floor — then the regulators should be vented to the outdoors directly. Regulators equipped with approved vent-limiting means may terminate within the chase because by design they cannot release an unsafe amount of gas.

A.3.5.3 Only vertical chases are recognized by the coverage. It is believed that welded joints for a horizontal gas line would be preferable to a horizontal chase.

3.6 Gas Pipe Turns

Changes in direction of gas pipe shall be made by the use of fittings, factory bends, or field bends.

3.6.1 Metallic Pipe.

Metallic pipe bends shall comply with the following:

(1) Bends shall be made only with bending equipment and procedures intended for that purpose.
(2) All bends shall be smooth and free from buckling, cracks, or other evidence of mechanical damage.
(3) The longitudinal weld of the pipe shall be near the neutral axis of the bend.

The neutral axis of the bend is the area where the metal is not stretched or compressed during the bending operation. It lies at the sides of the pipe at the approximate midpoint between the inside and the outside of the bend.

(4) Pipe shall not be bent through an arc of more than 90 degrees.
(5) The inside radius of a bend shall be not less than 6 times the outside diameter of the pipe.

3.6.2 Plastic Pipe.

Plastic pipe bends shall comply with the following:

(1) The pipe shall not be damaged, and the internal diameter of the pipe shall not be effectively reduced.

(2) Joints shall not be located in pipe bends.

(3) The radius of the inner curve of such bends shall not be less than 25 times the inside diameter of the pipe.

(4) Where the piping manufacturer specifies the use of special bending equipment or procedures, such equipment or procedures shall be used.

3.6.3* Mitered Bends.

Mitered bends shall be permitted subject to the following limitations:

(1) Miters shall not be used in systems having a design pressure greater than 50 psi (340 kPa). Deflections caused by misalignments up to 3 degrees shall not be considered as miters.

(2) The total deflection angle at each miter shall not exceed 90 degrees.

A.3.6.3 Care should be taken in making mitered joints to provide proper root opening and alignment and full weld penetration.

3.6.4 Elbows.

Factory-made welding elbows or transverse segments cut therefrom shall have an arc length measured along the crotch of at least 1 in. (25 mm) for pipe sizes 2 in. and larger.

Section 3.6 opens with a general permission to bend pipe and then restricts the methods as necessary to ensure an acceptable outcome. Metal pipe bends must be made using equipment for that purpose.

The allowance of misalignments of up to 3 degrees in mitered bends without classifying the misalignment as a miter is included so that unavoidable misalignments in pipe are not a cause for rejection of welded joints in pipe.

The requirement of leaving at least a 1-in. (25-mm) crotch in 2-in. (50-mm) and larger weld fittings that have been cut down exists so that the segment will retain its shape and allow a proper weld.

3.7 Drips and Sediment Traps

3.7.1 Provide Drips Where Necessary.

For other than dry gas conditions, a drip shall be provided at any point in the line of pipe where condensate could collect. Where required by the authority having jurisdiction or the serving gas supplier, a drip shall also be provided at the outlet of the meter. This drip shall be so installed as to constitute a trap wherein an accumulation of condensate will shut off the flow of gas before it will run back into the meter.

In the 1999 edition, the phrase "For other than dry gas conditions" was added at the beginning of the requirement. This phrase clarifies that drips are needed when "other than dry gas conditions exist" and are not necessary when only dry gas is expected to be present. Note that sediment traps are still required in 3.7.3 and 5.5.7.

Drips collect condensate from gases that contain condensable products (usually water) and are required by the code if required by the serving gas supplier or the authority having jurisdiction. Usually, drips are located near the outlet of the meter or the service entrance and at other locations where condensate could collect. The drip is installed at the bottom of a downward-flowing line by placing a tee at the bottom of the line, installing a nipple and cap in the run of the tee, and continuing the pipe run out the side of the tee. Drips must be located so that they can be emptied to prevent liquid from causing an obstruction to the flow of gas. Drip locations also must be protected from freezing.

Many old manufactured gas installations used low-pressure mains that were sealed with jute (a natural fiber) packing. This packing must be kept wet to prevent leakage. As cities were converted to natural gas, these mains were converted to natural gas service, and water foggers and other means were added to ensure packing integrity. Some of the water can pass through the meter into the house piping, and drip traps are still required in areas served by these old low-pressure mains or where water in the gas can condense prior to reaching the meter. Drip traps also can be required if the natural gas supplier expects condensate in the gas received from the transmission company. Drip traps are not required when the fuel gas is LP-Gas. (See Exhibit 3.8.)

Sediment traps are identical to drip traps, but they are intended for removal of debris from construction or corrosion.

3.7.2 Location of Drips.

All drips shall be installed only in such locations that they will be readily accessible to permit cleaning or emptying. A drip shall not be located where the condensate is likely to freeze.

3.7.3 Sediment Traps. *(See 5.5.7.)*

Sediment traps, sometimes confused with drip traps, are installed to collect solid foreign particles to prevent such material from entering close-fitting parts or small passageways (e.g., valves and orifices). While some gases can contain foreign solids, this situation is not likely to be a problem in either utility gas or LP-Gas because of the methods and equipment used in handling these products. Dirt and pipe material cuttings that are placed unavoidably in the system itself (usually during construction) and are present in limited amounts are the target of sediment traps. Thus, sediment traps seldom need to be opened for service or cleaning. See 5.5.7 for a depiction of a sediment trap.

Evidence exists that sediment, particularly copper sulfide, has been a significant factor in problems with gas appliance controls in the field. *Testing of Gas Fired Water Heater Combination Control Valves,* which was a study by the U.S. Consumer

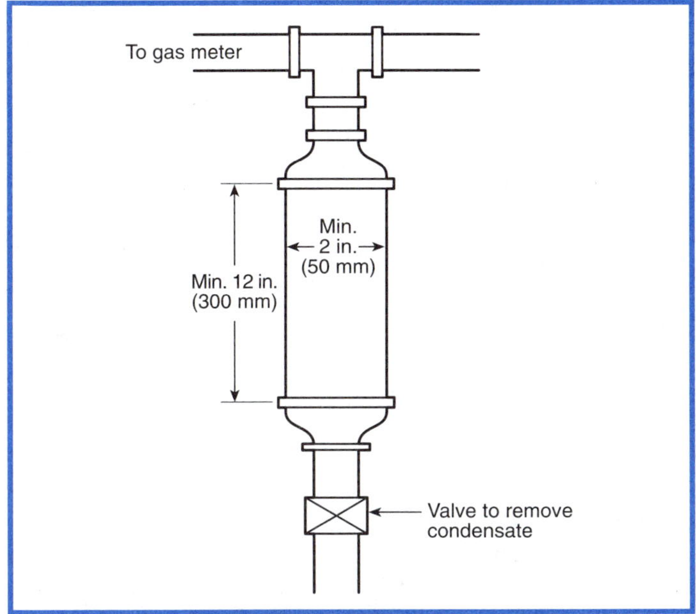

Exhibit 3.8 *Typical drip trap design. (Courtesy of J. J. Drechsler.)*

Product Safety Commission on various sediment trap configurations, found that the configuration shown in 5.5.7 is of significant benefit in removing sediment from the gases. Many appliance manufacturers are incorporating sediment traps in their appliances, and a number of the ANSI Z21 standards for gas appliances require the installation of a sediment trap.

In light of this information, 5.5.7 requires the installation of a sediment trap at the time of installation of most appliances if the appliance is not equipped with one already.

3.8 Outlets

3.8.1 Location and Installation.

(1) The outlet fittings or piping shall be securely fastened in place.
(2) Outlets shall not be located behind doors.
(3) Outlets shall be located far enough from floors, walls, patios, slabs, and ceilings to permit the use of wrenches without straining, bending, or damaging the piping.
(4) The unthreaded portion of gas piping outlets shall extend not less than 1 in. (25 mm) through finished ceilings or indoor or outdoor walls.

(5) The unthreaded portion of gas piping outlets shall extend not less than 2 in. (50 mm) above the surface of floors or outdoor patios or slabs.

(6) The provisions of 3.8.1(4) and (5) shall not apply to listed quick-disconnect devices of the flush-mounted type or listed gas convenience outlets. Such devices shall be installed in accordance with the manufacturers' installation instructions.

These provisions state what is common sense. Minimum projections of pipe are specified, in the case of walls and ceilings, to allow a wrench to grip the pipe. In the case of floors, the extra projection will tend to protect the threads from flooding and mechanical damage.

Since gas convenience outlets are listed as an installed assembly, these concerns are addressed by their listing, and the provisions of this paragraph need not apply.

3.8.2 Cap All Outlets.

(a) Each outlet, including a valve or cock outlet, shall be closed gastight with a threaded plug or cap immediately after installation and shall be left closed until the gas utilization equipment is connected thereto. When equipment is disconnected from an outlet and the outlet is not to be used again immediately, it shall be closed gastight.

Outlets shall not be closed with tin caps, wooden plugs, corks, or by other improvised methods.

Exception No. 1: Laboratory equipment installed in accordance with 5.5.2(a) shall be permitted.

Exception No. 2: The use of a listed quick-disconnect device with integral shutoff or listed gas convenience outlet shall be permitted.

(b) Equipment shutoff valves installed in fireplaces shall be removed and the piping capped gastight where the fireplace is used for solid fuel burning.

A plug or cap is required for all gas openings. Closing a valve is not enough to satisfy this requirement, because the valve can be opened inadvertently or accidentally. No temporary or makeshift closure is permitted, because anything except a proper plug or cap could leak.

Listed, quick-disconnect devices and gas convenience outlets are permitted to go uncapped or unplugged, because they are required by their listing to have valves that are integral with the device that will automatically shut off the gas either prior to or during the disconnect.

When burning solid fuel, the removal of equipment shutoff valves from fireplaces is required, because experience has shown that these valves will leak if they are subjected to the heat of a wood fire. Whenever possible, all gas piping in a fireplace that is going to be used to burn solid fuel should also be removed as a preventive measure. Leaving the gas piping in the fireplace subjects the piping to extreme temperatures, which over time can cause the piping connections to leak.

An exception is made for laboratory equipment, such as Bunsen burners, that are connected and disconnected from hose-end valves as a normal, everyday operation.

3.9 Branch Pipe Connection

When a branch outlet is placed on a main supply line before it is known what size pipe will be connected to it, the outlet shall be of the same size as the line that supplies it.

This requirement has been misinterpreted to mean that the branch must be the same size as the largest main supply line. For example, if the main supply line starts out as a 2-in. (50-mm) line and is reduced to a 1-in. (25-mm) line downstream, a branch downstream of the size reduction must be 2 in. (50 mm). This is not the intent. In this case, a 1-in. (25-mm) branch is sufficient, as 1 in. is the size of the main supply line where the branch is made. Remember, when any modification is made to a fuel distribution system, the system must be recalculated to be certain that it can provide the necessary fuel without an excessive pressure drop. And if it cannot, then larger or additional pipe must be installed.

3.10 Manual Gas Shutoff Valves

(Also see 5.5.4.)

3.10.1 Valves at Regulators.

An accessible gas shutoff valve shall be provided upstream of each gas pressure regulator. Where two gas pressure regulators are installed in series in a single gas line, a manual valve shall not be required at the second regulator.

3.10.2 Valves Controlling Multiple Systems.

The reason for requiring a shutoff valve for each line is to allow for the system to be separated for repair or maintenance. In a case where only part of the system has to be shut down for repair, the remaining system can continue to operate. Shutoff valves must be plainly marked so that in the future, one can determine which line services which unit or building.

The requirement of a shutoff valve for each building provides emergency responders with the ability to shut off the flow of gas to a building involved in fire.

(a) *Accessibility of Gas Valves.* Main gas shutoff valves controlling several gas piping systems shall be readily accessible for operation and installed so as to be protected from physical damage. They shall be marked with a metal tag or other permanent means attached by the installing agency so that the gas piping systems supplied through them can be readily identified.

(b) *Shutoff Valves for Multiple House Lines.* In multiple tenant buildings supplied through a master meter, or through one service regulator where a meter is not provided, or where meters or service regulators are not readily accessible from the equipment location, an individual shutoff valve for each apartment or tenant line shall be provided at a convenient point of general accessibility.

In a common system serving a number of individual buildings, shutoff valves shall be installed at each building.

3.10.3 Emergency Shutoff Valves.

An exterior shutoff valve to permit turning off the gas supply to each building in an emergency shall be provided. The emergency shutoff valves shall be plainly marked as such and their locations posted as required by the authority having jurisdiction.

In natural gas installations, this shutoff valve is part of the utility's service and meter installation. In LP-Gas systems, this valve is normally at the propane tank. These installations are presumed to be obvious enough to meet the requirement of being "plainly marked." The requirement of posting the location is not treated uniformly by different utilities and fire departments. When an installation is not obvious (e.g., an underground propane tank or a meter/regulator concealed by vegetation), a sign or other marking is needed where required by the authority having jurisdiction. The intent of this provision is to enable emergency responders to shut off the gas supply to a building in the event of a fire. The local fire department is often the authority having jurisdiction for this requirement.

3.11 Prohibited Devices

No device shall be placed inside the gas piping or fittings that will reduce the cross-sectional area or otherwise obstruct the free flow of gas, except where proper allowance in the piping system design has been made for such a device and where approved by the authority having jurisdiction.

An example of an acceptable device is an orifice plate (part of an orifice meter set assembly), which can be used to measure gas flow to a large gas user, such as a boiler. Obviously, such a device should be sized adequately to allow for the pressure drop it produces.

3.12 Systems Containing Gas–Air Mixtures Outside the Flammable Range

Where gas–air mixing machines are employed to produce mixtures above or below the flammable range, they shall be provided with stops to prevent adjustment of the mixture to within or approaching the flammable range.

3.13 Systems Containing Flammable Gas–Air Mixtures

3.13.1 Required Components.

A central premix system with a flammable mixture in the blower or compressor shall consist of the following components:

(1) Gas-mixing machine in the form of an automatic gas–air proportioning device combined with a downstream blower or compressor
(2) Flammable mixture piping, minimum Schedule 40 NPS
(3) Automatic firecheck(s)
(4) Safety blowout(s) or backfire preventers for systems utilizing flammable mixture lines above $2^1/_2$ in. nominal pipe size or the equivalent

3.13.2 Optional Components.

The following components shall also be permitted to be utilized in any type central premix system:

(1) Flowmeter(s)
(2) Flame arrester(s)

3.13.3 Additional Requirements.

Gas-mixing machines shall have nonsparking blowers and shall be so constructed that a flash-back will not rupture machine casings.

3.13.4* Special Requirements for Mixing Blowers.

A mixing blower system shall be limited to applications with minimum practical lengths of mixture piping, limited to a maximum mixture pressure of 10-in. water column (25 Pa) and limited to gases containing no more than 10 percent hydrogen.

The blower shall be equipped with a gas-control valve at its air entrance so arranged that gas is admitted to the airstream, entering the blower in proper proportions for correct combustion by the type of burners employed, the said gas-control valve being of either the zero governor or mechanical ratio valve type that controls the gas and air adjustment simultaneously. No valves or other obstructions shall be installed between the blower discharge and the burner or burners.

A.3.13.4 The mixing blower is acknowledged as a special case because of its inability to tolerate control valves or comparable restrictions between mixing blower and burner(s). With these limitations, mixing blower installations are not required to utilize safety blowouts, backfire preventers, explosion heads, flame arresters, or automatic firechecks that introduce pressure losses.

3.13.5 Installation of Gas-Mixing Machines.

(a)* The machine shall be located in a large, well-ventilated area or in a small detached building or cutoff room provided with room construction and explosion vents in accordance with sound engineering principles. Such rooms or belowgrade installations shall have adequate positive ventilation.

A.3.13.5(a) For information on venting of deflagrations, see NFPA 68, *Guide for Venting of Deflagrations.*

(b) Where gas-mixing machines are installed in well-ventilated areas, the type of electrical equipment shall be in accordance with NFPA 70, *National Electrical Code®*, for general service conditions unless other hazards in the area prevail.

Where gas-mixing machines are installed in small detached buildings or cutoff rooms, the electrical equipment and wiring shall be installed in accordance with NFPA 70, *National Electrical Code,* for hazardous locations (Articles 500 and 501, Class 1, Division 2).

(c) Air intakes for gas-mixing machines using compressors or blowers shall be taken from outdoors whenever practical.

(d)* Controls for gas-mixing machines shall include interlocks and a safety shutoff valve of the manual reset type in the gas supply connection to each machine arranged to automatically shut off the gas supply in the event of high or low gas pressure. Except for open burner installations only, the controls shall be interlocked so that the blower or compressor will stop operating following a gas supply failure. Where a system employs pressurized air, means shall be provided to shut off the gas supply in the event of air failure.

A.3.13.5(d) Additional interlocks might be necessary for safe operation of equipment supplied by the gas-mixing machine.

(e) Centrifugal gas-mixing machines in parallel shall be reviewed by the user and equipment manufacturer before installation, and means or plans for minimizing these effects of downstream pulsation and equipment overload shall be prepared and utilized as needed.

Prior to the 1992 edition, the code required that a gas-mixing machine be located in a large, well-ventilated area or in a cutoff room that was provided with room construction and explosion vents in accordance with NFPA 68, *Guide for Venting of Deflagrations.* The 1992 edition was revised to require that the design be in accordance with sound engineering principles, because NFPA 68 is a guide and is not appropriate as a mandatory reference. NFPA 68 is still recommended as an excellent reference for design information.

3.13.6 Use of Automatic Firechecks, Safety Blowouts, or Backfire Preventers.

Automatic firechecks and safety blowouts or backfire preventers shall be provided in piping systems distributing flammable air–gas mixtures from gas-mixing machines to protect the piping and the machines in the event of flashback, in accordance with the following:

(a)* Approved automatic firechecks shall be installed upstream as close as practicable to the burner inlets following the firecheck manufacturers' instructions.

A.3.13.6(a) Two basic methods are generally used. One calls for a separate firecheck at each burner, the other a firecheck at each group of burners. The second method is generally more practical if a system consists of many closely spaced burners.

An approved automatic firecheck should be installed as near as practical upstream from a flame arrester used for local protection where test burners or lighting torches are employed.

(b) A separate manually operated gas valve shall be provided at each automatic firecheck for shutting off the flow of gas–air mixture through the firecheck after a flashback has occurred. The valve shall be located upstream as close as practical to the inlet of the automatic firecheck.

CAUTION
These valves shall not be reopened after a flashback has occurred
until the firecheck has cooled sufficiently to prevent reignition
of the flammable mixture and has been reset properly.

(c) A safety blowout or backfiring preventer shall be provided in the mixture line near the outlet of each gas-mixing machine where the size of the piping is larger than $2^1/_2$ in. NPS, or equivalent, to protect the mixing equipment in the event of an explosion passing through an automatic firecheck. The manufacturers' instructions shall be followed when installing these devices, particularly after a disc has burst.

The discharge from the safety blowout or backfire preventer shall be located or shielded so that particles from the ruptured disc cannot be directed toward personnel. Wherever there are interconnected installations of gas-mixing machines with safety blowouts or backfire preventers, provision shall be made to keep the mixture from other machines from reaching any ruptured disc opening. Check valves shall not be used for this purpose.

(d) Explosion heads (rupture disc) shall be permitted to be provided in large-capacity premix systems to relieve excessive pressure in pipelines. They shall be located at and vented to a safe outdoor location. Provisions shall be provided for automatically shutting off the supply of the gas–air mixture in the event of rupture.

Any installation of this type can require the advice and assistance from an experienced person or organization to achieve a satisfactory, safe project.

3.14 Electrical Bonding and Grounding

(a) Each aboveground portion of a gas piping system upstream from the equipment shutoff valve shall be electrically continuous and bonded to any grounding electrode, as defined by NFPA 70, *National Electrical Code.*

(b) Gas piping shall not be used as a grounding conductor or electrode.

This is an electrical requirement to provide consistency between the *National Fuel Gas Code* and NFPA 70, *National Electrical Code (NEC)*.

Article 250 of the *NEC* defines a grounding electrode as a rod driven into the ground, the steel frame of a building, buried metal water piping, or a concrete encased electrode. It further requires that if more than one of these examples are used, the multiple-grounding electrodes must be bonded together.

The purpose of these requirements is to avoid any electrical potential buildup that could produce a spark. In addition, a difference in electrical potential between an ungrounded piping system and a grounded metal object could pose an electrical shock hazard to a person who comes into contact with the two systems. If the gas piping is maintained at ground potential, these problems are prevented. Bonding the gas piping to the other grounded components inside the structure will maintain ground potential. If the interior gas piping system is connected to an underground metallic line that is cathodically protected, a dielectric union must be installed between the two portions of the system to prevent the outside piping from being "grounded out" by the bonding required by this section.

3.15 Electrical Circuits

Electrical circuits shall not utilize gas piping or components as conductors.

Exception: Low-voltage (50 V or less) control circuits, ignition circuits, and electronic flame detection device circuits shall be permitted to make use of piping or components as a part of an electric circuit.

Section 3.15 continues the prohibition on electrical circuits that could develop a significant spark as well as a severe shock hazard. This section recognizes that if high-voltage electricity is conducted by the gas pipe and the gas pipe is disconnected, the person disconnecting the pipe can get a severe shock.

The allowance of high-voltage applications is limited to ignition and flame-detection applications, both of which have low exposure levels because the entire circuit is within the gas utilization equipment.

3.16 Electrical Connections

(a) All electrical connections between wiring and electrically operated control devices in a piping system shall conform to the requirements of NFPA 70, *National Electrical Code. (See Section 3.14.)*

(b) Any essential safety control depending on electric current as the operating medium shall be of a type that will shut off (fail safe) the flow of gas in the event of current failure.

Electrical requirements for fuel gas systems are presented here as they apply to piping systems and in Section 5.6 as they apply to appliances. The reference to the *National Electrical Code* is made to guide those who are not familiar with electrical requirements. The requirement for essential safety controls covers items such as control valves installed in a gas line that are operated in an emergency. Such valves are sometimes installed in commercial and industrial buildings to stop the flow of gas when a fire alarm system activates. These values are not prohibited, but they must stop the flow of gas in the event of failure of electric current. If such valves are installed in gas piping systems serving manually operated appliances, such as ranges, provisions should be made so that the flow of gas cannot be restarted with gas valves in the open position.

References Cited in Commentary

The following publication is available from CSA International Inc., 8501 East Pleasant Valley Road, Cleveland, OH 44131.

ANSI/AGA LC 1-1997, *Standard for Interior Fuel Gas Piping Systems Using Corrugated Stainless Steel Tubing.*

The following publications are available from the National Fire Protection Association, 1 Batterymarch Park, P.O. Box 9101, Quincy, MA 02269-9101.

NFPA 51B, *Standard for Fire Prevention During Welding, Cutting, and Other Hot Work,* 1999 edition.
NFPA 68, *Guide for Venting of Deflagrations,* 1998 edition.
NFPA 70, *National Electrical Code,*® 1999 edition.

The following publication is available from the U.S. Consumer Product Safety Commission, 4330 East-West Highway, Bethesda, MD 20814-4408.

Testing of Gas Fired Water Heater Combination Control Valves, US CPSC, Contract No. CPSC-C-84-1167.

Inspection, Testing, and Purging

Chapter 4 covers the next step in the process after installation of the gas piping system — testing and placing the pipe in service. It is divided into the following three sections:

- Pressure Testing and Inspection, which contains the requirements for testing new and modified gas piping systems. This chapter covers the piping system only, and does not include appliances and gas connectors (in Section 4.1).
- System and Equipment Leakage Test, which contains the requirements for the leak test that is conducted before turning on the gas in a new or modified piping system; when leakage is suspected; and when the gas meter is replaced. The test is also conducted when an out-of-gas condition occurs (in Section 4.2).
- Purging, or removing air from gas piping systems that are new or have been modified (in Section 4.3).

4.1 Pressure Testing and Inspection

This section is the essence of the code. Before being put into operation, piping installations must be inspected and tested to determine that the total project complies with the code. The piping installation includes all fixed piping from the point of delivery to, and including, the equipment shutoff valves. Appliances and appliance connectors are not part of the fixed piping system and are not included in this test. The pressure test normally is conducted once, prior to initial operation of the piping system, and again only when the piping system is modified. This section is not intended to be used when a piping system is checked for leaks after initial testing. Section 4.2 covers these leak tests.

4.1.1* General.

(a) Prior to acceptance and initial operation, all piping installations shall be inspected and pressure tested to determine that the materials, design, fabrication, and installation practices comply with the requirements of this code.

(b) Inspection shall consist of visual examination, during or after manufacture, fabrication, assembly, or pressure tests as appropriate. Supplementary types of nondestructive inspection techniques, such as magnetic-particle, radiographic, and ultrasonic, shall not be required unless specifically listed herein or in the engineering design.

(c) In the event repairs or additions are made following the pressure test, the affected piping shall be tested.

Exception: Minor repairs or additions, provided the work is inspected and connections are tested with a noncorrosive leak-detecting fluid or other leak-detecting methods approved by the authority having jurisdiction.

(d) A piping system shall be tested as a complete unit or in sections. Under no circumstances shall a valve in a line be used as a bulkhead between gas in one section of the piping system and test medium in an adjacent section, unless two valves are installed in series with a valved "telltale" located between these valves. A valve shall not be subjected to the test pressure unless it can be determined that the valve, including the valve closing mechanism, is designed to safely withstand the pressure.

(e) Regulator and valve assemblies fabricated independently of the piping system in which they are to be installed shall be permitted to be tested with inert gas or air at the time of fabrication.

A.4.1.1 Because it is sometimes necessary to divide a piping system into test sections and install test heads, connecting piping, and other necessary appurtenances for testing, it is not required that the tie-in sections of pipe be pressure-tested. Tie-in connections, however, should be tested with a noncorrosive leak detection fluid after gas has been introduced and the pressure has been increased sufficiently to give some indications whether leaks exist.

The test procedure used should be capable of disclosing all leaks in the section being tested and should be selected after giving due consideration to the volumetric content of the section and to its location.

Under no circumstances should a valve in a line be used as a bulkhead between gas in one section of the piping system and test medium in an adjacent section, unless two valves are installed in series with a valved "telltale" located between these valves. A valve should not be subjected to the test pressure unless it can be determined that the valve, including the valve closing mechanism, is designed to safely withstand the test pressure.

Minor editorial changes were made in the 1999 edition to clarify the intent of the committee. Note that "exotic" test methods are not required by the code, but they might be required in the contract between the installer and owner.

Clearly the code's intent is that all piping should be tested. However, the code also recognizes that testing minor areas or testing the minimum amount of material between two substantial areas that have been tested already is not needed, provided that the work is reasonably and carefully inspected and connections are tested with a leak-detecting solution or other leak-detecting methods. The requirement to use a noncorrosive, leak-detecting fluid recognizes that some users were employing solutions that contained ammonia and other corrosive compounds. The Exception to 4.1.1(c) permits the use of "other leak-detecting methods." This addition permits portable gas detectors to be used for such purposes.

A piping system might have to be tested in sections for any of several reasons, but the use of a valve by itself to separate the gas in one part of the system from the test medium in another part is not permitted. The first reason for this requirement is that the valve might not close completely, so testing could indicate a leak when there is none. In addition, the valve could be opened inadvertently, allowing the test medium to contaminate the gas, which could lead to operational difficulties with the appliance or could result in a flammable gas–air mixture in the piping system. Using two valves with a valved telltale that is left open during the test is an acceptable method of separating the gas from a test medium.

Paragraph 4.1.1(e) permits regulator and valve assemblies that are assembled and tested elsewhere to be used. By implication, any tested subassembly should be acceptable without retesting.

4.1.2 Test Medium.

The test medium shall be air or an inert gas. OXYGEN SHALL NEVER BE USED.

The prohibition on use of oxygen as a test medium is based on the fact that a mixture of fuel gas and pure oxygen is a much greater hazard than a mixture of fuel gas and air. In the 1999 edition, "inert gas" replaced "nitrogen" and "argon" in this paragraph to clarify that any inert gas can be used for the test.

4.1.3 Test Preparation.

(a) Pipe joints, including welds, shall be left exposed for examination during the test.

Exception: If the pipe end joints have been previously tested in accordance with this code, they shall be permitted to be covered or concealed.

(b) Expansion joints shall be provided with temporary restraints, if required, for the additional thrust load under test.

Manufacturers' information can be helpful in planning a system to secure expansion joints for a pressure test.

(c) Appliances and equipment that are not to be included in the test shall be either disconnected from the piping or isolated by blanks, blind flanges, or caps. Flanged joints at which blinds are inserted to blank off other equipment during the test shall not be required to be tested.

(d) Where the piping system is connected to appliances, equipment, or equipment components designed for operating pressures of less than the test pressure, such appliances, equipment, or equipment components shall be isolated from the piping system by disconnecting them and capping the outlet(s).

> This paragraph recognizes that the 3-psi (21-kPa) minimum test pressure (see 4.1.4) can damage appliances that operate at about $^1/_2$ psi (14-in. water column) (3 kPa). In cases where the piping to the appliance has been isolated according to 4.1.3, after the test the piping has to be reconnected. The piping between the shutoff or cap and the appliance must be checked to be sure there are no leaks. Leak solution or electronic leak detectors can be used to comply. See Section 4.2 for addition details.

(e) Where the piping system is connected to appliances, equipment, or equipment components designed for operating pressures equal to or greater than the test pressure, such appliances and equipment shall be isolated from the piping system by closing the individual equipment shutoff valve(s).

(f) All testing of piping systems shall be done with due regard for the safety of employees and the public during the test. Bulkheads, anchorage, and bracing suitably designed to resist test pressures shall be installed if necessary. Prior to testing, the interior of the pipe shall be cleared of all foreign material.

> All joints must be left exposed until testing is complete so that leakage, if found, can be located. Cleaning foreign material from pipe systems can be done with compressed air prior to connecting any appliances.

4.1.4 Test Pressure.

(a) Test pressure shall be measured with a manometer or with a pressure measuring device designed and calibrated to read, record, or indicate a pressure loss due to leakage during the pressure test period. The source of pressure shall be isolated before the pressure tests are made.

(b) The test pressure to be used shall be no less than $1\,^1/_2$ times the proposed maximum working pressure, but not less than 3 psi (20 kPa), irrespective of design pressure. Where the test pressure exceeds 125 psi (862 kPa), the test pressure shall not exceed a value that produces a hoop stress in the piping greater than 50 percent of the specified minimum yield strength of the pipe.

(c) Test duration shall be not less than $^1/_2$ hour for each 500 ft^3 (14 m^3) of pipe volume or fraction thereof. When testing a system having a volume less than 10 ft^3 (0.28 m^3) or a system in a single-family dwelling, the test duration shall be permitted to be reduced to 10 minutes. For piping systems having a volume of more than 24,000 ft^3 (680 m^3), the duration of the test shall not be required to exceed 24 hours.

> Note that 10 ft^3 (0.28 m^3) of volume in a gas system is equivalent to approximately 8000 ft (2400 m) of $^1/_2$-in. copper tubing. Most systems will not require more than a test duration of ten minutes.

The 1996 edition of the code eliminated an exception for single-stage LP-Gas systems that allowed pressure testing at approximately 9-in. (2-kPa) water column pressure. Single-stage LP-Gas systems were allowed by NFPA 58, *Liquefied Petroleum Gas Code,* until the 1995 edition. In that edition, two-stage pressure regulation was mandated for building piping systems.

4.1.5 Detection of Leaks and Defects.

(a) The piping system shall withstand the test pressure specified without showing any evidence of leakage or other defects. Any reduction of test pressures as indicated by pressure gauges shall be deemed to indicate the presence of a leak unless such reduction can be readily attributed to some other cause.

Other possible causes include sharp drops in temperature. The likelihood of this phenomenon's causing a pressure change is practically zero, except on very large systems with long test times. A pressure drop almost always means a leak. When trying to locate leaks, remember that the test apparatus itself can be the culprit.

(b) The leakage shall be located by means of an approved gas detector, a noncorrosive leak detection fluid, or other approved leak detection methods. **Matches, candles, open flames, or other methods that provide a source of ignition shall not be used.**

Repairs are not permitted to fittings and pipe. (See 2.6.5.)
In the 1999 edition, the requirement for the indefinite retention of records of the inspection and testing of piping systems was deleted. This was done because the committee could see no benefit in retaining these records after the piping system is demonstrated to be leak-free. There may be local requirements for retention of construction documents, which should be followed.

Where leakage or other defects are located, the affected portion of the piping system shall be repaired or replaced and retested. *[See 4.1.1(c).]*

4.2 System and Equipment Leakage Test

This section describes a test that checks for leakage throughout the system of piping and appliances. It is equally applicable to new and existing gas systems. For systems of new piping, it is conducted after the pressure test described in Section 4.1. Normally, the test is conducted when one of the following occurs:

- A system of new gas or modified gas piping is placed into service for the first time.
- Gas leakage is suspected.
- A gas meter is replaced.

- An appliance or appliance connector is replaced.
- An out-of-gas condition occurs.

The test differs from the test in Section 4.1 in that it requires no special preparations.

Over the years the pressure test and leak check have been confused with each other. A pressure test is required for new piping installations and additions to piping installations, while a leak check is required whenever the gas system is initially placed into service or when the gas is turned back on after being turned off.

The medium used for a leak check is fuel gas at its normal supply pressure. The gas is applied to the total system (i.e., piping, equipment, and equipment connections and valves). See Appendix D, "Suggested Method for Checking for Leakage," for a recommended leak check procedure. The procedure in Appendix D is only one possible procedure. Other test methods can be used.

4.2.1 Test Gases.

Fuel gas shall be permitted to be used for leak checks in piping systems that have been tested in accordance with Section 4.1.

This is a new section in the 1999 edition. It clarifies that fuel gas may be used to test for leaks in piping systems after the system has been tested in accordance with 4.1.

4.2.2 Before Turning Gas On.

Before gas is introduced into a system of new gas piping, the entire system shall be inspected to determine that there are no open fittings or ends and that all manual valves at outlets on equipment are closed and all unused valves at outlets are closed and plugged or capped.

Note that this requirement, numbered 4.2.1 in previous editions, is now applicable only to new systems. Its applicability to existing systems was removed in the 1999 edition based on input from the propane industry recognizing that it is often not possible to enter a building when gas service is restored after a shutoff. This condition occurs much more frequently when propane is the fuel, since the propane tank might have been routinely emptied with use, and the homeowner might not be home during delivery.

4.2.3* Test for Leakage.

Immediately after the gas is turned on into a new system or into a system that has been initially restored after an interruption of service, the piping system shall be tested for leakage. If leakage is indicated, the gas supply shall be shut off until the necessary repairs have been made.

Note that this requirement, numbered 4.2.2 in previous editions, has been revised in the 1999 edition. It is applicable to both new and existing systems. It now requires that a leak check be done following an interruption of service as well as for new systems. This is necessary to be certain that the system does not have any leaks. Leaks might have been present and not noticed, or during the service interruption the piping system might have been disturbed, creating a leak. Whenever the opportunity to perform a leak test presents itself, it should be done to ensure a safe system.

The requirement states that if leakage is indicated by the test, repairs must be made. The committee in its deliberations and review of many proposals and comments on this subject stated that "it recognizes that the standards for gas valves and gas controls allow a very small amount of gas leakage, which is not a safety problem. The leakage tests traditionally used for natural gas and propane vary and should not be mandated by the code."

A.4.2.3 See Appendix D for a suggested method.

4.2.4 Placing Equipment in Operation.

Gas utilization equipment shall not be placed in operation until after the piping system has been tested in accordance with 4.2.3 and purged in accordance with 4.3.2.

This requirement was numbered 4.2.3 in previous editions. Even though the gas piping system has been purged at this point, caution should be used when lighting equipment. A chance always exists that not all of the air in the system is gone, and delayed ignition can result. Wearing safety glasses and keeping faces, hands, and arms away from combustion chambers are reasonable precautions.

4.3* Purging

A.4.3 The processes of purging a gas pipeline of fuel gas and replacing the fuel gas with air or charging a gas pipeline that is full of air with fuel gas require that a significant amount of combustible mixture not be developed within the pipeline or released within a confined space.

4.3.1 Removal from Service.

When gas piping is to be opened for servicing, addition, or modification, the section to be worked on shall be turned off from the gas supply at the nearest convenient point and the line pressure vented to the outdoors or to ventilated areas of sufficient size to prevent accumulation of flammable mixtures.

If this section exceeds the lengths shown in Table 4.3.1, the remaining gas shall be displaced with an inert gas.

Table 4.3.1 Length of Gas Line Requiring Purging for Servicing or Modification

Nominal Pipe Size (in.)	Minimum Length of Piping Requiring Purging (ft)
$2\frac{1}{2}$	50
3	30
4	15
6	10
8 or larger	Any length

For SI units, 1 ft = 0.305 m.

4.3.2 Placing in Operation.

When piping full of air is placed in operation, the air in the piping shall be displaced with fuel gas, provided the piping does not exceed the length shown in Table 4.3.2. The air can be safely displaced with fuel gas provided that a moderately rapid and continuous flow of fuel gas is introduced at one end of the line and air is vented out at the other end. The fuel gas flow shall be continued without interruption until the vented gas is free of air. **The point of discharge shall not be left unattended during purging.** The vent shall then be closed.

If the piping exceeds the lengths shown in Table 4.3.2, the air in the piping shall be displaced with an inert gas, and the inert gas shall then be displaced with fuel gas.

Table 4.3.2 Length of Piping Requiring Purging Before Being Placed in Operation

Nominal Pipe Size (in.)	Minimum Length of Piping Requiring Purging (ft)
3	30
4	15
6	10
8 or larger	Any length

For SI units, 1 ft = 0.305 m.

4.3.3 Discharge of Purged Gases.

The open end of piping systems being purged shall not discharge into confined spaces or areas where there are sources of ignition unless precautions are taken to perform this operation in a safe manner by ventilation of the space, control of purging rate, and elimination of all hazardous conditions.

4.3.4 Placing Equipment in Operation.

After the piping has been placed in operation, all equipment shall be purged and then placed in operation, as necessary.

The goal of this section is to provide instructions to minimize the possibility of developing a combustible mixture either in the piping or while gas is exiting from the piping during purging. Accordingly, if a system is to be shut down for service, removing the gas by filling the piping with a nonflammable atmosphere is necessary. The general method for purging gas is to vent the gas outdoors and then to fill the line with inert gas. If the pipe volume is small, inert gas does not need to be used, which simplifies the process. This is due to the small volume of potentially flammable mix that could be produced.

Placing the system back in service presents the same possibility of a flammable mix being formed within the pipe system. First, larger systems must be filled with inert gas, then with fuel gas. Again, smaller systems can be purged directly with fuel gas, which simplifies the procedure. Regardless of the method, the purge outlet must be attended during the procedure. Although discharge of purged gases is allowed indoors with appropriate restrictions, outdoor discharge eliminates any associated hazard and should be the preferred method when practical.

Reference Cited in Commentary

The following publication is available from the National Fire Protection Association, 1 Batterymarch Park, P.O. Box 9101, Quincy, MA 02269-9101.

NFPA 58, *Liquefied Petroleum Gas Code,* 1998 edition.

Equipment Installation

Chapter 5 covers requirements for equipment installation and is applicable to all gas-fired equipment. Along with Chapter 6, Chapter 5 provides all the code requirements for equipment installation. Chapter 6 provides additional requirements for specific types of equipment that are primarily used in residential and commercial installations. Chapter 5 includes references to the Chapter 7, "Venting of Equipment," to remind the installer that appliances must be properly vented.

Section 5.1, General, provides the following general requirements covering many appliances and installations:

- Approval of appliances and accessories (in 5.1.1)
- Addition of equipment to existing systems (in 5.1.2)
- Type of gas being used (in 5.1.3)
- Safety shutoff devices for LP-Gas appliances (in 5.1.4)
- Use of air or oxygen under pressure (in 5.1.5)
- Protection of equipment from the products of combustion (in 5.1.6)
- Building structural members (in 5.1.7)
- Flammable vapors and installation in residential garages, commercial garages, and aircraft hangers (in 5.1.8 through 5.1.11)
- Protection of equipment (in 5.1.12)
- Venting of flue gases (in 5.1.13)
- Extra devices or attachments (in 5.1.14)
- Adequacy of pipe sizing and prevention of strain on piping (in 5.1.15 and 5.1.16)
- Installation and venting of appliance pressure regulators, and bleed lines for diaphragm valves (In 5.1.17 through 5.1.19)
- Combination of equipment (in 5.1.20)
- Installation instructions (in 5.1.21)
- Protection of outdoor equipment (in 5.1.22)

Section 5.2, Accessibility and Clearance, provides requirements for installers to allow for access of equipment for maintenance and clearance to combustible construction.

Section 5.3, Air for Combustion and Ventilation, provides the following requirements for sufficient air for gas to burn properly, dilution of products of combustion, and ventilation of appliances rooms and spaces:

- General requirements (in 5.3.1)
- Equipment in unconfined spaces (in 5.3.2) (refer to the definition of a space, unconfined, in Section 1.7)
- Equipment in confined spaces (in 5.4.3) (refer to the definition of a space, confined, in Section 1.7)
- Specially engineered installations (in 5.3.4)
- Louvers and grills (in 5.3.5)
- Combustion air ducts (in 5.36)

Section 5.4, Equipment on Roofs, provides requirements for roof installations.

Section 5.5, Equipment Connections to Building Piping, provides the following requirements for connection of equipment to the piping system:

- Connection materials (in 5.5.1)
- Use of gas hose connectors (in 5.5.2)
- Connection of movable equipment (in 5.5.3)
- Equipment shutoff valves (in 5.5.4)
- Quick connection devices and gas convenience outlets (in 5.5.5 and 5.5.6)
- Sediment traps (in 5.5.7)
- Installation of piping so as not to interfere with equipment (in 5.5.8)

Section 5.6, Electrical, includes requirements for electrical connections, ignition and control devices, circuits, and continuous power for electrically controlled appliances.

Section 5.7, Room Temperature Thermostats, provides requirements for location of thermostats for thermostatically controlled space conditioning appliances.

5.1 General

5.1.1* Appliances, Accessories, and Equipment to Be Approved.

Gas appliances, accessories, and gas utilization equipment shall be approved. Approved shall mean "acceptable to the authority having jurisdiction."

Acceptance of unlisted gas utilization equipment and accessories shall be on the basis of a sound engineering evaluation. In such cases, the equipment shall be safe and suitable for the proposed service and shall be recommended for the service by the manufacturer.

The authority having jurisdiction, which can be the local building or fire official, insurer, property owner, or commanding officer (in the military), will approve the materials, equipment, and method of construction as the result of investigations and tests or by reason of accepted principles or tests by national authorities, technical organizations, and scientific organizations. The second paragraph provides criteria for approving unlisted equipment.

Appendix B, Coordination of Gas Utilization Equipment Design, Construction, and Maintenance, is provided to assist an installer in developing criteria for acceptance of nonconventional gas systems that have been properly designed and installed. This appendix is especially useful when applied to industrial systems.

A.5.1.1 The American Gas Association, the American National Standards Institute, Inc., and the National Fire Protection Association do not approve, inspect, or certify any installations, procedures, equipment, or materials. In determining acceptability of installations or procedures, the authority having jurisdiction can base acceptance on compliance with AGA, ANSI, NFPA, or other appropriate standards. In the absence of such standards, said authority can require evidence of proper installation, procedure, or use. The authority having jurisdiction can also refer to the listings or labeling practices *(see Section 1.7)* of an organization concerned with product evaluations and is in a position to determine compliance with appropriate standards for the current production of listed items. Additional information regarding the coordination of gas utilization equipment design, construction, and maintenance can be found in Appendix B.

5.1.2 Added or Converted Equipment.

When additional or replacement equipment is installed or an appliance is converted to gas from another fuel, the location in which the equipment is to be operated shall be checked to verify the following:

(a) Air for combustion and ventilation is provided where required, in accordance with the provisions of Section 5.3. If existing facilities are not adequate, they shall be upgraded to Section 5.3 specifications.

(b) The installation components and equipment meet the clearances to combustible material provisions of 5.2.2.

It shall be determined that the installation and operation of the additional or replacement equipment does not render the remaining equipment unsafe for continued operation.

(c) The venting system is constructed and sized in accordance with the provisions of Chapter 7. If the existing venting system is not adequate, it shall be upgraded to comply with Chapter 7.

These requirements are included in the code to ensure that a thorough review of the installation procedures is made when an appliance is installed, replaced, or converted to gas from another fuel. References specify the need to verify that air for combustion and ventilation is adequate, that the clearance to combustible material is sufficient, and that the venting system meets existing requirements. If any of the requirements are not met, the portion not meeting code requirements must be upgraded to present code standards.

Paragraph 5.1.2(c) emphasizes that installations are not "grandfathered" when the appliance is replaced or converted. In many instances, the location has undergone renovations since the original installation. In the case of new equipment, the design and operation of the new equipment may be different from the existing appliance. New, more energy efficient equipment has different installation requirements, and new equipment can have different requirements for clearance to combustible materials.

The requirement of 5.1.2(c) is especially important because the venting requirements have changed significantly since the 1988 edition of the code. In most cases the venting requirements of the new appliance must meet the new tables in the code because many of the newer appliances are more efficient and produce different vent gases.

5.1.3 Type of Gas(es).

It shall be determined whether the gas utilization equipment has been designed for use with the gas to which it will be connected. No attempt shall be made to convert the equipment from the gas specified on the rating plate for use with a different gas without consulting the installation instruction, the serving gas supplier, or the equipment manufacturer for complete instructions.

Equipment rating plates will specify for which gas a particular appliance is specifically designed. Some appliances are not designed to be converted, while others include parts and installation instructions for conversion from one gas to another that specify a specific conversion kit part number. If there are no such instructions, contact the manufacturer or serving gas supplier to determine whether the unit can be converted safely and how to make the conversion. If the burners are satisfactory for use with both gases, changing only the burner and pilot orifice(s) and regulator will be necessary. [Natural gas appliances usually operate at 2 in. to 5 in. (0.5 kPa to 1.2 kPa) water column manifold pressure; LP-Gas appliances operate at 10 in. to 11 in. (2.5 kPa to 2.7 kPa) water column manifold pressure.] A complete (100%) shutoff-type safety device generally is required for LP-Gases. Refer to Appendix F, "Flow of Gas Through Fixed Orifices," for additional information on orifice sizing.

5.1.4 Safety Shutoff Devices for Unlisted LP-Gas Equipment Used Indoors.

Unlisted gas utilization equipment for use with undiluted liquefied petroleum gases and installed indoors shall be equipped with safety shutoff devices of the complete shutoff type.

A safety shutoff device of the complete shutoff type is one that will stop the flow of gas to both the main burner(s) and the pilot burner in the event of pilot failure.

5.1.5 Use of Air or Oxygen Under Pressure.

Where air or oxygen under pressure is used in connection with the gas supply, effective means such as a back-pressure regulator and relief valve shall be provided to prevent air or oxygen from passing back into the gas piping. Where oxygen is used, installation shall be in accordance with NFPA 51, *Standard for the Design and Installation of Oxygen-Fuel Gas Systems for Welding, Cutting, and Allied Processes.*

A backup of air or oxygen into a gas pipe can result in a flammable mixture in the pipe. This mixture can be extremely dangerous and must be avoided to prevent an explosion within the gas pipe. The code requires that when oxygen is used, the installation must be in accordance with NFPA 51, *Standard for the Design and Installation of Oxygen-Fuel Gas Systems for Welding, Cutting, and Allied Processes.*

In most cases in which air or oxygen is used, it is at a greater pressure than the fuel gas. If the air or oxygen enters the gas supply, the air or oxygen could travel all the way back to the gas supply source. If this occurs in a natural gas distribution system, it could result in a very dangerous situation in which the gas supply to other customers contains air or oxygen.

Installations where the use of air or oxygen combined with natural gas or propane is likely to be found include jewelry stores that do repairs, muffler shops, and other metalworking facilities.

5.1.6* Protection of Gas Equipment from Fumes or Gases Other Than Products of Combustion.

Where corrosive or flammable process fumes or gases are present, means for their safe disposal shall be provided. Such fumes or gases include carbon monoxide, hydrogen sulfide, ammonia, chlorine, and halogenated hydrocarbons.

Gas appliances installed in beauty shops, barber shops, or other facilities where chemicals that generate corrosive or flammable products such as aerosol sprays are routinely used shall be located in an equipment room separate or partitioned off from other areas with provisions for combustion and dilution air from outdoors.

Exception: This requirement shall not apply to direct vent equipment that is constructed and installed so all air for combustion is obtained from the outside atmosphere and all flue gases are discharged to the outside atmosphere.

Corrosive or flammable process fumes or gases can be disposed safely only by a separate exhaust system and an adequate supply of fresh makeup air. Providing a system of isolation for the gas-fired appliance can be necessary to ensure that uncontaminated air is available for its combustion process.

For many years, hydrocarbons treated with halogen were used extensively in aerosol spray cans containing various household and commercial products. These products are still in use, though their widespread use has been curtailed. However, chlorine and fluorine compounds are still present in many products (i.e., water softener salt, laundry bleaches, detergents, adhesives, paints, varnishes, paint strippers, waxes, and plastics). When burned, these compounds form acids that corrode the heat exchanger and vent system. The only way to avoid the subsequent problem of excessive corrosion of the heat exchanger surface and vent system is to isolate the heating or water-heating appliance so that it uses only outside air for combustion, ventilation, and draft hood dilution. Direct-vent equipment does not need to meet these requirements because it is designed to obtain all air for combustion from the outside atmosphere and discharge flue gases into it as well, provided that the combustion air is not located in a contaminated area (i.e., swimming pool, dryer vent, etc.).

A.5.1.6 Halogenated hydrocarbons are particularly injurious and corrosive after contact with flames or hot surfaces.

5.1.7 Building Structural Members.

(a) Structural members of a building shall not pass through gas utilization equipment having an operating temperature in excess of 500°F (260°C).

(b) Structural members passing through gas utilization equipment having an operating temperature of 500°F (260°C) or less shall be of noncombustible material. Building columns, girders, beams, or trusses shall not be installed within equipment, unless insulation and ventilation are provided to avoid all deterioration in strength and linear expansion of the building structure in either a vertical or a horizontal direction.

(c) Gas utilization equipment shall be furnished either with load distributing bases or with a sufficient number of supports to prevent damage to either the building structure or equipment.

(d) At the locations selected for installation of gas utilization equipment, the dynamic and static load-carrying capacities of the building structure shall be checked to determine whether they are adequate to carry the additional loads. The equipment shall be supported and shall be connected to the piping so as not to exert undue stress on the connections.

Local building codes should be reviewed for the installation requirements of certain equipment. Some codes will specify equipment location and, in many instances, have specific requirements for rooftop installations.

5.1.8 Flammable Vapors.

Gas appliances shall not be installed in areas where the open use, handling, or dispensing of flammable liquids occurs, unless the design, operation, or installation reduces the potential of ignition of the flammable vapors. Gas utilization equipment installed in compliance with 5.1.9, 5.1.10, or 5.1.11 shall be considered to comply with the intent of this provision.

Note that this requirement was revised in the 1999 edition for clarity. The previously used phrase "flammable vapors are likely to be present" was replaced with the phrase "areas where the open use, handling, or dispensing of flammable liquids occurs."

Automotive painting facilities, for example, probably would be best served by the use of direct-vent equipment or an indirect steam or hot water coil in a makeup air unit with a remote boiler to ensure that the boiler is not in the proximity of the flammable vapors that are generated by the painting process.

Liquids that readily generate flammable vapors include gasoline, mineral spirits, acetone, denatured alcohol, oil-based paint and lacquer thinners, and camp stove fuel. For information on safe, recommended storage enclosures for these materials, see NFPA 30, *Flammable and Combustible Liquids Code.*

The installer must verify that the room in which an appliance is being installed is not an area where flammable vapors will be routinely present. While the future uses (or misuses) of a room cannot be predicted, gasoline, paint thinners, and other household solvents must be verified as not being stored and used in the room. A small spill of gasoline or other flammable liquid will create an invisible vapor cloud near the floor, because the vapors generated usually are heavier than air. This cloud can be drawn into an appliance through its combustion air intake, ignite, and flash back to the liquid pool. Many water heaters come with warning labels that describe this hazard to the installer and consumer.

Residential gas water heaters designed to resist the ignition of flammable vapors are likely to become available in the near future (see commentary following 5.1.9). These new water heaters, listed ANSI Z21.10.1, will be required to meet a flammable vapor ignition test as part of their listing. Their installation requirement under this section is different from the installation requirement for residential garages under 5.1.9. While 5.1.9 mandates that all gas utilization equipment installed in residential garages be installed so that their burners and burner ignition devices be at least 18 in. (460 mm) above the floor, 5.1.8 requires that gas appliances not be installed in areas where flammable vapors are likely to be present (except for residential garages) "unless the design, operation, and installation are such to eliminate the probable ignition of flammable vapors." Residential water heaters listed to the revised ANSI Z21.10.1 standard (that contains the flammable vapor ignition test) may possibly be installed in areas where flammable vapors are likely to be present without an 18-in. (460-mm) elevation or other means to prevent ignition. Their installation in these locations should be in accordance with the manufacturer's installation instruction.

5.1.9 Installation in Residential Garages.

(a) Gas utilization equipment in residential garages shall be installed so that all burners and burner ignition devices are located not less than 18 in. (450 mm) above the floor.

(b) Such equipment shall be located or protected so it is not subject to physical damage by a moving vehicle.

Household solvents, oil-based paint and thinners, and gasoline for use as fuel for autos, lawn mowers, and so forth, are often stored in residential garages. Spills can occur when these products are dispensed from their storage containers or during automobile maintenance. These flammable liquids vaporize readily, and their vapors remain near the floor. Available evidence indicates that elevating burners and ignition devices 18 in. (460 mm) or more above the floor locates them high enough so the vapors will not be ignited. (See Exhibit 5.1.)

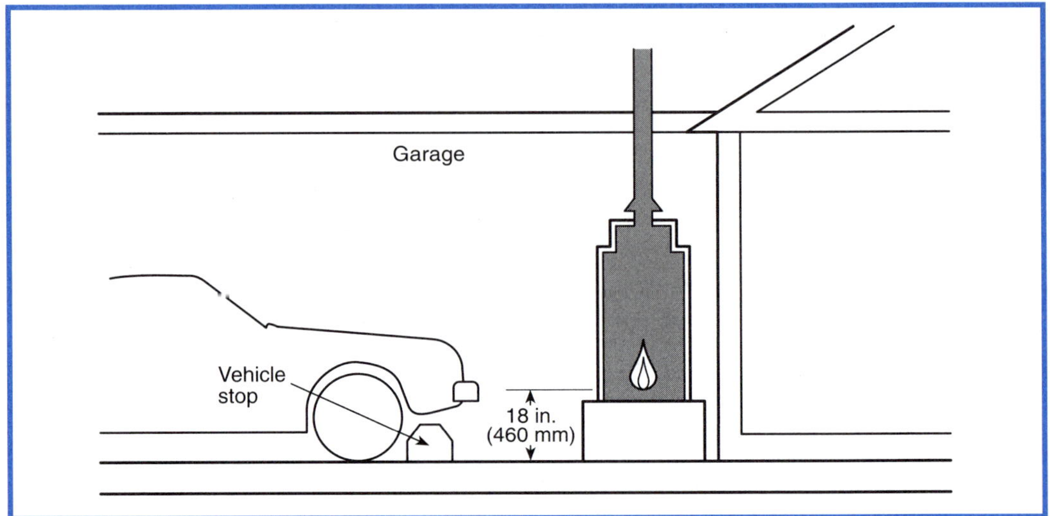

Exhibit 5.1 *Typical installation in residential garages. (Courtesy of American Gas Association.)*

(c) When appliances are installed in a separate, enclosed space having access only from outside of the garage, such equipment may be installed at floor level, providing the required combustion air is taken from the exterior of the garage.

The new paragraph (c) was added in the 1999 edition to allow for an appliance to be installed in a room that is constructed in a garage, but does not have a door that opens into the garage. This absence of a door prevents any flammable vapors that may be present in the garage from interacting with the combustion air for the appliance. If the room has the ability to open into the garage, appliances cannot be installed on the floor because the door can be left open.

Much concern has been voiced over the past several years regarding the ignition of flammable vapors by gas appliances, and specifically residential water heaters. Water heater manufacturers have responded by implementing a public service awareness campaign, aimed at children, on the hazards of flammable vapors. They have also proposed changes to ANSI Z21.10.1, *Gas Water Heaters — Volume I — Storage, Water Heaters with Input Ratings Above 75,000 Btu per Hour or Less,* that were accepted by the ANSI Z21 committee in May 1999. These changes will require that all residential water heaters pass a flammable vapor ignition resistance test in order to be certified. The editor is advised that the new edition of ANSI Z21.10.1 will go into effect in the near future. These products are in evaluation testing in selected installations. Their design has not been divulged by the manufacturers, and no details are available for publication at this writing. See Supplement 4 of this handbook for information on the new test procedure that was added to ANSI Z21.10.1.

Note that 5.1.9 requires that all gas utilization equipment installed in garages have their burners and burner ignition devices 18 in. (460 mm) or more above the floor. Therefore, unless the code is amended, the new water heaters will need to be installed with their burners and burner ignition devices 18 in. (460 mm) or more above the floor.

5.1.10 Installation in Commercial Garages.

(a) *Parking Structures.* Gas utilization equipment installed in enclosed, basement, and underground parking structures shall be installed in accordance with NFPA 88A, *Standard for Parking Structures.*

The installation of gas utilization equipment in any enclosed, basement, and underground parking structures, both with and without fuel-dispensing facilities, must meet the requirements of NFPA 88A, *Standard for Parking Structures.*

NFPA 88A, *Standard for Parking Structures,* requires that all gas-fueled heating equipment in enclosed parking garages be approved and be located at least 18 in. (460 mm) above the floor or be protected by a partition that is at least 18 in. (460 mm) high (see 4-2.2 of NFPA 88A).

An additional requirement of NFPA 88A is that continuous, mechanical ventilation must be provided in enclosed, basement, and underground parking garages that store flammable liquids. The fuels covered by NFPA 88A include CNG, LNG, LP-Gas; all Class I flammable liquids, including gasoline; and Class II flammable liquids, including diesel fuel, fuel oil, and kerosene.

While installation of heaters at least 8 ft (2.5 m) above the floor to provide clearance for passenger vehicles is not required for parking garages in NFPA 88A, as it is in NFPA 88B for repair garages, it is a good practice for overhead-type heaters. Consult the equipment manufacturer's instructions for the specific requirements for the installation of overhead radiant heaters to avoid overheating vehicles that could be parked underneath them.

(b) *Repair Garages.* Gas utilization equipment installed in repair garages shall be installed in a detached building or room, separated from repair areas by walls or partitions, floors, or floor ceiling assemblies that are constructed so as to prohibit the transmission of vapors and having a fire resistance rating of not less than 1 hour, and that have no openings in the wall separating the repair area within 8 ft (2.5 m) of the floor. Wall penetrations shall be firestopped. Air for combustion purposes shall be obtained from outside the building. The heating room shall not be used for the storage of combustible materials.

Exception No. 1: Overhead heaters where installed not less than 8 ft (2.5 m) above the floor shall be permitted.

Exception No. 2: Heating equipment for vehicle repair areas where there is no dispensing or transferring of Class I or Class II flammable or combustible liquids or liquefied petroleum gas shall be installed in accordance with NFPA 30A, Automotive and Marine Service Station Code.

This section requires that gas utilization equipment that is installed in repair garages must be either in a detached building or in a room separated from the repair garage, with air for combustion and ventilation obtained from outside the building.

Exceptions are made for unit heaters that are installed not less than 8 ft (2.5 m) above the floor and for repair garages where there is no dispensing or transferring of flammable liquids or liquid petroleum gas. This paragraph is in agreement with NFPA 88B, *Standard for Repair Garages,* which has the following requirements:

3-2 Heating

3-2.1 General.

3-2.1.1* Heating equipment shall be of an approved type. Improvised furnaces, salamanders, or space heaters shall not be permitted.

3-2.1.2* When compressed natural gas (CNG) or liquefied natural gas (LNG) fueled vehicles are repaired or stored, open flame heaters or heating equipment with exposed surfaces having a temperature in excess of 750°F (399°C) shall not be permitted.

3-2.1.3 Fuels used shall be of the type and quality specified by the manufacturer of the heating appliance. Crankcase trainings shall not be used in oil-fired units.

Exception: Heating equipment using crankcase trainings in accordance with NFPA 31, Standard for the Installation of Oil-Burning Equipment.

3-2.1.4 Heating equipment shall be installed to conform with NFPA 90A, *Standard for the Installation of Air Conditioning and Ventilating Systems;* NFPA 31, *Standard for the Installation of Oil-Burning Equipment;* NFPA 54, *National Fuel Gas Code;* NFPA 211, *Standard for Chimneys, Fireplaces, Vents, and Solid Fuel-Burning Appliances;* and NFPA 82, *Standard on Incinerators and Waste and Linen Handling Systems and Equipment,* as applicable, except as hereinafter specifically provided.

3-2.2 Location of Heating Equipment.

Heating equipment shall be installed in a detached building or room. It shall be separated from repair areas by walls or partitions and floor or floor-ceiling assemblies that are constructed so as to prohibit the transmission of vapors and having a fire resistance rating of not less than 1 hour, with no openings in the wall separating the repair area within 8 ft (2.4 m) of the floor. Wall penetrations shall be firestopped. Air for combustion purposes shall be obtained from outside the building. The heating room shall not be used for storage of combustible materials, except for fuel storage as permitted by the standards referenced in 3-2.1.4.

Exception No. 1: Unit heaters, where installed in accordance with 3-2.3. Ventilation requirements of this Section do not apply.

Exception No. 2: Heating equipment for vehicle repair areas where there is no dispensing or transferring of Class I or II flammable or combustible liquids, or liquefied petroleum gas, shall be installed in accordance with NFPA 30A, Automotive and Marine Service Station Code.

3-2.3 Suspended Unit Heaters.

3-2.3.1 Approved suspended unit heaters shall be located not less than 8 ft (2.4 m) above the floor and installed in accordance with the conditions of their approval.

3-2.3.2 A distance shall be maintained between the heater and its vent and any adjacent combustible material (which is part of the building or its contents) in conformance with NFPA 54, *National Fuel Gas Code,* or NFPA 31, *Standard for the Installation of Oil-Burning Equipment.*

3-2.4 Warm Air Heating.

3-2.4.1 Return air openings in motor vehicle repair or parking areas shall be not less than 18 in (.5 m) above floor level measured to the bottom of the openings.

3-2.4.2 Recirculated air shall not be taken from any floors below grade level.

5.1.11 Installation in Aircraft Hangars.

Heaters in aircraft hangars shall be installed in accordance with NFPA 409, *Standard on Aircraft Hangars.*

5.1.12 Gas Equipment Physical Protection.

Where it is necessary to locate gas utilization equipment close to a passageway traveled by vehicles or equipment, guardrails or bumper plates shall be installed to protect the equipment from damage.

See Exhibit 5.1.

5.1.13 Venting of Flue Gases.

Gas utilization equipment shall be vented in accordance with the provisions of Chapter 7.

5.1.14 Extra Device or Attachment.

No device or attachment shall be installed on any gas utilization equipment that could in any way impair the combustion of gas.

Refer to the equipment manufacturer's instructions to determine whether or not there are any accessories approved for use on a particular piece of gas utilization equipment.

5.1.15 Adequate Capacity of Piping.

When additional gas utilization equipment is being connected to a gas piping system, the existing piping shall be checked to determine whether it has adequate capacity. *(See Section 2.4.)* If inadequate, the existing system shall be enlarged as necessary, or separate gas piping of adequate capacity shall be run from the point of delivery to the equipment.

5.1.16 Avoiding Strain on Gas Piping.

Gas utilization equipment shall be supported and so connected to the piping as not to exert undue strain on the connections.

5.1.17 Gas Appliance Pressure Regulators.

Where the gas supply pressure is higher than that at which the gas utilization equipment is designed to operate or varies beyond the design pressure limits of the equipment, a gas appliance pressure regulator shall be installed.

Normally, the gas utilization equipment will come equipped with a pressure regulator to accommodate the gas pressure that is provided by the serving gas utility. Fuel gases in residential structures are usually at a nominal pressure of 4 in. to 11 in. (1 kPa to 2.7 kPa) of water column, and most appliance regulators can accommodate up to a 14-in. (3.5-kPa) water column of inlet pressure [approximately 1/2 psi (3.4-kPa)]. Commercial and industrial natural gas applications can have piping systems that deliver gas at a pressure of several psi to the proximity of the appliance, requiring the appliance installer to connect a "pounds to inches" regulator ahead of the gas appliance.

Propane systems generally are provided with two regulators at the tank and outside the building. The second-stage regulator will deliver gas downstream from the regulator at a nominal 11-in. (2.7-kPa) water column pressure. If a propane appliance is not equipped with a regulator, consult the manufacturer's instructions to ensure that the equipment is suitable for use at the nominal 11 in. (2.5 kPa to 2.7 kPa) of pressure from the LP-Gas system tank.

5.1.18 Venting of Gas Appliance Pressure Regulators.

(a) Gas appliance pressure regulators requiring access to the atmosphere for successful operation shall be equipped with vent piping leading outdoors or, if the regulator vent is an integral part of the equipment, into the combustion chamber adjacent to a continuous pilot, unless constructed or equipped with a vent limiting means to limit the escape of gas from the vent opening in the event of diaphragm failure.

Any gas that is relieved by a gas pressure regulator must be vented safely to the outdoors to prevent the accumulation of a potentially dangerous concentration of gas indoors. As a minimum, the size of the vent line should not be smaller than the vent connection on the regulator. Regulator vent lines must be made using piping materials permitted by Section 2.6. [See 2.8.4(a)(1).] Plastic piping materials are not permitted. When the regulator vent is an integral part of the gas-fired appliance, it can be vented adjacent to the continuous burning pilot in the combustion chamber.

Many gas appliance pressure regulators are equipped with a vent-limiting device that restricts the flow of gas to under 2.5 ft^3/hr (0.07 m^3/hr) for natural gas and 1 ft^3/hr (0.028 m^3/hr) for LP-Gas or propane gas. This leakage rate is considered not likely to create a dangerous concentration of gas.

The code provides neither restrictions nor guidance to those who manifold the vents of multiple regulators together. Information on sizing of discharge piping from pressure regulators can be found in American Petroleum Institute Recommended Practice RP 520, *Sizing, Selection, and Installation of Pressure-Relieving Devices in Refineries, Part II, Installation.* Prohibiting termination of bleed lines inside a building is imperative, because if the rubber diaphragm fails, unburned gas will be released into the building, creating a potentially hazardous condition.

(b) Vent limiting means shall be employed on listed gas appliance pressure regulators only.

(c) In the case of vents leading outdoors, means shall be employed to prevent water from entering this piping and also to prevent blockage of vents by insects and foreign matter.

Outdoor vents from regulators can become blocked by insects, insect nests, ice, or freezing rain. This blockage usually is prevented by pointing the end of the vent line down, flaring it approximately 40 degrees, and screening the opening.

(d) Under no circumstances shall a regulator be vented to the gas utilization equipment flue or exhaust system.

(e) In the case of vents entering the combustion chamber, the vent shall be located so the escaping gas will be readily ignited by the pilot and the heat liberated thereby will not adversely affect the normal operation of the safety shutoff system. The terminus of the vent shall be securely held in a fixed position relative to the pilot. For manufactured gas, the need for a flame arrester in the vent piping shall be determined.

(f) A vent line(s) from a gas appliance pressure regulator and a bleed line(s) from a diaphragm type valve shall not be connected to a common manifold terminating in a combustion chamber. Vent lines shall not terminate in positive-pressure-type combustion chambers.

Common manifolding of both a bleed line from a diaphragm-type valve and a vent line from a gas appliance regulator that terminates in a combustion chamber represents a potential hazard, because the gas vented by the diaphragm valve could be drawn into the regulator through the regulator vent line until an explosive air–gas mixture is reached. Ignition of this mixture by the continuous pilot could produce a small explosion that could extinguish the pilot and rupture the pressure regulator diaphragm. This rupture would then allow unburned gas to flow continuously through the vent line into the combustion chamber, causing an overrate condition for the main burner and pilot and subjecting the gas controls to excessive gas pressures.

The prohibition of vent lines terminating in positive-pressure-type combustion chambers recognizes that more gas utilization equipment is being designed for positive pressure combustion chambers, which can result in backpressure on the diaphragm of regulators and diaphragm valves and can cause overgassing or failure of equipment to operate.

5.1.19 Bleed Lines for Diaphragm-Type Valves.

(a) Diaphragm-type valves shall be equipped to convey bleed gas to the outside atmosphere or into the combustion chamber adjacent to a continuous pilot.

(b) In the case of bleed lines leading outdoors, means shall be employed to prevent water from entering this piping and also to prevent blockage of vents by insects and foreign matter.

(c) Bleed lines shall not terminate in the gas utilization equipment flue or exhaust system.

(d) In the case of bleed lines entering the combustion chamber, the bleed line shall be located so the bleed gas will be readily ignited by the pilot and the heat liberated thereby will not adversely affect the normal operation of the safety shutoff system. The terminus of the bleed line shall be securely held in a fixed position relative to the pilot. For manufactured gas, the need for a flame arrester in the bleed line piping shall be determined.

(e) A bleed line(s) from a diaphragm-type valve and a vent line(s) from a gas appliance pressure regulator shall not be connected to a common manifold terminating in a combustion chamber. Bleed lines shall not terminate in positive-pressure-type combustion chambers.

For the same reasons given in the commentary following 5.1.18(f), the bleed line from the diaphragm valve must not be interconnected by the manifold to the vent line from a gas pressure regulator. (See Exhibits 5.2 and 5.3.)

5.1.20 Combination of Equipment.

Any combination of gas utilization equipment, attachments, or devices used together in any manner shall comply with the standards that apply to the individual equipment.

A list of standards is included in Chapter 12 and Appendix L of this code. For further information, consult the gas utilization equipment manufacturer.

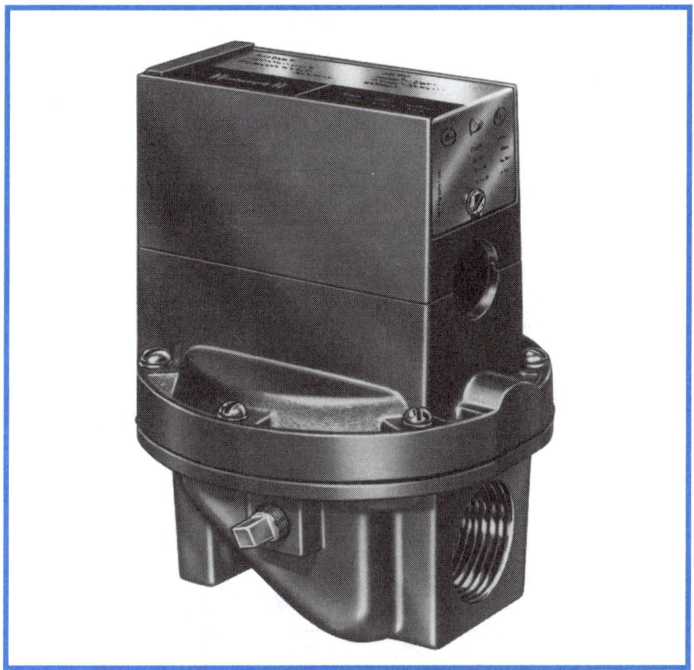

Exhibit 5.2 *Diaphragm valve. (Courtesy of Honeywell, Inc.)*

5.1.21 Installation Instructions.

The installing agency shall conform with the equipment manufacturers' recommendations in completing an installation. The installing agency shall leave the manufacturers' installation, operating, and maintenance instructions in a location on the premises where they will be readily available for reference and guidance of the authority having jurisdiction, service personnel, and the owner or operator.

> Leaving the manufacturer's installation, service and maintenance instructions, and owner's/user's manual on the job site is important for future reference by the authority having jurisdiction, service personnel, and the owner or user. This material includes specific installation and service requirements for the equipment not found in this or local codes.

5.1.22 Protection of Outdoor Equipment.

Gas utilization equipment not listed for outdoor installation but installed outdoors shall be provided with protection to the degree that the environment requires. Equipment listed for outdoor installation shall be permitted to be installed without protection in accordance with the provisions of their listing. *(See 5.2.1.)*

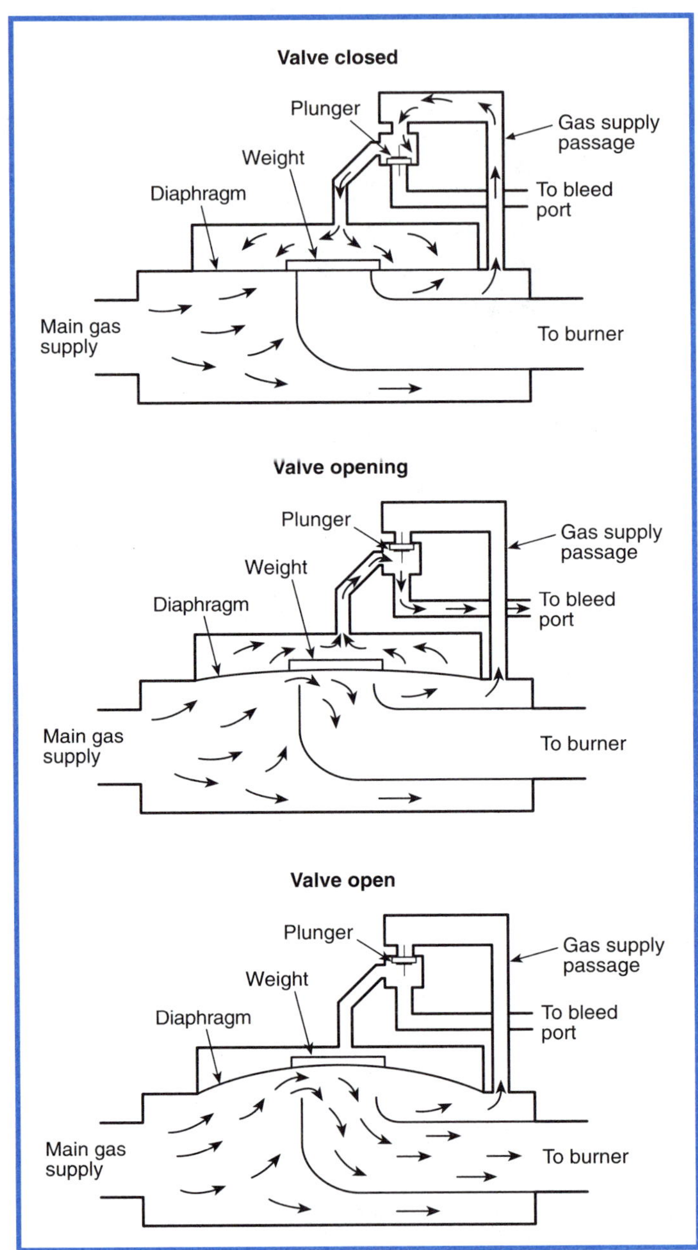

Exhibit 5.3 *Operation of diaphragm valve. (Courtesy of Honeywell, Inc.)*

5.2 Accessibility and Clearance

5.2.1 Accessibility for Service.

All gas utilization equipment shall be located with respect to building construction and other equipment so as to permit access to the gas utilization equipment. Sufficient clearance shall be maintained to permit cleaning of heating surfaces; the replacement of filters, blowers, motors, burners, controls, and vent connections; the lubrication of moving parts where necessary; the adjustment and cleaning of burners and pilots; and the proper functioning of explosion vents, if provided. For attic installation, the passageway and servicing area adjacent to the equipment shall be floored.

This paragraph highlights the importance of keeping service clearances of equipment in mind when it is installed. Follow manufacturer's instructions and/or consult local building codes for width and length of passageways to equipment installed in attics and crawl spaces. Generally, passageways must be floored and limited to not less than 24 in. (0.6 m) wide or to the width of the equipment for removal and 20 ft (6 m) in length. In addition, there should be at least 30 in. (0.76 m) clearance in front of the equipment to provide adequate service space and access to the electrical controls and other operating components. Gas utilization equipment approved for closet installation can have reduced clearance if service is accomplished through a door or removable panel and the reduced clearance is not less than that specified on the equipment rating plate or in Chapter 6, "Installation of Specific Equipment," of this code. (See 5.4.3 for access to equipment on roofs.)

5.2.2 Clearance to Combustible Materials.

Gas utilization equipment and their vent connectors shall be installed with clearances from combustible material so their operation will not create a hazard to persons or property.

Minimum clearances between combustible walls and the back and sides of various conventional types of equipment and their vent connectors are specified in Chapters 6 and 7. *(Reference can also be made to NFPA 211, Standard for Chimneys, Fireplaces, Vents, and Solid Fuel-Burning Appliances.)*

For listed gas utilization equipment, the manufacturer's instructions must be consulted for proper clearance to combustibles. For additional information, see Chapter 6, "Installation of Specific Equipment." (See Exhibit 5.4.)

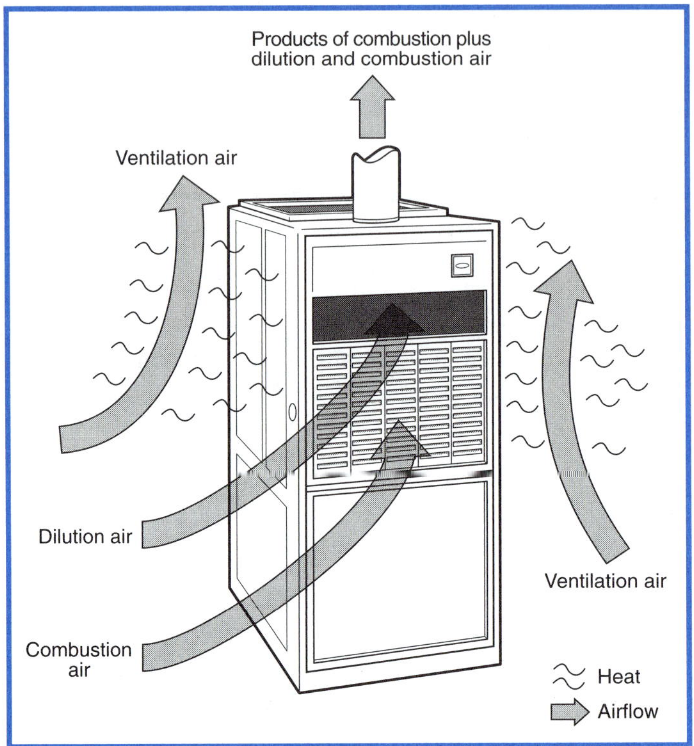

Exhibit 5.4 *Illustration of combustion, dilution, and ventilation air. (Courtesy of Ross Burnside.)*

5.2.3 Installation on Carpeting.

Equipment shall not be installed on carpeting, unless the equipment is listed for such installation.

This is a new paragraph in the 1999 edition. It was formerly part of 5.2.2, but was relocated to a separate paragraph so that it will not be overlooked.

5.3* Air for Combustion and Ventilation

The importance of combustion, dilution, and ventilation air cannot be overemphasized. Typical gas-fired natural draft (draft hood) furnaces require approximately 30 ft^3 (0.8 m^3) of air (i.e., combustion, vent dilution, and ventilation) for every cubic foot of gas burned. While modern fan-assisted combustion system furnaces do not need dilution air, they still require approximately 20 ft^3 (0.6 m^3) of air for each cubic foot of gas burned. This amount of air can range between 2000 ft^3/hr and 3000 ft^3/hr

(56.6 m³/hr – 85 m³/hr) for each 100,000 Btu/hr (29.3 kW) of gas input. This air is needed to support the combustion and venting process of the appliance and to provide ventilation cooling for the casing and internal controls. This section will help you determine the proper method for meeting these requirements. (See Exhibit 5.5.)

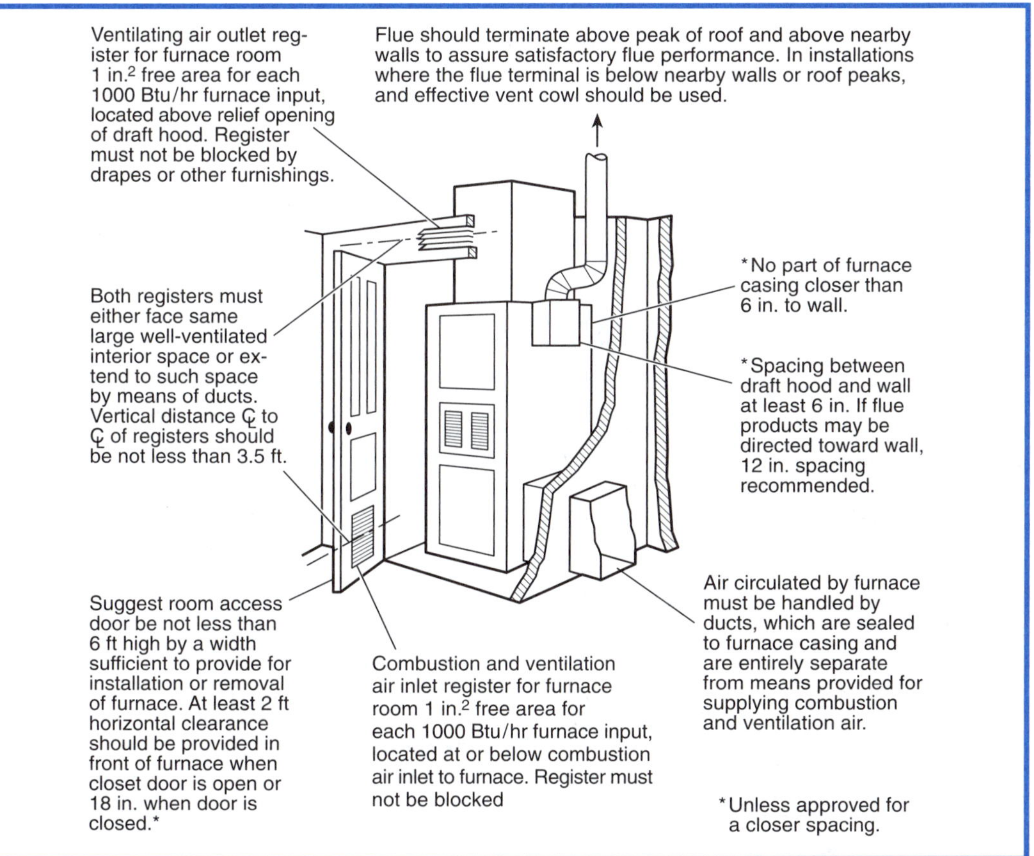

Exhibit 5.5 *Illustration showing air openings necessary to supply air for combustion when appliance is installed in confined space. (From NFPA 54 and ANSI Z21.30, 1950 edition.)*

As noted in the history of this code, the first gas safety standard, ASA K-2, was published in 1927. The requirements for air for combustion were very brief: "No appliance designed to burn gas at a rate greater than an ordinary lighting burner shall be installed in a room which is not adequately ventilated."

Not much was changed when American Standard ASA Z27 and NBFU (NFPA) Pamphlet 54 were issued in 1933: "Ventilation: No appliance shall be installed in a room in which the facilities for ventilation do not permit the proper combustion of gas under normal conditions of use."

The increased use of gas for heating and the popularity of "closet furnaces" following World War II resulted in AGA Research Bulletin #53, "The Effects of Confined Space Installation on Central Gas Space Heating Equipment Performance," published in 1947.

The study resulted in a requirement, included in the 1950 edition of the standards (ASA Z21.30 and NFPA 54), for two air openings into the confined space, each sized at 1 in.^{2} per 1000 Btu/hr (650 mm^2/kW) and communicating with an area having "adequate infiltration." (See Exhibit 5.5.)

If the confined space was within a building of "unusually tight construction," then air for combustion and ventilation was to be obtained from outdoors (e.g., crawl or attic space), and the size of the openings could be reduced by 50 percent.

In subsequent years, additional work by AGA Laboratories, which was published in AGA Research Bulletin #67, "Combustion and Ventilation Air Supply to Gas Equipment in Small Rooms," and work by the University of Illinois Engineer Experiment Station, published in Bulletin #427, "Outdoor Air Supply and Ventilation of a Furnace Closet Used with a Warm Air Heating System," reaffirmed the need for two openings in a confined space, and they provided more detailed information for the standards. In 1969, the most significant change in this section was to require that the openings be located within 12 in. (300 mm) of the top and 12 in. (300 mm) of the bottom of the enclosure. Consequently, five detailed drawings were added to Section 5.3 in 1959 for clarification. (One was dropped in the 1980 version.) An additional drawing was added in the 1999 edition, and all five drawings were relocated to Appendix A in the 1999 edition. The present requirement is that the minimum dimension of the air openings be not less than 3 in. (80 mm).

Significant changes in the design of gas appliances have occurred since the initial development of combustion air requirements. Increases in efficiency and the use of fan-assisted combustion systems prompted the Gas Research Institute to initiate a study of combustion air requirements. Report Number GRI 93/0316, "Analysis of Combustion Air Openings to the Outdoors: Preliminary Results," was issued in 1994. This report found that a single opening to the outdoors also could be used if certain qualifications were met [i.e., that the opening be within 12 in. (300 mm) of the ceiling and minimum specified clearances be provided around the appliance]. (See 5.3.3.)

The use of two openings, high and low, to the outdoors still presents practical problems, especially in an era when building envelopes are being tightened. In the 1996 edition, an alternate allowing one opening (located within 12 in. of the ceiling) was added, based on the GRI report cited above. The 1999 edition of the code adds language that recognizes that the two openings to the outdoors may be reduced in size if the appliance is in an indoor space that can contribute part of the needed air through infiltration. (See 5.3.4.) This concept has been in allowed in several states for some time.

A.5.3. *Special Conditions Created by Mechanical Exhausting or Fireplaces.* Operation of exhaust fans, ventilation systems, clothes dryers, or fireplaces can create conditions requiring special attention to avoid unsatisfactory operation of installed gas utilization equipment.

5.3.1 General.

(a) The provisions of Section 5.3 shall apply to gas utilization equipment installed in buildings that require air for combustion, ventilation, and dilution of flue gases.

Exception No. 1: Direct vent equipment that is constructed and installed so that all air for combustion is obtained directly from the outdoors and all flue gases are discharged to the outdoors.

Direct-vent appliances do not require air for combustion. They do however require air for ventilation. Gas appliances can become hot when they are operating. Although they do not need air for combustion, they do need air to keep the equipment room from overheating. If the appliances are operating for a long period of time, they can cause the room to overheat and can damage electrical components.

For an example of a direct-vent appliance, see Exhibit 6.30.

Exception No. 2: Enclosed furnaces that incorporate an integral total enclosure and use only outdoor air for combustion and dilution of flue gases.

(b) Equipment shall be located so as not to interfere with proper circulation of combustion, ventilation, and dilution air. Where normal infiltration does not provide the necessary air, outdoor air shall be introduced.

(c) In addition to air needed for combustion, process air shall be provided as required for cooling of equipment or material, controlling dew point, heating, drying, oxidation, dilution, safety exhaust, odor control, and air for compressors.

This requirement generally is specified by the equipment manufacturer or by the building designer.

(d) In addition to air needed for combustion, air shall be supplied for ventilation, including all air required for comfort and proper working conditions for personnel.

The local building code should be consulted, because the fresh air requirements for many occupancies are in excess of the fresh air requirements for combustion, ventilation, and dilution of vent gases.

(e) A draft hood or a barometric draft regulator shall be installed in the same room or enclosure as the equipment served so as to prevent any difference in pressure between the hood or regulator and the combustion air supply.

This paragraph is the same as 7.12.6 in Chapter 7, "Venting of Equipment," of this code. It is repeated to emphasize the need to have the draft hood relief opening and the combustion air supply in the same pressure zone.

(f) Air for combustion, ventilation, and dilution of flue gases for gas utilization equipment vented by natural draft shall be obtained by application of one of the methods covered in 5.3.3 and 5.3.4.

(g) Air requirements for the operation of exhaust fans, kitchen ventilation systems, clothes dryers, and fireplaces shall be considered in determining the adequacy of a space to provide combustion air requirements.

The operation of exhaust fans, kitchen ventilators, clothes dryers, fireplaces, and other systems that exhaust air from the dwelling can deprive the gas appliance of adequate combustion air or, in extreme cases, pull air in through the vent of natural draft appliances. This condition results in the combustion products spilling from the draft hood relief opening into the room in which the appliance is located. Coupled with a lack of combustion and ventilation air, incomplete combustion (i.e., high levels of carbon monoxide) and possible flame rollout can occur.

Section 5.3 contains provisions intended to provide adequate combustion and ventilation air for gas appliances based on an air infiltration rate of one-half air changes per hour. Because of increased interest in energy conservation, consumers are attempting to reduce heat losses by adding weatherstripping, caulking, and insulation to their residences. Although these measures are effective in reducing energy consumption, they also can reduce the air infiltration rate of the building to levels below those anticipated in the code. In these situations, additional air must be provided for proper operation of the equipment. In these tight buildings the operation of exhaust fans can create negative pressure in buildings. Air will enter the building through any opening available, and the vent of a gas appliance may provide the easiest available opening. An example of this is shown in Exhibit 5.6.

5.3.2* Equipment Located in Unconfined Spaces.

Equipment located in buildings of unusually tight construction *(see Section 1.7)* shall be provided with air for combustion, ventilation, and dilution of flue gases using the methods described in 5.3.3(b) or 5.3.4.

An unconfined space, which is illustrated in Exhibit 5.7, is defined in Section 1.7, Definitions, as follows:

Space, Unconfined. For purposes of this Code, a space whose volume is not less than 50 ft^3 per 1000 Btu/hr (4.8 m^3 per kW) of the aggregate input rating of all appliances installed in that space. Rooms communicating directly with the space in which the appliances are installed, through openings not furnished with doors, are considered a part of the unconfined space.

The requirement in 5.3.2 is sometimes referred to as the 1/20 rule, because 1000 Btu/hr ÷ 50 ft^3 = 20 Btu/hr per ft^3 or, for each cubic foot of room volume, you can install 20 Btu/hr.

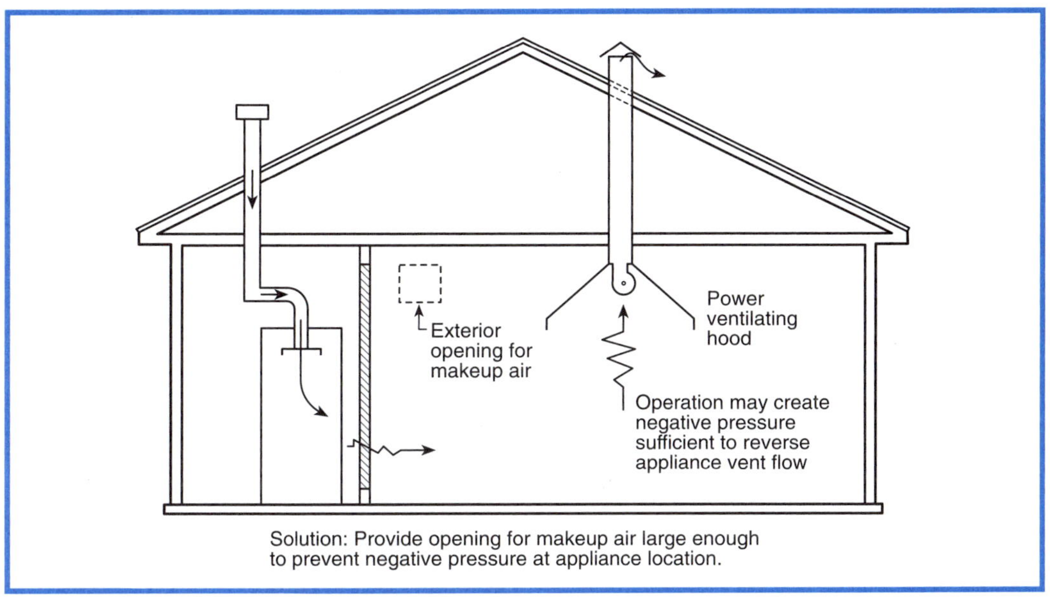

Exhibit 5.6 *Mechanical exhausting.*

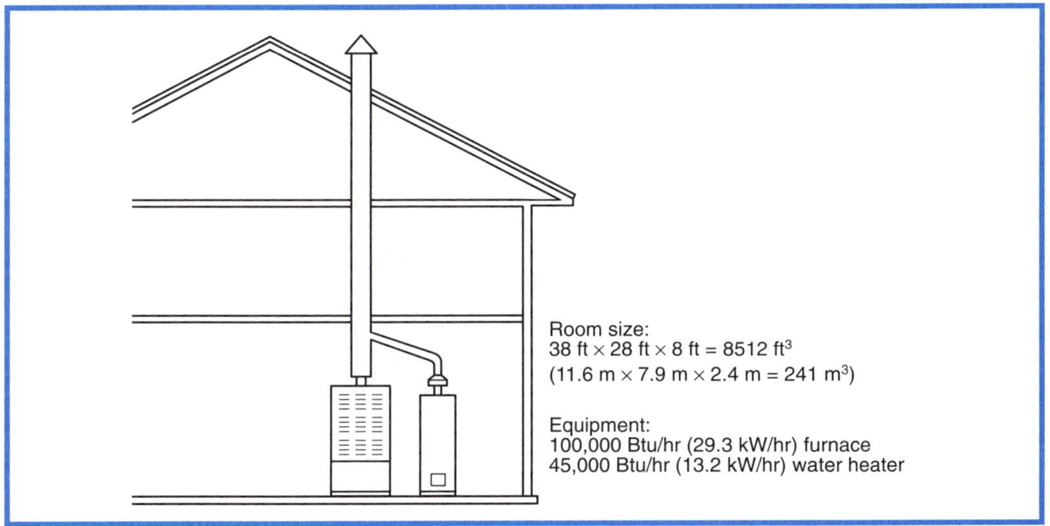

Exhibit 5.7 *Unconfined space. (Courtesy of Carrier Corporation.)*

The requirement for additional air for combustion, ventilation, and dilution recognizes that the definition of confined space came from research on buildings of
normal construction tightness. When unusually tight construction is used, additional

air is usually required, and the code provides guidance to local code authorities for determining when outside combustion and ventilation air is necessary.

Example:

$$\frac{\left(\dfrac{8512 \text{ ft}^3}{100{,}000 \text{ Btu/hr} + 45{,}000 \text{ Btu/hr}}\right)}{1000} = \frac{58.7 \text{ ft}^3}{1000 \text{ Btu/hr}}$$

or

$$\frac{241 \text{ m}^3}{425 \text{ kW}} = 5.7 \text{ m}^3/\text{kW}$$

Since 58.7 ft^3 per 1000 Btu/hr (5.7 m^3/kW) is greater than 50 ft^3 per 1000 Btu/hr (4.8 m^3/kW) of the total appliance input, this space by definition is unconfined. Another way to look at this situation is

$$\frac{8512 \text{ ft}^3}{(50 \text{ ft}^3 / 1000 \text{ Btu/hr})} \times 1000 \text{ Btu/hr} = 170{,}240 \text{ Btu/hr}$$

or

$$\frac{241 \text{ m}^3}{4.8 \text{ m}^3/\text{kW}} = 50.2 \text{ kW}$$

This result will equal the total input that can be installed in the space and still qualify as an unconfined space. In this example, the total input was 145,000 Btu/hr (42.5 kW). Therefore, there is no requirement for outside air.

Remember to look for possible building depressurization (negative pressure compared to the outside atmosphere) caused by fireplaces or mechanical exhausting [e.g., clothes dryers, bathroom fans, kitchen range or oven hoods, attic fans, and basement (radon gas control) fans]. If these are a factor, the use of additional sources of outside air for combustion and ventilation could be needed.

A.5.3.2 In unconfined spaces in buildings of other than unusually tight construction *(see Section 1.7)*, infiltration can be adequate to provide air for combustion, ventilation, and dilution of flue gases.

Unusually tight construction is defined in Section 1.7, Definitions. Note that the definition was revised in the 1999 edition by the addition of a fourth criterion in a new (4), and now reads as follows:

Unusually Tight Construction. Construction where (1) walls and ceilings exposed to the outside atmosphere have a continuous water vapor retarder with a rating of 1 perm (6×10^{-11} kg/Pa-sec-m^2) or less with openings gasketed or sealed; and (2) weatherstripping has been added on openable windows and doors; and (3) caulking or sealants are applied to areas such as joints around window and door frames, between sole plates and floors, between wall–ceiling joints, between wall panels, at penetrations for plumbing, electrical, and gas lines, and at other openings; and

(4) the building has an average air infiltration of less than 0.35 air changes per hour.

The concept of providing additional outside air to an unusually tight building was in the first edition of the code in 1974. The concept of "unusually tight construction" was introduced in 1988, and a separate definition was added to the code in the 1988 edition. Since that time, the techniques defining unusually tight construction have become commonplace and can now be described as normal in most new residential construction. A decade of experience with these newly constructed buildings did not demonstrate that widespread problems exist with combustion air obtained from the indoors through air infiltration. The air change per hour (ACH) criteria were added to provide guidance when it is determined that a building contains the first three construction details, but there still may be enough air infiltration to supply combustion air. This criterion recognizes that there may be enough combustion air available due to the inherent features of a building's overall design and construction, such as the amount of fenestration area, exterior wall area, number of envelope penetrations, how well a building is actually constructed, and so forth.

The air infiltration of a building occurs mainly as a result of pressure differences imposed by winds on the building or pressure-induced differences that result in a building being at a lower pressure than the atmosphere. Lower pressure in a building, compared to the atmosphere, is caused by the operation of exhaust fans, gas vents, and the relative temperature in the building compared to the outdoors. (See Exhibit 5.8.) The air change rate of 0.35 was selected to correspond to the definition of an unconfined space (50 ft^3 per 1000 Btu/hr), which is known to provide sufficient air for combustion and ventilation.

See Supplement 5, "Procedure to Estimate Infiltration Rate for Residential Structures," for further discussion of unusually tight construction and combustion air.

5.3.3 Equipment Located in Confined Spaces.

This subsection was revised in the 1996 edition with the addition of a second method of providing air from the outside using a single opening. Additional revisions in the 1999 edition add a third method that permits reduced outdoor opening sizes when indoor air can partially meet the combustion air needs.

(a)* *All Air from Inside the Building.* The confined space shall be provided with two permanent openings communicating directly with other spaces of sufficient volume so that the combined volume of all such spaces meets the criteria for an unconfined space. The total input of all gas utilization equipment installed in the combined spaces shall be used to determine the required minimum volume. Each opening shall have a minimum free area of not less than 1 in.^{2}/1000 Btu/hr (220 mm^2/kW) of the total input rating of all gas utilization equipment in the confined space, but not less than 100 in.^{2} (0.06 m^2). One opening shall commence within 12 in. (300 mm) of the top, and one opening shall commence within 12 in. (300 mm)

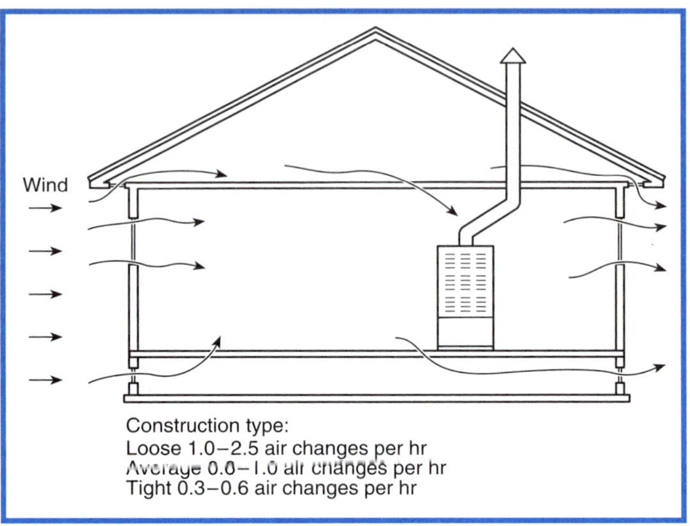

Exhibit 5.8 *Air infiltration and construction type. (Courtesy of Carrier Corporation.)*

of the bottom, of the enclosure *[see Figure A.5.3.3(a)]*. The minimum dimension of air openings shall be not less than 3 in. (8 cm).

A confined space is defined as "a space whose volume is less than 50 ft^3 per 1000 Btu/hr (4.8 m^3 per kW) of the aggregate input rating of all appliances installed in that space."

In the previous example in 5.3.2, consider that we were discussing a basement, and now we want to finish it into a family room. We divide the basement in half or into 4256 ft^3 (121 m^3). Is the space now confined or unconfined?

$$\frac{\left(\dfrac{4256 \text{ ft}^3}{100,000 \text{ Btu/hr} + 45,000 \text{ Btu/hr}}\right)}{1000 \text{ Btu/hr}} = \frac{29.4 \text{ ft}^3}{1000 \text{ Btu/hr}}$$

or

$$\frac{121 \text{ m}^3}{42.5 \text{ kW}} = 2.8 \text{ m}^3/\text{kW}$$

Since 29.4 ft^3/1000 Btu/hr (2.8 m^3/kW) is less than 50 ft^3 per 1000 Btu/hr (4.8 m^3/kW), by definition we have a confined space and must make provisions for combustion and ventilation air. In this case, if we did not have doors (only openings) between the two spaces, it would have still qualified as unconfined. However, since this room will be a family room, it will most likely have doors. The room will now require two permanent openings, each within 12 in. (300 mm) of the ceiling and floor. Each opening must have a free area of not less than 1 in.^{2} per 1000 Btu/hr (220 m^3/kW) of the total appliance input installed in the space, and the minimum dimen-

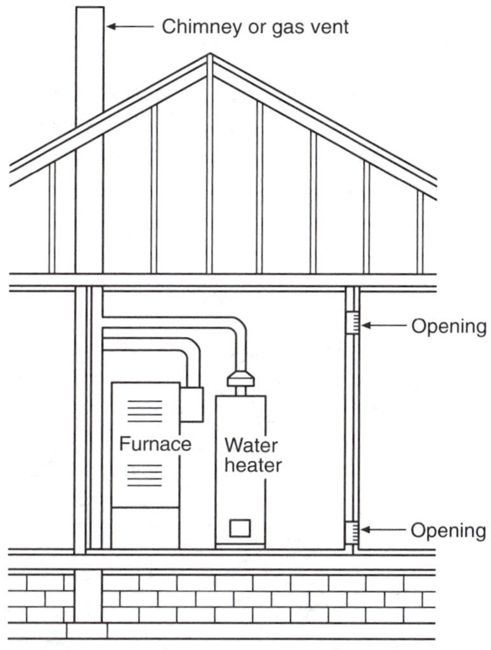

Figure A.5.3.3(a) *Equipment located in confined spaces; all air from inside the building. [See 5.3.3(a).]*

sion of the opening must not be less than 3 in. (80 mm). Also, the opening is limited to a minimum free area of 100 in.^{2} (0.0645 m^2), regardless of appliance input.

The openings at the top and bottom of the enclosure are necessary to provide adequate ventilation within the enclosure. Heat loss from the exposed appliance jacket, plenum, vent connector, and so forth, can cause a temperature increase in the enclosure. Since heated air will rise, natural circulation will bring in cooler air at the bottom opening and allow the heated air to escape from the top opening. To be effective, the openings must be within 12 in. (300 mm) of the top and bottom of the enclosure and have a minimum dimension of 3 in. (80 mm).

A.5.3.3(a) See Figure A.5.3.3(a).

(b) *All Air from Outdoors.* The confined space shall communicate with the outdoors in accordance with method 1 or 2, which follow. The minimum dimension of air openings shall not be less than 3 in. (80 mm). Where ducts are used, they shall be of the same cross-sectional area as the free area of the openings to which they connect.

(1) Two permanent openings, one commencing within 12 in. (300 mm) of the top and one commencing within 12 in. (300 mm) of the bottom, of the enclosure shall be provided.

The openings shall communicate directly, or by ducts, with the outdoors or spaces that freely communicate with the outdoors.

a.* Where directly communicating with the outdoors or where communicating to the outdoors through vertical ducts, each opening shall have a minimum free area of 1 in.^{2} per 4000 Btu/hr (550 mm^2/kW) of total input rating of all equipment in the enclosure. *[See Figures A.5.3.3(b)1a1 and A.5.3.3(b)1a2.]*

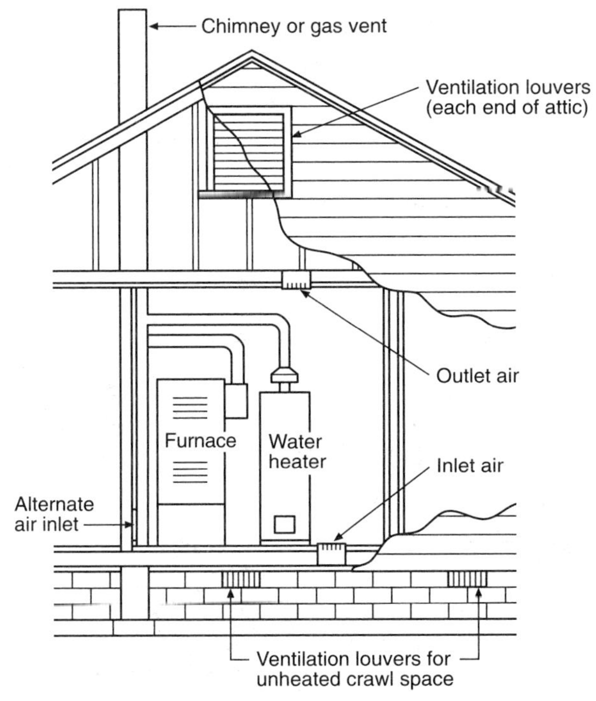

Figure A.5.3.3(b)1a1 *Equipment located in confined spaces; all air from outdoors — inlet air from ventilated crawl space and outlet air to ventilated attic. [See 5.3.3(b).]*

A.5.3.3(b)1a See Figure A.5.3.3(b)1a1.

A.5.3.3(b)1a2 See Figure A.5.3.3(b)1a2.

b.* Where communicating with the outdoors through horizontal ducts, each opening shall have a minimum free area of not less than 1 in.^{2} per 2000 Btu/hr (1100 mm^2/kW) of total input rating of all equipment in the enclosure. *[See Figure A.5.3.3(b)1b.]*

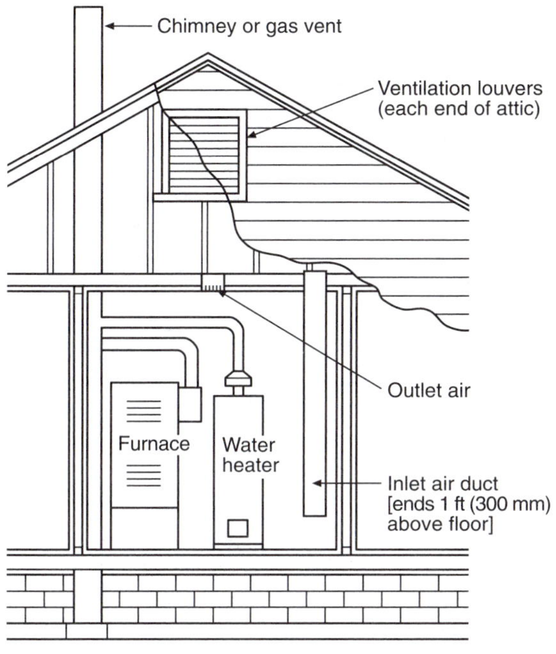

Figure A.5.3.3(b)1a2 *Equipment located in confined spaces; all air from outdoors through ventilated attic. [See 5.3.3(b).]*

A.5.3.3(b)1b See Figure A.5.3.3(b)1b.

(2)* One permanent opening, commencing within 12 in. (300 mm) of the top of the enclosure, shall be permitted where the equipment has clearances of at least 1 in. (25 mm) from the sides and back and 6 in. (160 mm) from the front of the appliance. The opening shall directly communicate with the outdoors or shall communicate through a vertical or horizontal duct to the outdoors or spaces that freely communicate with the outdoors *[see Figure A.5.3.3(b)2]* and shall have a minimum free area of

a. 1 in.^{2}/3000 Btu/hr (700 mm^2 per kW) of the total input rating of all equipment located in the enclosure, and
b. Not less than the sum of the areas of all vent connectors in the confined space.

A.5.3.3(b)2 See Figure A.5.3.3(b)2.

When all combustion and ventilation air is taken from outdoors, either directly or by ducts, one of two methods is permitted. The first method requires two openings or two ducts, either vertical or horizontal, commencing within 12 in. (300 mm) of the

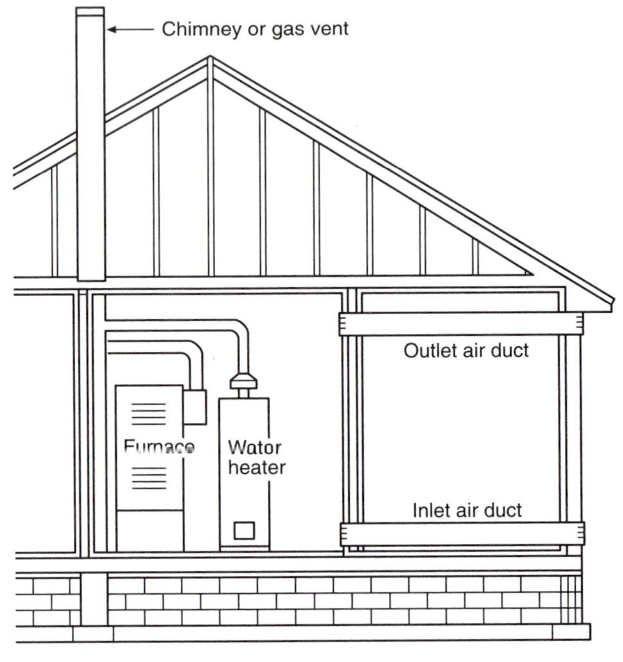

Figure A.5.3.3(b)1b   Equipment located in confined spaces; all air from outdoors. [See 5.3.3(b).]

ceiling and floor. The minimum dimension of the opening or duct is limited to 3 in. (80 mm), and the ducts must be the same area as the openings to which they connect. Vertical ducts or openings are sized at a minimum free area of 1 in.^{2} per 4000 Btu/hr (550 mm^2/kW) of the total appliance input installed in the space. Horizontal ducts require a minimum free area of 1 in.^{2} per 2000 Btu/hr (1100 mm^2/kW) of the total appliance input installed in the space. These openings or ducts can connect to spaces (e.g., crawl or attic) that freely communicate with the outdoors.

The second method addresses the concern to reduce the combustion and ventilation air requirements to reduce freezing water pipes and limit low temperatures in equipment rooms. Gas Research Institute Report Number GRI 93/0316 "Analysis of Combustion Air Openings to the Outdoors: Preliminary Results" identified our second method. One permanent opening or a vertical or horizontal duct can be used, provided that it directly communicates to the outdoors or connects through spaces (e.g., crawl or attic) that freely communicate with the outdoors. The opening or duct must be within 12 in. (300 mm) of the ceiling of the enclosure. Also, the study requires that the gas appliance(s) have a minimum clearance of 1 in. (25 mm) on each side and back and 6 in. (150 mm) in the front. The opening or duct is sized at a

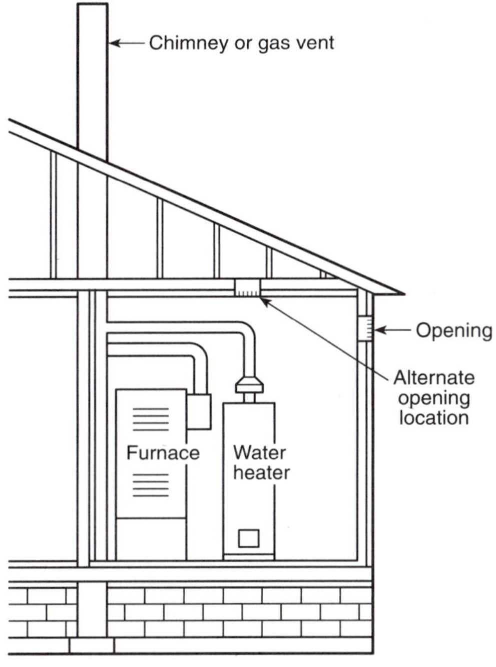

Figure A.5.3.3(b)2 Equipment located in confined spaces; single combustion air opening, all air from the outdoors. [See 5.3.3(b).]

minimum free area of 1 in.^{2} per 3000 Btu/hr (730 mm^2/kW) of the aggregate input rating of all appliances installed in the space, and the free area must not be less than the sum of the areas of all the vent connectors installed in the space. Exhibit 5.9 illustrates an example.

Example:

$$\frac{\left(\dfrac{4200 \text{ ft}^3}{100{,}000 \text{ Btu/hr} + 45{,}000 \text{ Btu/hr}}\right)}{1000 \text{ Btu/hr}} = \frac{28.9 \text{ ft}^3}{1000 \text{ Btu/hr}}$$

or

$$\frac{119 \text{ m}^3}{42.5 \text{ kW}} = 2.7 \text{ m}^3/\text{kW}$$

Because 28.9 ft^3 (2.7 m^3/kW) is less than the 50 ft^3 per 1000 Btu/hr (4.8 m^3/kW) of the total appliance input, the space is confined. Our appliances have a clearance of 1 in. (25.4 mm) to the sides and back and 6 in. (150 mm) to the front. Therefore, we can use a single opening to the attic, which freely communicates to the outdoors. We

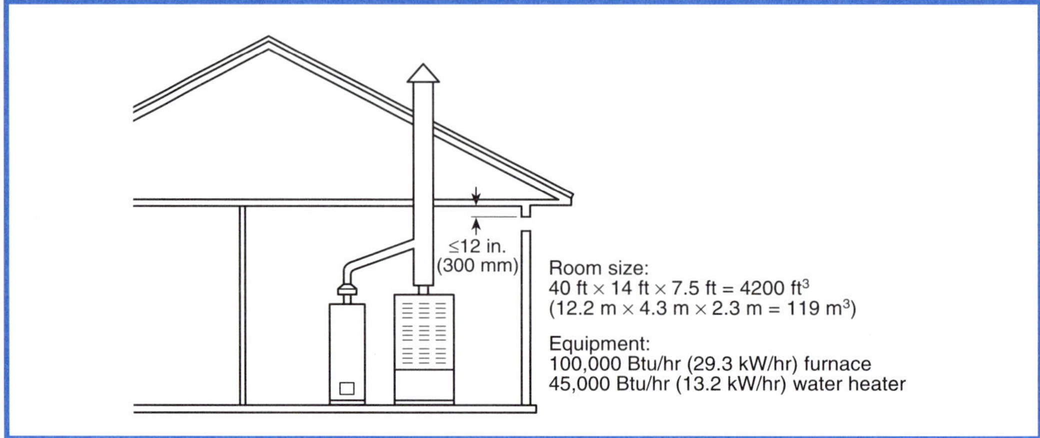

Exhibit 5.9 *Single air opening into a confined space. (Courtesy of Carrier Corporation.)*

size our opening at 1 in.^{2} per 3000 Btu/hr (730 mm^2/kW) for the total appliance input rating of the space.

$$\frac{100{,}000 \text{ Btu/hr} + 45{,}000 \text{ Btu/hr}}{3000 \dfrac{\text{Btu/hr}}{\text{in.}^2}} = 48.3 \text{ in.}^2$$

or
$$42.5 \text{ kW} \times 730 \text{ mm}^2/\text{kW} = 31{,}000 \text{ mm}^2$$

The opening net free area is limited to 48.3 in.^{2} (31,000 mm^2). Another requirement to be met is that the opening size should not be less than the sum of the total installed appliance vent connector area. In this case the furnace has a 4-in. or 12.2 in.^{2} (100 mm or 7870 mm^2) vent connector and the water heater has a 3-in. or 7.1 in.^{2} (76 mm or 0.00458 m^2) vent connector. The sum of our vent connectors is 12.2 in.^{2} + 7.1 in.^{2} = 19.3 in.^{2} (7870 mm^2 + 4500 mm^2 = 12,370 mm^2). Therefore, our opening size of 48.3 in.^{2} (31,000 mm^2) is satisfactory for this application. (See Table G.2.4 in the code for vent connector areas.)

(c) *Combination of air from the indoors and from the outdoors.* Where the building in which the fuel-burning appliances are located is not of unusually tight construction and the communicating interior spaces containing the fuel-burning appliances comply with all requirements of 5.3.3(a) except for the volumetric requirement of 5.3.3(a), required combustion and dilution air shall be obtained by opening the room to the outdoors utilizing a combination of indoor and outdoor air prorated in accordance with 5.3.3(c)6. Openings connecting the interior spaces shall comply with 5.3.3(a). The ratio of interior spaces shall comply with 5.3.3(c)5. The number, location, and ratios of openings connecting the space with the outdoor air shall comply with the following (see also sample calculation in Appendix L):

This new 5.3.3 (c) adds language that allows for the more flexible use of the traditional high/low openings to the outdoors. Basically, the openings to the outdoors may be reduced in size in consideration of infiltration available to the appliances from the local indoor communicating space. See commentary following Section 5.3 for more information.

This new provision is not needed if the appliance installation meets the requirements of an unconfined space. (See Section 5.3.2.) However, many installations fail to meet the 50 ft^2 per 1000 Btu/hr (4.8 m^3/kW) rule, but still are in a room or space that is a substantial fraction of what is needed to comply. The new language allows the two high/low openings to the outdoors to be reduced in proportion to the fraction that the communicating volume complies with the definition of an unconfined space. Note that it cannot be used to reduce the size of openings to other rooms or spaces within the building or to reduce the size of a single (high wall) opening to the outdoors [5.3.3 (b) 2].

For example, if the communicating volume amounted to 25 ft^3 per 1000 Btu/hr (2.4 m^3 per kW), which is 50 percent of what is needed, then the openings to the outdoors could be reduced 50 percent. This situation is shown schematically by Exhibit 5.10. The new provision is easy to apply if the following principles are understood:

(1) The volume of room that directly holds the appliances is included in the total communicating volume. All adjoining rooms that comply with Section 5.3.3 also may be considered to be part of the communicating volume. This means that it may be necessary to provide two high/low holes between rooms that meet the requirements of 5.3.3(a).
(2) The size of the high/low holes are based on the normal sizes found Section 5.3.3(b), depending on whether the openings are horizontal or vertical.
(3) The reduction is available only for the traditional two high/low opening method [Section 5.3.3(b)(1)]. It may not be used for the single high opening method [Section 5.5.5(b)(2)].
(4) The calculation should begin by calculating the volume needed qualify as be an unconfined space, the actual communicating volume, and the normal free area of the openings to the outdoors as required by 5.3.3(b).
(5) The reduced free area can then be calculated using the following equation:

$$\text{Reduced free area} = (\text{normal free area})\left(1 - \frac{\text{actual communicating volume}}{\text{confined space volume}}\right)$$

Example:

This example continues with installation options for the furnace and water heater used previously. The homeowner rejects the idea of openings into the new family room. The dimension of the room is 38 ft × 18 ft × 8 ft (11.6 m × 5.5 m × 2.4 m), or a total volume of 5472 ft^2 (153 m^3). The appliance's total input rate is 145,000 Btu/hr (42.5 kW). This input rate requires a volume of 7250 ft^3 (203 m^3) to meet the

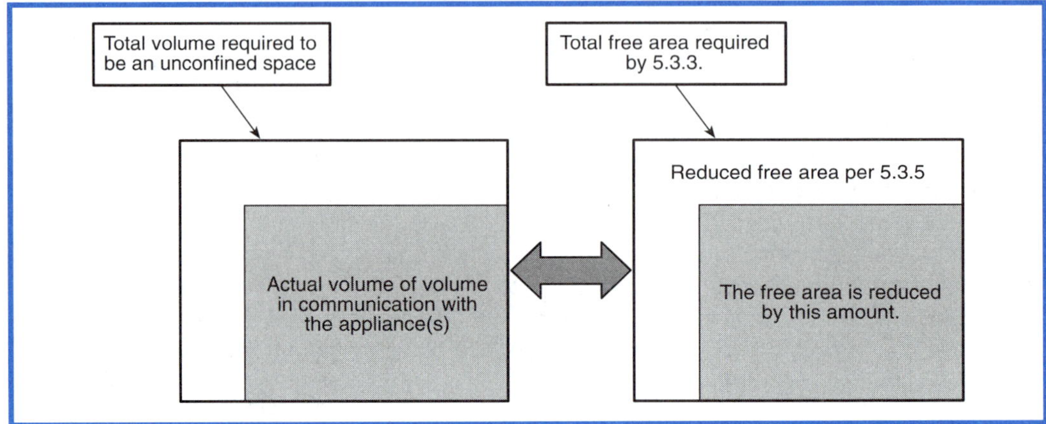

Exhibit 5.10 Free area reduction

definition of an unconfined space. Therefore, two openings to the outdoors, each 36.25 in.^{2}, are required (145,000 Btu/hr per 1 in.^{2}/4000 ft^3), or a little more than 6.02 in. × 6.02 in. Assume a 6 in. × 6 in. opening will be used. In this case, the openings are in direct communication with the outdoors, and no ducts are used. The volume of the two rooms (5472 ft^3) is only 72.5 percent of the volume required to be an unconfined space (7250 ft^3).

To use the new reduced outdoor air opening option, the installer must first size the usual two high/low openings to the outdoors in accordance with 5.3.3(a). The free area of each opening must be at least 36.25 in.^{2} The reduced sized openings may now be calculated using the preceeding equation:

$$\text{Reduced area} = 36.26 \text{ in}^2 \left(1 - \frac{5472 \text{ ft}^3}{7250 \text{ ft}^3}\right) = 36.25 \ (1 - 0.792) = 36.25 \ (0.208) = 7.54 \text{ in.}^2$$

A 7.54 in.^{2} opening is provided by an opening 2.5 in.^{2} × 2.5 in.^{2}(0.0016 m^2 × 0.0016 m^2). Note that the minimum opening size permitted by 5.3.3(b) is 3 in., so the opening is reduced from 6 in. × 6 in. (150 mm × 150 m), a reduction in area of 75 percent.

Note that the area reduction may not be workable in cold climates, where cold outside temperatures can cause freezing of water pipes. This is of special concern if there is a water heater present. While the reduced area of openings will reduce the cold air infiltration into the room, the option of one, high wall opening in 5.3.3(b)2 may be preferred in cold climates. Although the opening is larger (in this case, 48.3 in. sq, or about 7 in. × 7 in.), the elimination of the low opening can lead to higher temperature at the floor.

Another example of combined indoor and outdoor air can be found in Appendix L.

(1) *Number and Location of Openings.* At least two openings shall be provided, one within 1 ft (305 mm) of the ceiling of the room and one within 1 ft (305 mm) of the floor.

(2) *Ratio of Direct Openings.* Where direct openings to the outdoors are provided in accordance with 5.3.3(b), method 1a, the ratio of direct openings shall be the sum of the net free areas of both direct openings to the outdoors, divided by the sum of the required areas for both such openings as determined in accordance with 5.3.3(b), method 1a.

(3) *Ratio of Horizontal Openings.* Where openings connected to the outdoors through horizontal ducts are provided in accordance with 5.3.3(b), method 1b, the ratio of horizontal openings shall be the sum of the net free areas of both such openings, divided by the sum of the required areas for both such openings as determined in accordance with Section 5.3.3(b), method 1b.

(4) *Ratio of Vertical Openings.* Where openings connected to the outdoors through vertical ducts are provided in accordance with Section 5.3.3(b), method 1a, the ratio of vertical openings shall be the sum of the net free areas of both such openings, divided by the sum of the required areas for both such openings as determined in accordance with 5.3.3(b), method 1a.

(5) *Ratio of Interior Spaces.* The ratio of interior spaces shall be the available volume of all communicating spaces, divided by the required volume as determined in accordance with 5.3.3(a).

(6) *Prorating of Indoor and Outdoor Air.* In spaces that utilize a combination of indoor and outdoor air, the sum of the ratios of all direct openings, horizontal openings, vertical openings, and interior spaces shall equal or exceed 1.

5.3.4 Specially Engineered Installations.

The requirements of 5.3.3 shall be permitted to be waived where special engineering, approved by the authority having jurisdiction, provides an adequate supply of air for combustion, ventilation, and dilution of flue gases.

Special engineering can be necessary for large commercial and industrial installations where consideration must be given to the use and type of structure in which the equipment is located and where the requirements of 5.3.3 are difficult to apply. It is imperative that special engineering be done only by engineers thoroughly familiar with all aspects of the subject. Equipment manufacturers can often provide assistance.

The free area required for makeup air called for in 5.3.3 assumes that the only driving force that is moving air through the opening is the pressure reduction caused by the operation of the vent. Large commercial and industrial equipment frequently operate with forced air fans, which provide greater driving force than "natural" vents. This greater force moves more air through a given size opening, permitting a smaller opening than allowed by the requirements of 5.3.3.

With an emphasis on efficiency, many builders and home owners are sealing their homes from air infiltration. This tight construction can lead to many problems, such as reduced combustion air and increased indoor air pollution. The use of heat recovery ventilators (HRV) is one method to minimize heat loss due to infiltration and to maintain sufficient air for combustion and ventilation. See Exhibit 5.11 for an illustration of a heat recovery ventilator. This device draws fresh, outside air into the dwelling and exhausts stale, indoor air to the outdoors. These two air streams pass through a heat exchanger, warming the outdoor air and cooling the indoor air, thereby minimizing energy loss. These HRVs can make up for a lack of combustion and ventilation air. For each cubic foot of gas burned, a natural draft (draft hood) furnace requires about 30 ft^3/hr or 0.5 ft^3/min (0.8 m^3/hr or 0.24 1/sec) and a fan-assisted combustion system furnace needs about 20 ft^3/hr or 0.33 ft^3/min (0.6 m^3/hr or 0.16 1/sec) of air. Therefore, a 100,000-Btu/hr (29.3-kW) natural draft (draft hood) furnace would require 50 ft^3/min (23.6 l/sec) to meet its combustion and ventilation air requirements. If an HRV is used to supply combustion and ventilation air, it must be interlocked with the gas appliance(s) to ensure that it is operating prior to the main burner operation. If the HRV fails to operate or its air flow becomes restricted during gas appliance operation, the main burners must not operate.

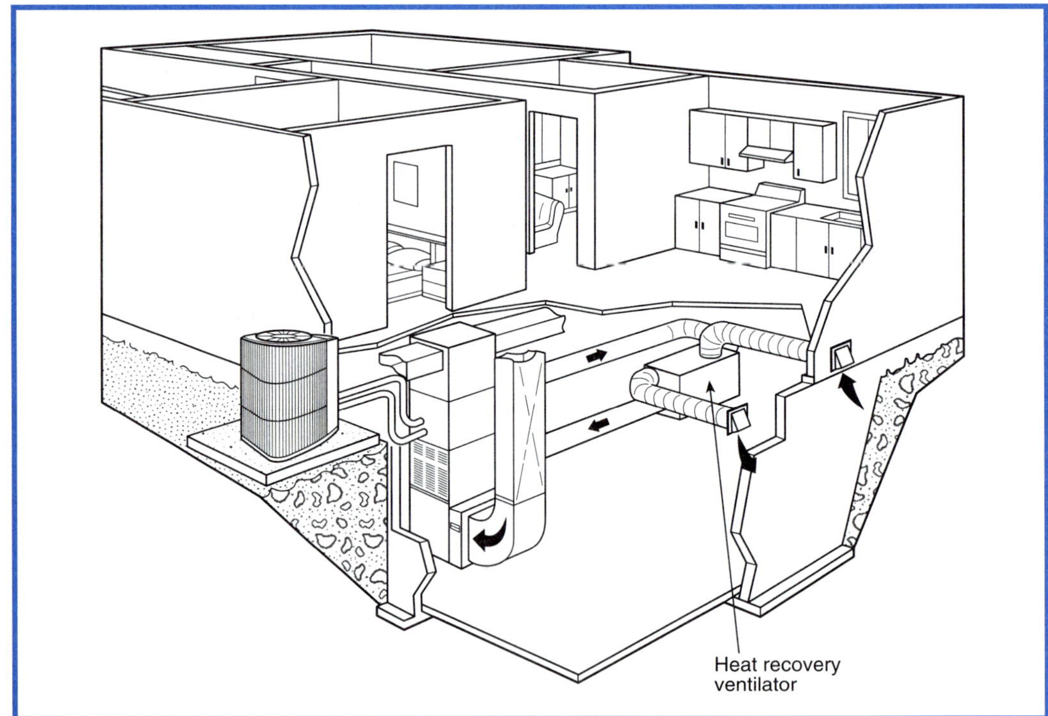

Exhibit 5.11 *HRV installation with forced air system. (Courtesy of Carrier Corporation.)*

5.3.5 Louvers and Grilles.

In calculating free area in 5.3.3, the required size of openings for combustion, ventilation, and dilution air shall be based on the net free area of each opening. If the free area through a design of louver or grille is known, it shall be used in calculating the size opening required to provide the free area specified. If the design and free area are not known, it shall be assumed that wood louvers will have 20–25 percent free area and metal louvers and grilles will have 60–75 percent free area. Louvers and grilles shall be fixed in the open position.

Exception: Louvers interlocked with the equipment so they are proven in the full open position prior to main burner ignition and during main burner operation. Means shall be provided to prevent the main burner from igniting should the louver fail to open during burner startup and to shut down the main burner if the louvers close during burner operation.

Louvers will prevent the entrance of rain and snow. All outside openings should be screened to be no smaller than 1/4 in. (6.4 mm). When calculating the free area of louvers, one must consider the restricting effect of the louver to the opening size. For example, louvers made of wood can have a free area of only 25 percent and need an opening size that is four times larger for the required free area. (See Exhibit 5.12)

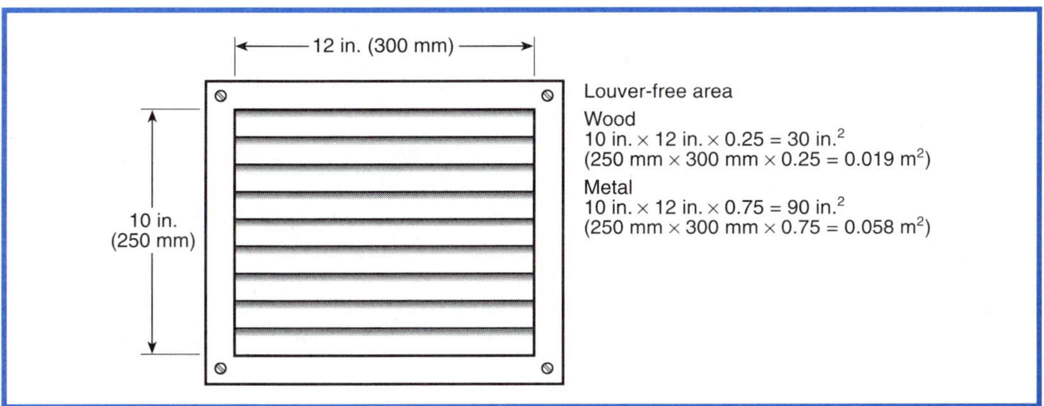

Exhibit 5.12 *Louver free area. (Courtesy of Carrier Corporation.)*

The exception was added in the 1996 edition of the code to permit the operation of closable louvers. Louvers can be fixed in the open position, or they must be interlocked with the equipment to ensure that they are open during main burner operation. They also must prevent main burner ignition if they fail to open. If automatic means are used to open and close the louvers, the louvers must be proven to be open during burner ignition and operation. If the louvers fail to open or close during operation, the main burners must not operate.

Louvers must be in the open position to ensure adequate combustion and ventilation air. Without adequate air, there is a strong possibility of incomplete combustion, which will produce toxic gases such as carbon monoxide. Carbon monoxide is an odorless, tasteless, invisible gas that is poisonous.

5.3.6 Combustion Air Ducts.

This new subsection on combustion air ducting was added in the 1999 edition. It specifies that the combustion air duct must have adequate resistance to corrosion and be dedicated to providing combustion air for one confined space. The high and low opening ducts to the confined space must be separate over their entire run. The last requirement, that the combustion air duct must not slope down and away from the appliances, prevents a "trap" being formed by the duct, which could impede the flow of the duct.

Combustion air ducts shall comply with the following:

(1) Ducts shall be of galvanized steel or an equivalent corrosion-resistant material.

Exception: Within dwellings units, unobstructed stud and joist spaces shall not be prohibited from conveying combustion air, provided that not more than one fireblock is removed.

(2) Ducts shall terminate in an unobstructed space, allowing free movement of combustion air to the appliances.
(3) Ducts shall serve a single space.
(4) Ducts shall not serve both upper and lower combustion air openings where both such openings are used. The separation between ducts serving upper and lower combustion air openings shall be maintained to the source of combustion air.
(5) Ducts shall not be screened where terminating in an attic space.
(6) Horizontal upper combustion air ducts shall not slope downward toward the source of combustion air.

5.4 Equipment on Roofs

5.4.1 General.

(a) Gas utilization equipment on roofs shall be designed or enclosed so as to withstand climatic conditions in the area in which they are installed. If enclosures are provided, each enclosure shall permit easy entry and movement, shall be of reasonable height, and shall have at least a 30-in. (760-mm) clearance between the entire service access panel(s) of the equipment and the wall of the enclosure.

(b) Roofs on which equipment is to be installed shall be capable of supporting the additional load or shall be reinforced to support the additional load.

(c) All access locks, screws, and bolts shall be of corrosion-resistant material.

A major portion of gas-fired equipment for rooftop installations is now certified for outdoor installation with its own integral vent system. An enclosure, if provided for a conventional furnace, must follow the guidelines given in 5.4.1(a). Any permanent structure on the roof must be designed in accordance with local building codes.

5.4.2 Installation of Equipment on Roofs.

(a) Gas utilization equipment shall be installed in accordance with its listing and the manufacturer's installation instructions.

(b) Equipment shall be installed on a well-drained surface of the roof. At least 6 ft (1.8 m) of clearance shall be available between any part of the equipment and the edge of a roof or similar hazard, or rigidly fixed rails or guards at least 42 in. (1.1 m) in height shall be provided on the exposed side.

Exception: Parapets or other building structures at least 42 in. (1.1 m) in height shall be permitted to be utilized in lieu of rails or guards.

(c) All equipment requiring an external source of electrical power for its operation shall be provided with (1) a readily accessible electrical disconnecting means within sight of the equipment that will completely deenergize the equipment, and (2) a 120-V ac grounding-type receptacle outlet on the roof adjacent to the equipment. The receptacle outlet shall be on the supply side of the disconnect switch.

(d) Where water stands on the roof at the equipment or in the passageways to the equipment, or where the roof is of a design having a water seal, a suitable platform, walkway, or both shall be provided above the water line. Such platform(s) or walkway(s) shall be located adjacent to the equipment and control panels so that the equipment can be safely serviced where water stands on the roof.

Equipment installed on sloping roofs with a pitch greater than 4 in. in 12 in. should be mounted on a suitable platform. A service platform of at least 30 in. (760 mm) in depth should be provided on the side(s) of the unit that has access panels for service and maintenance. (See Exhibit 5.13.) Local building codes can have more restrictive requirements for clearances to the edge of the roof and for the location and placement of guard rails.

When equipment is installed on sloping roofs, the installer must provide a means of access to the equipment in all anticipated weather conditions normal in the area. This requirement may mean that walkways and railings must be provided. To walk on a sloped roof when it is icy or snow covered is unsafe.

5.4.3 Access to Equipment on Roofs.

(a) Gas utilization equipment located on roofs or other elevated locations shall be accessible.

Some building codes can require that equipment on roofs be "readily accessible," which means that permanent access to the roof must be provided. In most cases, single-story dwellings can use a portable ladder to provide access to the roof.

Exhibit 5.13 *Sample of roof installation. (Courtesy of BDP Company.)*

(b) Buildings of more than 15 ft (4.6 m) in height shall have an inside means of access to the roof, unless other means acceptable to the authority having jurisdiction are used.

Fifteen feet (4.6 m) is a somewhat arbitrary figure, but it does imply that all two-story buildings must have a "ready" means of access to the roof. Some jurisdictions permit an exterior ladder but, in the interest of security, will require either that the lower portion be removable and be kept on the premises or that a locking door be placed over the lower 8 ft (2.4 m) of the ladder. Because local building codes vary on roof access requirements, consult a local building official if questions arise to ensure meeting both the *National Fuel Gas Code* requirements and local requirements.

(c) The inside means of access shall be a permanent or foldaway inside stairway or ladder, terminating in an enclosure, scuttle, or trapdoor. Such scuttles or trapdoors shall be at least 22 in. × 24 in. (560 mm × 610 mm) in size, shall open easily and safely under all conditions, especially snow, and shall be constructed so as to permit access from the roof side unless deliberately locked on the inside.

At least 6 ft (1.8 m) of clearance shall be available between the access opening and the edge of the roof or similar hazard, or rigidly fixed rails or guards at least 42 in. (1.1 m) in height shall be provided on the exposed side; parapets or other building structures at least 42 in. (1.1 m) in height shall be permitted to be utilized in lieu of guards or rails.

If access from the roof scuttle is to a sloping roof or to a roof that is designed to hold water, then a walkway with a level 30-in. (760-mm) working space can be required to provide ready access to the equipment. If the unit is closer than 6 ft (1.8 m) to the edge of the roof, guards, such as railings or parapets, must be provided to ensure safety of service personnel.

(d) Proper permanent lighting shall be provided at the roof access. The switch for such lighting shall be located inside the building near the access means leading to the roof.

5.4.4 Additional Provisions.

(Also see 5.1.22, 5.2.1, and 7.3.4.)

These references are 5.1.22, Protection of Outdoor Equipment; 5.2.1, Accessibility for Service; and 7.3.4, Mechanical Draft Systems.

5.5 Equipment Connections to Building Piping

5.5.1 Connecting Gas Equipment.

Gas utilization equipment shall be connected to the building piping in compliance with 5.5.4 by one of the following:

(1) Rigid metallic pipe and fittings.
(2) Semirigid metallic tubing and metallic fittings. Aluminum alloy tubing shall not be used in exterior locations.

Semirigid metallic tubing is permitted to be used to connect appliances to building piping systems. Traditionally, copper tubing was the only material available. With the introduction of corrugated stainless steel tubing (CSST), a second material is now available. If there are any questions on the use of CSST as a connector, the CSST manufacturer's installation instructions may provide information on the terms of the product's listing.
See Supplement 2, "Corrugated Stainless Steel Tubing Gas Piping Systems."

(3) Listed connectors used in accordance with the terms of their listing that are completely in the same room as the equipment.

Gas connectors are listed in accordance with ANSI Z21.24, *Metal Connectors for Gas Appliances,* or ANSI Z21.69, *Standard for Connectors for Moveable Gas Appliances.*

(4) Listed gas hose connectors in accordance with 5.5.2.
(5) Gas-fired food service (commercial cooking) equipment listed for use with casters or otherwise subject to movement for cleaning, and other large and heavy gas utilization equipment that can be moved, shall be connected in accordance with the connector man-

ufacturer's installation instructions using a listed appliance connector complying with ANSI Z21.69, *Standard for Connectors for Movable Gas Appliances.*

(6) In 5.5.1(2), (3), and (5), the connector or tubing shall be installed so as to be protected against physical and thermal damage. Aluminum alloy tubing and connectors shall be coated to protect against external corrosion where they are in contact with masonry, plaster, or insulation or are subject to repeated wettings by such liquids as water (except rain water), detergents, or sewage.

5.5.2 Use of Gas Hose Connectors.

Listed gas hose connectors shall be used in accordance with the terms of their listing and as follows:

(a) *Indoor.* Indoor gas hose connectors shall be permitted to be used with laboratory, shop, or ironing equipment that requires mobility during operation. An equipment shutoff valve shall be installed where the connector is attached to the building piping. The connector shall be of minimum length and shall not exceed 6 ft (1.8 m). The connector shall not be concealed and shall not extend from one room to another or pass through wall partitions, ceilings, or floors.

(b) *Outdoor.* Outdoor gas hose connectors shall be permitted to be used to connect portable outdoor gas-fired equipment. An equipment shutoff valve, a listed quick-disconnect device, or a listed gas convenience outlet shall be installed where the connector is attached to the supply piping and in such a manner so as to prevent the accumulation of water or foreign matter. This connection shall only be made in the outdoor area where the equipment is to be used.

Indoor gas hose connectors can be used only to connect equipment requiring mobility, such as laboratory, shop, and ironing equipment. The connector must be listed

---

Formal Interpretation 80-2

Reference: 3-8.2, 5-5.2

Question 1: Is 5.5.2(a) intended to require that all such [gas valve] outlets have a flexible hose and a mobile laboratory appliance permanently attached to it?

Answer: No.

Question 2: Is it the intent of 3-8.2(a) or 5-5.2(a) to preclude the use of such [gas valve] outlets as convenience outlets in a laboratory for connecting mobile laboratory equipment?

Answer: No.

Issue Edition: 1980

Date: December 1983

---

and connected to a listed manual shutoff valve immediately before the gas hose. Plugging or capping the outlet of the shutoff valve when the equipment is disconnected is not necessary. When a gas hose is used outdoors, it can be connected to a listed, quick-disconnect device instead of to a manual shutoff valve. Formal Interpretation 80-2 clarifies the use of gas hose connectors.

5.5.3 Connection of Portable and Mobile Industrial Gas Equipment.

(a) Portable industrial gas utilization equipment or equipment requiring mobility or subject to vibration shall be permitted to be connected to the building gas piping system by the use of flexible hose suitable and safe for the conditions under which it can be used.

(b) Industrial gas utilization equipment requiring mobility shall be permitted to be connected to the rigid piping by the use of swivel joints or couplings that are suitable for the service required. Where swivel joints or couplings are used, only the minimum number required shall be installed.

(c) Industrial gas utilization equipment subject to vibration shall be permitted to be connected to the building piping system by the use of all metal flexible connectors suitable for the service required.

(d) Where flexible connections are used, they shall be of the minimum practical length and shall not extend from one room to another or pass through any walls, partitions, ceilings, or floors. Flexible connections shall not be used in any concealed location. They shall be protected against physical or thermal damage and shall be provided with gas shutoff valves in readily accessible locations in rigid piping upstream from the flexible connections.

5.5.4 Equipment Shutoff Valves and Connections.

Gas utilization equipment connected to a piping system shall have an accessible, approved manual shutoff valve with a nondisplaceable valve member, or a listed gas convenience outlet, installed within 6 ft (1.8 m) of the equipment it serves. Where a connector is used, the valve shall be installed upstream of the connector. A union or flanged connection shall be provided downstream from this valve to permit removal of controls.

Shutoff valves serving decorative gas appliances shall be permitted to be installed in fireplaces if listed for such use.

The manual shutoff valve required is needed to isolate the equipment for servicing, removal, or replacement without shutting off the gas supply to other appliances. This valve is not a primary emergency shutoff valve. The shutoff valve located outside the building provides for that need. (See 3.10.3.) The manual shutoff may be used to isolate certain equipment from the building piping during pressure testing under 4.1.3. Because the equipment shutoff valve is not an emergency shutoff valve, it does not have to be readily accessible.

The equipment shutoff valve must be an approved manual shutoff valve and, if a connector is used, must be installed before the connector and within 6 ft (1.8 m) of the appliance. All shutoff valves that are 1 in. and smaller are required to be listed. (See Section 2.12.)

The code was changed in the 1992 edition to permit shutoff valves that are serving decorative gas appliances to be installed in fireplaces if listed for such use. (See Exhibit 5.14.)

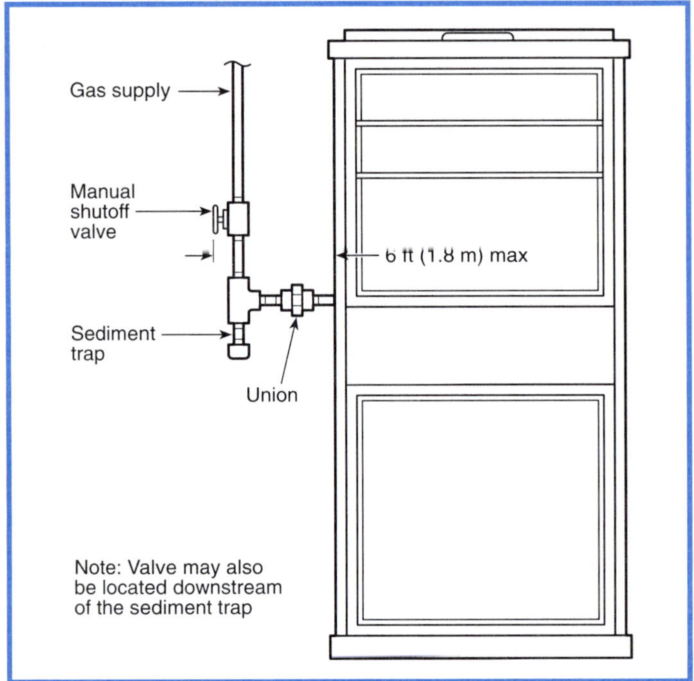

Exhibit 5.14 *Shutoff valve location. (Courtesy of BDP Company.)*

5.5.5 Quick-Disconnect Devices.

Gas utilization equipment connectors shall be permitted to be connected to the building piping by means of a listed quick-disconnect device and, where installed indoors, an approved manual shutoff valve with a nondisplaceable valve member shall be installed upstream of the quick-disconnect device.

A listed, quick-disconnect device often is used as one of the end fittings of a listed gas appliance connector or listed gas hose. The device functions as a manual gas shutoff in that, when it is manually disconnected, it is designed to shut off the flow of gas before the disconnection is made. The more common designs use a spring-loaded poppet valve, which is internal to the device. When installed indoors, a manual shutoff valve is required upstream to provide the additional safety of a positive and visual shutoff for the gas. The manual shutoff valve must be an approved gas valve.

5.5.6* Gas Convenience Outlets.

Gas utilization equipment shall be permitted to be connected to the building piping by means of a listed gas convenience outlet, in conjunction with a listed appliance connector, used in accordance with the terms of their listings.

This paragraph permits the use of a new category of devices, gas convenience outlets, which are being developed by several manufacturers. The gas convenience outlets must be listed under AGA 7-90, *Requirements for Gas Convenience Outlets,* for testing and certification. (See Exhibit 5.15.)

Exhibit 5.15 *A gas convenience outlet and fitting. (Courtesy of NAHB.)*

A.5.5.6 For information on gas convenience outlets, see *Requirements for Gas Convenience Outlets,* AGA 7-90.

5.5.7 Sediment Trap.

If a sediment trap is not incorporated as a part of the gas utilization equipment, a sediment trap shall be installed as close to the inlet of the equipment as practicable at the time of equipment installation. The sediment trap shall be either a tee fitting with a capped nipple in the bottom outlet as illustrated in Figure 5.5.7 or other device recognized as an effective sediment trap. Illuminating appliances, ranges, clothes dryers, decorative appliances for installation in vented fireplaces, gas fireplaces, and outdoor grills shall not be required to be so equipped.

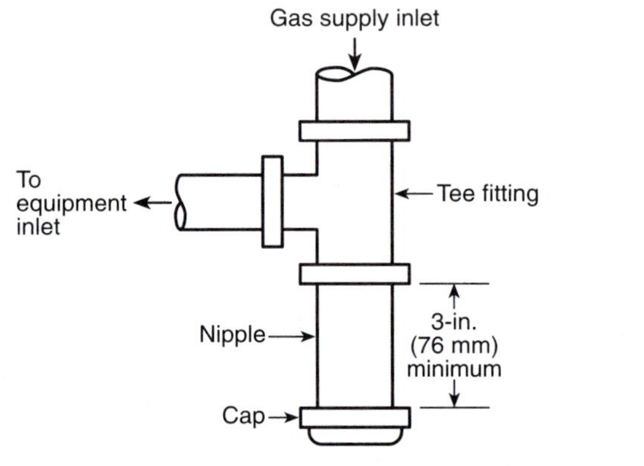

Figure 5.5.7 *Method of installing a tee fitting sediment trap.*

Refer to the commentary on 3.7.3, and note that this requirement was revised in the 1999 edition to be more specific, with the addition of decorative appliances for installation in vented fireplaces and decorative vented appliances to the list of appliances not requiring a sediment trap.

5.5.8 Installation of Piping.

Piping shall be installed in a manner not to interfere with inspection, maintenance, or servicing of the gas utilization equipment.

5.6 Electrical

5.6.1 Electrical Connections.

Electrical connections between gas utilization equipment and the building wiring, including the grounding of the equipment, shall conform to NFPA 70, *National Electrical Code®*.

Article 422 in NFPA 70, *National Electrical Code*, covers the requirement for appliances, and Article 440 covers the installation of appliances that also contain hermetic refrigeration motor compressors, which are found in many combination "gas-electric" rooftop units that provide heating and cooling. Article 250, Parts E and F, cover grounding requirements for equipment.

5.6.2 Electrical Ignition and Control Devices.

Electrical ignition, burner control, and electrical vent damper devices shall not permit unsafe operation of the gas utilization equipment in the event of electrical power interruption or when the power is restored.

5.6.3 Electrical Circuit.

The electrical circuit employed for operating the automatic main gas-control valve, automatic pilot, room temperature thermostat, limit control, or other electrical devices used with the gas utilization equipment shall be in accordance with the wiring diagrams supplied with the equipment.

Appliance wiring diagrams are included with appliances, usually inside one of the access panels and generally shown in the installation instructions. Most manufacturers will also show the necessary wiring connections for the installation of optional accessories, such as external vent dampers, humidifiers, air cleaners, and other auxiliary controls. Accessory installation instructions should be left with the equipment for future reference.

5.6.4 Continuous Power.

All gas utilization equipment using electrical controls shall have the controls connected into a permanently live electrical circuit—that is, one that is not controlled by a light switch. Central heating equipment shall be provided with a separate electrical circuit.

The *National Electrical Code* requires a separate electrical circuit and overcurrent protection for central heating equipment, as well as a means for disconnecting the power source at or within sight from the appliance. This separate circuit also can include circulating water pumps, valves, humidifiers, electronic air cleaners, and other accessories normally associated with the equipment. Certain appliances and installations also are permitted to be connected with a flexible cord that has a grounding-type attachment plug. The grounding-type attachment plug also serves as a disconnecting means for future service and maintenance.

5.7 Room Temperature Thermostats

5.7.1 Locations.

Room temperature thermostats shall be installed in accordance with the manufacturers' instructions.

Room thermostats generally are located on inside walls to avoid drafts from the heating or cooling system supply registers and should avoid heat sources, such as pipes in walls, lights, or TVs, which can affect their operation. Usually, thermostats are installed 4 ft to 5 ft (1.2 m to 1.5 m) above the floor and can include supplemental switches to control the functions of the system. Consult the equipment manufacturers' instructions for additional details. (See Exhibit 5.16.)

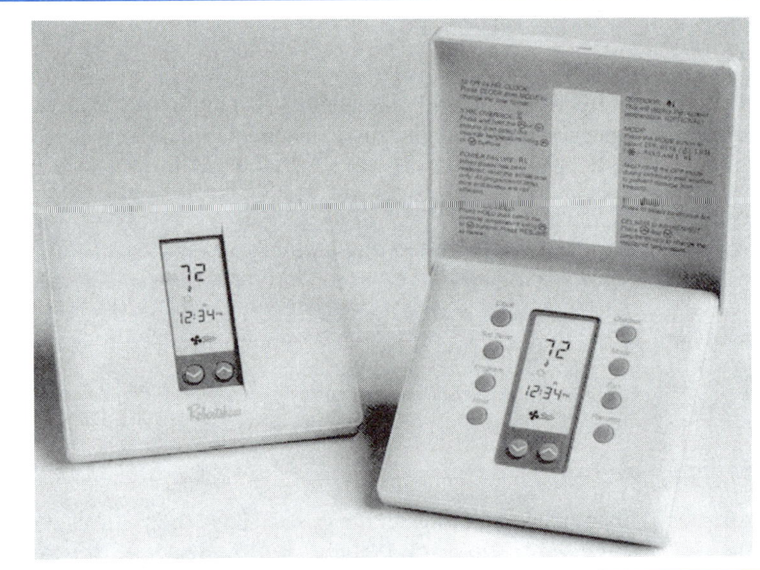

Exhibit 5.16 *A clock thermostat. (Courtesy of Robertshaw Controls.)*

Each appliance should have its own thermostat unless the equipment design specifically states that one thermostat can serve more than one appliance. If a thermostat serves more than one appliance, it can cause the appliance to operate erratically. Each heating appliance should have its own thermostat so that it can properly heat the area it is designed to heat. If it were connected to additional units, it would cause the different areas to become underheated or overheated, because it could not accurately record the temperature for both areas.

5.7.2 Drafts.

Any hole in the plaster or panel through which the wires pass from the thermostat to the gas utilization equipment being controlled shall be sealed so as to prevent drafts from affecting the thermostat.

This requirement increases the system's overall efficiency by reducing unnecessary equipment operating time.

References Cited in Commentary

The following publication is available from the American Petroleum Institute, 1220 L Street, NW, Washington, DC 20005.

API RP 520 PT II-1994, *Sizing, Selection, and Installation of Pressure-Relieving Devices in Refineries.*

The following publication is available from the Gas Research Institute, 8600 W. Bryn Mawr Avenue, Chicago, IL 60631.

Report Number GRI 93/0316, "Analysis of Combustion Air Openings to the Outdoors: Preliminary Results," 1994.

The following publications are available from the International Approval Services, CSA International, 8501 East Pleasant Valley Road, Cleveland, OH 44131.

AGA 7-90, *Requirements for Gas Convenience Outlets.*
ANSI Z21.10.1, *Volume I — Gas Water Heaters — with Input Ratings Above 75,000 Btu per Hour or Less,* 1993.
ANSI Z21.24-1993, *Metal Connectors for Gas Appliances.*
ANSI Z21.69-1992, *Standard for Connectors for Moveable Gas Appliances.*

The following publications are available from the National Fire Protection Association, 1 Batterymarch Park, P.O. Box 9101, Quincy, MA 02269-9101.

NFPA 30, *Flammable and Combustible Liquids Code,* 1996 edition.
NFPA 51, *Standard for the Design and Installation of Oxygen-Fuel Gas Systems for Welding, Cutting, and Allied Processes,* 1997 edition.
NFPA 70, *National Electrical Code®,* 1999 edition.
NFPA 88A, *Standard for Parking Structures,* 1998 edition.
NFPA 88B, *Standard for Repair Garages,* 1997 edition.

Installation of Specific Equipment

Chapter 6 covers requirements for installing specific gas-fired equipment and is applicable to nonindustrial equipment. Specific requirements are provided for most nonindustrial equipment.

Section 6.1, General, provides the overall requirement that listed equipment be installed in accordance with the terms of their listing. This requirement recognizes that the installation instructions for listed equipment provide clearances based on testing of the specific appliance. These clearances are in many cases lower than those required for unlisted equipment of the same type. It also covers restrictions on the installation of gas-fired equipment in bedrooms and bathrooms.

Sections 6.2 through 6.30 provide requirements for the following specific appliances:

- Gas-fired air-conditioning equipment (in Section 6.2)
- Central heating furnaces and boilers (in Section 6.3)
- Clothes dryers (in Section 6.4)
- Conversion burners (in Section 6.5)
- Decorative appliances for installation in vented fireplaces and vented gas fireplaces (in Sections 6.6 and 6.7)
- Air heaters (in Sections 6.8 and 6.9)
- Duct and floor furnaces (in Sections 6.10 and 6.11)
- Food service equipment and household cooking appliances (in Sections 6.12 – 6.15, 6.20, and 6.21)
- Illuminating appliances (in Section 6.16)
- Commercial-industrial and domestic incinerators (in Sections 6.17 and 6.18)
- Infrared heaters (in Section 6.19)
- Pool heaters (in Section 6.22)
- Gas refrigerators (in Section 6.23)

- Room (space) heaters (in Section 6.24)
- Stationary gas engines (in Section 6.25)
- Gas-fired toilets (in Section 6.26)
- Unit heaters (in Section 6.27)
- Wall furnaces (in Section 6.28)
- Water heaters (in Section 6.29)
- Compressed natural gas (CNG) vehicular fuel systems (in Section 6.30)

Section 6.31, Appliances for Installation in Manufactured Housing, is new in the 1999 edition and makes the code applicable to installation and replacement of gas appliances in manufactured housing after the first consumer sale.

6.1 General

(a) This chapter is applicable primarily to nonindustrial-type gas utilization equipment and installations and, unless specifically indicated, does not apply to industrial-type equipment and installations. Listed gas utilization equipment shall be installed in accordance with their listing and the manufacturers' instructions, or as elsewhere specified in this chapter. Unlisted equipment shall be installed as specified in this chapter as applicable to the equipment.

For additional information concerning particular gas equipment and accessories, including industrial types, reference can be made to the standards listed in Chapter 11 and Appendix M.

Chapter 6 of this code is, for the most part, directly descended from former ASA Z21.30 and NFPA 54, which covered residential and some commercial installations. Generalized coverage can be written for such equipment, because it usually is subjected to similar installation conditions. The installation conditions for industrial equipment and other commercial equipment are often unique and vary widely, so generalized installation coverage is not practical. However, the references in Chapter 11 and Appendix M will assist in determining the applicability of industrial-type gas equipment and accessories.

The reader's attention is drawn particularly to the following statement in 6.1(a): "Listed gas utilization equipment shall be installed in accordance with their listing and the manufacturers' instructions...." Although not every section in Chapter 6 that covers the various types of equipment contains this statement, many sections do; this requirement applies in all cases in which the equipment has been listed by a testing organization acceptable to the authority having jurisdiction.

(b)* Gas utilization equipment shall not be installed so its combustion, ventilation, and dilution air are obtained only from a bedroom or bathroom unless the bedroom or bathroom is an unconfined space. *(See 5.3.2 and Section 1.7, Definitions.)*

This requirement was added to the 1984 edition after considerable discussion on whether or not equipment should be permitted to be installed in bedrooms and bathrooms. Originally, the National Fuel Gas Code Committee's position was that the installation of most gas utilization equipment in bedrooms or bathrooms, or in closets or small spaces having access only through a bedroom or bathroom, be prohibited for the following reasons:

(1) Such rooms are usually kept closed, are small in size, and may not provide adequate air for combustion, draft hood dilution, and ventilation.
(2) This ruling would be consistent with other codes.
(3) There is a possibility of moisture and lint accumulation within the appliance.
(4) The potential for contact with hot surfaces and for ignition of fabric is increased.
(5) Proper ventilation for the products of combustion can be lacking.
(6) Hair spray and other products that can accelerate corrosion of the equipment are present.

The wording was changed in the 1988 edition of the code to provide a more positive statement on small spaces, to permit the installation of direct-vent appliances in bedrooms and bathrooms, and to provide for access to furnace closets (which can draw outside air for combustion and ventilation) located in or having access through bedrooms. Remember that the definition of an unconfined space is "a space whose volume is not less than 50 ft^3 per 1000 Btu/hr (4.8 m^3 per kW) of the aggregate input rating of all appliances installed in that space. Rooms communicating directly with the space in which the appliances are installed, through openings not furnished with doors, are considered a part of the unconfined space."

Bedrooms and bathrooms usually have doors; therefore, appliances that take air for combustion and ventilation from the room in which they are installed must not be installed in these rooms. A.6.1(b) draws attention to 6.6.1, 6.7.1, 6.24.1, and 6.29.1, which prohibit the installation of specific gas utilization equipment in bedrooms and bathrooms whether the space is confined or not.

In 1994, the code was amended to permit the installation of listed, wall-mounted room heaters with limited input ratings and oxygen depletion safety shut-off systems in bedrooms and bathrooms, which comply as unconfined spaces, as long as these heaters are acceptable to the authority having jurisdiction.

The addition of Exception No. 2 to 6.30.1 (6.29.1 in the 1999 edition) in the 1996 edition of the code further clarifies this subject. The exception permits the installation of appliances in closets that open into bedrooms and bathrooms when both the closet is equipped with a self-closing, solid door and air for combustion and ventilation is supplied from the outdoors.

A.6.1(b) Also see Prohibited Installations, 6.6.1, 6.7.1, 6.24.1, and 6.29.1.

6.2 Air-Conditioning Equipment (Gas-Fired Air Conditioners and Heat Pumps)

The installation requirements for listed gas furnaces, boilers, and air-conditioning equipment have been revised in the 1999 edition, and Section 6.2 is revised to be consistent with the revised Table 6.2.3(a). The table now covers clearances for unlisted equipment, and listed equipment is covered in Chapter 6. The requirements of this section have been correspondingly revised to require the installation of listed gas air-conditioning equipment in accordance with the terms of the listing. This section is now consistent with the installation requirements for other listed appliances.

6.2.1 Independent Gas Piping.

Gas piping serving heating gas utilization equipment shall be permitted to also serve cooling equipment where heating and cooling equipment cannot be operated simultaneously. *(See Section 2.4.)*

In accordance with Section 2.4, the gas piping serving either the heating or cooling equipment must be sized on the basis of the larger gas demand of the two pieces of equipment. The diversity provision in 2.4.2 permits the pipe size to be based on the larger, rather than the combined, gas demand once it is determined that the heating and cooling equipment cannot be operated simultaneously.

6.2.2 Connection of Gas Engine-Powered Air Conditioners.

To protect against the effects of normal vibration in service, gas engines shall not be rigidly connected to the gas supply piping.

Because an engine is prone to vibration, it is important that it be secured to a firm foundation so that it cannot move. An appliance shutoff valve must be located on the fixed pipe upstream of the connector. During installation, it is important that the connector be oriented so that there is no stress on the connector and so that any vibration does not adversely affect the connector.

6.2.3 Clearances for Indoor Installation.

(a) Listed air-conditioning equipment installed in rooms that are large in comparison with the size of the equipment shall be installed with clearances per the terms of their listing and the manufacturer's instructions. *(See Table 6.2.3(a) and Section 1.7 for definition.)*

Note that the installation requirements have been revised in the 1999 edition to require that clearances for listed gas air-conditioning equipment installed in rooms large in comparison with the size of the equipment be installed in accordance with the terms of their listing. Note that the following definition appears in Section 1.7:

Table 6.2.3(a) Clearances to Combustible Material for Unlisted Furnaces, Boilers, and Air Conditioners Installed in Rooms That Are Large in Comparison with the Size of Equipment

	Minimum Clearance (in.)					
	Above and Sides of Plenum	**Top of Boiler**	**Jacket Sides and Rear**	**Front**	**Draft Hood and Barometric Draft Regulator**	**Single-Wall Vent Connector**
I Automatically fired, forced air or gravity system, equipped with temperature limit control that cannot be set higher than 250°F (121°C).	6		6	18	6	18
II Automatically fired heating boilers — steam boilers operating at not over 15 psi (103 kPa) and hot water boilers operating at 250°F (121°C) or less	6	6	6	18	18	18
III Central heating boilers and furnaces, other than in I or II	18	18	18	18	18	18
IV Air conditioning equipment	18	18	18	18	18	18

Note: See 6.2.3 for additional requirements for air-conditioning equipment and 6.3.1 for additional requirements for central heating boilers and furnaces.

Room Large in Comparison with Size of Equipment. Rooms having a volume equal to at least 12 times the total volume of a furnace or air-conditioning appliance and at least 16 times the total volume of a boiler. Total volume of the appliance is determined from exterior dimensions and is to include fan compartments and burner vestibules, when used. When the actual ceiling height of a room is greater than 8 ft (2.4 m), the volume of the room is figured on the basis of a ceiling height of 8 ft (2.4 m).

(b) Air-conditioning equipment installed in rooms that are NOT large (such as alcoves and closets) in comparison with the size of the equipment shall be listed for such installations and installed in accordance with the manufacturer's instructions. Listed clearances shall not be reduced by the protection methods described in Table 6.2.3(b), regardless of whether the enclosure is of combustible or noncombustible material.

Table 6.2.3(b) Reduction of Clearances with Specified Forms of Protection

Type of protection applied to and covering all surfaces of combustible material within the distance specified as the required clearance with no protection [see Figures 6.3.1(b)1 through 6.3.1(b)3]	Where the required clearance with no protection from appliance, vent connector, or single wall metal pipe is									
	36 in.		18 in.		12 in.		9 in.		6 in.	
	Allowable Clearances with Specified Protection (Inches) Use Col. 1 for clearances above appliance or horizontal connector. Use Col. 2 for clearances from appliance, vertical connector, and single-wall metal pipe.									
	Above Col. 1	Sides and Rear Col. 2	Above Col. 1	Sides and Rear Col. 2	Above Col. 1	Sides and Rear Col. 2	Above Col. 1	Sides and Rear Col. 2	Above Col. 1	Sides and Rear Col. 2
(a) 3^1/$_2$-in. thick masonry wall without ventilated air space	—	24	—	12	—	9	—	6	—	5
(b) ^{1}/$_2$-in. insulation board over 1-in. glass fiber or mineral wool batts	24	18	12	9	9	6	6	5	4	3
(c) 0.024 sheet metal over 1-in. glass fiber or mineral wool batts reinforced with wire on rear face with ventilated air space	18	12	9	6	6	4	5	3	3	3
(d) 3^1/$_2$-in. thick masonry wall with ventilated air space	—	12	—	6	—	6	—	6	—	6
(e) 0.024 sheet metal with ventilated air space	18	12	9	6	6	4	5	3	3	2
(f) ^{1}/$_2$-in. thick insulation board with ventilated air space	18	12	9	6	6	4	5	3	3	3
(g) 0.024 sheet metal with ventilated air space over 0.024 sheet metal with ventilated air space	18	12	9	6	6	4	5	3	3	3
(h) 1-in. glass fiber or mineral wool batts sandwiched between two sheets 0.024 sheet metal with ventilated air space.	18	12	9	6	6	4	5	3	3	3

Notes Applicable to Table 6.2.3(b)

For SI units, 1 in. = 25.4 mm.

Notes:

1. Reduction of clearances from combustible materials shall not interfere with combustion air, draft hood clearance and relief, and accessibility of servicing.

2. All clearances shall be measured from the outer surface of the combustible material to the nearest point on the surface of the appliance, disregarding any intervening protection applied to the combustible material.

3. Spacers and ties shall be of noncombustible material. No spacer or tie shall be used directly opposite the appliance or connector.

4. Where all clearance reduction systems use a ventilated air space, adequate provision for air circulation shall be provided as described. *(See Figures 6.3.1(b)2 and 6.3.1(b)3.)*

5. There shall be at least 1 in. (25 mm) between clearance reduction systems and combustible walls and ceilings for reduction systems using a ventilated air space.

6. If a wall protector is mounted on a single flat wall away from corners, adequate air circulation shall be permitted to be provided by leaving only the bottom and top edges or only the side and top edges open with at least a 1-in. (25-mm) air gap.

7. Mineral wool batts (blanket or board) shall have a minimum density of 8 lb/ft^3 (128 kg/m) and a minimum melting point of 1500°F (816°C).

8. Insulation material used as part of a clearance reduction system shall have a thermal conductivity of 1.0 Btu in./ft^2/hr-°F (0.144 W/m-K) or less.

9. There shall be at least 1 in. (25 mm) between the appliance and the protector. In no case shall the clearance between the appliance and the combustible surface be reduced below that allowed in Table 6.2.3(b).

10. All clearances and thicknesses are minimum; larger clearances and thicknesses are acceptable.

11. Listed single-wall connectors shall be permitted to be installed in accordance with the terms of their listing and the manufacturer's instructions.

Note that equipment installed in alcoves and closets must be listed specifically for such installations. Also note that the last sentence prohibits clearance reduction for appliances installed in rooms *not* large in comparison with the size of the equipment. This prohibition recognizes that in these small rooms, closets, and alcoves, the minimum clearances are needed for air circulation to keep appliances from overheating.

(c) Unlisted air-conditioning equipment shall be installed with clearances from combustible material of not less than 18 in. (460 mm) above the equipment and at the sides, front, and rear, and 9 in. (230 mm) from the draft hood.

(d) Air-conditioning equipment (listed and unlisted) installed in rooms that are large in comparison with the size of the equipment shall be permitted to be installed with reduced clearances to combustible material provided the combustible material or equipment is protected as described in Table 6.2.3(b) *[see 6.2.3(b)]*.

This new requirement was added in the 1999 edition. It specifically permits the clearance reduction systems described in Table 6.2.3(b) to be used with all (listed and unlisted) gas-fired air-conditioning equipment.

(e) Where the plenum is adjacent to plaster on metal lath or noncombustible material attached to combustible material, the clearance shall be measured to the surface of the plaster or other noncombustible finish where the clearance specified is 2 in. (50 mm) or less.

When used alone, plaster or other insulating material that is applied to combustible material, such as wooden studs, is not considered to be adequate protection for combustible construction. For this reason, the clearance must be measured to the surface of the plaster or other finish if the specified clearance is 2 in. (51 mm) or less.

(f) Listed air-conditioning equipment shall have the clearance from supply ducts within 3 ft (0.9 m) of the plenum be not less than that specified from the plenum. No clearance is necessary beyond this distance.

This new requirement clarifies clearance to supply ducts near the equipment, which are to be provided the same clearance as the plenums to which they connect.

6.2.4 Assembly and Installation.

Air-conditioning equipment shall be installed in accordance with the manufacturer's instructions. Unless the equipment is listed for installation on a combustible surface such as a floor or roof, or unless the surface is protected in an approved manner, it shall be installed on a surface of noncombustible construction with noncombustible material and surface finish and with no combustible material against the underside thereof.

Prior to the 1988 edition, the *National Fuel Gas Code* contained a footnote referring the reader to Appendix E of the 1976 edition of the *National Building Code* for additional information on methods of installation on combustible floors. The 1976 edition of the *National Building Code* was the last edition of that code written by the American Insurance Association. After 1976, the project was taken over by the Building Officials and Code Administrators International. This appendix no longer appears in the *National Building Code*. A part of Appendix E is reprinted here for the information of the user of this handbook. By referring to the requirements for room heaters, the *National Building Code* covered the installation of residential air-conditioning equipment, clothes dryers, and water heaters, and stated as follows:

Placement
(a) Room heaters, except as permitted by the provisions of Sections (b) through (e), shall be placed on the ground; or on approved limited-combustible assemblies having a fire resistance rating of not less than 2 hours, with floors constructed of noncombustible material; or on concrete slabs or masonry arches that do not have combustible materials attached to the underside. Any floor covering on floor-ceiling assemblies, slabs, or arches shall extend not less than 6 in (15 cm) beyond the appliances on all sides, and, where solid fuel is used, they shall extend not less than 18 in (46 cm) at the front or side where ashes are removed.

(b) Room heaters, that are tested and listed by a nationally recognized testing laboratory for installation on floors constructed of combustible materials, are permitted to be placed on floors, other than as required by the provisions of Section (a), provided they are installed in accordance with the requirements of the listing and conditions of approval.

(c) Room heaters, which are set on legs or simulated legs that provide not less than 4 in (10 cm) open space under the base of the appliance, are permitted to be placed on floors other than as required by the provisions of Section (a), provided the floor under the appliance is protected with sheet metal of not less than No. 24 gage, or by other noncombustible material that will reflect heat and is durable. Where solid fuel is used, the protection shall extend not less than 18 in (46 cm) beyond the appliance at the front or side where ashes are removed. With radiating type gas burning room heaters which make use of metal, asbestos or ceramic material to direct radiation to the front of the device, the floor protection shall extend out at the front not less than 36 in (91 cm) when the heater has not been tested and listed by a nationally recognized testing laboratory for installation on a combustible floor.

(d) Room heaters which are set on legs that provide not less than 18 in (46 cm) open space under the base of the appliance, or which have no burners within 18 in (46 cm) of the floor, are permitted to be placed on floors other than as required by the provisions of Section (a) without special floor protection, provided there is at least one sheet metal baffle not less than No. 24 gage between the burners and the floor.

(e) Room heaters, other than those described in the provisions of Sections (b), (c), and (d), are permitted to be placed on floors other than as required by the provisions of Section (a), provided the floor under the appliance is protected with hollow masonry not less than 4 in (10 cm) thick covered with sheet metal not less than No. 24 gage. The masonry shall be laid with ends unsealed and joints matched in such a way as to provide a free circulation of air from side to side through the masonry. Where solid fuel is used, the floor, for 18 in. (46 cm) beyond the front of the appliance or side where ashes are removed, shall be protected with sheet metal not less than No. 24 gage or with other noncombustible materials that will not allow hot ashes to ignite the floor or floor covering.

NOTE: In Appendix E of the *National Building Code,* a sheet metal thickness of 24 gauge is specified. In the current *National Fuel Gas Code*, a dimension of 0.0195 in. (0.5 mm) is used in place of the 24 gauge that was specified in prior editions. According to the *Chemical Engineers' Handbook,* fifth edition, 1973, 24 U.S. Standard Gauge (for sheet or plate metal and wrought iron) measures 0.0250 in. thick. The use of gauge to specify metal thickness, instead of an actual measurement, is due both to the numerous gauges that are used for specific applications (e.g., wire, sheet, hoop, etc.) and to differing allowances for variation in the manufacturing process. The 0.0195-in. (0.5-mm) dimension currently used is a minimum thickness at any point in the sheet.

6.2.5 Plenums and Air Ducts.

A plenum supplied as a part of the air-conditioning equipment shall be installed in accordance with the manufacturer's instructions. Where a plenum is not supplied with the equipment, any fabrication and installation instructions provided by the manufacturer shall be followed. The method of connecting supply and return ducts shall facilitate proper circulation of air.

Where the air conditioner is installed within a room not large in comparison with the size of the equipment, the air circulated by the equipment shall be handled by ducts that are sealed to the casing of the equipment and that separate the circulating air from the combustion and ventilation air.

In addition to the reference standards NFPA 90A, *Standard for the Installation of Air-Conditioning and Ventilating Systems,* and NFPA 90B, *Standard for the Installation of Warm Air Heating and Air-Conditioning Systems,* the inspection authority is referred to the following standards. Underwriters Laboratories Inc. standards UL 181, *Standard for Safety Factory-Made Air Ducts and Air Connectors,* and UL 263, *Standard for Safety Fire Tests of Building Construction and Materials,* are the primary standards for factory-made air ducts. These standards divide factory-made nonmetallic ducts into three classes, depending on their surface burning characteristics and flame spread characteristics, and also they specify where they can be used. An additional standard for the construction of metal ducts is the Sheet Metal and Air Conditioning Contractors National Association (SMACNA) standard SMACNA 1985, *HVAC Duct Construction Standards — Metal and Flexible,* for low-pressure and high-pressure duct construction. Low-pressure metal ducts generally are those types that handle air distribution systems with internal pressures of up to 2 in. (0.5 kPa) water column, either positive or negative. These standards also show the construction methods for high-pressure ducts of up to 10 in. (2.5 kPa) water column and include methods for joining the sections and hanging the ducts to ensure compliance with most building codes. In addition, the standards contain methods for sealing duct sections to ensure that they are essentially airtight from the appliance to the discharge grille or diffuser.

6.2.6* Refrigeration Coils.

(See 6.3.7 and 6.3.8.)

A.6.2.6 Reference can be made to ANSI/NFPA 90A, *Standard for the Installation of Air-Conditioning and Ventilating Systems,* or NFPA 90B, *Standard for the Installation of Warm Air Heating and Air-Conditioning Systems.*

6.2.7 Switches in Electrical Supply Line.

Means for interrupting the electrical supply to the air-conditioning equipment and to its associated cooling tower (if supplied and installed in a location remote from the air conditioner) shall be provided within sight of and not over 50 ft (15 m) from the air conditioner and cooling tower.

In accordance with the provisions of NFPA 70, *National Electrical Code*®, Section 440-14, the location of the disconnecting means in the electrical supply to the air-conditioning unit and its associated cooling tower must be located within sight of and be readily accessible from the air-conditioning unit or the cooling tower. Indoor units that can form part of the system must also have a means for disconnect, and each branch circuit must be provided with adequate branch circuit short-circuit and ground-fault protection. In most instances, the branch-circuit breaker cannot be used as a separate disconnecting means unless it meets the conditions specified above and is also listed by its manufacturer as a short-circuit and ground-fault protection device, as an HACR breaker, and as a disconnecting means. Make certain that disconnect devices on outdoor units are also approved for outdoor installation or are in a suitable, weatherproof enclosure.

6.3 Central Heating Boilers and Furnaces

The installation requirements for listed gas furnaces, boilers, and air-conditioning equipment have been revised in the 1999 edition, and this section has been revised to be consistent with the revised Table 6.2.3(a). The table now covers clearances for unlisted equipment, and listed central heating boilers and furnaces are now covered in Section 6.3. The requirements of this section have been correspondingly revised to require the installation of listed gas air-conditioning equipment in accordance with the terms of the listing. The revised requirements are now consistent with the installation requirements for other listed appliances. Note that 6.3.1 (a) through (j) have been revised to incorporate the content of the former notes to Table 6.2.3(a), making the information part of the requirements rather than additional information.

Consult Section 1.7, Definitions, for the different types of central heating boilers and furnaces. Note that this code specifically covers boilers operating at low pressures and temperatures not exceeding 250°F (121°C) and that there are several different types of furnaces, including upflow furnaces, horizontal furnaces, downflow furnaces, forced-air furnaces with cooling units, gravity furnaces, direct-vent furnaces, and enclosed central furnaces. (See Exhibits 6.1 through 6.6.)

6.3.1 Clearance.

Central heating boilers and furnaces and their vent connectors must be installed so that continued or intermittent operation will not create a hazard. By its nature, heat-

Exhibit 6.1 *Central furnace with cooling unit (outdoor installation). (Courtesy of BDP Company.)*

Exhibit 6.2 *Hot water boiler. (Courtesy of Burnham Corporation.)*

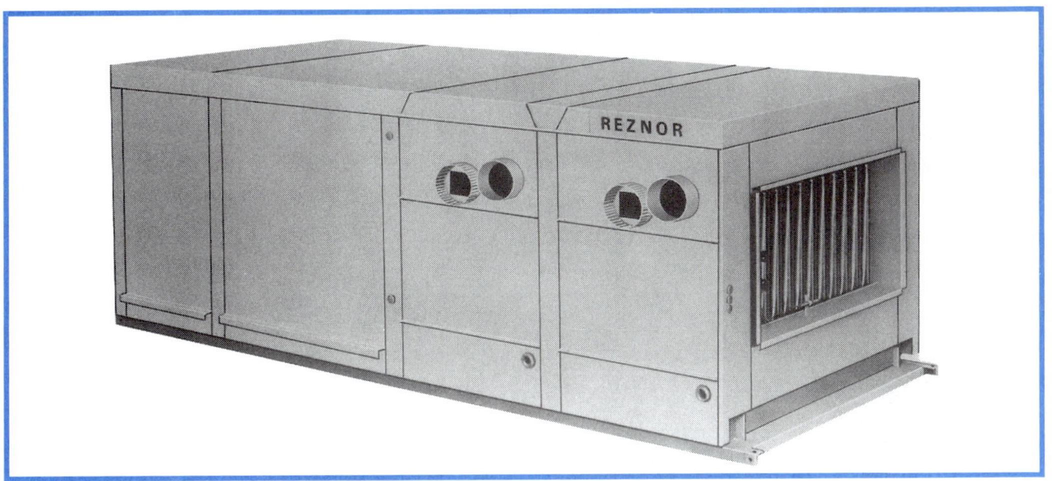

Exhibit 6.3 *Horizontal airflow furnace. (Courtesy of Thomas & Betts.)*

ing equipment gets warm, and even hot, during operation. If adequate clearances and ventilation around the appliance are not provided, a potential for fire and decreased service life of internal components exists.

(a) Listed central heating furnaces and low-pressure boilers installed in rooms that are large in comparison with the size of the equipment shall be installed with clearances per the terms of their listing and the manufacturer's instructions. *(See Section 1.7 for definition.)*

Note that the term "room large in comparison with size of the equipment" is defined as follows in Section 1.7:

> **Room Large in Comparison with Size of Equipment.** Rooms having a volume equal to at least 12 times the total volume of a furnace or air-conditioning appliance and at least 16 times the total volume of a boiler. Total volume of the appliance is determined from exterior dimensions and is to include fan compartments and burner vestibules, when used. When the actual ceiling height of a room is greater than 8 ft (2.4 m), the volume of the room is figured on the basis of a ceiling height of 8 ft (2.4 m).

(b) Central heating furnaces and low-pressure boilers installed in rooms that are NOT large (such as alcoves and closets) in comparison with the size of the equipment shall be listed for such installations. Listed clearances shall not be reduced by the protection methods described in Table 6.2.3(b) and illustrated in Figures 6.3.1(b)1 through 6.3.1(b)3, regardless of whether the enclosure is of combustible or noncombustible material.

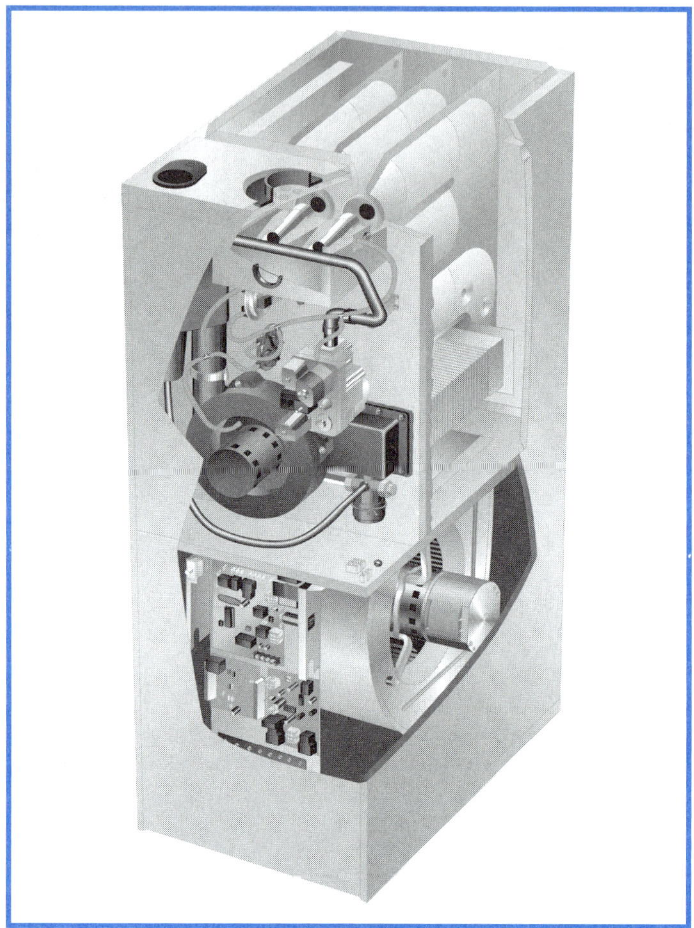

Exhibit 6.4 *Upflow furnace. (Courtesy of Carrier Corporation.)*

Central heating furnaces and boilers that are approved for installation in alcoves and closets will be specifically listed and marked. Note that the listed clearance must not be reduced by the application of methods given in Table 6.2.3(b). A typical rating plate for an appliance that is approved for closet installation at the specified clearances is pictured in Exhibit 6.7.

Note that the last sentence prohibits clearance reduction for appliances installed in rooms NOT large in comparison with the size of the equipment. Such small rooms, closets, and alcoves require the minimum clearances for air circulation to keep appliances from overheating.

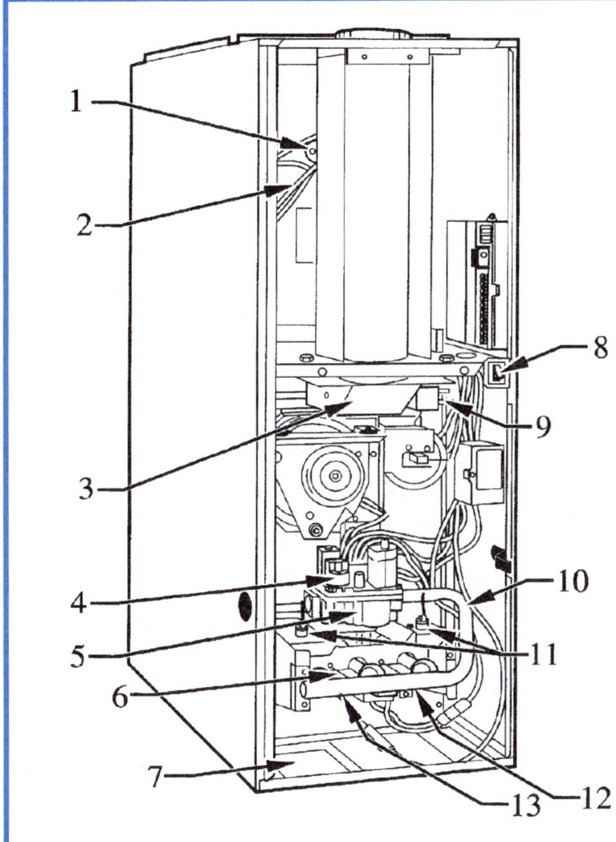

Downflow Horizontal Furnace Components

1. Manual-reset auxiliary limit switch (when used)
2. Blower and blower motor
3. Relief box
4. Gas valve control knob or electric switch (on/off)
5. Gas valve
6. Gas burner
7. Rating plate
8. Blower door safety switch
9. Blocked vent safeguard tube and switch
10. Gas manifold
11. Manual reset limit switch (2)
12. Hot surface ignitor
13. Flame sensor

Exhibit 6.5 Downflow furnace. (Courtesy of Payne Heating & Cooling.)

(c) Unlisted central heating furnaces and low-pressure boilers installed in rooms that are large in comparison with the size of the equipment shall be installed with clearances not less than those specified in Table 6.2.3(a).

(d) Central heating furnaces and low-pressure boilers (listed and unlisted) installed in rooms that are large in comparison with the size of the equipment shall be permitted to be installed with reduced clearances to combustible material provided the combustible material or equipment is protected as described in Table 6.2.3(b)*[see 6.3.1(b)]*.

Where the installation is in a room large in comparison with the size of the equipment, the clearance reduction systems of Table 6.2.3(b) can always be used.

(e) Front clearance shall be sufficient for servicing the burner and the furnace or boiler.

Space must be provided for both routine and major service work.

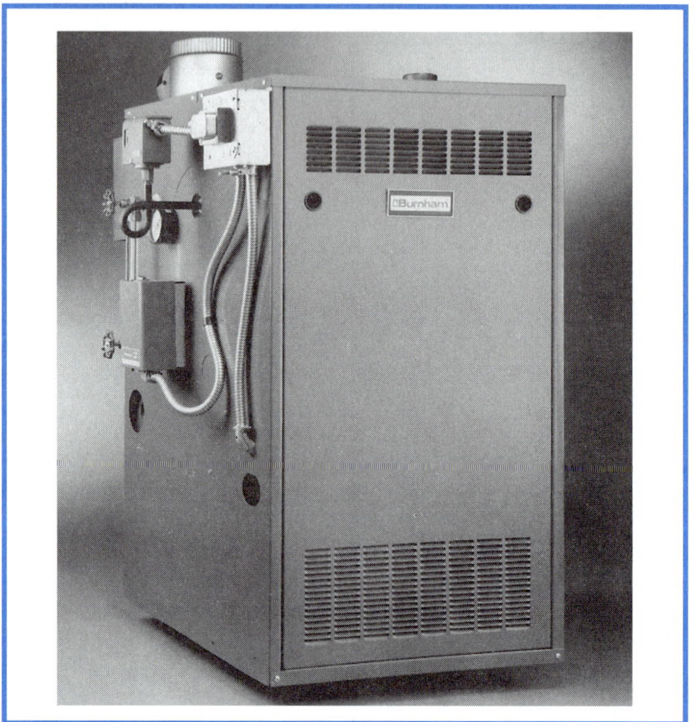

Exhibit 6.6 *Steam boiler. (Courtesy of Burnham Corporation.)*

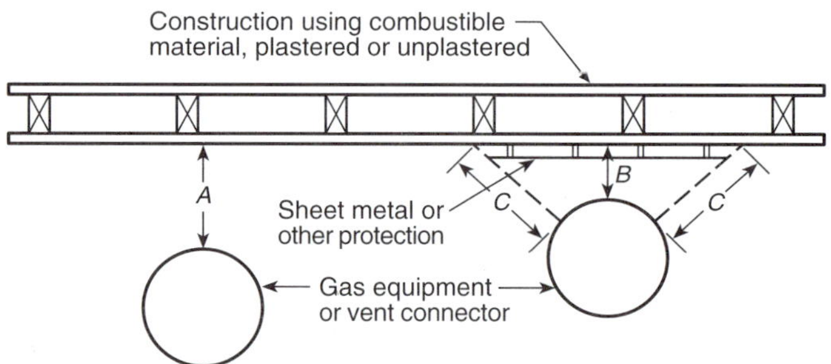

A equals the clearance with no protection specified in Tables 6.2.3(a) and 7.4.1 and in the sections applying to various types of equipment.

B equals the reduced clearance permitted in accordance with Table 6.2.3(b) The protection applied to the construction using combustible material shall extend far enough in each direction to make *C* equal to *A*.

Figure 6.3.1(b)1 *Extent of protection necessary to reduce clearances from gas equipment or vent connectors.*

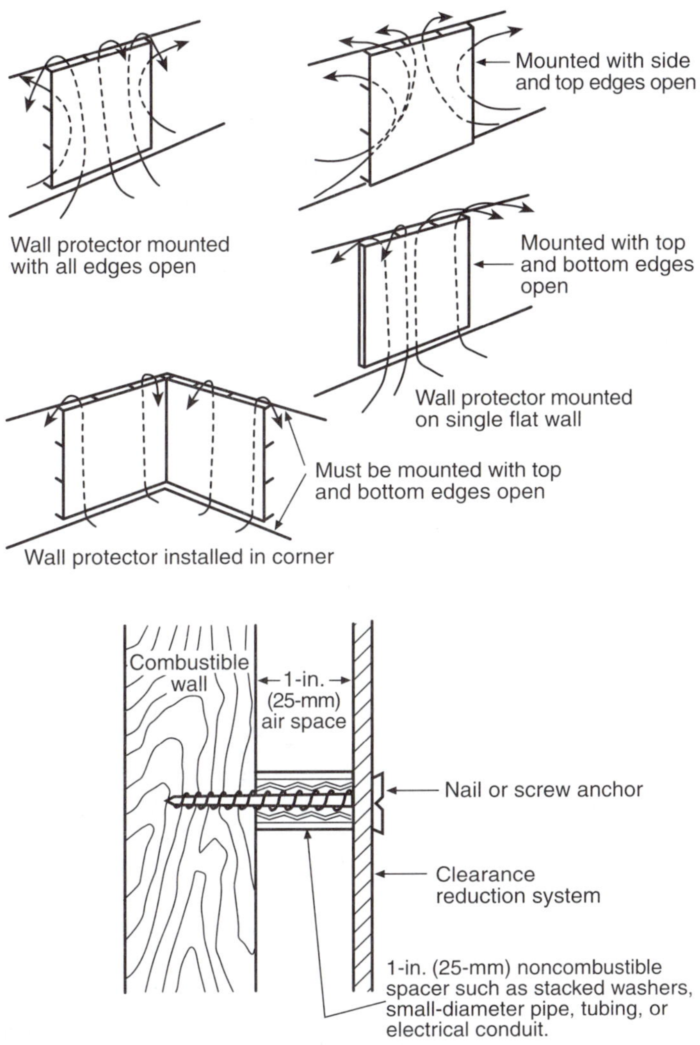

Figure 6.3.1(b)2   Wall protector clearance reduction system.

(f) Where the plenum is adjacent to plaster on metal lath or noncombustible material attached to combustible material, the clearance shall be measured to the surface of the plaster or other noncombustible finish where the clearance specified is 2 in. (50 mm) or less.

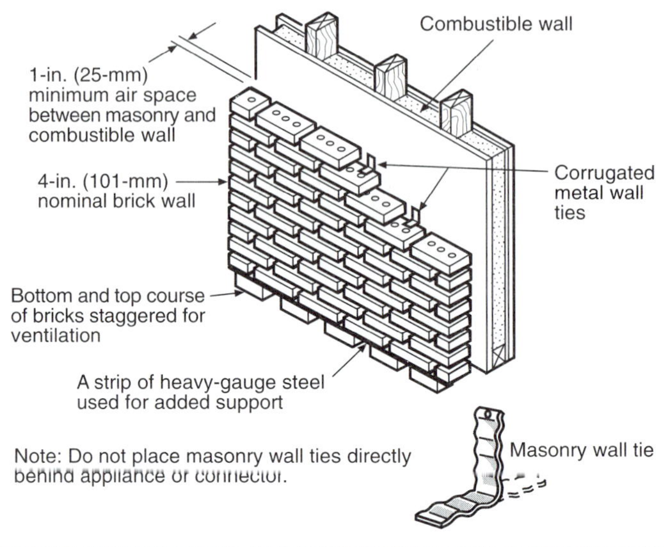

Figure 6.3.1(b)3 Masonry clearance reduction system.

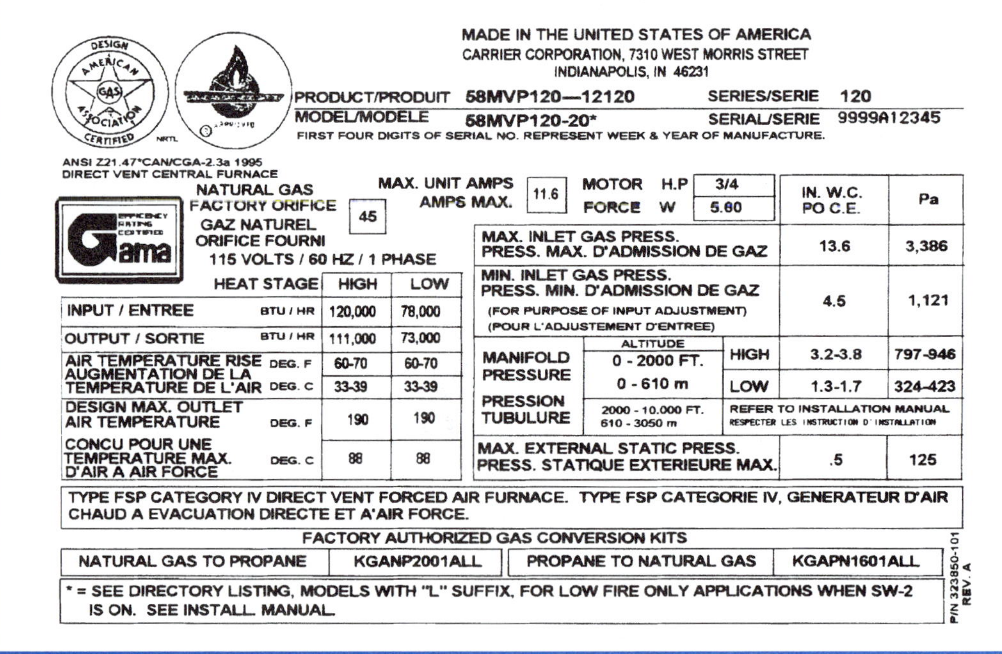

MADE IN THE UNITED STATES OF AMERICA
CARRIER CORPORATION, 7310 WEST MORRIS STREET
INDIANAPOLIS, IN 46231

PRODUCT/PRODUIT 58MVP120—12120   SERIES/SERIE 120
MODEL/MODELE 58MVP120-20*   SERIAL/SERIE 9999A12345
FIRST FOUR DIGITS OF SERIAL NO. REPRESENT WEEK & YEAR OF MANUFACTURE.

ANSI Z21.47*CAN/CGA-2.3a 1995
DIRECT VENT CENTRAL FURNACE

NATURAL GAS FACTORY ORIFICE **45**
GAZ NATUREL ORIFICE FOURNI
115 VOLTS / 60 HZ / 1 PHASE

MAX. UNIT AMPS / AMPS MAX. **11.6**
MOTOR H.P **3/4**
FORCE W **5.60**

HEAT STAGE	HIGH	LOW
INPUT / ENTREE BTU / HR	120,000	78,000
OUTPUT / SORTIE BTU / HR	111,000	73,000
AIR TEMPERATURE RISE DEG. F / AUGMENTATION DE LA TEMPERATURE DE L'AIR	60-70	60-70
DEG. C	33-39	33-39
DESIGN MAX. OUTLET AIR TEMPERATURE DEG. F	190	190
CONCU POUR UNE TEMPERATURE MAX. D'AIR A AIR FORCE DEG. C	88	88

		IN. W.C. PO C.E.	Pa
MAX. INLET GAS PRESS. PRESS. MAX. D'ADMISSION DE GAZ		13.6	3,386
MIN. INLET GAS PRESS. PRESS. MIN. D'ADMISSION DE GAZ (FOR PURPOSE OF INPUT ADJUSTMENT) (POUR L'ADJUSTEMENT D'ENTREE)		4.5	1,121

MANIFOLD PRESSURE / PRESSION TUBULURE	ALTITUDE		IN. W.C. PO C.E.	Pa
	0 - 2000 FT. / 0 - 610 m	HIGH	3.2-3.8	797-946
		LOW	1.3-1.7	324-423
	2000 - 10.000 FT. 610 - 3050 m	REFER TO INSTALLATION MANUAL RESPECTER LES INSTRUCTION D'INSTALLATION		

	IN. W.C. PO C.E.	Pa
MAX. EXTERNAL STATIC PRESS. PRESS. STATIQUE EXTERIEURE MAX.	.5	125

TYPE FSP CATEGORY IV DIRECT VENT FORCED AIR FURNACE. TYPE FSP CATEGORIE IV, GENERATEUR D'AIR CHAUD A EVACUATION DIRECTE ET A'AIR FORCE.

FACTORY AUTHORIZED GAS CONVERSION KITS

NATURAL GAS TO PROPANE	KGANP2001ALL	PROPANE TO NATURAL GAS	KGAPN1601ALL

* = SEE DIRECTORY LISTING, MODELS WITH "L" SUFFIX, FOR LOW FIRE ONLY APPLICATIONS WHEN SW-2 IS ON. SEE INSTALL. MANUAL.

P/N 323850-101
REV. A

Exhibit 6.7 Typical nameplate. (Courtesy of Carrier Corporation.)

Sheetrock, lath, or plaster applied to combustible materials is not adequate protection to prevent ignition of the combustible construction. If clearance reduction is required, use one of the systems in Table 6.2.3(b).

(g) The clearance to this equipment shall not interfere with combustion air, draft hood clearance and relief, and accessibility for servicing. *(See 5.2.1, Section 5.3, and 7.12.8.)*

As noted previously, the clearance required for circulation of combustion air, ventilation of the compartment, and clearance for the draft hood relief opening will be specified on the furnace or boiler rating plate. Under no conditions should this clearance be reduced, because any reduction can affect the safe and satisfactory performance of the appliance.

(h) Listed central heating furnaces shall have the clearance from supply ducts within 3 ft (0.9 m) of the plenum be not less than that specified from the plenum. No clearance is necessary beyond this distance.

(i) Unlisted central heating furnaces with temperature limit controls that cannot be set higher than 250°F (121°C) shall have the clearance from supply ducts within 6 ft (1.8 m) of the plenum be not less than 6 in. (150 mm). No clearance is necessary beyond this distance.

(j) Central heating furnaces other than those listed in 6.3.1(h) or (i) shall have clearances from the supply ducts of not less than 18 in. (0.46 m) from the plenum for the first 3 ft (0.9 m), then 6 in. (150 mm) for the next 3 ft (0.9 m) and 1 in. (25 mm) beyond 6 ft (1.8 m).

6.3.2 Assembly and Installation.

A central heating boiler or furnace shall be installed in accordance with the manufacturer's instructions and shall be installed on a floor of noncombustible construction with noncombustible flooring and surface finish and with no combustible material against the underside thereof, or on fire-resistive slabs or arches having no combustible material against the underside thereof.

Exception No. 1: Appliances listed for installation on a combustible floor.

Exception No. 2: Installation on a floor protected in an approved manner.

Many furnaces and boilers, particularly those intended for residential use, are listed for installation on combustible flooring. Furnaces and boilers listed for installation on a noncombustible flooring also can be listed for optional installation on a combustible flooring if a listed accessory base is added. An example of such a device is shown for a downflow-type furnace (see Exhibit 6.8). The base must be used and installed as shown when the furnace is installed on a combustible flooring.

Prior to the 1988 edition, the *National Fuel Gas Code* contained a footnote referring the reader to Appendix E of the 1976 edition of the *National Building Code* for information on installation of central heating boilers and furnaces not listed for installation on combustible floors. The 1976 edition of the *National Building Code*

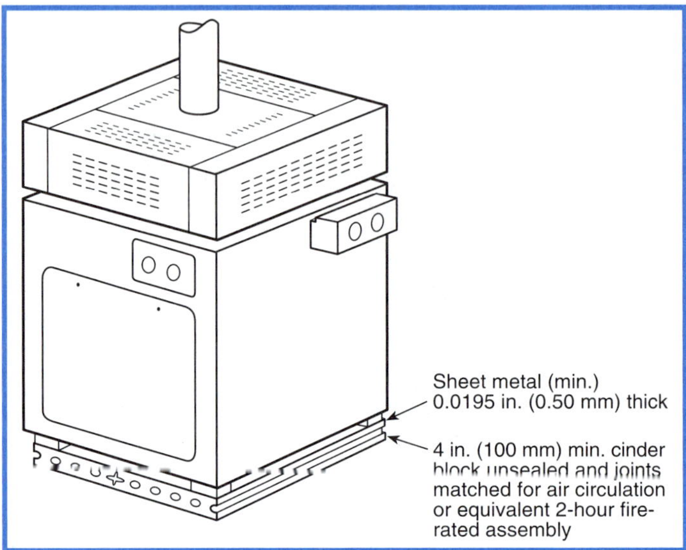

Exhibit 6.8 Installation of a central heating boiler on a combustible floor. (Courtesy of Teledyne Laars.)

was the last edition of that code written by the American Insurance Association. After 1976, the project was taken over by the Building Officials and Code Administrators International. Appendix E no longer appears in the *National Building Code,* but part of it is reprinted here for the information of the user of this handbook. The *National Building Code* covered the installation of residential type boilers and central furnaces as follows:

(a) Residential type boilers and central furnaces, except as permitted by the provisions of Section (b) through (g), shall be placed on concrete bases supported on compacted soil, crushed rock, or gravel; on approved limited-combustible floor-ceiling assemblies with noncombustible floors and having a fire resistance rating of not less than 2 hours; or on concrete slabs or masonry arches, that do not have any combustible materials attached to the underside. Any floor covering on floor-ceiling assemblies, slabs, or arches shall be of noncombustible materials. Concrete bases, concrete slabs, masonry arches, and floor-ceiling assemblies and their supports shall be designed and constructed to support the appliances and shall extend not less than 12 in (30 cm) beyond the appliance on all sides, and, where solid fuel is used, the base or floor shall extend not less than 18 in (46 cm) at the front or side where ashes are removed.

(b) Residential type boilers and central furnaces, that are tested and listed by a nationally recognized testing laboratory for installation on floors constructed of combustible materials, are permitted to be placed on floors other than as required by the provisions of Section (a), provided they are installed in accordance with the requirements of the listing and conditions of approval.

(c) Central furnaces are permitted to be placed on floors other than as required by the provisions of Section (a), provided they are designed so that the fan chamber occupies the entire area beneath the firing chamber and forms a well ventilated air space, between the firing chamber and the floor, not less than 18 in. (46 cm) in height, with at least one sheet metal baffle not less than No. 24 gage between the firing chamber and the floor.

(d) Residential type boilers of the water-base type are permitted to be placed on floors, other than as required by the provisions of Section (a), provided the water chamber extends under the firing chamber or under the ash pit and the firebox. Where solid fuel is used, the floor shall be covered as required by the provisions of Section (h).

(e) Residential type boilers and central furnaces, which are set on legs that provide an open space of not less than 4 in. (10 cm) under the base of the appliance, are permitted to be placed on the floors, other than as required by the provisions of Section (a), provided the appliance is designed so the flames or hot gases do not come in contact with its base, and further provided the floor under the appliance is protected with asbestos millboard not less than $^1/_4$ in. (6.4 mm) thick, covered with sheet metal not less than No. 24 gage. The floor protection shall extend not less than 6 in. (15 cm) beyond the appliance on all sides. Where solid fuel is used, the floor shall be covered as required by the provisions of Section (h).

(f) Residential type boilers and central furnaces are permitted to be placed on floors, other than as required by the provisions of Section (a), provided the appliance is designed so the flames or hot gases do not come in contact with its base, and further provided the floor under the appliance is protected with hollow masonry units not less than 4 in. (10 cm) thick, covered with sheet metal not less than No. 24 gage. The masonry shall be laid with ends unsealed and joints matched to provide a free circulation of air through the core spaces of the masonry. Where solid fuel is used, the floor shall be covered as required by the provisions of Section (h).

(g) Residential type boilers and central furnaces, which are designed so the flames or hot gases are in contact with the base, are permitted to be placed on floors, other than as required by the provisions of Section (a), provided the floor under the appliance is protected by two courses of 4 in (10 cm) hollow masonry units, that are laid with ends unsealed and joints matched to provide a free circulation of air through the core spaces of the masonry. The masonry shall be covered with a steel plate not less than $^3/_{16}$ in. (4.8 mm) thick. Where solid fuel is used, the floor shall be covered as required by the provisions of Section (h).

(h) Where appliances which burn solid fuel are placed on floors other than as required by the provisions of Section (a), the floor, for not less than 18 in (46 cm) beyond the front of the appliance or side where ashes are removed, shall be covered with asbestos millboard not less than $^1/_4$ in. (6.4 mm) thick, covered with sheet metal not less than No. 24 gage, or with materials providing equivalent protection to the floor and durability from hot ashes.

(i) Downflow type central furnaces shall not be placed on floors other than specified in Section (a) unless the appliance rests upon hollow masonry units not less than 4 in. (10 cm) thick. Masonry units shall be laid with the ends unsealed and joints matched to provide a free circulation of air through the core spaces of the masonry. Downflow type furnaces, which are tested and listed by a nationally recognized testing laboratory for installation on a floor constructed of combustible material, are permitted to be mounted in accordance with the requirements of the listing and conditions of approval. Downflow type furnaces shall be installed so that there are no open passages in the floor through which flame or hot gases from a fire originating in the space below the floor can travel to the room above.

NOTE: In Appendix E of the *National Building Code,* a sheet metal thickness of 24 gauge is specified. In the current *National Fuel Gas Code,* a dimension of 0.0195 in. (0.5 mm) is used in place of the 24 gauge that was specified in prior editions. According to the *Chemical Engineers' Handbook,* fifth edition, 1973, 24 U.S. Standard Gauge (for sheet or plate metal and wrought iron) measures 0.0250 in. thick. The use of gauge to specify metal thickness, instead of an actual measurement, is due both to the numerous gauges that are used for specific applications (e.g., wire, sheet, hoop, etc.) and to differing allowances for variation in the manufacturing process. The 0.0195-in. (0.5-mm) dimension currently used is a minimum thickness at any point in the sheet.

6.3.3 Temperature- or Pressure-Limiting Devices.

Steam and hot water boilers, respectively, shall be provided with approved automatic limiting devices for shutting down the burner(s) to prevent boiler steam pressure or boiler water temperature from exceeding the maximum allowable working pressure or temperature.

Safety limit controls shall not be used as operating controls.

Paragraph 6.3.3 covers both automatic temperature-limiting devices and automatic pressure-limiting devices for installation on steam and hot water boilers. If these devices are provided by the manufacturer, they must be installed in accordance with the boiler manufacturer's instructions. If provided by the installer, they must be installed in accordance with the control manufacturer's instructions. Neither the safety pressure control nor the temperature-limiting control can be used as an "operating control" for the appliance. In all cases, refer to the manufacturer's installation instructions for the necessary control wiring to the room thermostat or operating control specified in the piping system. A supplemental, low-temperature control can be used if permitted by the appliance manufacturer.

6.3.4 Low Water Cutoff.

Hot water boilers installed above the radiation level and all steam boilers shall be provided with an automatic means to shut off the fuel supply to the burner(s) if the boiler water level drops to the lowest safe water line.

The requirement for a low water cutoff in water boilers is applicable to installations in which the boiler is installed at a higher elevation than the lowest radiator, baseboard, or other device used to transfer heat from the circulating fluid to the room air. In this configuration, a leak at the lowest radiator (or other device) could drain the boiler without shutting down the burner. This situation would probably damage the boiler because boilers are not designed to be operated dry. An identical requirement appears in the appliance standard for boilers, ANSI Z21.13, *Gas-Fired Low-Pressure Steam and Hot Water Boilers,* and the conversion burner installation standard, ANSI Z21.8, *Installation of Domestic Gas Conversion Burner.*

In some cases, the authority having jurisdiction has allowed using check valves or vacuum relief valves or locating the pipes above the radiation at a point near the ceiling so that it creates a hydraulic trap, similar to a "Hartford Loop." This decision should be left to the authority having jurisdiction.

6.3.5* Steam Safety and Pressure Relief Valves.

Steam and hot water boilers shall be equipped, respectively, with listed or approved steam safety or pressure relief valves of appropriate discharge capacity and conforming with ASME requirements. A shutoff valve shall not be placed between the relief valve and the boiler or on discharge pipes between such valves and the atmosphere.

(1) Relief valves shall be piped to discharge near the floor.
(2) The entire discharged piping shall be at least the same size as the relief valve discharge piping.
(3) Discharge piping shall not contain threaded end connection at its termination point.

These safety devices generally are provided by, or specified by, the steam or hot water boiler manufacturer and must not be altered. The control should be installed at the location specified in the manufacturer's installation instructions. The prohibition of valves in piping to or from these safety devices prevents the devices from being defeated. In addition, follow the termination and size requirements for added safety.

Three specific requirements [6.3.5(1) to 6.3.5(3)] were added in the 1996 edition. These requirements are additional to the minimum requirements for pressure safety and they ensure the following:

(1) In the unlikely event of discharge from a pressure relief valve, the discharge does not create an additional hazard to people in the area.

(2) The safe operation of pressure relief devices is not compromised by downstream flow restrictions.

(3) A valve or cap is not installed unintentionally.

A.6.3.5 For details of requirements on low-pressure heating boiler safety devices, refer to ASME *Boiler and Pressure Vessel Code,* Section IV, "Rules for Construction of Heating Boilers."

6.3.6 Plenums and Air Ducts.

In addition to the reference standards NFPA 90A, *Standard for the Installation of Air-Conditioning and Ventilating Systems,* and NFPA 90B, *Standard for the Installation of Warm Air Heating and Air-Conditioning Systems,* the inspection authority is referred to the following standards. Underwriters Laboratories Inc. standards UL 181, *Standard for Safety Factory-Made Air Ducts and Air Connectors,* and UL 263, *Standard for Safety Fire Tests of Building Construction and Materials,* are the primary standards for factory-made air ducts. These standards divide factory-made nonmetallic ducts into three classes, depending on their surface burning characteristics and flame spread characteristics, and also they specify where these ducts can be used. An additional standard for the construction of metal ducts is the Sheet Metal and Air Conditioning Contractors National Association (SMACNA) standard SMACNA 1985, *HVAC Duct Construction Standards — Metal and Flexible,* for low-pressure and high-pressure duct construction. Low-pressure metal ducts generally are those types that handle air distribution systems with internal pressures of up to 2 in. (0.5 kPa) water column, either positive or negative. These standards also show the construction methods for high-pressure ducts of up to 10 in. (2.5 kPa) water column and include methods for joining the sections and hanging the ducts to ensure compliance with most building codes. In addition, the standards contain methods for sealing duct sections to ensure that they are essentially airtight from the appliance to the discharge grille or diffuser.

(a) Plenums and air ducts shall be installed in accordance with NFPA 90A, *Standard for the Installation of Air Conditioning and Ventilating Systems,* or NFPA 90B, *Standard for the Installation of Warm Air Heating and Air Conditioning Systems.*

(b) A plenum supplied as a part of a furnace shall be installed in accordance with the manufacturer's instructions.

(c)* Where a plenum is not supplied with the furnace, any fabrication and installation instructions provided by the manufacturer shall be followed. The method of connecting supply and return ducts shall facilitate proper circulation of air.

A.6.3.6(c) Reference can be made to NFPA 90A, *Standard for the Installation of Air Conditioning and Ventilating Systems,* or to NFPA 90B, *Standard for the Installation of Warm Air Heating and Air Conditioning Systems.*

(d) Where a furnace is installed so supply ducts carry air circulated by the furnace to areas outside the space containing the furnace, the return air shall also be handled by a duct(s) sealed to the furnace casing and terminating outside the space containing the furnace.

Installing the furnace as required in 6.3.6(d) will separate the furnace's circulating air system physically from its combustion system. The reason for this physical separation is that if the circulating air is carried in ducts out of, but not back to, the space in which a forced-air furnace is installed, flow in the chimney or vent can be reversed, even if the space is fairly large or the doors to the space are louvered. In a small utility room, this reversal of the vent flow is almost a certainty. When the flow is reversed, combustion products will enter the circulating air system through the open return air duct connection on the furnace and will circulate throughout the dwelling. For this reason, the central furnace standard, ANSI Z21.47, *Gas-Fired Central Furnaces,* requires an interlock to prevent furnace operation if the blower (filter) access panel is not in place. (See Exhibit 6.9.)

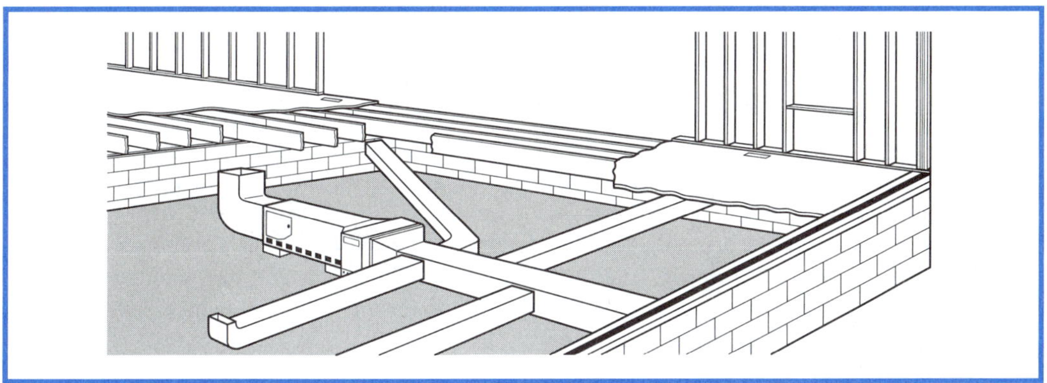

Exhibit 6.9 *Typical installation of a horizontal furnace showing plenum and ducts. (Courtesy of BDP Company.)*

6.3.7 Refrigeration Coils.

(a) A refrigeration coil shall not be installed in conjunction with a forced air furnace where circulation of cooled air is provided by the furnace blower, unless the blower has sufficient capacity to overcome the external static resistance imposed by the duct system and cooling coil and the air throughput necessary for heating or cooling, whichever is greater.

The general requirement for a blower of sufficient capacity to overcome the additional external static pressure resistance imposed by the addition of a cooling coil is to assume that the blower capacity on the furnace must be rated for air delivery of at least 0.5 in. (120 Pa) water column static pressure. This requirement will provide enough blower capacity for a nominal 0.15 in. (37 Pa) water column pressure for the supply duct system, 0.30 in. (75 Pa) water column for the cooling coil, and 0.05 in. (12.5 Pa) water column for the return air duct system and filters.

(b) Furnaces shall not be located upstream from cooling units, unless the cooling unit is designed or equipped so as not to develop excessive temperature or pressure.

Direct-expansion cooling coils with a specified condensing unit are listed as part of the air-conditioning unit assembly, and the requirements for listing generally will include the protective devices on the system necessary to prevent the development of excessive temperatures and pressures in the cooling coil when it is installed downstream from the furnace. In direct expansion systems, the vapor in the cooling coil will expand into the interconnected piping and other portions of the air-conditioning unit. A chilled water coil will be protected by provisions for expansion of the water in the piping system and remote chiller.

(c) Refrigeration coils shall be installed in parallel with or on the downstream side of central furnaces to avoid condensation in the heating element, unless the furnace has been specifically listed for downstream installation. With a parallel flow arrangement, the dampers or other means used to control flow of air shall be sufficiently tight to prevent any circulation of cooled air through the furnace.

Only a furnace specifically constructed and designed for installation downstream from a cooling coil should be installed in this location, and the heat exchanger and burners must be of corrosion-resistant material. If the furnace is listed for downstream installation, then the tests conducted on the furnace will show that there is no danger of condensate dripping on the burners and pilots during the cooling cycle. When a parallel flow system is designed, there should be face and bypass dampers in front of both the cooling coil and the heating unit to ensure that there is no recirculation of chilled air back over the heat exchanger.

(d) Means shall be provided for disposal of condensate and to prevent dripping of condensate on the heating element.

A typical coil for installation in a plenum above an upflow furnace is shown in Exhibit 6.10. The "A-type" coil is constructed so that the condensate forming on the coil will travel down the edge of the coil fins to the drain pan connection. Note also that this particular coil has a secondary drain connection to provide for disposal of condensate in the event that the primary drain connection becomes plugged. Condensate drains should follow local plumbing or mechanical codes.

6.3.8 Cooling Units Used with Heating Boilers.

(a) Boilers, where used in conjunction with refrigeration systems, shall be installed so that the chilled medium is piped in parallel with the heating boiler with appropriate valves to prevent the chilled medium from entering the heating boiler.

Exhibit 6.10 *Air-conditioning refrigeration coil with drain pan. (Courtesy of Bryant Heating & Cooling Systems.)*

(b) Where hot water heating boilers are connected to heating coils located in air-handling units where they can be exposed to refrigerated air circulation, such boiler piping systems shall be equipped with flow control valves or other automatic means to prevent gravity circulation of the boiler water during the cooling cycle.

6.4 Clothes Dryers

Clothes dryers covered by this code are classified into the following two types:

Type 1—Primarily for residential use

Type 2—Designed for use by businesses frequented by the public

Both types can be coin-operated. (See Section 1.7, Definitions, and Exhibit 6.11.)

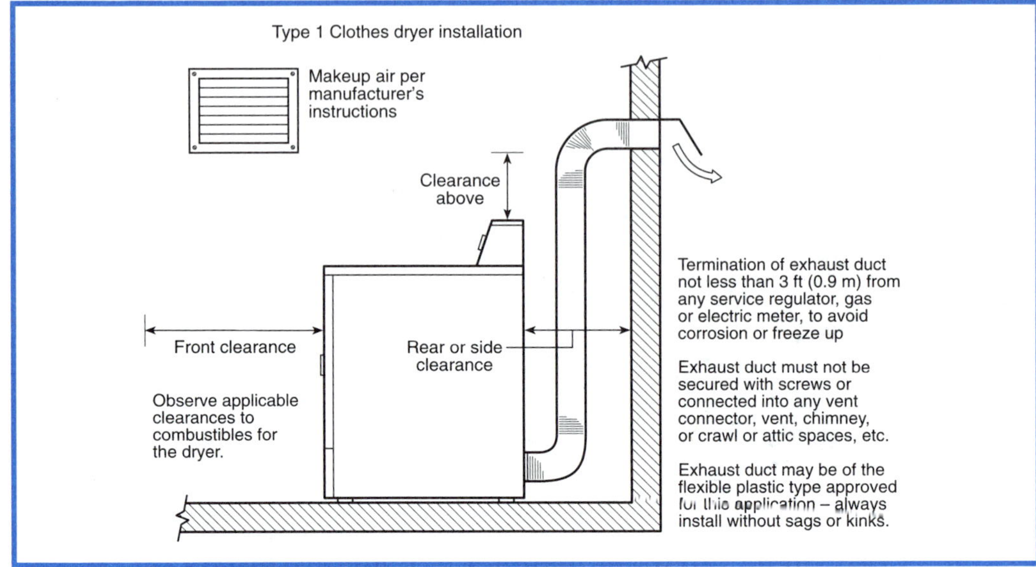

Exhibit 6.11 *Clothes dryer installation requirements.*

6.4.1 Clearance.

(a) Listed Type 1 clothes dryers shall be installed with a minimum clearance of 6 in. (15 cm) from adjacent combustible material, except that clothes dryers listed for installation at lesser clearances shall be permitted to be installed in accordance with their listing. Type 1 clothes dryers installed in closets shall be specifically listed for such installation.

Check the manufacturer's instructions or rating plate on the unit for specified clearance from combustible materials. A drawing of the typical installation of a Type 1 clothes dryer in an alcove or closet is shown in Exhibit 6.12. Clearance must be provided for adequate air supply and ease of installation and service. All dryers must be exhausted to the outdoors. Again, be sure to follow the minimum installation clearances defined by the appliance manufacturer. Consult Section 1.7, Definitions, for the difference between a Type 1 and Type 2 dryer.

(b) Listed Type 2 clothes dryers shall be installed with clearances of not less than shown on the marking plate and in the manufacturers' instructions. Type 2 clothes dryers designed and marked "For use only in noncombustible locations" shall not be installed elsewhere.

(c) Unlisted clothes dryers shall be installed with clearances to combustible material of not less than 18 in. (460 mm). Combustible floors under unlisted clothes dryers shall be protected in an approved manner.

Refer to the commentary on 6.2.4 for additional information on installation on combustible floors.

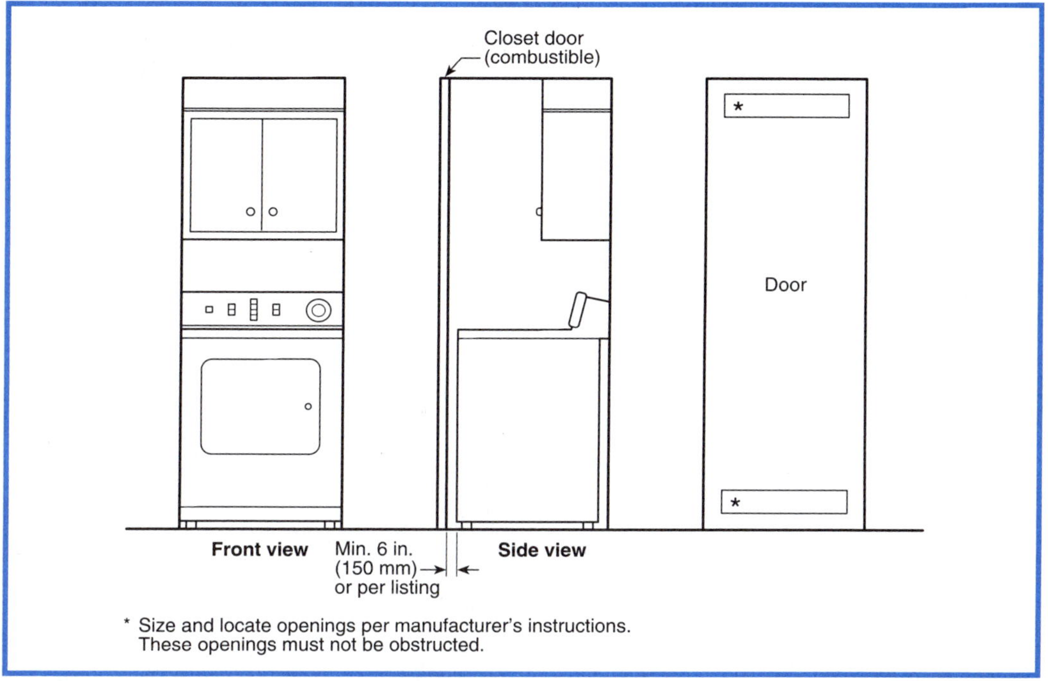

Exhibit 6.12 Installation of a Type 1 clothes dryer in an alcove or closet. (Courtesy of Gas Vent Institute.)

6.4.2 Exhausting to the Outdoors.

(a) Type 1 and Type 2 clothes dryers shall be exhausted to the outside air.

Exhausting of clothes dryers sometimes is confused with venting of appliances. Clothes dryers are exhausted so that both the products of combustion and water removed from the clothing are combined and removed under positive pressure from the dryer and the space in which it is installed. Venting of an appliance, such as a furnace or water heater, is the removal of the products of combustion from the appliance and the space in which it is installed. Because of the significant difference between the exhaust gas from a clothes dryer and the vent gas from a furnace or water heater (in both temperature and moisture content), the removal methods are given different names and have different requirements.

Follow the manufacturer's instructions for the exhaust duct installation and the location of the termination. Note again that dryers installed in closets, bathrooms, or bedrooms should be exhausted to the outdoors.

In the 1999 edition, a change was made to require exhausting of Type 1 clothes dryers to the outdoors. This change parallels a similar change in the equipment standard, ANSI Z21.10.1, *Gas Water Heaters — Volume 1 — Storage Water Heaters with Input Ratings of 75,000 Btu per Hour or Less.* Now NFPA 54 requires that both Type 1 and Type 2 clothes dryers always be exhausted to the outdoors. Refer to 6.4.5 for the requirements for exhaust ducts for Type 2 clothes dryers.

6.4.3 Provisions for Make-Up Air.

(a) Make-up air shall be provided for Type 1 clothes dryers in accordance with the manufacturers' installation instructions.

The quantity of make-up air required for a Type 1 clothes dryer exceeds that of other gas appliances. Generally, five to ten times the air required for an appliance of similar heat input should be allowed. If several gas appliances are installed in a confined space, as defined by the code, or in a limited space (e.g., a utility room or service porch), provisions should be made for adequate combustion and ventilation air. Again, Type 1 dryers that can be installed in closets will be identified by the manufacturer's installation instructions.

(b) Provision for make-up air shall be provided for Type 2 clothes dryers, with a minimum free area *(see 5.3.5)* of 1 in.^{2} (6.5 m^2) for each 1000 Btu/hr (2200 mm^2/kW) total input rating of the dryer(s) installed.

Type 2 dryers are required to have at least 1 in.^{2} of make-up air opening per 1000 Btu/hr (22 cm^2/kW) input rating. In large rooms, this requirement can be fulfilled by installing a separate ventilation inlet duct. A ventilation system normally will be provided in a commercial public laundry.

6.4.4 Exhaust Ducts for Type 1 Clothes Dryers.

(a) A clothes dryer exhaust duct shall not be connected into any vent connector, gas vent, chimney, crawl space, attic, or other similar concealed space.

(b) Ducts for exhausting clothes dryers shall not be assembled with screws or other fastening means that extend into the duct and that would catch lint and reduce the efficiency of the exhaust system.

The manufacturer's instructions will specify the type of material that can be used for a duct assembly. Many manufacturers recommend that only single wall metal pipe or flexible metal ducts be used and not flexible plastic or other types of ducts. Note that sheet metal screws or other fastenings that can extend into the duct and catch lint are prohibited and that all duct connections and fittings must be made by either external clamps or suitable duct tape.

6.4.5 Exhaust Ducts for Type 2 Clothes Dryers.

(a) Exhaust ducts for Type 2 clothes dryers shall comply with 6.4.4.

(b) Exhaust ducts for Type 2 clothes dryers shall be constructed of sheet metal or other noncombustible material. Such ducts shall be equivalent in strength and corrosion resistance to ducts made of galvanized sheet steel not less than 0.0195 in. (0.5 mm) thick.

(c) Type 2 clothes dryers shall be equipped or installed with lint-controlling means.

(d) Exhaust ducts for Type 2 clothes dryers shall have a clearance of at least 6 in. (150 mm) to combustible material.

Exception: Exhaust ducts for Type 2 clothes dryers shall be permitted to be installed with reduced clearances to combustible material, provided the combustible material is protected as described in Table 6.2.3(b).

(e) Where ducts pass through walls, floors, or partitions, the space around the duct shall be sealed with noncombustible material.

(f) Multiple installations of Type 2 clothes dryers shall be made in a manner to prevent adverse operation due to back pressures that might be created in the exhaust systems.

6.4.6 Multiple Family or Public Use.

All clothes dryers installed for multiple-family or public use shall be equipped with approved safety shutoff devices and shall be installed as specified for a Type 2 clothes dryer under 6.4.5.

6.5 Conversion Burners

Installation of conversion burners shall conform to ANSI Z21.8, *Standard for Installation of Domestic Gas Conversion Burners.*

Exhibit 6.13 is an example of a gas conversion burner.

6.6 Decorative Appliances for Installation in Vented Fireplaces

Decorative appliances for installation in vented fireplaces are self-contained, free-standing, fuel-gas-burning appliances designed for installation only in vented fireplaces. They rely on the fireplace venting to vent their products of combustion. (See Exhibit 6.14.)

Their primary function lies in the aesthetic effect of their flames. There are four different types of decorative appliances, described as follows:

(1) *Coal Basket.* An open-flame appliance with a metal basket filled with simulated coals.

(2) *Fireplace Insert.* An open-flame, radiant-type appliance mounted in a decorative metal panel that covers the fireplace opening. This appliance has a provision for venting to the fireplace chimney.

Exhibit 6.13 *A gas conversion burner. (Courtesy of Wayne.)*

Exhibit 6.14 *A decorative gas appliance installed in a fireplace. (Courtesy of Desa International.)*

(3) *Gas Log.* An open-flame appliance consisting of a metal frame or base supporting simulated logs.

(4) *Radiant Appliance.* An open-flame appliance designed primarily to convert the
energy in fuel gas to radiant heat by the means of refractory or similar radiating
materials. A radiant heater typically has no external jacket.

6.6.1* Prohibited Installations.

Decorative appliances for installation in vented fireplaces shall not be installed in bathrooms
or bedrooms unless the appliance is listed and the bedroom or bathroom is an unconfined
space. *(See 5.3.2 and Section 1.7.)*

> Note the requirement of 6.1(b) that gas utilization equipment can only be installed
> in bedrooms or bathrooms if the room is an unconfined space. Since many homes
> now have bedrooms and bathrooms large enough to meet the unconfined space
> requirement, the code now permits these installations if the decorative appliance is
> listed.

A.6.6.1 For information on decorative appliances for installation in vented fireplaces, see
ANSI Z21.60, *Decorative Gas Appliances for Installation in Solid-Fuel Burning Fireplaces.*

> A.6.6.1 and A.6.7.1 assist code users in identifying decorative appliances for instal-
> lation in vented fireplaces and vented decorative appliances. Because the appliance
> nameplate lists the standard to which the appliance is listed, the identification can be
> made easily.

6.6.2 Installation.

A decorative appliance for installation in a vented fireplace shall be installed only in a vented
fireplace having a working chimney flue and constructed of noncombustible materials. These
appliances shall not be thermostatically controlled.

(1) A listed decorative appliance for installation in a vented fireplace shall be installed in
accordance with its listing and the manufacturer's instructions.

(2) An unlisted decorative appliance for installation in a vented fireplace shall be installed in
a fireplace having a permanent free opening, based on appliance input rating and chimney height, equal to or greater than that specified in Table 6.6.2.

6.6.3 Fireplace Screens.

A fireplace screen shall be installed with a decorative appliance for installation in a vented
fireplace.

> This subsection requires a screen to be installed to provide protection from inadvert-
> ent contact with the appliance.

Table 6.6.2 Free Opening Area of Chimney Damper for Venting Flue Gases from Unlisted Decorative Appliances for Installation in Vented Fireplaces

Chimney Height (ft)	Minimum Permanent Free Opening (in.^{2})[*]						
	8	13	20	29	39	51	64
	Appliance Input Rating (Btu/hr)						
6	7,800	14,000	23,200	34,000	46,400	62,400	80,000
8	8,400	15,200	25,200	37,000	50,400	68,000	86,000
10	9,000	16,800	27,600	40,400	55,800	74,400	96,400
15	9,800	18,200	30,200	44,600	62,400	84,000	108,800
20	10,600	20,200	32,600	50,400	68,400	94,000	122,200
30	11,200	21,600	36,600	55,200	76,800	105,800	138,600

For SI units, 1 ft = 0.305 m; 1 in.^{2} = 645 mm^2; 1000 Btu/hr = 0.293 kW.

[*]The first six minimum permanent free openings (8 in.^{2} to 51 in.^{2}) correspond approximately to the cross-sectional areas of chimneys having diameters of 3 in. through 8 in., respectively. The 64-in.^{2} opening corresponds to the cross-sectional area of standard 8 in. × 8 in. chimney tile.

6.7 Gas Fireplaces, Vented

A vented gas fireplace is a vented appliance whose only function lies in the aesthetic effect of its flames (see Exhibit 6.15). These appliances were formerly called vented decorative appliances. The name change follows an identical name change in the standard used for listing these appliances, ANSI Z21.50, *Vented Gas Fireplaces.* Follow the manufacturer's installation instructions for the installation of listed appliances and the recommendations in 6.7.2, Chapter 7, "Venting of Equipment," and Table 6.6.2 for the protection of combustibles and the essential venting requirements.

6.7.1* Prohibited Installations.

Vented gas fireplaces shall not be installed in bathrooms or bedrooms unless the appliance is listed and the bedroom or bathroom is an unconfined space. *(See 5.3.2 and Section 1.7.)*

Exception: Direct-vent gas fireplaces.

This requirement was revised in the 1996 edition of the code to permit the installation of vented decorative appliances in bedrooms and bathrooms when the bedroom or bathroom is an unconfined space. The revision brings this requirement into agreement with the requirements for decorative appliances installed in vented fireplaces.

Exhibit 6.15 *A vented gas fireplace. (Courtesy of Superior Fireplace.)*

See the commentary following 6.6.1. In 1999, an exception was added to clarify that direct-vent gas fireplaces can be installed whether or not the space is confined.

A.6.7.1 For information on vented gas fireplaces, see ANSI Z21.50, *Vented Gas Fireplaces.*

Refer to the commentary following A.6.6.1.

6.7.2 Installation.

(a) Listed vented gas fireplaces shall be installed in accordance with their listing and the manufacturers' instructions. They shall be permitted to be installed in or attached to combustible material where so listed.

(b) Unlisted vented gas fireplaces shall not be installed in or attached to combustible material. They shall have a clearance at the sides and rear of not less than 18 in. (460 mm). Combustible floors under unlisted vented gas fireplaces shall be protected in an approved manner. Unlisted appliances of other than the direct vent type shall be equipped with a draft hood and shall be properly vented in accordance with Chapter 7.

Exception: Appliances that make use of metal, asbestos, or ceramic material to direct radiation to the front of the appliance shall have a clearance of 36 in. (910 mm) in front and, if constructed with a double back of metal or ceramic, shall be permitted to be installed with a clearance of 18 in. (460 mm) at the sides and 12 in. (300 mm) at the rear.

Refer to the commentary on 6.2.4 and 6.3.2 for information on the installation of gas utilization equipment on combustible floors.

(c) Panels, grilles, and access doors that are required to be removed for normal servicing operations shall not be attached to the building.

(d) Direct-vent gas fireplaces shall be installed with the vent-air intake terminal in the outdoors and in accordance with the manufacturers' instructions.

6.7.3 Combustion and Circulating Air.

Combustion and circulating air shall be provided in accordance with Section 5.3.

Size the combustion air requirements based on whether the appliance is located within a confined or an unconfined space. In an unconfined space, infiltration can be expected to provide the necessary air for combustion. When the appliance is located in a confined space, provisions must be made for combustion air to come either from outdoors or from an adjoining, unconfined space. (See Section 5.3.)

6.8 Direct Make-up Air Heaters

6.8.1 Installation.

Direct make-up air heaters shall not be used to supply any area containing sleeping quarters.

The installation of a direct-fired make-up air heater must be in accordance with its listing and the general requirements outlined in Section 6.8. These heaters are intended for commercial and industrial applications.

6.8.2 Clearance from Combustible Material.

The following clearances shall not interfere with combustion air, accessibility for operation, and servicing:

(1) Listed direct make-up air heaters shall be installed in accordance with their listings and the manufacturers' instructions.
(2) Unlisted direct make-up air heaters shall be installed with clearances to combustible material of not less than 18 in. (460 mm). Combustible floors under unlisted floor-mounted heaters shall be protected in an approved manner.

Refer to the commentary on 6.2.4 and 6.3.2 for information on the installation of gas utilization equipment on combustible floors.

6.8.3 Outside Air.

All air handled by a direct make-up air heater, including combustion air, shall be brought in from outdoors. Indoor air shall be permitted to be added to the outdoor airstream after the outdoor airstream has passed the combustion zone.

> Because the air for combustion mixes with the fuel and the products of combustion are exhausted directly into the building, it is important that the exhaust contain only a minimal amount of carbon monoxide. The manufacturers design the appliance to burn cleanly with unlimited fresh air. If building air were to be recycled before the combustion zone, there would be no way to be certain that the recycled air has enough oxygen and no CO or other toxic substances.

6.8.4 Outside Louvers.

If outside louvers of either the manual or automatic type are used, they shall be proved in the open position before the main burners operate.

6.8.5 Controls.

(a) Listed direct make-up air heaters shall be equipped with airflow sensing devices, safety shutoff devices, operating temperature controls, and thermally actuated temperature limit controls in accordance with the terms of their listings.

(b) Unlisted direct make-up air heaters shall be equipped with all of the following:

(1) Airflow sensing devices so designed and installed as to shut off the gas to the main burners upon failure of either combustion air or main air supply. Controls actuated by failure of the power supply to the blower motor shall not be permitted to meet this requirement.
(2) Combustion safeguards, including manual reset safety shutoff devices.
(3) Operating temperature controls and thermally actuated manual reset temperature limit controls, the latter of which shall not permit the discharge air temperature to exceed 150°F (66°C).

6.8.6 Input Ratings.

Unlisted direct make-up air heaters shall have input ratings such that the ratio of gas input by volume to the total volume of gas–air mixture discharged will not exceed 0.2 percent.

6.8.7 Atmospheric Vents and Gas Reliefs or Bleeds.

Direct make-up air heaters with valve train components equipped with atmospheric vents or gas reliefs or bleeds shall have their atmospheric vent lines or gas reliefs or bleeds lead to a safe point outdoors. Means shall be employed on these lines to prevent water from entering and to prevent blockage by insects and foreign matter. An atmospheric vent line shall not be required to be provided on a valve train component equipped with a listed vent limiter.

6.8.8 Relief Opening.

The design of the installation shall include provision to permit make-up air heaters to operate at rated capacity by taking into account the structure's designed infiltration rate, providing properly designed relief openings or an interlocked power exhaust system, or a combination of these methods.

(1) The structure's designed infiltration rate and the size of relief openings shall be determined by approved engineering methods.
(2) Relief openings shall be permitted to be louvers or counterbalanced gravity dampers. Motorized dampers or closable louvers shall be permitted to be used, provided they are verified to be in their full open position prior to main burner operation.

An exhaust system or properly designed relief openings must be provided to exhaust the air that is brought in; otherwise, the buildup of static pressure within the building (area) could reduce the amount of airflow through the heater.

6.8.9 Purging.

The blower of an unlisted direct make-up heater and the exhaust system shall be operated to effect at least four air changes of the combustion chamber before the gas to the main burners is ignited.

6.9 Direct Gas-Fired Industrial Air Heaters

Safety requirements are based on ANSI Z83.18, *Direct Gas-Fired Industrial Air Heaters,* first edition, and the installation requirements herein.

Direct gas-fired industrial air heaters are intended for use in industrial and commercial occupancies, such as factories and warehouses. They must not be installed in areas of public assembly or in areas containing sleeping quarters.

6.9.1 Application.

Direct gas-fired industrial air heaters shall be listed in accordance with ANSI Z83.18, *Standard for Direct Gas-Fired Industrial Air Heaters.*

Exception: Unlisted equipment shall be permitted to be installed where the installation is approved by the authority having jurisdiction and a site-specific equipment evaluation is performed by an independent testing agency based on the applicable requirements of ANSI Z83.18, Standard for Direct Gas-Fired Industrial Air Heaters.

This requirement was revised in the 1999 edition with the addition of the exception, which was relocated from 6.9.3(b). The relocation more clearly illustrates the committee's intent.

6.9.2 Prohibited Installations.

(a) Direct gas-fired industrial air heaters shall not use recirculation of room air in buildings that contain flammable solids, liquids, or gases; explosive materials; or substances that can become toxic when exposed to flame or heat.

(b) Direct gas-fired industrial air heaters shall not be installed in any area containing sleeping quarters.

6.9.3 Installation.

(a) Listed direct gas-fired industrial air heaters shall be permitted to be installed in accordance with their listing and the manufacturers' instructions.

(b) Unlisted direct gas-fired industrial air heaters shall be installed according to the manufacturers' instructions.

(c) Direct gas-fired industrial air heaters shall be installed only in industrial or commercial occupancies.

(d) Direct gas-fired industrial air heaters shall be permitted to provide fresh air ventilation.

The installation requirements were revised in the 1999 edition by revising (b) to resolve confusion over installation of unlisted direct gas-fired heaters. Unlisted heater requirements are covered in the exception to 6.9.1, and (b) requires that any installation recommendations resulting from the evaluation mandated in the exception to 6.9.1 be followed.

While direct gas-fired industrial air heaters are permitted to provide fresh air for ventilation, they are not to be used for this purpose in an area containing sleeping quarters.

6.9.4 Clearance from Combustible Materials.

(a) Listed direct gas-fired industrial air heaters shall be installed with a clearance from combustible material of not less than that shown on the marking plate and in the manufacturers' instructions.

(b) Clearances to combustible materials for unlisted direct gas-fired industrial air heaters shall be determined to be acceptable during evaluation testing. *[See 6.9.3(b).]*

6.9.5* Air Supply.

Air to direct gas-fired industrial air heaters shall be taken from the building, ducted directly from outdoors, or a combination of both.

(a) Direct gas-fired industrial air heaters shall incorporate a means to supply outside ventilation air to the space at a rate of not less than 4 ft^3/min/1000 Btu/hr (0.38 m^3/min/kW) of rated input of the heater. If a separate means is used to supply ventilation air, an interlock shall be provided so as to lock out the main burner operation until the mechanical means is verified.

(b) If outside air dampers or closing louvers are used, they shall be verified to be in the open position prior to main burner operation.

A.6.9.5 Recirculation of room air can be hazardous in the presence of flammable solids, liquids, gases, explosive materials (e.g., grain dust, coal dust, gun powder), and substances (e.g., refrigerants, aerosols) that can become toxic when exposed to flame or heat.

6.9.6 Atmospheric Vents or Gas Reliefs or Bleeds.

Direct gas-fired industrial air heaters with valve train components equipped with atmospheric vents, gas reliefs, or bleeds shall have their atmospheric vent lines and gas reliefs or bleeds lead to a safe point outdoors.

Means shall be employed on these lines to prevent water from entering and to prevent blockage by insects and foreign matter. An atmospheric vent line shall not be required to be provided on a valve train component equipped with a listed vent limiter.

6.9.7 Relief Opening.

The design of the installation shall include adequate provision to permit direct gas-fired industrial air heaters to operate at rated capacity by taking into account the structure's designed infiltration rate, providing properly designed relief openings, an interlocked power exhaust system, or a combination of these methods.

(a) The structure's designed infiltration rate and the size of relief openings shall be determined by approved engineering methods.

(b) Relief openings shall be permitted to be louvers or counterbalanced gravity dampers. Motorized dampers or closable louvers shall be permitted to be used, provided they are verified to be in their full open position prior to main burner operation.

An exhaust system or properly designed relief openings must be provided to exhaust the air that is brought into an area. Otherwise, the buildup of static pressure within the building (area) could reduce the amount of airflow through the heater and affect its combustion process.

6.10 Duct Furnaces

Listed duct furnaces are available for either indoor installation only, outdoor installation only, or both. For the installation of indoor furnaces, the general requirements of this section must be followed. For the installation of outdoor furnaces, the instructions provided by the manufacturer and the conditions of listing must be followed. Certain duct furnaces also can be constructed of corrosion-resistant material and will be listed for installation downstream from a refrigeration coil. Note in 6.10.6 that duct furnaces can be installed downstream from evaporative coolers or air washers if the heat exchanger is made of corrosion-resistant material.

6.10.1 Clearances.

(a) Listed duct furnaces shall be installed with clearances of at least 6 in. (150 mm) between adjacent walls, ceilings, and floors of combustible material and the furnace draft hood, except that furnaces listed for installation at lesser clearances shall be permitted to be installed in accordance with their listings. In no case shall the clearance be such as to interfere with combustion air and accessibility. *(See 5.2.1 and Section 5.3.)*

Refer to the commentary on 6.2.4 and 6.3.2 for information on the installation of gas utilization equipment on combustible floors.

(b) Unlisted duct furnaces shall be installed with clearances to combustible material in accordance with the clearances specified for unlisted furnaces and boilers in Table 6.2.3(a). Combustible floors under unlisted duct furnaces shall be protected in an approved manner.

6.10.2 Erection of Equipment.

Duct furnaces shall be erected and firmly supported in accordance with the manufacturers' instructions.

6.10.3 Access Panels.

The ducts connected to duct furnaces shall have removable access panels on both the upstream and downstream sides of the furnace.

6.10.4 Location of Draft Hood and Controls.

The controls, combustion air inlet, and draft hoods for duct furnaces shall be located outside the ducts. The draft hood shall be located in the same enclosure from which combustion air is taken.

6.10.5 Circulating Air.

Where a duct furnace is installed so that supply ducts carry air circulated by the furnace to areas outside the space containing the furnace, the return air shall also be handled by a duct(s) sealed to the furnace casing and terminating outside the space containing the furnace.

The duct furnace shall be installed on the positive-pressure side of the circulating air blower.

The second paragraph clearly states that all duct furnaces be installed on the positive-pressure side of the circulation air blower. Prior to the 1996 edition, the text could be interpreted to mean that this was not required in all cases.

6.10.6 Duct Furnaces Used with Refrigeration Systems.

(a) A duct furnace shall not be installed in conjunction with a refrigeration coil where circulation of cooled air is provided by the blower.

Exception: Where the blower has sufficient capacity to overcome the external static resistance imposed by the duct system, furnace, and the cooling coil and the air throughput necessary for heating or cooling, whichever is greater.

(b) Duct furnaces used in conjunction with cooling equipment shall be installed in parallel with or on the upstream side of cooling coils to avoid condensation within heating elements. With a parallel flow arrangement, the dampers or other means used to control the flow of air shall be sufficiently tight to prevent any circulation of cooled air through the unit.

When a duct furnace is installed in a parallel arrangement with a cooling coil, it must be installed with suitable, tight-fitting face dampers in front of both the duct furnace and the cooling coil to divert the airflow as necessary. Recirculated cold air over the heat exchanger will cause corrosion and premature failure

Exception: Where the duct furnace has been specifically listed for downstream installation.

(c) Where duct furnaces are to be located upstream from cooling units, the cooling unit shall be so designed or equipped as to not develop excessive temperatures or pressures.

(d) Duct furnaces shall be permitted to be installed downstream from evaporative coolers or air washers if the heating element is made of corrosion-resistant material. Stainless steel, ceramic-coated steel, and an aluminum-coated steel in which the bond between the steel and the aluminum is an iron–aluminum alloy are considered to be corrosion resistant. Air washers operating with chilled water that deliver air below the dew point of the ambient air at the equipment are considered as refrigeration systems.

6.10.7 Installation in Commercial Garages and Aircraft Hangars.

Duct furnaces installed in garages for more than three motor vehicles or in aircraft hangars shall be of a listed type and shall be installed in accordance with 5.1.10 and 5.1.11.

6.11 Floor Furnaces

A floor furnace is a completely self-contained unit, suspended from the floor of the space being heated. It takes its air for combustion from areas other than the heated space. A floor furnace can be equipped with a fan to provide the primary means for circulation of air, or it can be a gravity-type furnace that depends on the circulation of air by convection. (See Exhibit 6.16.) A gravity furnace that is equipped with a booster-type fan is considered also to be a gravity flow furnace when the fan does not restrict materially free circulation of air and when the fan is not in operation.

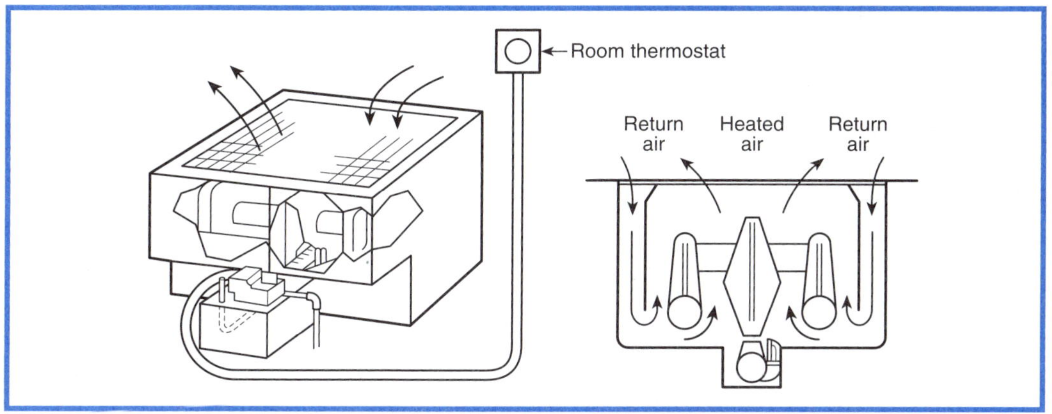

Exhibit 6.16 Installation of a floor furnace. (Courtesy of the American Gas Association.)

6.11.1 Installation.

(a) Listed floor furnaces shall be permitted to be installed in accordance with their listing and the manufacturers' instructions.

(b) Unlisted floor furnaces shall not be installed in combustible floors.

(c) Thermostats controlling floor furnaces shall not be located in a room or space that can be separated from the room or space in which the register of the floor furnace is located.

6.11.2 Temperature Limit Controls.

(a) Listed automatically operated floor furnaces shall be equipped with temperature limit controls in accordance with the terms of their listing.

(b) Unlisted automatically operated floor furnaces shall be equipped with a temperature limit control arranged to shut off the flow of gas to the burner in the event the temperature at the warm air outlet register exceeds 350°F (177°C) above room temperature.

Historically, floor furnaces did not have controls to limit the temperature at the warm air register. Some years ago, the standard for floor furnaces (ANSI Z21.86, *Gas-Fired Gravity Space Heating Equipment*) was revised to require additional controls to limit register temperatures, minimizing the burn hazard potential for the occupants.

6.11.3 Combustion and Circulating Air.

Combustion and circulating air shall be provided in accordance with Section 5.3.

Since the burners and draft hood are generally below the floor, there must be adequate ventilation of the underfloor space to provide air for combustion, draft hood dilution, and ventilation. Normal construction methods in houses with crawl spaces usually will provide more than adequate ventilation and combustion air, but finding such spaces blocked in an effort to maintain warmer floors is not uncommon.

6.11.4 Placement.

The following provisions apply to furnaces that serve one story.

(a) *Floors.* Floor furnaces shall not be installed in the floor of any doorway, stairway landing, aisle, or passageway of any enclosure, public or private, or in an exitway from any such room or space.

(b) *Walls and Corners.* The register of a floor furnace with a horizontal warm air outlet shall not be placed closer than 6 in. (150 mm) to the nearest wall. A distance of at least 18 in. (460 mm) from two adjoining sides of the floor furnace register to walls shall be provided to eliminate the necessity of occupants walking over the warm air discharge. The remaining sides shall be permitted to be placed not closer than 6 in (150 mm) to a wall. Wall register models shall not be placed closer than 6 in. (150 mm) to a corner.

(c) *Draperies.* The furnace shall be placed so that a door, drapery, or similar object cannot be nearer than 12 in. (300 mm) to any portion of the register of the furnace.

6.11.5 Bracing.

The space provided for the furnace shall be framed with doubled joists and with headers not lighter than the joists.

In locating the floor furnace, the installer must avoid cutting beams that support floor joists and must provide double joists and headers for the opening where floor joists are cut to install the furnace.

6.11.6 Support.

Means shall be provided to support the furnace when the floor register is removed.

6.11.7 Clearance.

The lowest portion of the floor furnace shall have at least a 6-in. (150-mm) clearance from the general ground level. Where these clearances are not present, the ground below and to the sides shall be excavated to form a "basin-like" pit under the furnace so that the required clearance is provided beneath the lowest portion of the furnace. A 12-in. (300-mm) clearance shall be provided on all sides except the control side, which shall have an 18-in. (460-mm) clearance.

Exception: Where the lower 6-in. (150-mm) portion of the floor furnace is sealed by the manufacturer to prevent entrance of water, the clearance shall be permitted to be reduced to not less than 2 in. (50 mm).

6.11.8 Access.

The space in which any floor furnace is installed shall be accessible by an opening in the foundation not less than 24 in. × 18 in. (600 mm × 460 mm) or by a trapdoor not less than 24 in. × 24 in. (600 mm × 600 mm) in any cross section thereof, and a passageway not less than 24 in. × 18 in. (600 mm × 460 mm) in any cross section thereof.

6.11.9 Seepage Pan.

Where the excavation exceeds 12 in. (300 mm) in depth or water seepage is likely to collect, a watertight copper pan, concrete pit, or other suitable material shall be used, unless adequate drainage is provided or the equipment is sealed by the manufacturer to meet this condition. A copper pan shall be made of not less than 16-oz/ft^2 (4.9-kg/m^2) sheet copper. The pan shall be anchored in place so as to prevent floating, and the walls shall extend at least 4 in. (100 mm) above the ground level with at least 6 in. (150 mm) clearance on all sides, except the control side, which shall have at least 18 in. (460 mm) clearance.

6.11.10 Wind Protection.

Floor furnaces shall be protected, where necessary, against severe wind conditions.

6.11.11 Upper Floor Installations.

Listed floor furnaces shall be permitted to be installed in an upper floor, provided the furnace assembly projects below into a utility room, closet, garage, or similar nonhabitable space. In such installations, the floor furnace shall be enclosed completely (entirely separated from the nonhabitable space) with means for air intake to meet the provisions of Section 5.3, with access for servicing, the minimum furnace clearances of 6 in. (150 mm) to all sides and bottom, and with the enclosure constructed of portland cement plaster or metal lath or other noncombustible material.

6.11.12 First Floor Installation.

Listed floor furnaces installed in the first or ground floors of buildings shall not be required to be enclosed unless the basements of these buildings have been converted to apartments or sleeping quarters, in which case the floor furnace shall be enclosed as specified for upper floor installations and shall project into a nonhabitable space.

6.12 Food Service Equipment, Floor-Mounted

The titles of this section and of Section 6.13 correlate with the terminology of ANSI Z83.11/CGA 1.8, *Gas Food Service Equipment,* which covers this equipment. For examples of this equipment, see Exhibits 6.17 through 6.20.

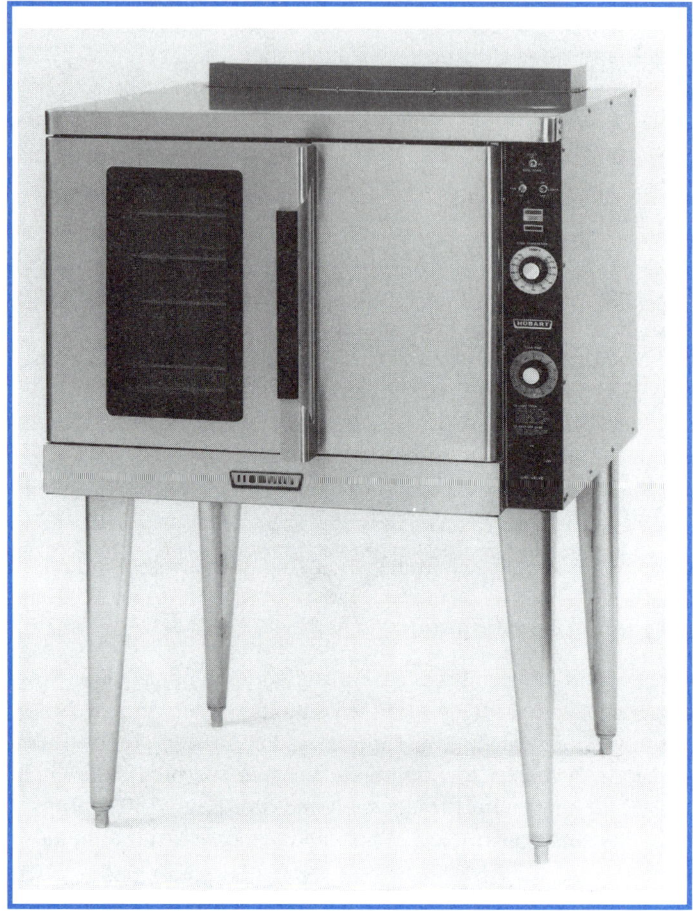

Exhibit 6.17 *Gas convection oven. (Courtesy of Hobart.)*

6.12.1 Clearance for Listed Equipment.

Listed floor-mounted food service equipment, such as ranges for hotels and restaurants, deep fat fryers, unit broilers, gas-fired kettles, steam cookers, steam generators, and baking and roasting ovens, shall be installed at least 6 in. (150 mm) from combustible material except that at least a 2-in. (50-mm) clearance shall be maintained between a draft hood and combustible material. Floor-mounted food service equipment listed for installation at lesser clearances shall be permitted to be installed in accordance with its listing and the manufacturer's instructions. Equipment designed and marked "For use only in noncombustible locations" shall not be installed elsewhere.

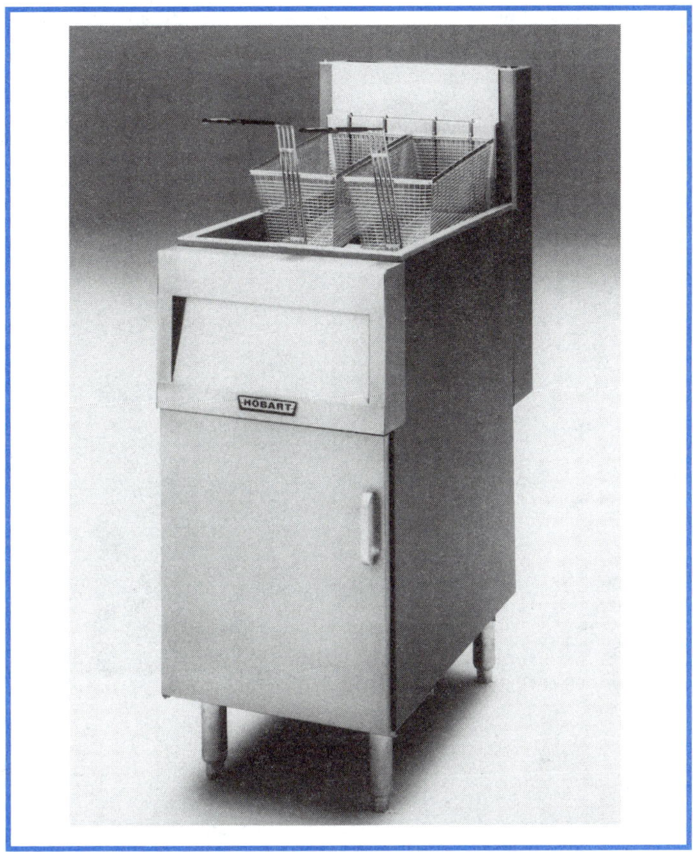

Exhibit 6.18 *Deep fat fryer. (Courtesy of Hobart.)*

6.12.2 Clearance for Unlisted Equipment.

Unlisted floor-mounted food service equipment shall be installed to provide a clearance to combustible material of not less than 18 in. (460 mm) at the sides and rear of the equipment and from the vent connector and not less than 48 in. (1.2 m) above cooking tops and at the front of the equipment.

Exception No. 1: Unlisted floor-mounted food service equipment shall be permitted to be installed in rooms, but not in partially enclosed areas such as alcoves, with reduced clearances to combustible material, provided the combustible material or the equipment is protected as described in Table 6.2.3(b).

Exception No. 2: Unlisted floor-mounted food service equipment shall be permitted to be installed in rooms, but not in partially enclosed areas such as alcoves, with reduced clearance

Exhibit 6.19 *Commercial gas range. (Courtesy of Vulcan-Hart.)*

of 6 in. (150 mm) to combustible material provided the wall or combustible material is protected by sheet metal not less than 0.0152 in. (0.4 mm) thick, fastened with noncombustible spacers that are spaced at not less than 2-ft (0.6-m) vertical and horizontal intervals to provide a clearance of $1^1/_2$ in. (38 mm) from such wall or material. Such protection shall extend at least 12 in. (300 mm) beyond the back, side, top, or any other part of the equipment, and the space between the sheet metal and wall or combustible material shall be open on both sides and top and bottom to permit circulation of air.

6.12.3 Mounting on Combustible Floor.

(a) Listed floor-mounted food service equipment that is listed specifically for installation on floors constructed of combustible material shall be permitted to be mounted on combustible floors in accordance with its listing and the manufacturer's instructions.

Exhibit 6.20 *Dry hot food table. (Courtesy of Foodservice Equipment Co., Inc.)*

(b) Floor-mounted food service equipment that is not listed for mounting on a combustible floor shall be mounted in accordance with 6.12.4 or be mounted in accordance with one of the following:

(1) Where the equipment is set on legs that provide not less than 18 in. (460 mm) open space under the base of the equipment or where it has no burners and no portion of any oven or broiler within 18 in. (460 mm) of the floor, it shall be permitted to be mounted on a combustible floor without special floor protection, provided there is at least one sheet metal baffle between the burner and the floor.

(2) Where the equipment is set on legs that provide not less than 8 in. (200 mm) open space under the base of the equipment, it shall be permitted to be mounted on combustible floors, provided the floor under the equipment is protected with not less than $^3/_8$-in. (9.5-mm) insulating millboard covered with sheet metal not less than 0.0195 in. (0.5 mm) thick. The preceding specified floor protection shall extend not less than 6 in. (150 mm) beyond the equipment on all sides.

(3) Where the equipment is set on legs that provide not less than 4 in. (100 mm) under the base of the equipment, it shall be permitted to be mounted on combustible floors, provided the floor under the equipment is protected with hollow masonry not less than 4 in.

(100 mm) in thickness covered with sheet metal not less than 0.0195 in. (0.5 mm) thick. Such masonry courses shall be laid with ends unsealed and joints matched in such a way as to provide for free circulation of air through the masonry.

(4) Where the equipment does not have legs at least 4 in. (100 mm) high, it shall be permitted to be mounted on combustible floors, provided the floor under the equipment is protected by two courses of 4-in. (100-mm) hollow clay tile, or equivalent, with courses laid at right angles and with ends unsealed and joints matched in such a way as to provide for free circulation of air through such masonry courses, and covered with steel plate not less than $^3/_{16}$ in. (4.8 mm) in thickness.

6.12.4 Mounting on Noncombustible Floor.

Listed floor-mounted food service equipment that is designed and marked "For use only in noncombustible locations" shall be mounted on floors of noncombustible construction with noncombustible flooring and surface finish and with no combustible material against the underside thereof, or on noncombustible slabs or arches having no combustible material against the underside thereof. Such construction shall in all cases extend not less than 12 in. (300 mm) beyond the equipment on all sides.

6.12.5 Combustible Material Adjacent to Cooking Top.

Any portion of combustible material adjacent to a cooking top section of a food service range, even though listed for close-to-wall installation, that is not shielded from the wall by a high shelf, warming closet, and so on, shall be protected as specified in 6.12.2 for a distance of at least 2 ft (0.6 m) above the surface of the cooking top.

6.12.6 For Use with Casters.

Floor-mounted equipment with casters shall be listed for such construction and shall be installed in accordance with their listing and the accompanying instructions for limiting the movement of the equipment to prevent strain on the connection.

6.12.7 Level Installation.

Floor-mounted food service equipment shall be installed level on a firm foundation.

6.12.8* Ventilation.

Means shall be provided to properly ventilate the space in which food service equipment is installed to permit proper combustion of the gas.

A.6.12.8 Where exhaust fans are used for ventilation, precautions might be necessary to avoid interference with the operation of the equipment.

6.13 Food Service Equipment Counter Appliances

6.13.1 Vertical Clearance.

A vertical distance of not less than 48 in. (1.2 m) shall be provided between the top of all food service hot plates and griddles and combustible material.

6.13.2 Clearance for Listed Appliances.

Listed food service counter appliances such as hot plates and griddles, food and dish warmers, and coffee brewers and urns, where installed on combustible surfaces, shall be set on their own bases or legs and shall be installed with a minimum horizontal clearance of 6 in. (150 mm) from combustible material, except that at least a 2-in. (50-mm) clearance shall be maintained between a draft hood and combustible material. Food service counter appliances listed for installation at lesser clearances shall be permitted to be installed in accordance with their listing and the manufacturers' instructions.

6.13.3 Clearance for Unlisted Appliances.

Unlisted food service hot plates and griddles shall be installed with a horizontal clearance from combustible material of not less than 18 in. (460 mm). Unlisted gas food service counter appliances, including coffee brewers and urns, waffle bakers, and hot water immersion sterilizers, shall be installed with a horizontal clearance from combustible material of not less than 12 in. (300 mm). Gas food service counter appliances shall be permitted to be installed with reduced clearances to combustible material provided as described in Table 6.2.3(b). Unlisted food and dish warmers shall be installed with a horizontal clearance from combustible material of not less than 6 in. (150 mm).

6.13.4 Mounting of Unlisted Appliances.

Unlisted food service counter appliances shall not be set on combustible material unless they have legs that provide not less than 4 in. (100 mm) of open space below the burners and the combustible surface is protected with insulating millboard at least $1/4$ in. (6.4 mm) thick covered with sheet metal not less than 0.0122 in. (0.3 mm) thick, or with equivalent protection.

6.14 Hot Plates and Laundry Stoves

Refer to commentary on 6.2.4 and 6.3.2 for information on the installation of gas utilization equipment on combustible floors.

(a) Listed domestic hot plates and laundry stoves installed on combustible surfaces shall be set on their own legs or bases. They shall be installed with minimum horizontal clearances of 6 in. (150 mm) from combustible material.

(b) Unlisted domestic hot plates and laundry stoves shall be installed with horizontal clearances to combustible material of not less than 12 in. (300 mm). Combustible surfaces under unlisted domestic hot plates and laundry stoves shall be protected in an approved manner. Refer to the Commentary on 6.2.4 and 6.3.2 for information on the installation of gas utilization equipment on combustible floors.

(c) The vertical distance between tops of all domestic hot plates and laundry stoves and combustible material shall be at least 30 in. (760 mm).

6.15 Household Cooking Appliances

Household cooking appliances are for domestic food preparation and provide surface cooking, oven cooking, broiling, or a combination of these functions. Broilers can be either enclosed or open, with the radiant heat source above the cooking surface in an enclosed broiler and below the cooking surface in an open broiler. Domestic cooking appliances can be either built in or floor mounted. (See Exhibit 6.21.)

Refer to the commentary on 6.2.4 and 6.3.2 for information on the installation of gas utilization equipment on combustible floors.

6.15.1 Floor-Mounted Units.

(a) *Clearance from Combustible Material.* The clearances specified as follows shall not interfere with combustion air, accessibility for operation, and servicing.

(1) Listed floor-mounted household cooking appliances, where installed on combustible floors, shall be set on their own bases or legs and shall be installed in accordance with their listing and the manufacturers' instructions.

Exception No. 1: Listed household cooking appliances with listed gas room heater sections shall be installed so that the warm air discharge side shall have a minimum clearance of 18 in. (460 mm) from adjacent combustible material. A minimum clearance of 36 in. (910 mm) shall be provided between the top of the heater section and the bottom of cabinets.

Exception No. 2: Household cooking appliances that include a solid or liquid fuel-burning section shall be spaced from combustible material and otherwise installed in accordance with the standards applying to the supplementary fuel section of the appliance.

(2) Unlisted floor-mounted household cooking appliances shall be installed with at least a 6-in. (150-mm) clearance at the back and sides to combustible material. Combustible floors under unlisted appliances shall be protected in an approved manner.

(b) *Vertical Clearance Above Cooking Top.* Household cooking appliances shall have a vertical clearance above the cooking top of not less than 30 in. (760 mm) to combustible material or metal cabinets.

Exhibit 6.21 *Household cooking range. (Courtesy of Maytag.)*

Exception: The clearance shall be permitted to be reduced to not less than 24 in. (610 mm) as follows:

(a) The underside of the combustible material or metal cabinet above the cooking top is protected with not less than $^1/_4$-in. (6.4-mm) insulating millboard covered with sheet metal not less than 0.0122 in. (0.3 mm) thick, or

(b) A metal ventilating hood of sheet metal not less than 0.0122 in. (0.3 mm) thick is installed above the cooking top with a clearance of not less than $^1/_4$ in. (6.4 mm) between the hood and the underside of the combustible material or metal cabinet, and the hood is at least as wide as the appliance and is centered over the appliance.

(c) A listed cooking appliance or microwave oven is installed over a listed cooking appliance and will conform to the terms of the upper appliance's listing and the manufacturer's instructions.

This text addresses the conflicts between the code and installation requirements for separate, listed microwave ovens installed above household cooking appliances.

Listed cooking appliances and microwave ovens placed above listed cooking appliances are now permitted, but they must be installed according to the terms of the upper appliance listing and the manufacturer's instructions. The instructions must be followed because heat from the lower cooking appliance can adversely affect the operation of the upper appliance.

(c) *Level Installation.* Cooking appliances shall be installed so that the cooking top or oven racks are level.

6.15.2 Built-In Units.

(a) *Installation.* Listed built-in household cooking appliances shall be installed in accordance with their listing and the manufacturer's instructions. Listed built-in household cooking appliances shall be permitted to be installed in combustible material unless otherwise marked.

The installation shall not interfere with combustion air, accessibility for operation, and servicing.

Unlisted built-in household cooking appliances shall not be installed in, or adjacent to, combustible material.

(b) *Vertical Clearance.* Built-in top (or surface) cooking appliances shall have a vertical clearance above the cooking top of not less than 30 in. (760 mm) to combustible material or metal cabinets.

Exception: The clearance shall be permitted to be reduced to not less than 24 in. (610 mm) as follows:

(a) The underside of the combustible material or metal cabinet above the cooking top is protected with not less than $^{1}/_{4}$-in. (6.4-mm) insulating millboard covered with sheet metal not less than 0.0122 in. (0.3 mm) thick, or

(b) A metal ventilating hood of sheet metal not less than 0.0122 in. (0.3 mm) thick is installed above the cooking top with a clearance of not less than $^{1}/_{4}$ in. (6.4 mm) between the hood and the underside of the combustible material or metal cabinet, and the hood is at least as wide as the appliance and is centered over the appliance.

(c) A listed cooking appliance or microwave oven is installed over a listed cooking appliance and will conform to the terms of the upper appliance's listing and the manufacturer's instructions.

(c) *Horizontal Clearance.* The minimum horizontal distance from the center of the burner head(s) of a listed top (or surface) cooking appliance to vertical combustible walls extending above the top panel shall be not less than that distance specified by the permanent marking on the appliance.

See the commentary on 6.15.1(b), Exception (b).

(d) *Level Installation.* Built-in household cooking appliances shall be installed so that the cooking top, broiler pan, or oven racks are level.

6.16 Illuminating Appliances

Exhibit 6.22 provides an example of an outdoor gas light.

Exhibit 6.22 *Outdoor gas light. (Courtesy of U.S. Gaslight.)*

6.16.1 Clearances for Listed Appliances.

Listed illuminating appliances shall be installed in accordance with their listing and the manufacturers' instructions.

6.16.2 Clearances for Unlisted Appliances.

(a) *Enclosed Type*.

(1) Unlisted enclosed illuminating appliances installed outdoors shall be installed with clearances in any direction from combustible material of not less than 12 in. (300 mm).
(2) Unlisted enclosed illuminating appliances installed indoors shall be installed with clearances in any direction from combustible material of not less than 18 in. (460 mm).

(b) *Open-Flame Type*.

(1) Unlisted open-flame illuminating appliances installed outdoors shall have clearances from combustible material not less than that specified in Table 6.16.2(b)1. The distance

from ground level to the base of the burner shall be at least 7 ft (2 m) where installed within 2 ft (0.6 m) of walkways. Lesser clearances shall be permitted to be used where acceptable to the authority having jurisdiction.

(2) Unlisted open-flame illuminating appliances installed outdoors shall be equipped with a limiting orifice or other limiting devices that will maintain a flame height consistent with the clearance from combustible material, as given in Table 6.16.2(b)1.

(3) Appliances designed for flame heights in excess of 30 in. (760 mm) shall be permitted to be installed if acceptable to the authority having jurisdiction. Such appliances shall be equipped with a safety shutoff device or automatic ignition.

(4) Unlisted open-flame illuminating appliances installed indoors shall have clearances from combustible material acceptable to the authority having jurisdiction.

Table 6.16.2(b)1 Clearances for Unlisted Outdoor Open-Flame Illuminating Appliances

Flame Height Above Burner Head (in.)	Minimum Clearance from Combustible Material (ft)*	
	Horizontal	Vertical
12	2	6
18	3	8
24	3	10
30	4	12

For SI units, 1 in. = 25.4 mm; 1 ft = 0.305 m.

*Measured from the nearest portion of the burner head.

6.16.3 Mounting on Buildings.

Illuminating appliances designed for wall or ceiling mounting shall be securely attached to substantial structures in such a manner that they are not dependent on the gas piping for support.

6.16.4 Mounting on Posts.

Illuminating appliances designed for post mounting shall be securely and rigidly attached to a post.

Posts shall be rigidly mounted. The strength and rigidity of posts greater than 3 ft (0.9 m) in height shall be at least equivalent to that of a $2^1/_2$-in. (64-mm) diameter post constructed of 0.064-in. (1.6-mm) thick steel or a 1-in. Schedule 40 steel pipe. Posts 3 ft (0.9 m) or less in height shall not be smaller than a $^3/_4$-in. Schedule 40 steel pipe.

Drain openings shall be provided near the base of posts where there is a possibility of water collecting inside them.

6.16.5 Gas Appliance Pressure Regulators.

Where a gas appliance pressure regulator is not supplied with an illuminating appliance and the service line is not equipped with a service pressure regulator, an appliance pressure regulator shall be installed in the line to the illuminating appliance. For multiple installations, one regulator of adequate capacity shall be permitted to be used to serve more than one illuminating appliance.

6.17 Incinerators, Commercial-Industrial

Commercial-industrial-type incinerators shall be constructed and installed in accordance with NFPA 82, *Standard on Incinerators, and Waste and Linen Handling Systems and Equipment.*

6.18 Incinerators, Domestic

6.18.1 Clearance.

(a) Listed incinerators shall be installed in accordance with their listing and the manufacturers' instructions, provided that, in any case, the clearance shall be sufficient to afford ready accessibility for firing, cleanout, and necessary servicing.

(b) The clearances to combustible material above a charging door shall be not less than 48 in. (1220 mm).

Exception No. 1: The clearance shall be permitted to be reduced to 24 in. (610 mm), provided the combustible material is protected with sheet metal not less than 0.0122 in. (0.3 mm) thick, spaced out 1 in. (25 mm) on noncombustible spacers, or equivalent protection. Such protection shall extend 18 in. (460 mm) beyond all sides of the charging door opening.

Exception No. 2: Listed incinerators designed to retain the flame during loading need not comply with this paragraph.

(c) Unlisted incinerators shall be installed with clearances to combustible material of not less than 36 in. (920 mm) at the sides, top, and back and not less than 48 in. (1.2 m) at the front, but in no case shall the clearance above a charging door be less than 48 in. (1.2 m). Unlisted wall-mounted incinerators shall be installed on a noncombustible wall communicating directly with a chimney.

(d) Domestic-type incinerators shall be permitted to be installed with reduced clearances to combustible material in rooms, provided the combustible material is protected as described in Table 6.2.3(b).

Exception: In partially enclosed areas, such as alcoves, clearances shall not be so reduced.

(e) Where a domestic-type incinerator that is refractory lined or insulated with heat insulating material is encased in common brick not less than 4 in. (100 mm) in thickness, the clearances shall be permitted to be reduced to 6 in. (150 mm) at the sides and rear, and the clearance at the top shall be permitted to be reduced to 24 in. (610 mm), provided that the construction using combustible material above the charging door and within 48 in. (1220 mm) is protected with sheet metal not less than 0.0122 in. (0.3 mm) thick, spaced out 1 in. (25 mm), or equivalent protection.

6.18.2 Mounting.

(a) Listed incinerators specifically listed for installation on combustible floors shall be permitted to be so installed.

(b) Unlisted incinerators shall be mounted on the ground or on floors of noncombustible construction with noncombustible flooring or surface finish and with no combustible material against the underside thereof, or on noncombustible slabs or arches having no combustible material against the underside thereof. Such construction shall extend not less than 12 in. (300 mm) beyond the incinerator base on all sides, except at the front or side where ashes are removed, where it shall extend not less than 18 in. (460 mm) beyond the incinerator.

Exception No. 1: Unlisted incinerators shall be permitted to be mounted on floors other than as specified in 6.18.2(b), provided the incinerator is so arranged that flame or hot gases do not come in contact with its base and, further, provided the floor under the incinerator is protected with hollow masonry not less than 4 in. (100 mm) thick, covered with sheet metal not less than 0.0195 in. (0.5 mm) thick. Such masonry course shall be laid with ends unsealed and joints matched in such a way as to provide a free circulation of air from side to side through the masonry. The floor for 18 in. (460 mm) beyond the front of the incinerator or side where ashes are removed, and 12 in. (300 mm) beyond all other sides of the incinerator, shall be protected with not less than $^{1}/_{4}$-in. (6.4-mm) insulating millboard covered with sheet metal not less than 0.0195 in. (0.5 mm) thick, or with equivalent protection.

Exception No. 2: Unlisted incinerators that are set on legs that provide not less than 4 in. (100 mm) open space under the base of the incinerator shall be permitted to be mounted on floors other than as specified in 6.18.2(b), provided the incinerator is such that flame or hot gases do not come in contact with its base and, further, provided the floor under the incinerator is protected with not less than $^{1}/_{4}$-in. (6.4-mm) insulating millboard covered with sheet metal not less than 0.0195 in. (0.5 mm) thick. The above specified floor protection shall extend not less than 18 in. (460 mm) beyond the front of the incinerator or side where ashes are removed and 12 in. (300 mm) beyond all other sides of the incinerator.

6.18.3 Draft Hood Prohibited.

A draft hood shall not be installed in the vent connector of an incinerator.

6.18.4 Venting.

Incinerators shall be vented in accordance with Sections 7.4, 7.5, 7.7, and 7.10.

6.19 Infrared Heaters

An infrared heater is a heater that directs infrared energy into a specific area. Such heaters can be either vented or unvented. (See Exhibits 6.23 and 6.24.)

Exhibit 6.23 *Infrared space heater. (Courtesy of Desa International.)*

6.19.1 Support.

Suspended-type infrared heaters shall be fixed in position independent of gas and electric supply lines. Hangers and brackets shall be of noncombustible material.

Heaters subject to vibration shall be provided with vibration-isolating hangers.

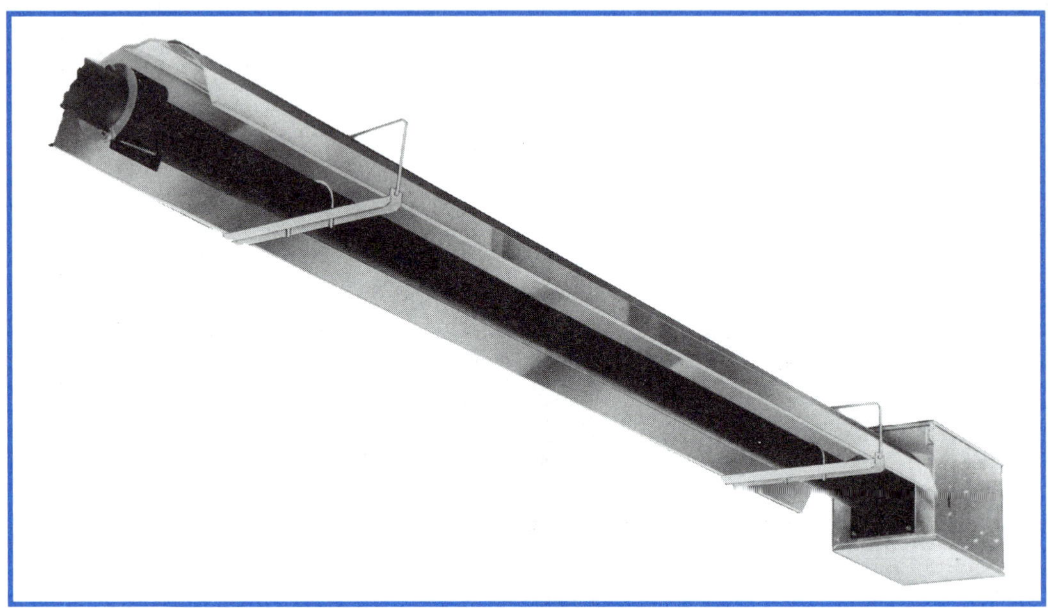

Exhibit 6.24 *An infrared heater. (Courtesy of Reznor.)*

Suitable materials for hanging infrared heaters are steel pipe, steel channels, or fabricated hangers of at least 16-gauge material. Consult heater manufacturer's installation instructions for specific recommendations. Additional bracing to protect against seismic forces is also recommended in areas with active seismicity.

6.19.2 Clearance.

(a) Listed heaters shall be installed with clearances from combustible material in accordance with their listing and the manufacturers' instructions.

(b) Unlisted heaters shall be installed in accordance with clearances from combustible material acceptable to the authority having jurisdiction.

(c) In locations used for the storage of combustible materials, signs shall be posted to specify the maximum permissible stacking height to maintain required clearances from the heater to the combustibles.

6.19.3 Combustion and Ventilation Air.

(a) Where unvented infrared heaters are used, natural or mechanical means shall be provided to supply and exhaust at least 4 ft^3/min/1000 Btu/hr (0.38 m^3/min/kW) input of installed heaters.

(b) Exhaust openings for removing flue products shall be above the level of the heaters.

6.19.4 Installation in Commercial Garages and Aircraft Hangars.

Overhead heaters installed in garages for more than three motor vehicles or in aircraft hangars shall be of a listed type and shall be installed in accordance with 5.1.10 and 5.1.11.

6.20 Open-Top Broiler Units

Open-top broiler units are similar to outdoor gas grills, except that they are designed for countertop installation indoors in conjunction with a ventilating hood. These units cook primarily by radiated heat.

6.20.1 Listed Units.

Listed open-top broiler units shall be installed in accordance with their listing and the manufacturers' instructions.

6.20.2 Unlisted Units.

Unlisted open-top broiler units shall be installed in accordance with the manufacturers' instructions but shall not be installed in combustible material.

6.20.3 Protection Above Domestic Units.

Domestic open-top broiler units shall be provided with a metal ventilating hood not less than 0.0122 in. (0.3 mm) thick with a clearance of not less than $^1/_4$ in. (6.4 mm) between the hood and the underside of combustible material or metal cabinets. A clearance of at least 24 in. (610 mm) shall be maintained between the cooking top and the combustible material or metal cabinet, and the hood shall be at least as wide as the open-top broiler unit and centered over the unit. Listed domestic open-top broiler units incorporating an integral exhaust system and listed for use without a ventilating hood need not be provided with a ventilating hood if installed in accordance with 6.15.1(b)1.

Most listed domestic broilers must be installed with at least a 24-in. (610-mm) clearance between the cooking top and the ventilating hood. Open-top broiler units that incorporate an integral exhaust system, which usually includes a fan and filter located below the broiler level, are designed to pull the flue gases and cooking vapors downward and discharge them to the outside. These units do not require a hood, but they must maintain the clearance of at least 24 in. (610 mm) above the cooking top.

6.20.4 Commercial Units.

Commercial open-top broiler units shall be provided with ventilation in accordance with NFPA 96, *Standard for Ventilation Control and Fire Protection of Commercial Cooking Operations.*

NFPA 96, *Standard for Ventilation Control and Fire Protection of Commercial Cooking Operations,* contains detailed requirements for the design and installation of hoods, duct systems, grease-removal devices, auxiliary equipment, and fire-extinguishing equipment.

6.21 Outdoor Cooking Appliances

An outdoor cooking-gas appliance as used in this code is a cooking appliance installed directly on a post that is provided by the manufacturer as a part of the appliance. (See Exhibit 6.25.)

Exhibit 6.25 *Outdoor gas grill. (Courtesy of Ducane.)*

6.21.1 Listed Units.

Listed outdoor cooking appliances shall be installed in accordance with their listing and the manufacturers' instructions.

6.21.2 Unlisted Units.

Unlisted outdoor cooking appliances shall be installed outdoors with clearances to combustible material of not less than 36 in. (910 mm) at the sides and back and not less than 48 in. (1220 mm) at the front. In no case shall the appliance be located under overhead combustible construction.

6.22 Pool Heaters

A pool heater is an appliance designed for heating nonpotable water stored at atmospheric pressure, such as water in swimming pools, therapeutic pools, and similar applications. It can be either a coil-type heater, where the heat exchanger consists primarily of water tubes whose inside diameters are less than $1\frac{1}{4}$ in. (3.2 cm), or an indirect-type heater, which uses water in a primary heat exchanger to transmit heat from the gas combustion process to the pool water by means of a secondary heat exchanger. Pool heaters are listed for either indoor installations, outdoor installations, or both. Those heaters that are installed indoors must be vented and have air for combustion and ventilation which is provided in accordance with Part 5, Equipment Installation. Pool heaters installed outdoors must have the vent termination and any protective cover installed as required by manufacturer's instructions. (See Exhibit 6.26.)

Exhibit 6.26 *A pool heater. (Courtesy of Teledyne Laars.)*

6.22.1 Location.

A pool heater shall be located or protected so as to minimize accidental contact of hot surfaces by persons.

6.22.2 Clearance.

(a) In no case shall the clearances be such as to interfere with combustion air, draft hood or vent terminal clearance and relief, and accessibility for servicing.

(b) A listed pool heater shall be installed in accordance with its listing and the manufacturer's instructions.

(c) An unlisted pool heater shall be installed with a minimum clearance of 12 in. (300 mm) on all sides and the rear. A combustible floor under an unlisted pool heater shall be protected in an approved manner.

> Refer to the commentary on 6.2.4 and 6.3.2 for information on installation of heaters and similar appliances on a combustible floor.

6.22.3 Temperature- or Pressure-Limiting Devices.

(a) An unlisted pool heater shall be provided with overtemperature protection or overtemperature and overpressure protection by means of an approved device(s).

(b) Where a pool heater is provided with overtemperature protection only and is installed with any device in the discharge line of the heater that can restrict the flow of water from the heater to the pool (such as a check valve, shutoff valve, therapeutic pool valving, or flow nozzles), a pressure relief valve shall be installed either in the heater or between the heater and the restrictive device.

6.22.4 Bypass Valves.

If an integral bypass system is not provided as a part of the pool heater, a bypass line and valve shall be installed between the inlet and outlet piping for use in adjusting the flow of water through the heater.

6.22.5 Venting.

A pool heater listed for outdoor installation shall be installed with the venting means supplied by the manufacturer and in accordance with the manufacturer's instructions. *(See 7.2.5, 7.2.6, 7.3.4, and Section 7.8.)*

6.23 Refrigerators

> Gas refrigerators were used widely in rural areas prior to the expansion of electricity to these locations. They are still used in remote areas that are not served by electric-

ity, in areas where the supply of electricity is interrupted frequently, by people whose religious beliefs do not permit the use of electricity (e.g., the Amish), and in recreational vehicles. Their use in recreational vehicles is not covered in this code, but it is covered in ANSI A119.2/NFPA 501C, *Standard on Recreational Vehicles.*

6.23.1 Clearance.

Refrigerators shall be provided with clearances for ventilation at the top and back in accordance with the manufacturers' instructions. If such instructions are not available, at least 2 in. (50 mm) shall be provided between the back of the refrigerator and the wall and at least 12 in. (300 mm) above the top.

6.23.2 Venting or Ventilating Kits Approved for Use with a Refrigerator.

If an accessory kit is used for conveying air for burner combustion or unit cooling to the refrigerator from areas outside the room in which it is located, or for conveying combustion products diluted with air-containing waste heat from the refrigerator to areas outside the room in which it is located, the kit shall be installed in accordance with the refrigerator manufacturer's instructions.

A gas refrigerator generally has a low enough input that it can be installed conveniently in the desired space, usually with no special provisions required for air for combustion and ventilation. The clearances required are similar to those required for electric refrigerators. If the manufacturer provides an accessory kit for venting and ventilation, it must be installed in accordance with the manufacturer's instructions.

6.24 Room Heaters

By design, room heaters are either unvented or vented. They can be either the circulating type or the radiant type. Some room heaters are equipped with circulating fans to provide more satisfactory heating performance. (See Exhibit 6.27.)

6.24.1* Prohibited Installations.

Unvented room heaters shall not be installed in bathrooms or bedrooms.

This provision clearly prohibits the installation of unvented room heaters in bedrooms and bathrooms. [See also the commentary after 6.1(b).] There are two reasons for this ban, as follows:

(1) These rooms are often small, confined spaces that provide inadequate air for combustion and ventilation and present a carbon monoxide or an asphyxiation hazard.

Exhibit 6.27 *Unvented room heater. (Courtesy of Desa International.)*

(2) No matter where an unvented room heater is located, in these small rooms the potential for accidental ignition of flammable materials (e.g., towels, curtains, and clothing) always exists.

Exception No. 1: Where approved by the authority having jurisdiction, one listed wall-mounted unvented room heater equipped with an oxygen depletion safety shutoff system shall be permitted to be installed in a bathroom provided that the input rating shall not exceed 6000 Btu/hr (1760 W/hr) and combustion and ventilation air is provided as specified in 6.1(b).

Exception No. 2: Where approved by the authority having jurisdiction, one listed wall-mounted unvented room heater equipped with an oxygen depletion safety shutoff system shall be permitted to be installed in a bedroom provided that the input rating shall not exceed 10,000 Btu/hr (2930 W/hr) and combustion and ventilation air is provided as specified in 6.1(b).

Two exceptions were added to the 1996 code that permit, with approval of the authority having jurisdiction, installation of listed, wall-mounted, unvented room heaters of limited capacity and with an oxygen depletion safety shutoff system in bedrooms and bathrooms, provided that they comply with 6.1(b).

If a heater of greater capacity than permitted in the exceptions is required or if the approval of the authority having jurisdiction is not received, heaters of the direct-vent type can be used. Direct-vent heaters take all of the air for combustion from outside the building and discharge the products of combustion back outside the building.

Also note that Section 6.1(b) specifies that gas utilization equipment must be installed so that combustion and ventilation air is not obtained from a bedroom or a bathroom unless the bedroom or bathroom is an unconfined space.

A.6.24.1 It is recommended that space heating appliances installed in all bedrooms or rooms generally kept closed be of the direct vent type. *(See Section 6.28.)*

6.24.2 Installations in Institutions.

Room heaters shall not be installed in institutions such as homes for the aged, sanitariums, convalescent homes, or orphanages.

6.24.3 Clearance.

A room heater shall be placed so as not to cause a hazard to walls, floors, curtains, furniture, doors when open, and so on, and to the free movements of persons within the room. Heaters designed and marked "For use in noncombustible fireplace only" shall not be installed elsewhere. Listed room heaters shall be installed in accordance with their listings and the manufacturers' instructions. In no case shall the clearances be such as to interfere with combustion air and accessibility.

Unlisted room heaters shall be installed with clearances from combustible material not less than the following:

(a) *Circulating Type.* Room heaters having an outer jacket surrounding the combustion chamber, arranged with openings at top and bottom so that air circulates between the inner and outer jacket, and without openings in the outer jacket to permit direct radiation, shall have clearance at sides and rear of not less than 12 in. (300 mm).

(b) *Radiating Type.* Room heaters other than those of the circulating type described in 6.24.3(a) shall have clearance at sides and rear of not less than 18 in. (460 mm), except that heaters that make use of metal, asbestos, or ceramic material to direct radiation to the front of the heater shall have a clearance of 36 in. (910 mm) in front and, if constructed with a double back of metal or ceramic, shall be permitted to be installed with a clearance of 18 in. (460 mm) at sides and 12 in. (300 mm) at rear. Combustible floors under unlisted room heaters shall be protected in an approved manner.

6.24.4 Wall-Type Room Heaters.

Wall-type room heaters shall not be installed in or attached to walls of combustible material unless listed for such installation.

6.25 Stationary Gas Engines.

The installation of gas engines shall conform with NFPA 37, *Standard for the Installation and Use of Stationary Combustion Engines and Gas Turbines.*

Prior to the 1999 edition, Section 6.25 provided coverage of sauna heaters. This coverage was deleted in the 1999 edition because gas-fired sauna heaters are no longer being made.

NFPA 37, *Standard for the Installation and Use of Stationary Combustion Engines and Gas Turbines,* applies to the installation and operation of stationary combustion engines and gas turbines of up to 7500 horsepower per unit. Engines that are used to drive fire pumps and those used in essential electrical systems in health care facilities are excluded because such engines are covered by NFPA 20, *Standard for the Installation of Centrifugal Fire Pumps,* and NFPA 99, *Standard for Health Care Facilities,* respectively.

For fuel gas piping supplying engines, NFPA 37 refers back to the *National Fuel Gas Code* for piping of up to 125 psi (862 kPa) and to ANSI B31.3, *Process Piping for Piping Above 125 psi (862 kPa).* Additional requirements for installation, engine protective devices, valves, and exhaust piping and chimneys (including clearance to combustibles) are included in NFPA 37.

6.26 Gas-Fired Toilets

Gas-fired toilets often are used at construction sites and in areas that are remote from water supplies (e.g., hunting and skiing cabins and office trailers). They also are used in areas without sewer service and areas where soils will not accept septic systems. (See Exhibit 6.28.)

6.26.1 Clearance.

A listed gas-fired toilet shall be installed in accordance with its listing and the manufacturer's instructions, provided that the clearance shall in any case be sufficient to afford ready accessibility for use, cleanout, and necessary servicing.

6.26.2 Mounting.

Listed gas-fired toilets specifically listed for installation on combustible floors shall be permitted to be so installed.

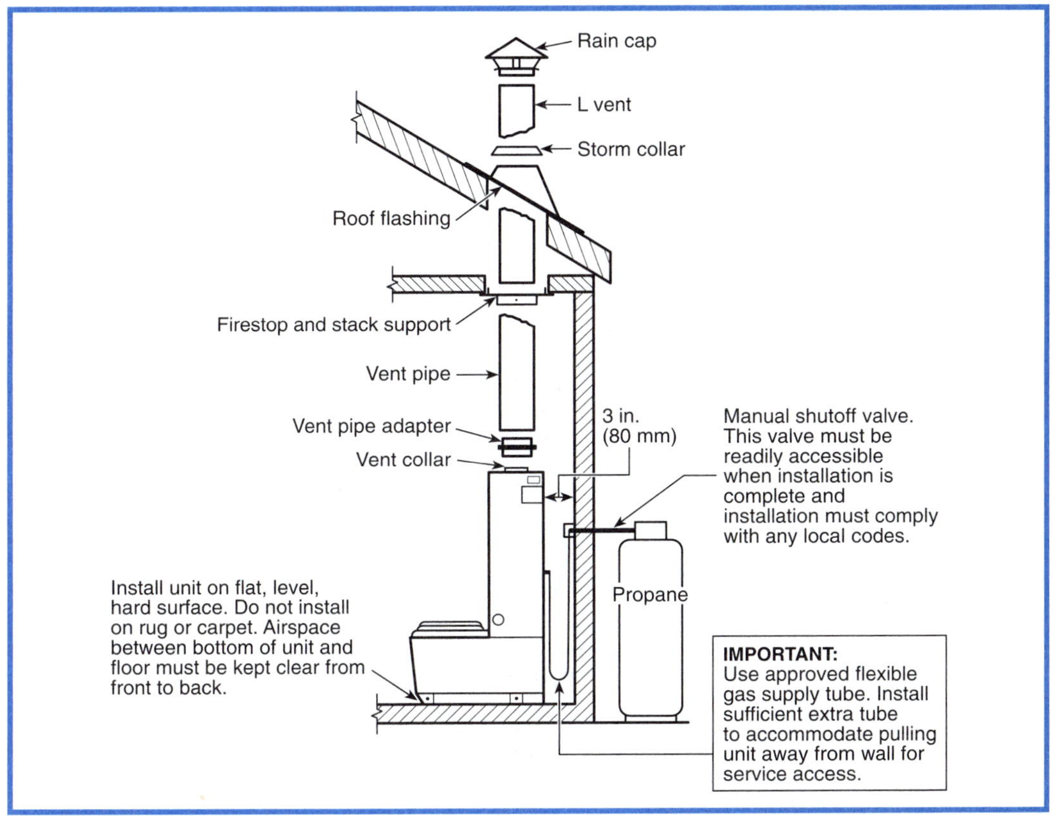

Exhibit 6.28 *Installation of a gas-fired toilet. (Courtesy of Storburn.)*

6.26.3 Installation.

Vents or vent connectors that are capable of being contacted during casual use of the room in which the toilet is installed shall be protected or shielded to prevent such contact.

6.27 Unit Heaters

A unit heater is a self-contained, automatically controlled, vented, fuel-gas-burning appliance that has an integral means for the circulation of air. (See Exhibit 6.29.) A high-static-pressure unit heater can deliver air against a pressure of 0.2 in (50 Pa) water column or greater static pressure. It is equipped with provisions for attaching an outlet air duct, and, if designed for indoor installation in a place remote from the space to be heated, it also will be equipped with provisions for attaching an inlet air duct. A low-static-pressure unit heater usually circulates air by means of an integral propeller fan, and it can be equipped with louvers or face extensions on the discharge side in accordance with the manufacturer's specifications. A low-static unit heater is intended for installation in the space to be heated and must not be used with circulating air ducts.

Exhibit 6.29 *A unit heater. (Courtesy of Reznor.)*

6.27.1 Support.

Suspended-type unit heaters shall be safely and adequately supported with due consideration given to their weight and vibration characteristics. Hangers and brackets shall be of noncombustible material.

Suspension methods for unit heaters generally are specified by the manufacturer, depending on the weight of the heater. Normally, steel pipe and the use of pipe couplings or bushings will provide satisfactory support and will enable the unit to be removed for service. Also, additional bracing to protect the heater against seismic forces is recommended in areas with active seismicity.

6.27.2 Clearance.

(a) *Suspended-Type Unit Heaters.*

(1) A listed unit heater shall be installed with clearances from combustible material of not less than 18 in. (460 mm) at the sides, 12 in. (300 mm) at the bottom, and 6 in. (150 mm)

above the top where the unit heater has an internal draft hood, or 1 in. (25 mm) above the top of the sloping side of a vertical draft hood.

Exception: A unit heater listed for reduced clearances shall be installed in accordance with its listing and the manufacturer's instructions.

(2) Unlisted unit heaters shall be installed with clearances to combustible material of not less than 18 in. (460 mm).

(3) Clearances for servicing shall be in accordance with the manufacturers' recommendations contained in the installation instructions.

(b) *Floor-Mounted-Type Unit Heaters.*

(1) A listed unit heater shall be installed with clearances from combustible material at the back and one side only of not less than 6 in. (150 mm). Where the flue gases are vented horizontally, the 6-in. (150-mm) clearance shall be measured from the draft hood or vent instead of the rear wall of the unit heater.

Exception: A unit heater listed for reduced clearances shall be installed in accordance with its listing and the manufacturer's instructions.

(2) Floor-mounted-type unit heaters shall be permitted to be installed on combustible floors if listed for such installation.

(3) Combustible floors under unlisted floor-mounted unit heaters shall be protected in an approved manner.

(4) Clearances for servicing shall be in accordance with the manufacturers' recommendations contained in the installation instructions.

Refer to the commentary following 6.22.2 for suggested methods.

6.27.3 Combustion and Circulating Air.

Combustion and circulating air shall be provided in accordance with Section 5.3.

Normally, unit heaters are installed in large commercial or industrial areas for spot heating; thus, adequate air for combustion and ventilation usually will be available. If the heater is in a confined space, consult Section 5.3 of this code for requirements for combustion and ventilation air.

6.27.4 Ductwork.

A unit heater shall not be attached to a warm air duct system unless listed and marked for such installation.

6.27.5 Installation in Commercial Garages and Aircraft Hangars.

Unit heaters installed in garages for more than three motor vehicles or in aircraft hangars shall be of a listed type and shall be installed in accordance with 5.1.10 and 5.1.11.

6.28 Wall Furnaces

A wall furnace is a self-contained vented appliance, complete with grilles or their equivalent and designed for incorporation in or permanent attachment to the structure of a building. It furnishes radiant heat or heated air, circulated by gravity or by a fan, directly into the space to be heated through openings in the casing.

Some appliances can be listed with accessory duct extensions, or boots, which do not extend more than 10 in. (250 mm) beyond the appliance casing for extension through walls of nominal thickness. These boots must be supplied by the manufacturer as an integral part of the appliance.

A direct-vent wall furnace is a system composed of an appliance, combustion air and flue gas connections between the appliance and outdoor atmosphere, and a vent cap supplied by the manufacturer. It is constructed so that all air for combustion is obtained from outdoors and all flue gases are discharged to the outdoors. (See Exhibits 6.30, 6.31, 6.32, and 6.33.) Note that the definition of a wall furnace (see Section 1.7, Definitions) excludes floor furnaces, unit heaters, and central furnaces.

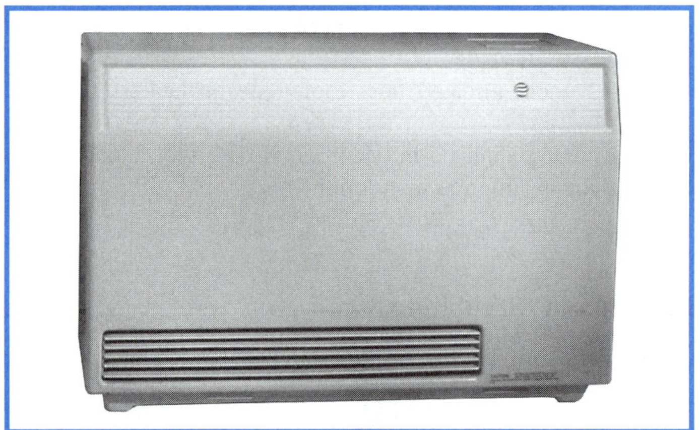

Exhibit 6.30 *A direct-vent wall furnace. (Courtesy of Empire Comfort Systems.)*

6.28.1 Installation.

(a) Listed wall furnaces shall be installed in accordance with their listing and the manufacturers' instructions. They shall be permitted to be installed in or attached to combustible material.

(b) Unlisted wall furnaces shall not be installed in or attached to combustible material.

(c) Vented wall furnaces connected to a Type B-W gas vent system listed only for a single story shall be installed only in single-story buildings or the top story of multistory build-

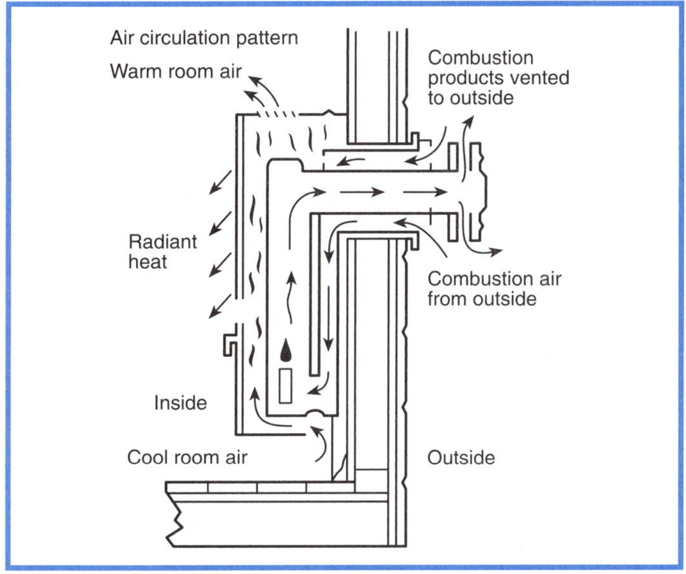

Exhibit 6.31 *Air circulation of a direct-vent wall furnace. (Courtesy of Williams.)*

Exhibit 6.32 *A counterflow wall furnace. (Courtesy of Empire Comfort Systems.)*

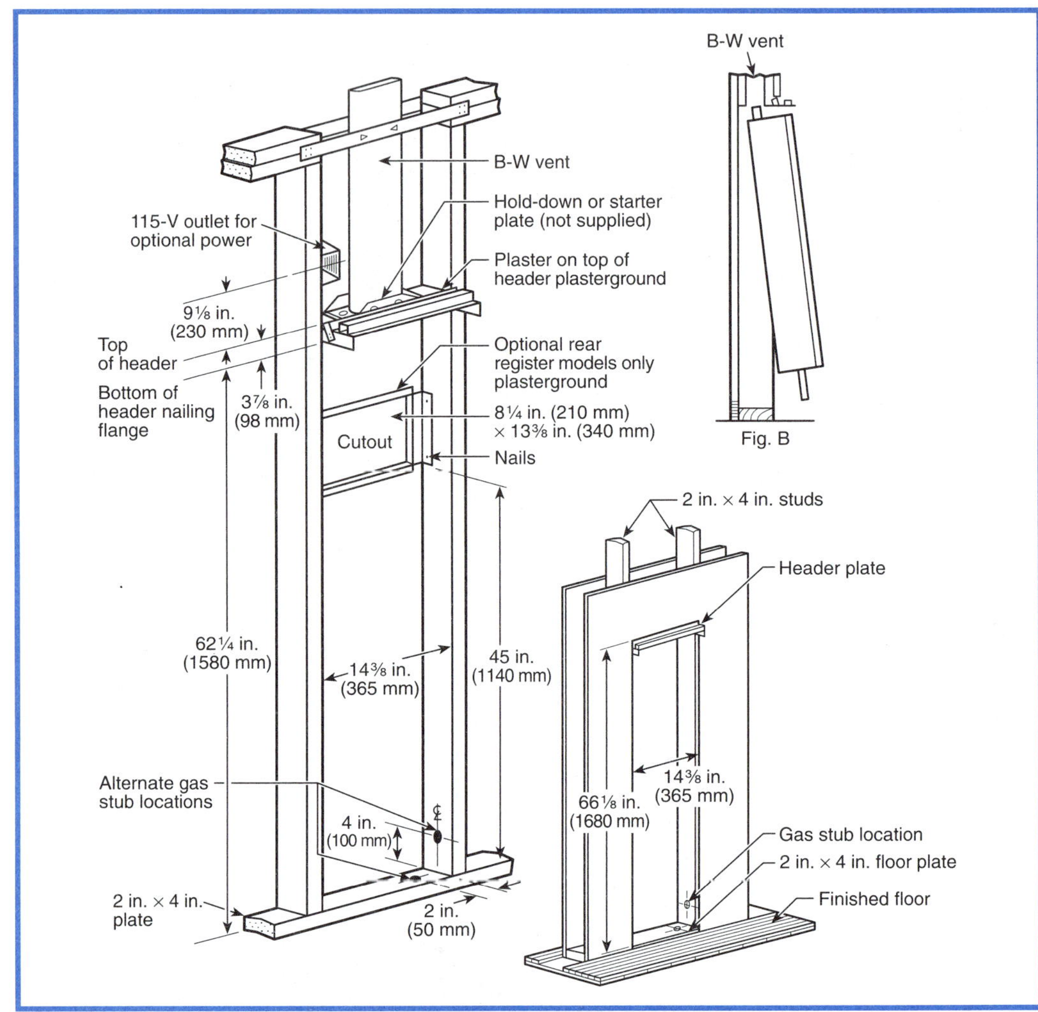

Exhibit 6.33 *Typical installation of a vented wall furnace. (Courtesy of Williams.)*

ings. Vented wall furnaces connected to a Type B-W gas vent system listed for installation in multistory buildings shall be permitted to be installed in single-story or multistory buildings. Type B-W gas vents shall be attached directly to a solid header plate that serves as a firestop at that point and that shall be permitted to be an integral part of the vented wall furnace. The stud space in which the vented wall furnace is installed shall be ventilated at the first ceiling level by installation of the ceiling plate spacers furnished with the gas vent. Firestop spacers shall be installed at each subsequent ceiling or floor level penetrated by the vent. *[See Figure 6.28.1(c) for Type B-W gas vent installation.]*

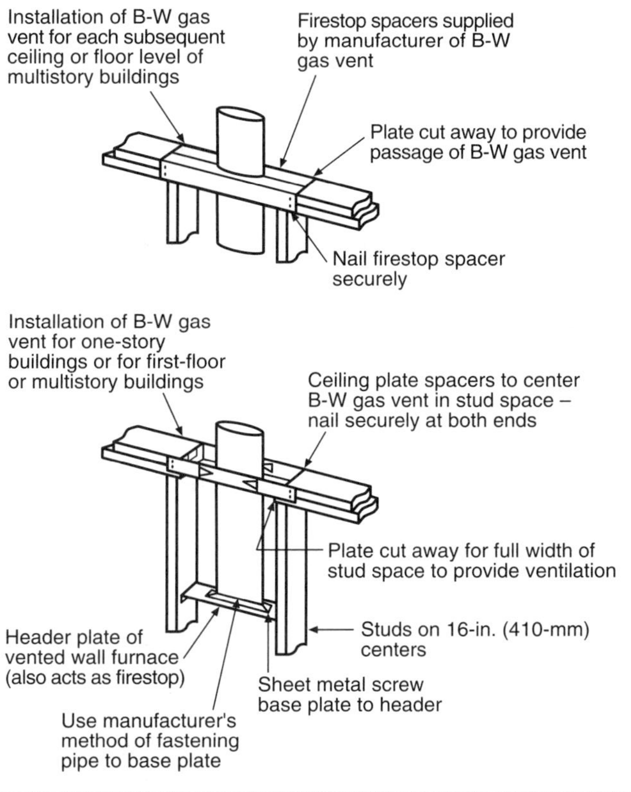

Figure 6.28.1(c) *Installation of Type B-W gas vents for vented wall furnaces.*

(d) Direct-vent wall furnaces shall be installed with the vent-air intake terminal in the outside atmosphere. The thickness of the walls on which the furnace is mounted shall be within the range of wall thickness marked on the furnace and covered in the manufacturers' installation instructions.

(e) Panels, grilles, and access doors that are required to be removed for normal servicing operations shall not be attached to the building.

For additional information on the venting of wall furnaces, consult Chapter 7 in this code.

6.28.2 Location.

Wall furnaces shall be located so as not to cause a hazard to walls, floors, curtains, furniture, or doors. Wall furnaces installed between bathrooms and adjoining rooms shall not circulate air from bathrooms to other parts of the building.

In addition to the location restrictions identified above, there must be adequate space in front of the wall furnace to service the controls.

6.28.3 Combustion and Circulating Air.

Combustion and circulating air shall be provided in accordance with Section 5.3.

6.29 Water Heaters

A water heater is an appliance for supplying hot water for domestic or commercial purposes. A direct-vent water heater is constructed and installed so that all air for combustion is obtained directly from the outside atmosphere and all flue gases are discharged to the outside atmosphere.

Questions continue to be raised on the installation of water heaters in manufactured housing. The 1999 edition of the code addresses this subject for the first time in Section 6.31, which makes the code applicable to the installation and replacement of gas-fired appliances in manufactured housing after the initial sale.

Small water heaters [e.g., 75,000 Btu/hr (22 kW) and less] are listed in accordance with ANSI Z21.10.1, *Gas Water Heaters — Volume I — Storage Water Heaters with Input Ratings of 75,000 Btu per Hour or Less,* which contains the requirements and tests for all storage water heaters of this size. Additional requirements are included in ANSI Z21.10.1 for water use heaters in manufactured housing. These requirements include provisions for securing the water heater to the vehicle structure, a requirement that the water heater be convertible for use with both natural gas and propane, special marking and installation requirements, and other construction requirements. Certification that the water heater is suitable for installation in manufactured housing means that the specific requirements for water heater construction in ANSI Z21.10.1 have been met. Both atmospheric-vent and direct-vent water heaters can be listed for installation in manufactured housing. While new manufactured housing is covered by standards of the federal government in the United States, replacement of appliances is covered by the building codes in effect at the site. The requirements of the *National Fuel Gas Code* now specifically apply to the installation of water heaters in manufactured housing (after consumer sale) and must be followed.

The decision to replace a direct-vent water heater in manufactured housing with a non-direct-vent type should be made carefully. It is necessary to comply with the provisions of Sections 5.3 and 6.31. The replacement appliance must be listed or have the approval of the local authority as required by Section 6.31. The proper clearance to combustibles must also be provided.

In addition to the water heater installation requirements in this section, the installer should be aware that requirements that prevent the installation of water heaters in certain locations where flammable vapors are or can be present are stated in Section 5.1 (See 5.1.8 through 5.1.11.) At the time of this writing, the ANSI Z21/83 committee has approved revisions to the next edition of Z21.10.1, *Gas Water Heaters – Volume I – Storage, Water Heaters with Input Ratings of 75,000 Btu per Hour or Less.* These revisions require that all new water heaters must pass a flammable vapor ignition resistance test in order to be listed. (Refer to the commentary following 5.1.8 and 5.1.9 for more information.)

6.29.1 Prohibited Installations.

(a) Water heaters shall not be installed in bathrooms, bedrooms, or any occupied rooms normally kept closed.

Exception No. 1: Direct-vent water heaters.

Exception No. 2: Water heaters shall be permitted to be installed in a closet located in a bathroom or bedroom where the closet has a weather-stripped solid door with a self-closing device and where all combustion air is obtained from the outdoors.

Exception No. 2 permits, with special requirements, the installation of water heaters in special equipment rooms or closets opening into bedrooms and bathrooms. The closet must be used only for a water heater, it must have a weather-tripped door with an automatic self-closing device, and all combustion air must be taken from outdoors per 5.3.3.

(b) Single-faucet automatic instantaneous water heaters, as permitted under 7.2.2 in addition to (a), shall not be installed in kitchen sections of light housekeeping rooms or rooms used by transients.

(c) See 5.1.8 for flammable vapors.

6.29.2 Location.

Water heaters of other than the direct-vent type shall be located as close as practical to the chimney or gas vent.

Direct-vent water heaters are vented the same way as other direct-vent gas appliances. For specific information, see Chapter 7, "Venting of Equipment," in the code. Direct-vent water heaters installed in confined spaces are not required to have air for combustion provided as specified in Section 5.3 of the code, but they do need ventilation air to keep the confined space from overheating.

6.29.3 Clearance.

(a) The clearances shall not be such as to interfere with combustion air, draft hood clearance and relief, and accessibility for servicing. Listed water heaters shall be installed in accordance with their listing and the manufacturers' instructions.

(b) Unlisted water heaters shall be installed with a clearance of 12 in. (300 mm) on all sides and rear. Combustible floors under unlisted water heaters shall be protected in an approved manner.

Refer to commentary on 6.2.4 and 6.3.2 for information on the protection of combustible floors.

6.29.4 Pressure-Limiting Devices.

A water heater installation shall be provided with overpressure protection by means of an approved device constructed, listed, and installed in accordance with the terms of its listing and the manufacturer's instructions.

The pressure setting of the device shall exceed the water service pressure and shall not exceed the maximum pressure rating of the water heater.

Water heaters are required to have overpressure protection to prevent internal overpressuring and failure of the heater. This protection is accomplished by installing on the heater a pressure relief valve with a nonadjustable setting of 75 psi–150 psi (520 kPa–1030 kPa), with 125 psi (862 kPa) being a commonly used pressure. Water heaters usually are designed and constructed with a 150 psi (1030 kPa) maximum operating pressure. The pressure relief valve setting must be greater than the water service pressure to prevent unnecessary operation of the valve. Most municipal water supplies have pressures of 80 psi (560 kPa) or less. In most cases, the overpressure protection device is provided and installed in accordance with the water heater manufacturer's installation instructions.

The pressure relief valve frequently is combined with a temperature relief valve that has a nonadjustable setting of 200°F–210°F (93.3°C–98.9°C). An extended temperature probe extends through the inlet of the relief valve, and the valve must be installed so that the temperature probe is immersed in water in the top 6 in. (150 mm) of a storage-type water heater. (See Exhibits 6.34, 6.35, and 6.36.)

6.29.5 Temperature-Limiting Devices.

Water heater installation or a hot water storage vessel installation shall be provided with overtemperature protection by means of an approved device constructed, listed, and installed in accordance with the terms of its listing and the manufacturers' instructions.

Water heaters are required to have overtemperature protection, which usually is accomplished by an energy cutoff (ECO) device, an independent, high-temperature switch that shuts off the gas valve. Modern units incorporate a switch, built into the thermostat, which can be a replaceable "fusible link" that cannot be reset. This fusible link is replaceable by a serviceman or, in the case of ECO operation, could require the replacement of the thermostat. Many water heater manufacturers require in their installation instructions that a combination temperature and pressure relief valve also be installed. (See the commentary on 6.29.4.)

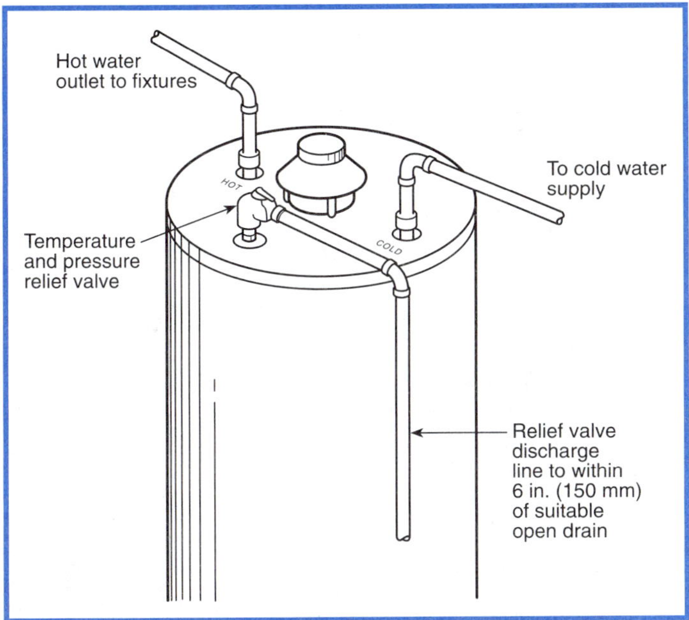

Exhibit 6.34 *Water heater. (Courtesy of Rheem.)*

Exhibit 6.35 *Temperature and pressure relief valve. (Courtesy of Watts Regulator Company.)*

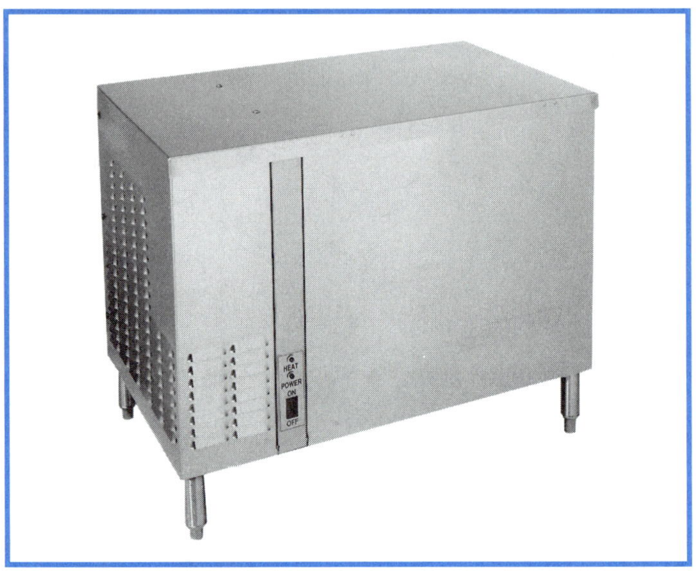

Exhibit 6.36 *Commercial instantaneous water heater. (Courtesy of Mighty Therm.)*

6.29.6 Temperature, Pressure, and Vacuum Relief Devices.

The installation of temperature, pressure, and vacuum relief devices or combinations thereof, and automatic gas shutoff devices, shall be in accordance with the terms of their listing and the manufacturers' instructions.

A shutoff valve shall not be placed between the relief valve and the water heater or on discharge pipes between such valves and the atmosphere.

The hourly Btu discharge capacity or the rated steam relief capacity of the device shall not be less than the input rating of the water heater.

Subsections 6.29.4 through 6.29.6 provide requirements for the installation of safety devices on water heaters. These devices must be installed in accordance with the manufacturer's instructions provided with the devices. Shutoff valves are prohibited in piping to or from relief valves to prevent their being compromised accidentally or intentionally. The discharge line (pipe) from such devices must be full sized and not reduced for its full length. Also, the end of the pipe should not be threaded. The lack of threads prevents the installation of a cap on the pipe.

6.29.7 Automatic Instantaneous Type: Cold Water Supply.

The water supply to an automatic instantaneous water heater that is equipped with a water flow–actuated control shall be such as to provide sufficient pressure to properly operate the control when water is drawn from the highest faucet served by the heater.

6.29.8 Circulating Tank Types.

(a) *Connection to Tank.* The method of connecting the circulating water heater to the tank shall provide proper circulation of water through the heater and permit a safe and useful temperature of water to be drawn from the tank. *[See Figure 6.29.8(a).]*

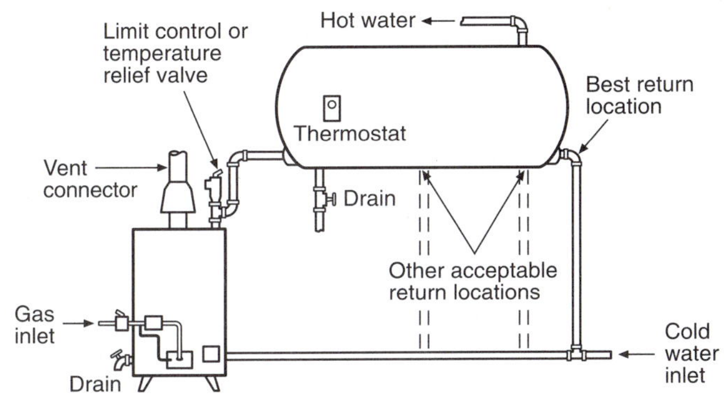

Figure 6.29.8(a) *Typical gravity circulating tank-type installation.*

(b) *Size of Water Circulating Piping.* The size of the water circulating piping shall conform with the size of the water connections of the heater.

(c) *Sediment Drain.* A suitable water valve or cock, through which sediment can be drawn off or the tank emptied, shall be installed at the bottom of the tank.

6.29.9* Antisiphon Devices.

Means acceptable to the authority having jurisdiction shall be provided to prevent siphoning in any water heater or any tank to which a circulating water heater that incorporates a cold water inlet tube is attached.

A.6.29.9 A hole near the top of a cold water inlet tube that enters the top of the water heater or tank is commonly accepted for this purpose.

6.30 Compressed Natural Gas (CNG) Vehicular Fuel Systems.

The installation of compressed natural gas (CNG) fueling (dispensing) systems shall conform with NFPA 52, *Standard for Compressed Natural Gas (CNG) Vehicular Fuel Systems.*

6.31 Appliances for Installation in Manufactured Housing.

Appliances installed in manufactured housing after the initial sale shall be listed for installation in manufactured housing, or approved, and shall be installed in accordance with the requirements of this code and the manufacturers' installation instructions.

Appliances installed in the living space of manufactured housing shall be in accordance with the requirements of 5.3.

This section was added in the 1999 edition to provide coverage in the *National Fuel Gas Code* for installation and replacement of appliances in manufactured housing after the initial sale. In the United States, the production of manufactured housing is regulated by the Federal Department of Housing and Urban Development (HUD). HUD has standards governing the production of manufactured housing that are applicable only to the manufacturer and to dealers prior to the first consumer sale. After the first consumer sale, the HUD standard is not applicable, and for many years no other codes or standards have covered installation and replacement of gas appliances in manufactured housing.

The committee recognized this lack and, after reviewing the existing code requirements, determined that they can be used for the safe installation of gas appliances in manufactured housing after the initial consumer sale.

References Cited in Commentary

The following publication is available from Engineering and Safety Service, 85 John Street, New York, NY 10038.

National Building Code, 1976 edition.

The following publications are available from CSA International, 8501 East Pleasant Valley Road, Cleveland, OH 44131.

ANSI B31.3, *Process Piping for Piping Above 125 psi (862 kPa),* 1993.
ANSI Z21.8, *Installation of Domestic Gas Conversion Burners,* 1994.
ANSI Z21.10.1, *Gas Water Heaters — Volume I — Storage Water Heaters with Input Ratings of 75,000 Btu per Hour or Less,* 1993.
ANSI Z21.13, *Gas-Fired Low-Pressure Steam and Hot Water Boilers,* 1994.
ANSI Z21.47, *Gas-Fired Central Furnaces,* 1993.
ANSI Z21.48, *Gas-Fired Gravity and Fan Type Floor Furnaces,* 1993.
ANSI Z21.50, *Vented Gas Fireplaces,* 1996.
ANSI Z83.11/CGA 1.8, *Gas Food Service Equipment,* 1996.
ANSI Z83.18, *Standard for Direct Gas-Fired Industrial Air Heaters,* 1992.
ANSI Z21.86, *Gas-Fired Gravity Space Heating Equipment,* 1998.

The following publication is available from McGraw-Hill, Inc., 1221 Avenue of the Americas, New York, NY 10020.

Perry, Robert H., and Cecil H. Chilton. *Chemical Engineers' Handbook,* fifth edition, 1973.

The following publications are available from the National Fire Protection Association, 1 Batterymarch Park, P.O. Box 9101, Quincy, MA 02269-9101.

NFPA 20, *Standard for the Installation of Stationary Pumps for Fire Protection,* 1999 edition.
NFPA 37, *Standard for the Installation and Use of Stationary Combustion Engines and Gas Turbines,* 1998 edition.
NFPA 70, *National Electrical Code®,* 1999 edition.
NFPA 90A, *Standard for the Installation of Air-Conditioning and Ventilating Systems,* 1999 edition.
NFPA 90B, *Standard for the Installation of Warm Air Heating and Air-Conditioning Systems,* 1999 edition.
NFPA 96, *Standard for Ventilation Control and Fire Protection of Commercial Cooking Operations,* 1998 edition.
NFPA 99, *Standard for Health Care Facilities,* 1999 edition.
ANSI A119.2/NFPA 501C, *Standard on Recreational Vehicles,* 1996 edition.

The following publication is available from the Sheet Metal and Air Conditioning Contractors National Association, P.O. Box 70, Merrifield, VA 22116.

SMACNA, *HVAC Duct Construction Standards — Metal and Flexible,* 1985.

The following publications are available from Underwriters Laboratories Inc., Publication Stock, 333 Pfingsten Road, Northbrook, IL 60062.

UL 181, *Standard for Safety Factory-Made Air Ducts and Air Connectors,* 1996.
UL 263, *Standard for Safety Fire Tests of Building Construction and Materials,* 1992.

Venting of Equipment

Chapter 7 covers venting equipment, including the construction and installation of venting equipment but not the sizing of the vents. This chapter and Chapter 10 comprise the venting portion of the *National Fuel Gas Code.* Note that Chapter 7 covers installation requirements; however, safe installation of venting systems requires engineering calculations to ensure that sufficient draft is present. The tables in Chapter 10 perform these calculations. They do not cover all vent sizes and configurations, and other methods can be used in lieu of the tables. The sections in Chapter 7 generally follow the procedures an installer would use to design or evaluate a venting system for use with a Category I, mid-efficiency furnace or boiler. The venting system is roughly divided into two parts. The main, vertical section of the venting system is generally a chimney or Type B vent. The main vent is attached to the appliance(s) by the vent connector, which often has a horizontal component. Each portion of the venting system has different requirements, which are detailed in the corresponding section of Chapter 7. The first four sections of Chapter 7 cover the following:

- General requirements (in Section 7.1)
- The specification for venting, which identifies appliances that must be vented and those that do not need to be vented (in 7.2)
- Design and construction of venting systems, including some general design and construction requirements and some more detailed design and construction requirements for mechanical draft systems, ventilating hoods, and exhaust systems (in Section 7.3)
- The type of venting system to be used for the different types of appliances, provided in a table (in Section 7.4)

The next three sections cover material and installation requirements for the three types of venting systems used for conventional venting systems. The driving force for venting is the heat of the vent gases in these conventional (Category I) venting systems. These sections cover the following:

- Masonry, metal, and factory-built chimneys, including requirements for construction, termination, sizing (by reference to Chapter 10), inspection, and cleaning of chimneys. Also included are detailed diagrams specifying proper chimney termination, making these requirements easily understood. Chimneys can be listed or unlisted (in Section 7.5).
- Gas vents, including requirements for sizing, termination, multiple equipment venting, and gas vent support and marking. Gas vents are always listed (in Section 7.6).
- Single-wall metal pipe, including requirements for construction, termination installation, sizing, support, and marking of single-wall metal pipe. Single-wall metal pipe is not listed (in Section 7.7).
- Venting system location, which provides requirements for locating the termination of vent systems exiting through the sides of buildings relative to other building openings (Section in 7.8)
- Condensation drain (in Section 7.9)

The next three sections cover material and installation requirements for the two types of connectors used in Category I venting systems. Vent connectors connect gas utilization equipment to vents. Vents are usually vertical and connectors are usually horizontal. These sections cover the following:

- Vent connectors for Category I gas utilization equipment, including requirements for installation, materials, sizing, and clearance to combustibles for vent connectors used with conventionally vented equipment (in Section 7.10)
- Vent connectors for Category II, Category III, and Category IV, which are the types of appliances that are not "utilization equipment." Vent connectors for these types of appliances are specified in Section 7.4 (in Section 7.11).
- Draft hoods and draft controls, which defines which gas utilization equipment requires draft hoods and their installation and covers other draft control devices and their installation (in Section 7.12)
- Manually operated dampers, which are limited to venting systems serving certain larger gas-fired equipment (in Section 7.13)
- Automatically operated vent dampers, which must be listed (in Section 7.14)
- Obstructions, which are not permitted except for (1) approved installations of listed draft regulators and safety controls and (2) heat recovery systems, which are generally used with larger equipment, such as industrial boilers (in Section 7.15)

7.1 General

This chapter recognizes that the choice of venting materials and the methods of installation of venting systems are dependent on the operating characteristics of the gas utilization equipment. The operating characteristics of vented gas utilization equipment can be categorized with respect to (1) positive or negative pressure within the venting system and (2) whether or not the equipment generates flue or vent gases that can condense in the venting system. See Section 1.7 for the definition of these vented appliance categories.

Although gas is a clean-burning fuel, the products of combustion must not be allowed to accumulate inside of a building. (See Exhibit 7.1.) Therefore, venting of most gas utilization equipment is required. (See 7.2.2 for a list of appliances not requiring venting.) A properly installed and maintained venting system should perform the following functions to provide for proper appliance function and the safety of building occupants:

(1) Convey all of the combustion products to the outside atmosphere
(2) Prevent damage to the gas equipment, vent, building, and furnishings from water vapor condensation in the flue gases
(3) Prevent overheating of walls, building structure, and other combustible materials that are installed with required clearance to the appliance and venting system
(4) Provide fast priming of natural draft venting systems to minimize spillage of combustion products into the building

The operation of a venting system may appear complicated, but remember that natural draft venting works because HOT AIR RISES. When gas is burned, the products of combustion are hotter than the ambient air and, therefore, rise. The venting system's job is to channel these combustion products out of the building. To do so, it

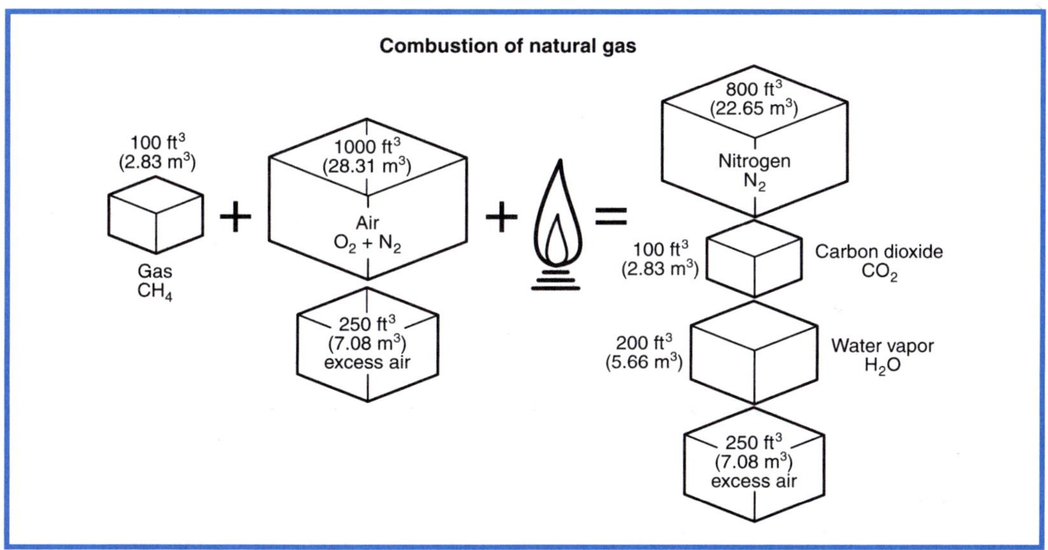

Exhibit 7.1 *Combustion of natural gas. (Courtesy of Gas Vent Institute.)*

must keep the combustion products as warm as possible in the vent. Heat maximizes the draft produced.

Prior to the 1980s when the Department of Energy's minimum appliance efficiency regulation mandated higher-efficiency appliances, gas furnaces had a seasonal efficiency of about 60 percent. This meant that approximately 40 percent of the heat of combustion remained in the vent gases (the flue loss), providing ample heat for venting and for controlling condensation. With higher-efficiency space-heating appliances, the old safety margin no longer applies, and extra care must be taken so that the vents operate properly.

The mid-efficiency appliances available today are known as Category I appliances and operate at an efficiency of about 80 percent. Some use a fan to assist the flow of combustion products through the appliance and are not equipped with a draft hood. Other appliances are equipped with a draft hood but reduce their total flue losses by using an automatic damper to cut off the flow in the vent during the off-cycle. Problems with early models and vent system failures were traced to condensation and resulted in the development of new venting tables, shown in Chapter 10, to provide proper sizing of vents for these appliances. For more details on these considerations, see Supplement 1 of this handbook.

Water heaters must also meet minimum efficiency requirements, but they are still typically draft hood equipped. Vent sizing for draft hood–equipped water heaters (and other appliances) connected to a dedicated venting system did not require an extensive reevaluation. However, the size of common vents serving water heaters and Category I fan-assisted appliances has changed; sizing is to be in accordance with the venting tables in Chapter 10.

High-efficiency condensing appliances have an efficiency of 90 percent or higher, which reduces vent gas temperatures to a point where the water vapor produced as a product of combustion condenses to liquid water in the appliance or in the vent. These condensing appliances carry a vented appliance category (see Section 1.7, Definitions) of Category IV. This type of appliance produces much cooler vent gases, resulting in water condensing in the vent. Venting must be accomplished with a fan because the vent gases are not hot enough to operate the natural draft vent. The vent materials used with these appliances must be able to resist the acidic condensate.

The advantage of these new appliances is that they reduce the amount of gas that is consumed with no loss in output. A mid-efficiency appliance uses one-third less gas than a conventional appliance, and a condensing appliance uses only one-half of the gas of a conventional appliance.

7.2 Specification for Venting

7.2.1 Connection to Venting Systems.

Except as permitted in 7.2.2 through 7.2.6, all gas utilization equipment shall be connected to venting systems.

7.2.2 Equipment Not Required to Be Vented.

(a) Listed ranges

(b) Built-in domestic cooking units listed and marked for optional venting

(c) Listed hot plates and listed laundry stoves

(d) Listed Type 1 clothes dryers *(see 6.4.4 for exhausting requirements)*

(e) A single listed booster-type (automatic instantaneous) water heater, when designed and used solely for the sanitizing rinse requirements of a dishwashing machine, provided that the equipment is installed, with the draft hood in place and unaltered, if a draft hood is required, in a commercial kitchen having a mechanical exhaust system; where installed in this manner, the draft hood outlet shall not be less than 36 in. (910 mm) vertically and 6 in. (150 mm) horizontally from any surface other than the equipment.

(f) Listed refrigerators

(g) Counter appliances

(h) Room heaters listed for unvented use *(See 6.24.1 and 6.24.2.)*

(i) Direct gas-fired makeup air heaters

(j) Other equipment listed for unvented use and not provided with flue collars

(k) Specialized equipment of limited input such as laboratory burners or gas lights

Where any or all of the equipment in 7.2.2(e) through (k) is installed so the aggregate input rating exceeds 20 Btu/hr/ft^3 (207 W/m^3) of room or space in which it is installed, one or more shall be provided with venting systems or other approved means for removing the vent gases to the outside atmosphere so the aggregate input rating of the remaining unvented equipment does not exceed the 20 Btu/hr/ft^3 (207 W/m^3) figure. Where the room or space in which the equipment is installed is directly connected to another room or space by a doorway, archway, or other opening of comparable size that cannot be closed, the volume of such adjacent room or space shall be permitted to be included in the calculations.

The equipment listed in this section does not require venting. Note that some of the appliances [listed in (e) through (k)] must be vented if the total input of all the appliances in this class exceeds 20 Btu/hr per ft^3 (207 W per m^3) of room volume. This ratio is equivalent to that in the definition of Space, Confined in Section 1.7, which refers to a ratio of 50 ft^3 per 1000 Btu/hr (4.8 m^3/kW). For further information see the commentary following 5.3.3.

Item (d), clothes dryers, was revised in the 1996 edition of the code to add a reference to Section 6.4.4. The addition was made to highlight that clothes dryers are exhausted, rather than vented, because the products of combustion and the moisture removed from the clothing are combined and exhausted together. This exhaust is powered to ensure proper operation. Clothes dryer exhausts and appliance vents must never be combined.

Certain appliances that are installed in unconfined spaces (see Section 1.7 for definition) are not required to be connected to a venting system for reasons such as their having limited input, or operating for short periods during infrequent intervals, or being in locations that have means to dilute the combustion products to safe levels. Listed, unvented appliances have been tested by the listing agency to determine that the burners, when properly adjusted, burn cleanly without producing unacceptable levels of dangerous combustion products. However, unvented appliances should be operated in rooms with ample air for combustion, and the burners should be maintained in proper adjustment. Except for decorative appliances, a yellow flame is indicative of improper combustion. Sooting or carbon buildups are another indication. Usually, this fault can be corrected by one of the following:

(1) Properly adjusting the burner
(2) Cleaning the burner
(3) Admitting adequate combustion air into the room

Hot plates, laundry stoves, ranges and other cooking units, and so forth, must never be used for room heating. Direct gas-fired makeup heaters are intended for industrial or commercial applications and are designed so that the products of combustion are diluted with large quantities of fresh air.

Unvented room heaters and fireplaces should be listed, installed, and operated strictly in accordance with the manufacturer's instructions, which are required to be furnished with each heater. Listed, unvented room heaters and fireplaces are equipped with an oxygen depletion safety shutoff system designed to shut off the gas supply to the heater before the oxygen in the surrounding air is reduced below a safe level. However, in all cases, and as specified in the manufacturer's instructions, the room in which the heater is located must be provided with means for supplying fresh air for combustion and ventilation.

7.2.3* Ventilating Hoods.

Ventilating hoods and exhaust systems shall be permitted to be used to vent gas utilization equipment installed in commercial applications *(see 7.3.5)* and to vent industrial equipment, particularly where the process itself requires fume disposal. *(See 5.1.6 and 5.1.8.)*

A common application of this provision is in restaurant kitchens, where very hot water is required for dishwashing for sanitary reasons. The gas supply to the booster is interlocked with the hood, usually with a normally closed solenoid valve in the gas line that is opened only when the hood is running. This interlock is needed to meet the requirements of 7.3.4 (d), and 7.3.5. Refer to commentary following 7.3.5 for more information on this application.

A.7.2.3   Information on the construction and installation of ventilating hoods can be obtained from NFPA 96, *Standard for Ventilation Control and Fire Protection of Commercial Cooking Operations.*

Information on the design and installation of ventilating hoods in industrial plants can be obtained from NFPA 91, *Standard for Exhaust Systems for Air Conveying of Vapors,*

Gases, Mists, and Noncombustible Particulate Solids. Information on ventilating hoods for restaurants and commercial kitchens may be obtained from NFPA 96, *Standard for Ventilation Control and Fire Protection of Commercial Cooking Operations.*

7.2.4 Well-Ventilated Spaces.

Where located in a large and well-ventilated space, industrial gas utilization equipment shall be permitted to be operated by discharging the flue gases directly into the space.

This unusual provision is an exception to the rule stated in 7.2.1 that all gas utilization equipment be connected to a venting system. It recognizes common, safe practice in metal-treating and other industries. Some of the equipment normally installed under the requirements of this paragraph are heat-treating furnaces, radiant tube burners, and pot heaters in foundries. Building ventilation must be designed to dilute flue gases to a safe level by means of natural draft or mechanical ventilation. In many of these installations, ventilation for fume removal or process reasons exceeds the ventilation level needed for safe removal of the products of combustion.

7.2.5 Direct-Vent Equipment.

Listed direct-vent gas utilization equipment shall be considered properly vented where installed in accordance with the terms of its listing, the manufacturers' instructions, and 7.8(c).

Direct-vent equipment is designed and manufactured to draw all air for combustion directly from the outdoors. In addition, all products of combustion are ejected directly to the outdoors, in the same pressure zone as the inlet. This design completely separates the products of combustion from the building in which the appliance is located.

The venting system is part of a direct-vent appliance and is considered to be part of its listing. When the appliance is listed, it is tested with inlet and outlet configurations taken from its installation directions. Therefore, the installation instructions must be carefully followed. Exhibit 7.2 shows one type of a direct-vent appliance, which uses a concentric vent and air supply. Some direct-vent appliances use independent ducts for the inlet and outlets.

Note that this section applies to listed direct-vent appliances, which are the most likely to be encountered. Section 7.8 provides requirements for the location of the vent terminations of direct-vent appliances.

7.2.6 Equipment with Integral Vents.

Gas utilization equipment incorporating integral venting means shall be considered properly vented where installed in accordance with its listing, the manufacturers' instructions, and 7.8(a) and (b).

An integral vent is a vent that is supplied with a gas appliance by the manufacturer.

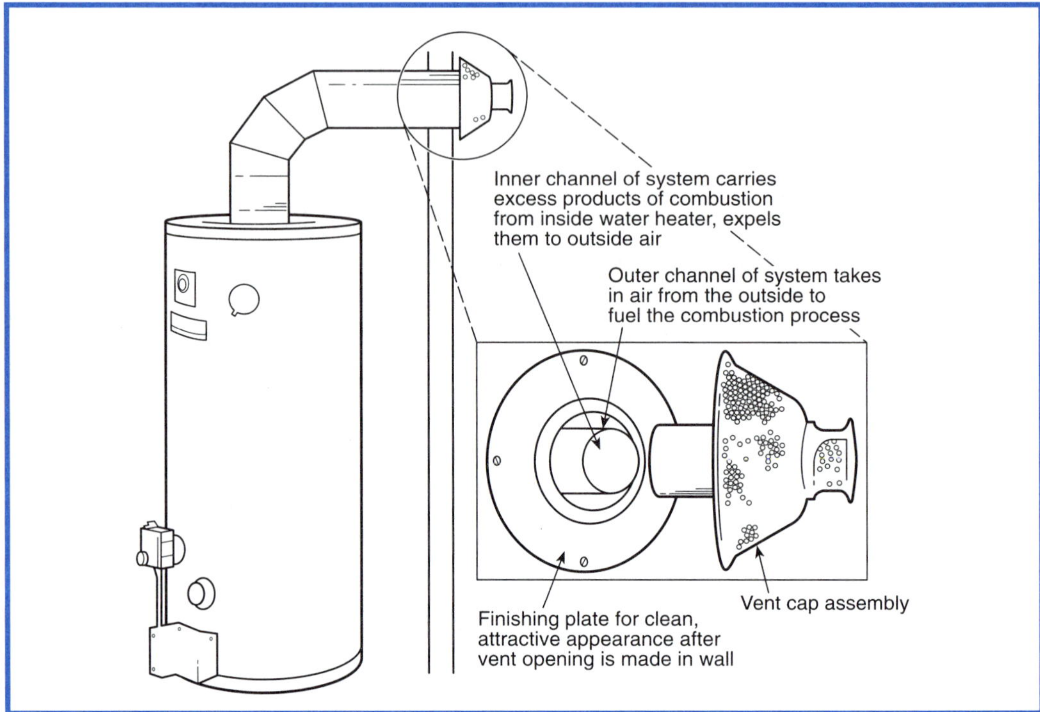

Exhibit 7.2 *Direct-vent water heater using a concentric vent and supply.*

7.3 Design and Construction

7.3.1 Minimum Safe Performance.

A venting system shall be designed and constructed so as to develop a positive flow adequate to remove flue or vent gases to the outside atmosphere.

7.3.2 Equipment Draft Requirements.

A venting system shall satisfy the draft requirements of the equipment in accordance with the manufacturer's instructions.

The principle on which natural draft vents operate is simple. Reduced to fundamentals, heat is the power that operates a natural draft vent or chimney. (See Exhibit 7.3.) Combustion gases rise in the chimney or vent only because they are hotter, and, therefore, lighter, than the surrounding air. The hotter the gases and the higher the vent, the more swiftly and powerfully they will rise. Conversely, the cooler the gases and the shorter the vent, the more sluggish their movement will be. The flue gases in the vent must remain hot enough over the length of the vent to provide a strong draft. If cooled enough, the upward motion stops altogether and combustion gases can spill into the building through the relief opening of the appliance draft hood.

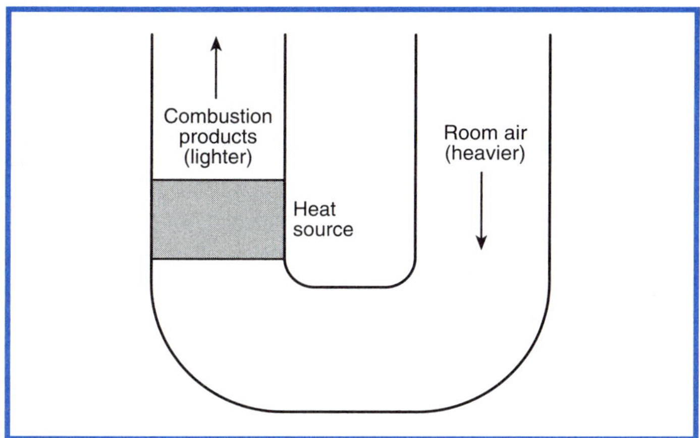

Exhibit 7.3 *Motive force in vents. (Courtesy of Gas Vent Institute.)*

When the combustion products and dilution air rise in the vent, this volume must be replaced by air from outside the building. This replacement air can be supplied through normal air infiltration (through small openings in the building walls), through outdoor openings purposely installed in outside walls, or by a mechanical air system. So, to ensure proper vent operation, the building's air tightness and other devices exhausting air from the building must be taken into account.

Fan-assisted combustion appliances cause some confusion because they are listed as Category I appliances, meaning that their vents operate by natural draft due to the heat of the vent gas. However, they have a fan to assist the flow of the combustion product through the appliance. The fan is necessary because modern, higher-efficiency appliances have a higher pressure drop through their heat exchangers. The pressure provided by the induced draft blower is carefully matched to the resistance of the heat exchanger. Once the combustion products exit the appliance, the natural buoyancy takes over in the vent. If the vent is designed using the tables of Chapter 10, the pressure in the vent system will be negative.

Some Category I furnaces are listed as Category I or Category III appliances, depending on the method used to install the system. Category III installation requires using the manufacturer's installation instructions.

Blockage of a natural draft vent may cause flue gases to spill into the building through the relief opening of the draft hood. Spillage can also be caused if the pressure in the vicinity of the appliance is much lower than the pressure outside the building (depressurization). Depressurization may be caused by mechanical exhausts, fireplaces, wind — anything that removes air from the building. Consequently, periodically verifying the performance of appliance venting systems is wise. This can be done readily with a natural draft system as follows (see Appendix H):

Operate the appliance for at least five minutes to allow the flue gases to heat the vent. Move a lighted match across the entire width of the draft hood relief opening (see Exhibit 7.4). If the match flame is blown downward or extinguished, have the venting system or chimney checked by a qualified agency. (See Section 1.4.) If the appliance is not equipped with a draft hood, you might not be able to perform the match test. In this case, consult the appliance instruction manual.

Good air circulation in adequate amounts is also vital for good venting and efficient appliance operation. Signs of improper operation include a yellow or wavering flame, discoloration around access doors, a pungent odor, or soot near the burner or vent area. If any of these conditions are observed, call a qualified agency to have the installation inspected.

Under extremely adverse conditions, carbon monoxide can be produced as a result of either improper venting of combustion products, insufficient fresh air to support the proper burning of gas, or improperly adjusted appliances. If symptoms of carbon monoxide poisoning are experienced—headache, yawning, ringing in ears, weariness, vomiting, or heart fluttering or throbbing—get fresh air promptly. Open windows and go outdoors. Shut off any gas appliance that you suspect is operating improperly and immediately have it checked by a qualified agency.

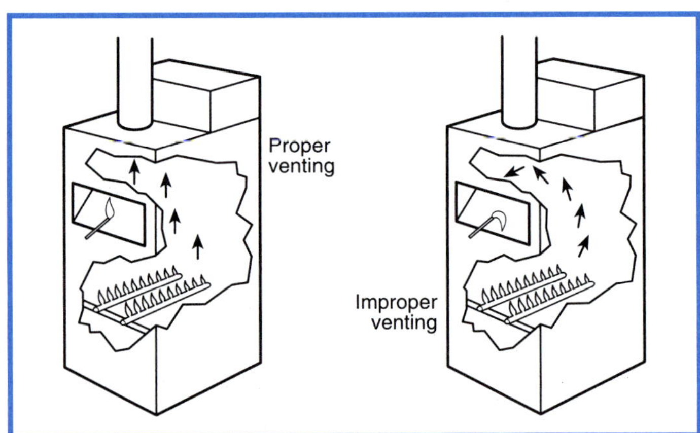

Exhibit 7.4 *Match test to determine proper and improper venting. (Courtesy of Gas Vent Institute.)*

7.3.3 Design and Construction.

Gas utilization equipment required to be vented shall be connected to a venting system designed and constructed in accordance with the provisions of Sections 7.4 through 7.15.

7.3.4 Mechanical Draft Systems.

Mechanical draft systems use the mechanical force from a blower to vent the combustion products. Natural buoyancy is not needed or significant. Mechanical draft systems can be roughly divided as being either forced or induced draft. A forced-draft system has the blower at the inlet of the venting system. An induced-draft system's blower is near the outlet. Typically, any section of vent downstream of the blower will be under a positive gauge pressure. Therefore, care must be taken to ensure that the vent pipe used is appropriate for positive-pressure combustion gases. Otherwise, leaks of the combustion products may occur. A mechanical draft water heater is shown in Exhibit 7.5.

Exhibit 7.5 *A mechanically vented water heater. (Courtesy of A. O. Smith Corporation.)*

(a) Gas utilization equipment requiring venting shall be permitted to be vented by means of mechanical draft systems of either forced or induced draft design.

Exception: Incinerators.

(b) Forced draft systems and all portions of induced draft systems under positive pressure during operation shall be designed and installed so as to prevent leakage of flue or vent gases into a building.

(c) Vent connectors serving equipment vented by natural draft shall not be connected into any portion of mechanical draft systems operating under positive pressure.

(d) Where a mechanical draft system is employed, provision shall be made to prevent the flow of gas to the main burners when the draft system is not performing so as to satisfy the operating requirements of the equipment for safe performance.

(e) The exit terminals of mechanical draft systems shall be not less than 7 ft (2.1 m) above grade where located adjacent to public walkways and shall be located as specified in 7.8(a) and (b).

Sections 7.8(a) and 7.8(b) specify the location of vent terminals in relation to air inlets, doors, and windows. The information is shown also in Figure A.7.8.

(f) Mechanical draft systems shall be installed in accordance with the terms of their listing and the manufacturers' instructions.

This paragraph provides the installer with guidance on the sizing of mechanical draft systems. This need is recognized because of the increased use of power vent kits to vent appliances. The manufacturer of a listed power vent kit provides sizing information in the instructions provided with the kit.

7.3.5* Ventilating Hoods and Exhaust Systems.

This section recognizes a safe, alternate method for venting of gas utilization equipment in commercial applications. A common application of this section is in restaurant kitchens, where a "booster" water heater is used to provide very hot water for sanitation. In this application, the water heater is operated only while the restaurant kitchen is in operation, and the range hood is used to vent both the range and the water heater. Range hoods are covered under NFPA 96, *Standard for Ventilation Control and Fire Protection of Commercial Cooking Operations,* which requires an interlock that operates when the airflow falls below a predetermined value. This circuit can be interlocked with the main gas valve of the water heater (or other gas appliance vented via the range hood) so that gas will not flow to the main burners unless a minimum airflow is achieved in the vent system.

(a) Ventilating hoods and exhaust systems shall be permitted to be used to vent gas utilization equipment installed in commercial applications.

(b) Where automatically operated gas utilization equipment is vented through a ventilating hood or exhaust system equipped with a damper or with a power means of exhaust, provisions shall be made to allow the flow of gas to the main burners only when the damper is open to a position to properly vent the equipment and when the power means of exhaust is in operation.

A.7.3.5 See A.7.2.3.

7.3.6 Circulating Air Ducts and Plenums.

No portion of a venting system shall extend into or pass through any circulating air duct or plenum.

7.4 Type of Venting System to Be Used

7.4.1

The type of venting system to be used shall be in accordance with Table 7.4.1.

To use Table 7.4.1, locate in the left-hand column the type of gas utilization equipment to be vented and read across the row to find the type(s) of venting system(s) that is permitted. For example, listed Category I equipment can be vented using a Type B gas vent, chimney, single-wall metal pipe, chimney lining system that is listed for gas venting, or a special gas vent listed for the equipment.

The table refers to appliance Category I through Category IV. The categories are based on vent temperature and pressure (see Table 7.1, Appliance Vent Categories). In the 1992 edition, the definitions were changed from a specific temperature break-

Table 7.1 Appliance Vent Categories

Appliance Category[3]	Vent Pressure	Temperature Above/ Below Defined in Relevant ANSI Standard	Comment
I	Nonpositive[1]	Above	Natural draft venting
II	Nonpositive[1]	Below	Materials must be corrosion resistant. Condensate must be drained.
III	Positive[2]	Above	Vent must be gas tight.
IV	Positive[2]	Below	Vent must be liquid tight and gas tight. Condensate must be drained.

[1] The term *nonpositive vent pressure* means that even if fans or blowers are used in the appliance or vent systems, venting is accomplished by natural draft. (The vent pressure is lower than the atmospheric pressure.)

[2] The term *positive vent pressure* means that fans, blowers, or other means are used to propel vent gases through the vent at above atmospheric pressure.

[3] The newer models of appliances will be identified as Category I, II, III, or IV on the nameplate on the appliance and will be stated as such in the manufacturer's installation instructions.

Table 7.4.1 Type of Venting System to Be Used

Gas Utilization Equipment	Type of Venting System
Listed Category I equipment	Type B gas vent (7.6)
Listed equipment equipped with draft hood	Chimney (7.5) Single-wall metal pipe (7.7)
Equipment listed for use with Type B gas vent	Listed chimney lining system for gas venting [7.5.1(c)]. Special gas vent listed for this equipment (7.4.3)
Listed vented wall furnaces	Type B-W gas vent (7.6, 6.28)
Category II equipment Catcgory III equipment Category IV equipment	As specified or furnishcd by manufacturers of listed equipment (7.4.2, 7.4.3)
Incinerators, outdoors	Single-wall metal pipe [7.7, 7.7.3(c)]
Incinerators, indoors Equipment that can be converted to use of solid fuel Unlisted combination gas- and oil-burning equipment Combination gas- and solid-fuel-burning equipment Equipment listed for use with chimneys only Unlisted equipment	Chimney (7.5)
Listed combination gas- and oil-burning equipment	Type L vent (7.6) or chimney (7.5)
Decorative appliancc in vented fireplace	Chimney [6.6.2(2)]
Gas-fired toilets	Single-wall metal pipe (7.7, 6.26.3)
Direct-vent equipment	See 7.2.5
Equipment with integral vent	See 7.2.6
Equipment in commercial and industrial installations	Chimney, ventilating hood, and exhaust system (7.3.5)

point for all appliances of 140°F (78°C) above dewpoint to the temperature specified in the relevant ANSI standard for each appliance.

The criteria in the ANSI Z21 standards for appliance categorization are based on a flue loss of 17 percent of total energy. The 17 percent flue loss is the same flue loss built into the vent sizing tables for fan-assisted appliances. In this way, the standards ensure that the appliance will work properly with the vent system.

In Table 7.1, the term nonpositive vent pressure means that the pressure in the vent will be lower than the surrounding atmosphere if the vent system meets the requirements of Chapter 7 and Chapter 10. The incorporation of a fan into the appliance does not always mean that the vent pressure is positive. If unsure, check the appliance nameplate or manufacturer's instructions for the venting category or check the vent pressure with a manometer or other pressure gauge when the appliance is operating.

7.4.2 Plastic Piping.

Approved plastic piping shall be permitted to be used for venting equipment listed for use with such venting materials.

Before the introduction of high-efficiency gas utilization equipment, plastic piping was prohibited as a vent material. High-efficiency (Category IV) appliances reduced vent temperatures, resulting in condensate formation. As accumulation of condensate became a source of corrosion of metal vents, plastic piping became the preferred material. Note that plastic vent materials can be used for listed gas utilization equipment only when specified in the manufacturer's instructions.

7.4.3 Special Gas Vent.

Special gas vent shall be listed and installed in accordance with the terms of the special gas vent listing and the manufacturers' instructions.

All special gas vents are listed vent materials. They are listed in accordance with UL 1738, *Standard for Safe Venting Systems for Gas-Burning Appliances, Categories II, III and IV.* Installation instructions for special gas vents include limitations on operating temperature, categories of appliance to be used with each vent, clearance to combustible materials, types of fittings and joint sealant to be used, and vent termination requirements.

Special attention should be given to the following areas:

(1) Proper support for the special vent to prevent sagging and to allow for expansion, contraction, and condensate drainage
(2) Proper cutting and cleaning of joints and fittings, and the use of recommended joint sealants (substitutes are not usually permitted)
(3) Construction of a condensate trap (see appliance manufacturer's instruction for special requirements)
(4) Wall penetrations (do not secure the pipe at a thimble as the pipe must be allowed to move to accommodate expansion and contraction)
(5) Insulation [do not insulate the vent pipe or the fittings of the inside of a wall thimble when polymeric (nonmetallic) vent materials are used]

Product Recall

Over the last ten years, a class of special gas vent known as "High Temperature Plastic Vent" (HTPV) was introduced to the market for use with mid-efficiency appliances. The field experience has been that these vent systems are prone to failure. The failure may occur because of improper installation practice and/or corrosion from acidic condensate.

At the time of this writing, an active product recall was under way, with the cooperation of the U.S. Consumer Product Safety Commission, appliance manufacturers, and the manufacturers of the vent pipe. The product recall covers furnaces that are horizontally vented and all boiler installations. If you encounter one of these vent systems, you should call (800) 758-3688 for information on how to proceed. This number is operated by the product manufacturers, and the editor is advised that it will be in operation until the recall is substantially complete. If the number is not in operation, questions can be referred to the furnace, boiler, or water heater manufacturer.

To determine whether the installation has an HTPV pipe system that is subject to this program, first check the vent pipes attached to the natural gas or propane furnaces or boilers. Vent pipes subject to this recall program can be identified as follows:

- The vent pipes are plastic.
- The vent pipes are colored gray or black.
- The vent pipes have the names "Plexvent," "Plexvent II," or "Ultravent" stamped on the vent pipe or printed on stickers placed on pieces used to connect the vent pipes together.

Also, check the location of those vent pipes. For furnaces, only HTPV systems that have vent pipes that go through the sidewalls of structures (horizontal systems) are subject to this program. Other plastic vent pipes, such as white PVC or CPVC, are not involved in this program.

7.5 Masonry, Metal, and Factory-Built Chimneys

This section applies to chimneys of all types. Not included are gas vents (including Type B vents), which are covered in Section 7.6, and single-wall metal pipe, which is covered in Section 7.7.

7.5.1 Listing or Construction.

(a) Factory-built chimneys shall be installed in accordance with their listing and the manufacturers' instructions. Factory-built chimneys used to vent appliances that operate at positive vent pressure shall be listed for such application.

Factory-built chimneys consist entirely of factory-built parts, such as chimney sections, supports, thimbles, flashings, caps, and other parts required to complete a particular installation. The parts are designed to be assembled with other parts of the same model without requiring field alteration or construction. All of the parts for a particular model as described in the chimney manufacturer's instructions are to be used.

The following three types of factory-built chimneys are available:

(1) Residential-type and building heating appliance chimneys that are intended for venting flue gases at a temperature not exceeding 1000°F (538°C) under continuous operating conditions

(2) 1400°F Type chimneys for venting flue gases at a temperature not exceeding 1400°F (760°C) under continuous operating conditions

(3) Medium-heat appliance chimneys for venting flue gases at a temperature not exceeding 1800°F (982°C)

The installation of chimneys serving appliances with vent temperatures over 100°F (538°C) is not covered in this code. See NFPA 211, *Standard for Chimneys, Fireplaces, Vents, and Solid Fuel-Burning Appliances,* which also describes the various types of heating appliances mentioned here.

Factory-built chimneys must be installed in accordance with the terms of their listing and the manufacturer's instructions, which are furnished with each chimney assembly and have been verified by the listing agency. The last sentence of the requirement was added in the 1999 edition to recognize that chimneys suitable for positive pressure are now available. Recently, the product standard for chimneys, UL 103, *Factory-Built Chimneys for Residential Type and Building Heating Appliances,* has added separate tests to verify the proper operation of factory-built chimneys that are designed for positive combustion-product pressure.

Category II, III, and IV appliances are not suitable for venting by the conventional method of natural draft venting. These appliances must not be connected to chimneys. They should be vented in accordance with the appliance manufacturer's instructions, which cover the type, size, and length of material to be used; how the venting system is to be installed; and the minimum clearance to combustible material. These instructions have been verified by the listing agency.

For descriptions of Type B, Type B-W, and Type L vents, refer to the commentary on Section 7.6. Diagrams of typical installations of residential-type chimneys are included in Exhibits 7.6 through 7.9.

(b) Metal chimneys shall be built and installed in accordance with NFPA 211, *Standard for Chimneys, Fireplaces, Vents, and Solid Fuel-Burning Appliances,* or local building codes.

(c)* Masonry chimneys shall be built and installed in accordance with NFPA 211, *Standard for Chimneys, Fireplaces, Vents, and Solid Fuel-Burning Appliances,* or local building codes and lined with approved clay flue lining, a listed chimney lining system, or other

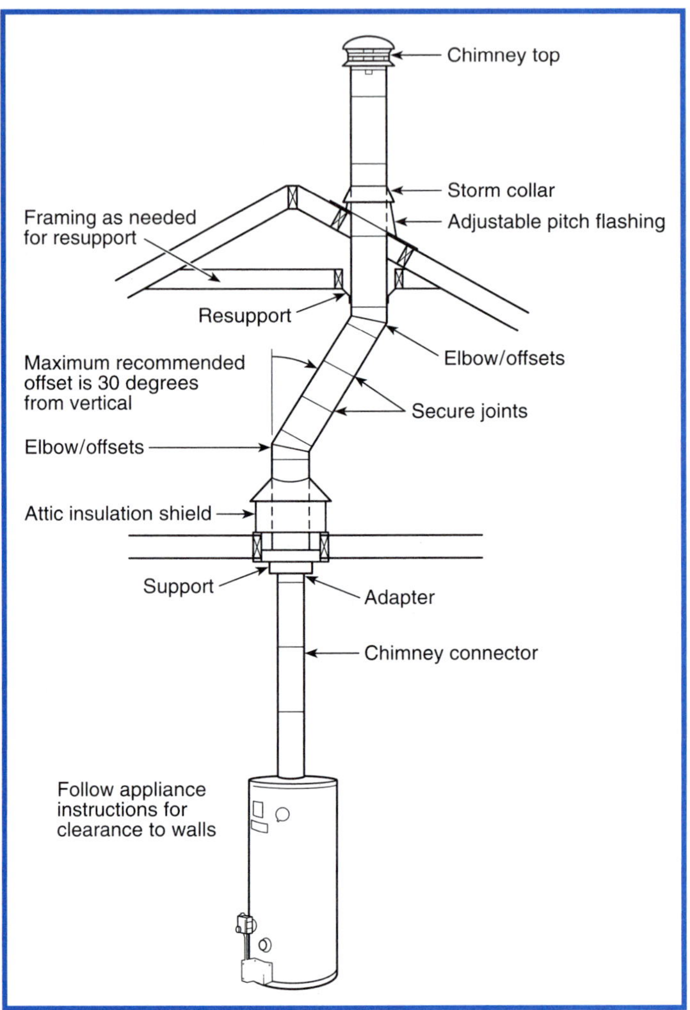

Exhibit 7.6 *Typical installation of a factory-built chimney. (Courtesy of Gas Vent Institute.)*

approved material that will resist corrosion, erosion, softening, or cracking from vent gases at temperatures up to 1800°F (982°C).

These requirements are for chimneys that can be used for all appliances permitted in 7.4.1 that can be vented using chimneys. The code user must be aware that Section 10.1, Additional Requirements to Single Appliance Vent Tables 10.1 Through 10.5, includes 10.1.9, and that Section 10.2, Additional Requirements to Multiple Appliance Vent Tables 10.6 Through 10.13(a) and (b), includes 10.2.18. Both 10.1.9 and

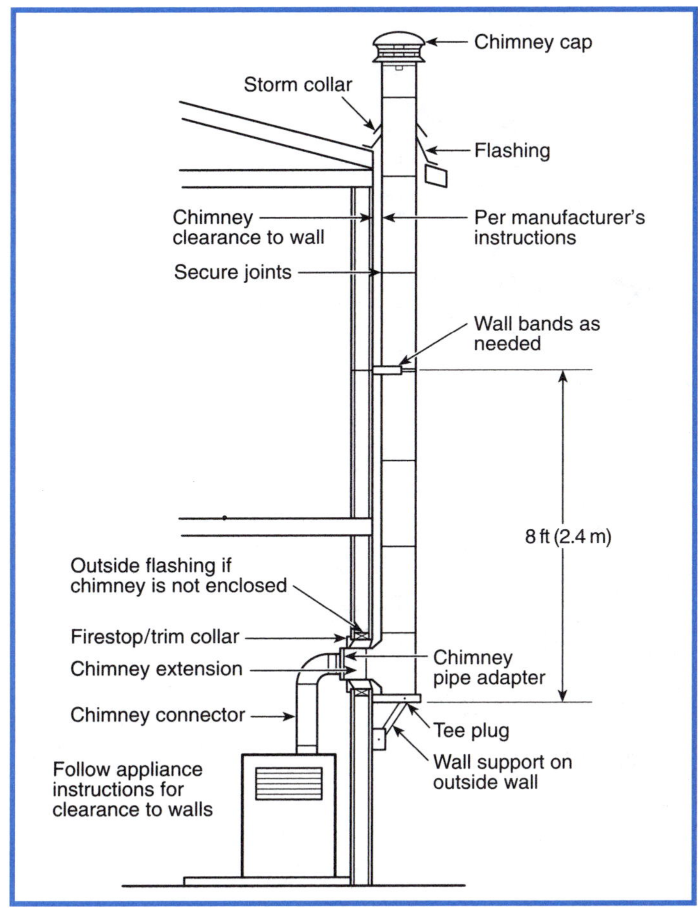

Exhibit 7.7 *Alternate installation of a factory-built chimney. (Courtesy of Gas Vent Institute.)*

10.2.18 place restrictions on the use of the sizing tables for chimneys that vent fan-assisted combustion appliances.

Exception: Masonry chimney flues serving listed gas appliances with draft hoods, Category I appliances, and other gas appliances listed for use with Type B vents shall be permitted to be lined with a chimney lining system specifically listed for use only with such appliances. The liner shall be installed in accordance with the liner manufacturer's instructions and the terms of the listing. A permanent identifying label shall be attached at the point where the connection is to be made to the liner. The label shall read "This chimney liner is for appliances that burn gas only. Do not connect to solid or liquid fuel-burning appliances or incinerators."

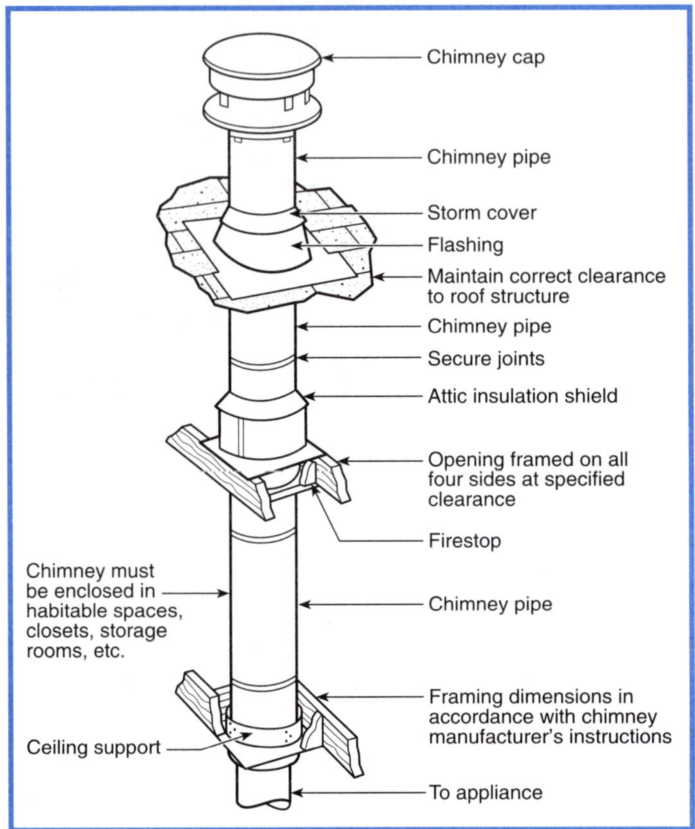

Exhibit 7.8 *Alternate installation of a factory-built chimney. (Courtesy of Gas Vent Institute.)*

The exception to 7.5.1(c) permits installation of masonry chimney liners serving appliances that do not exceed 1800°F (982°C). The labeling requirement notifies future installers of the vent liner's limitations.

A.7.5.1(c) For information on the installation of gas vents in existing masonry chimneys, see Section 7.6.

7.5.2 Termination.

(a) A chimney for residential-type or low-heat gas utilization equipment shall extend at least 3 ft (0.9 m) above the highest point where it passes through a roof of a building and at least 2 ft (0.6 m) higher than any portion of a building within a horizontal distance of 10 ft (3.0 m). *[See Figure 7.5.2(a).]*

Exhibit 7.9 *Hardware for a positive-pressure factory-built chimney. (Courtesy of Selkirk Metalbestos.)*

(b) A chimney for medium-heat equipment shall extend at least 10 ft (3.0 m) higher than any portion of any building within 25 ft (7.6 m).

(c) A chimney shall extend at least 5 ft (1.5 m) above the highest connected equipment draft hood outlet or flue collar.

(d) Decorative shrouds shall not be installed at the termination of factory-built chimneys except where such shrouds are listed and labeled for use with the specific factory-built chimney system and are installed in accordance with manufacturers' installation instructions.

This requirement was added in the 1999 edition to prohibit the installation of unlisted shrouds around factory-built chimneys, thereby affecting the flow of flue gases and trapping hot flue gases in close proximity to the combustible roof.

7.5.3 Size of Chimneys.

The effective area of a chimney venting system serving listed gas appliances with draft hoods, Category I appliances, and other appliances listed for use with Type B vents shall be in accordance with Chapter 10 or other approved engineering methods.

Information on the sizing of chimneys is provided in Chapter 10 in the form of tables and additional requirements. These venting requirements were developed by a research project funded by the Gas Research Institute and conducted by Battelle.

The requirements recognize that there is a significant difference between the operation of interior and exterior chimneys. An interior chimney is not exposed to

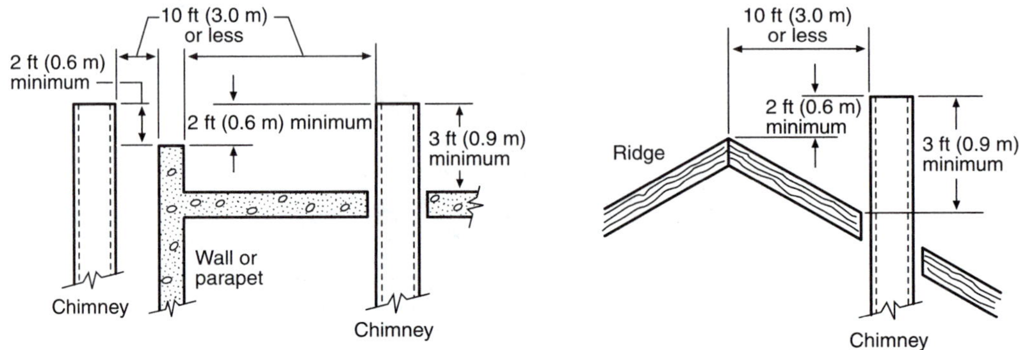

(a) Termination 10 ft (3.0 m) or Less from Ridge, Wall, or Parapet

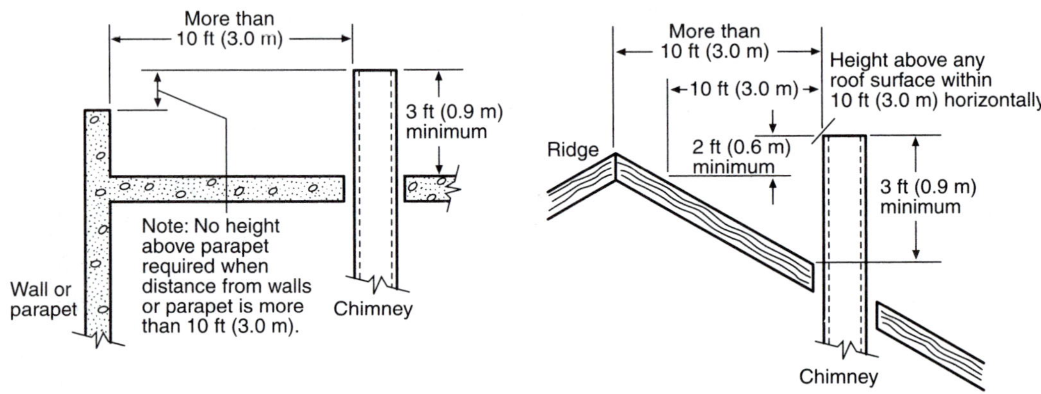

(b) Termination More Than 10 ft (3.0 m) from Ridge, Wall, or Parapet

Figure 7.5.2(a) *Typical termination locations for chimneys and single-wall metal pipes serving residential-type and low-heat equipment.*

the outdoors below the building's roof line. (Chimneys that pass through unheated attics or garages are not considered to be outdoors.) An interior chimney is largely isolated from weather changes. An exterior chimney is affected by the outside ambient temperature. Therefore, the requirements for exterior chimneys are keyed to the lowest temperature expected in different parts of the country.

These requirements for chimneys were developed using a flue loss of 17 percent for the heating appliance. Most real furnaces and boilers operate at a higher flue loss. Therefore, the code requirements are conservative. Recently, several manufacturers have produced customized venting requirements for masonry chimneys that are keyed to the exact performance of their appliances. In this case, their recommendations should be used. An example of a custom vent kit is shown in Exhibit 7.10.

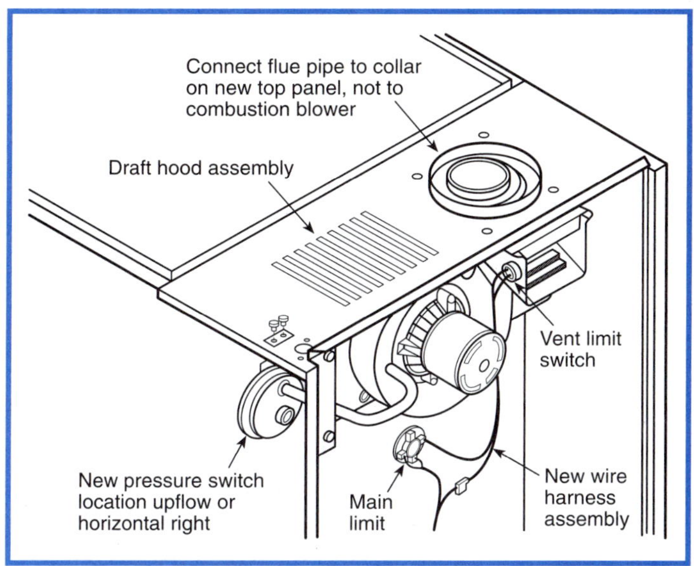

Exhibit 7.10 *A furnace with a field-installed modification utilizing customized venting instructions for venting using a masonry chimney. Note the draft hood to the right of the induced draft blower, which has been added. The kit includes the hardware and venting instructions. (Courtesy of ICP.)*

Exception No. 1: As an alternate method of sizing an individual chimney venting system for a single appliance with a draft hood, the effective areas of the vent connector and chimney flue shall be not less than the area of the appliance flue collar or draft hood outlet or greater than seven times the draft hood outlet area.

Exception No. 2: As an alternate method for sizing a chimney venting system connected to two appliances with draft hoods, the effective area of the chimney flue shall be not less than the area of the larger draft hood outlet plus 50 percent of the area of the smaller draft hood outlet, or greater than seven times the smaller draft hood outlet area.

These exceptions permit a simple, alternate, chimney-sizing method that limits the vent to a maximum size of seven times the smallest draft hood outlet area and a minimum size based on the draft hood outlet areas. The exceptions limit the maximum size of the vent connector and gas vent to minimize condensation in the gas vent caused by insufficient flow of hot vent gases to heat an excessively large gas vent. Excessive condensation can cause premature failure of a chimney.

Exception No. 2 was revised in the 1999 edition to limit the use of this exception to chimney venting systems serving only two appliances. Sample calculations using the venting tables in Chapter 10 and the alternate method showed that the alternate method results in vent sizes too small when four or more appliances are involved, and for some tall vents. The alternate always provides acceptable vent sizes for two appliances. Venting systems serving more than two appliances must use the vent sizing tables in Chapter 10, or other engineering methods.

Where an incinerator is vented by a chimney serving other gas utilization equipment, the gas input to the incinerator shall not be included in calculating chimney size, provided the chimney flue diameter is not less than 1 in. (25 mm) larger in equivalent diameter than the diameter of the incinerator flue outlet.

7.5.4 Inspection of Chimneys.

(a) Before replacing an existing appliance or connecting a vent connector to a chimney, the chimney passageway shall be examined to ascertain that it is clear and free of obstructions and shall be cleaned if previously used for venting solid or liquid fuel-burning appliances or fireplaces.

In the 1999 edition, the former 7.5.4 (a) was revised and split into two paragraphs, (a) and (b). The present 7.5.4 (c) and (d) are the former (b) and (c), unchanged. The revised 7.5.4 (a) covers inspection prior to the connection of a new appliance to a chimney that previously served appliances using fuels other than gas fuels.

(b) Chimneys shall be lined in accordance with NFPA 211, *Standard for Chimneys, Fireplaces, Vents, and Solid Fuel-Burning Appliances,* or local building codes.

Paragraph (b) was added in the 1999 edition to emphasize this important chimney construction requirement.

Exception: Existing chimneys shall be permitted to have their use continued when an appliance is replaced by an appliance of similar type, input rating, and efficiency.

This exception recognizes that not all appliances have been converted to the higher-efficiency fan-assisted combustion type at this time. Water heaters and space heaters are examples of appliances that have not been converted to fan-assisted combustion types.

Note that all chimneys must be lined for any appliance to be connected to it, unless the replacement appliance is of the same type, input rating, and efficiency [see 7.5.4(b), Exception]. Fan-assisted combustion appliances have been developed to meet the minimum energy efficiency requirements of PL100-12, *National Appliance Energy Conservation Act of 1987.* These appliances obtain a higher efficiency by extracting heat that formerly went up with the vent gases. Therefore, the vent gases are cooler, and condensation of water (a product of combustion) will occur during the initial period of appliance operation, making the lining of chimneys mandatory to prevent corrosion. This feature has resulted in revising Tables 10-1 through 10-4 and Tables 10-6 through 10-9 and adding new Tables 10-11 through 10-13.

These revised and new tables provide a method to properly size vents and chimneys for these new appliances, and they provide a method to size masonry chimneys that are exposed to the outdoors below the roof line.

For additional information on the venting of fan-assisted combustion appliances, see Supplement 1, "Development of Revised Venting Guidelines," in this handbook, which describes the work that went into the development of the new tables and the code requirements.

(c) Cleanouts shall be examined to determine they will remain tightly closed when not in use.

(d) When inspection reveals that an existing chimney is not safe for the intended application, it shall be repaired, rebuilt, lined, relined, or replaced with a vent or chimney to conform to NFPA 211, *Standard for Chimneys, Fireplaces, Vents, and Solid Fuel-Burning Appliances,* or local building codes, and shall be suitable for the equipment to be attached.

Some common chimney troubles and ways to detect and remedy them are shown in Exhibit 7.11.

7.5.5 Chimney Serving Equipment Burning Other Fuels.

(a) Gas utilization equipment shall not be connected to a chimney flue serving a separate appliance designed to burn solid fuel.

(b) Gas utilization equipment and equipment burning liquid fuel shall be permitted to be connected to one chimney flue through separate openings or shall be permitted to be connected through a single opening if joined by a suitable fitting located as close as practical to the chimney. If two or more openings are provided into one chimney flue, they shall be at different levels. If the gas utilization equipment is automatically controlled, it shall be equipped with a safety shutoff device.

(c)* A listed combination gas- and solid fuel–burning appliance equipped with a manual reset device to shut off gas to the main burner in the event of sustained backdraft or flue gas spillage shall be permitted to be connected to a single chimney flue. The chimney flue shall be sized to properly vent the appliance.

A.7.5.5(c) Reference can also be made to the chapter on chimney, gas vent, and fireplace systems of the *ASHRAE Handbook — HVAC Systems and Equipment.*

(d) A listed combination gas- and oil-burning appliance shall be permitted to be connected to a single chimney flue. The chimney flue shall be sized to properly vent the appliance.

7.5.6 Support of Chimneys.

All portions of chimneys shall be supported for the design and weight of the materials employed. Listed factory-built chimneys shall be supported and spaced in accordance with their listings and the manufacturers' instructions.

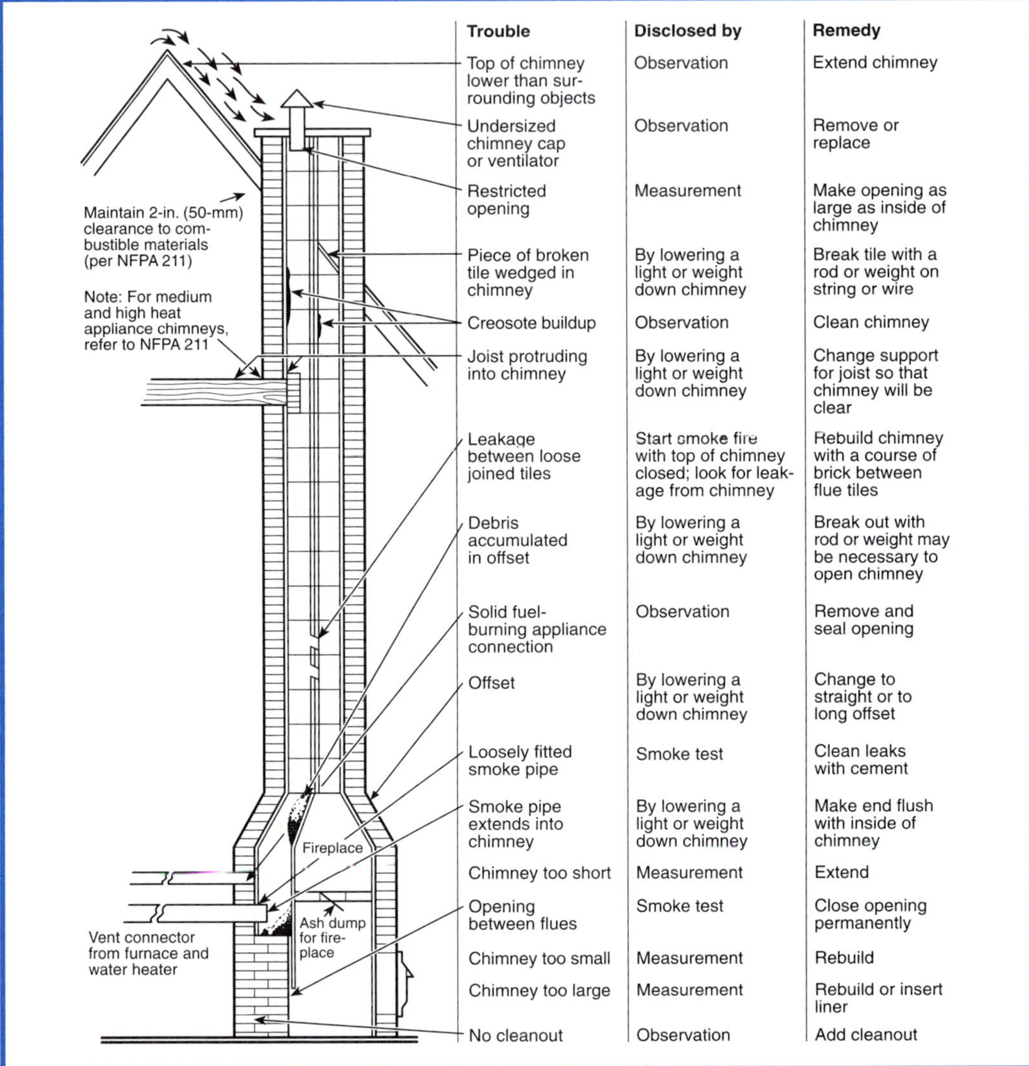

Trouble	Disclosed by	Remedy
Top of chimney lower than surrounding objects	Observation	Extend chimney
Undersized chimney cap or ventilator	Observation	Remove or replace
Restricted opening	Measurement	Make opening as large as inside of chimney
Piece of broken tile wedged in chimney	By lowering a light or weight down chimney	Break tile with a rod or weight on string or wire
Creosote buildup	Observation	Clean chimney
Joist protruding into chimney	By lowering a light or weight down chimney	Change support for joist so that chimney will be clear
Leakage between loose joined tiles	Start smoke fire with top of chimney closed; look for leakage from chimney	Rebuild chimney with a course of brick between flue tiles
Debris accumulated in offset	By lowering a light or weight down chimney	Break out with rod or weight may be necessary to open chimney
Solid fuel-burning appliance connection	Observation	Remove and seal opening
Offset	By lowering a light or weight down chimney	Change to straight or to long offset
Loosely fitted smoke pipe	Smoke test	Clean leaks with cement
Smoke pipe extends into chimney	By lowering a light or weight down chimney	Make end flush with inside of chimney
Chimney too short	Measurement	Extend
Opening between flues	Smoke test	Close opening permanently
Chimney too small	Measurement	Rebuild
Chimney too large	Measurement	Rebuild or insert liner
No cleanout	Observation	Add cleanout

Exhibit 7.11 *Some common chimney troubles — how to detect and remedy them. (Courtesy of American Radiator Company.)*

7.5.7 Cleanouts.

Where a chimney that formerly carried flue products from liquid or solid fuel-burning appliances is used with an appliance using fuel gas, an accessible cleanout shall be provided. The cleanout shall have a tight-fitting cover and be installed so its upper edge is at least 6 in. (150 mm) below the lower edge of the lowest chimney inlet opening.

This provision recognizes that when a gas heating unit or conversion burner is connected to a chimney that previously served for venting another fuel, after a period of operation on natural gas, carbon deposits or mortar materials can become dislodged from the chimney wall and drop to the base of the chimney. The accumulation of this debris, if great enough, can result in blockage of the venting system, sometimes with fatal results. Effective with the 1999 edition, the code requires a 6-in. (150-mm) drop leg and inspection/maintenance cleanout, which provides a sump to collect a considerable amount of debris before a potential accident occurs. The cleanout also enables home owners and service personnel to clean debris easily and to inspect conditions. This distance was reduced from 12 in. distance based on the experience of committee members that the requirement was excessive.

7.5.8 Space Surrounding Lining or Vent.

The remaining space surrounding a chimney liner, gas vent, special gas vent, or plastic piping installed within a masonry chimney flue shall not be used to vent another appliance.

This paragraph covers the installation of nonmetallic gas vents in an abandoned chimney or a chase with a gas vent in it. The use of a masonry chimney as a chase for vents is a practical solution to an existing unlined, oversized, or failed chimney. When installing vents in a chimney space, all of the vent material manufacturer's installation requirements must be followed. If required by the vent manufacturer, penetrating the masonry chimney to install supports can be necessary.

When using an abandoned chimney as a chase for multiple vents, take care to ensure that the vent materials are compatible and that vent failure does not result inadvertently. Category I or III gas appliance vent gas temperatures can exceed 300°F (14°C). When a Category IV gas furnace, vented with PVC pipe, is used in the chimney with a listed chimney liner on a Category I appliance or on a listed vent on a Category III gas appliance (e.g., furnace or water heater), a potential exists for exceeding the working temperature of the PVC pipe [140°F (60°C)]. Contact between the high-temperature Category I or III vent material and the PVC vent pipe can cause the PVC pipe to degrade or soften to the extent of failure. With the chimney opening sealed around the vent pipes, the products of combustion, which can escape from the failed PVC pipe, will be trapped in the chimney. These gases could eventually leak back into the living space or corrode the other nearby vents.

Exception: The insertion of another liner or vent within the chimney as provided in this code and the liner or vent manufacturer's instructions.

7.6 Gas Vents

(See Section 1.7.)

This section applies to gas vents of all types. Not included are chimneys, which are covered in Section 7.5, and single-wall metal pipe, which is covered in Section 7.7.

Note that the term "gas vent" is defined as a listed product. This section applies only to listed products.

Type B and Type B-W gas vents that currently are listed are of double-wall construction — basically, a pipe within a pipe, with airspace between the two walls. This construction reduces the heat loss from the vent gases, in much the same way that an insulated coffee bottle works.

Usually, the inner wall is made of aluminum to resist corrosion, and the outer wall is made of galvanized steel for strength. The parts of the vent — pipe sections, supports, spacers, caps, and flashings — are furnished for field erection to form a continuous passageway from the gas appliance to the terminus the roof, including the cap or roof assembly. These parts are manufactured by several companies that furnish the vent pipe and other related parts needed to erect a complete vent. The piping and parts produced by the different manufacturers are not interchangeable without the use of special, listed adapters.

Type B gas vents can be round or oval. Type B-W gas vents are always oval, because they are designed to be installed within a stud space. (See Exhibits 7.12 and 7.13.) Type L vents are similar in construction to Type B gas vents, except that the inner pipe of a Type L vent is stainless steel. Type L vents are intended for venting flue gases at higher temperatures than are normally produced by gas appliances, such as for oil furnaces listed for use with Type L vents.

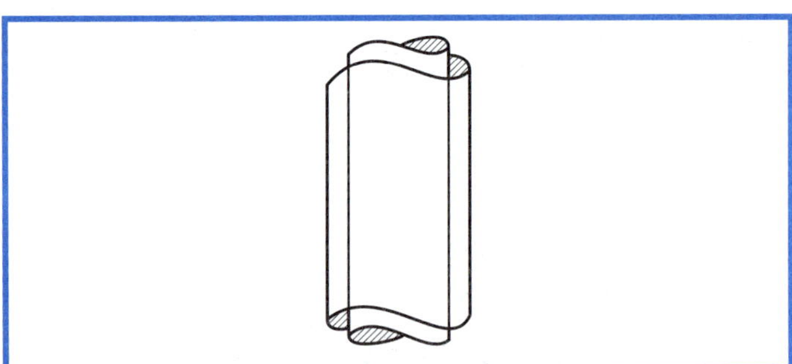

Exhibit 7.12 Construction of double-wall Type B gas vent. (Courtesy of Gas Vent Institute.)

7.6.1 Application.

Typical Type B and Type B-W gas vent installations are diagrammed in Exhibits 7.14 and 7.15, respectively.

(a) Gas vents shall be installed in accordance with the terms of their listings and the manufacturers' instructions.

(b) A Type B-W gas vent shall have a listed capacity not less than that of the listed vented wall furnace to which it is connected.

Exhibit 7.13 *Type B and Type B-W double-wall gas vent fittings. (Courtesy of Selkirk Metalbestos.)*

(c) A gas vent passing through a roof shall extend through the entire roof flashing, roof jack, or roof thimble and be terminated with a listed termination cap.

Paragraph (c) was revised in the 1999 edition to provide guidance to installers and inspectors to prevent improper installations. The word "entire" was added to ensure that vents terminate above the roof or flashing. It also reminds installers that a listed cap must be used.

(d) Type B or Type L vents shall extend in a generally vertical direction with offsets not exceeding 45 degrees, except that a vent system having not more than one 60-degree offset shall be permitted.

Any angle greater than 45 degrees from the vertical is considered horizontal. The total horizontal distance of a vent plus the horizontal vent connector serving draft hood-equipped appliances shall not be greater than 75 percent of the vertical height of the vent.

The code permits Category I equipment to be installed in accordance with the Chapter 10 tables, which incorporate Category I vent- sizing tables. These tables provide three columns for each set of vent horizontal and vertical runs: one column for draft hood–equipped appliances and two columns that specify both minimum and maximum capacities for fan-assisted appliances. Additional information on use of the tables can be found in Chapter 10. Information on the development of the tables can be found in Supplement 1, "Development of Revised Venting Guidelines," in this handbook. In lieu of using the tables in Chapter 10 for vent sizing, approved engineering methods are permitted. Additional information on sources of design information can be found in A.7.6.3(a).

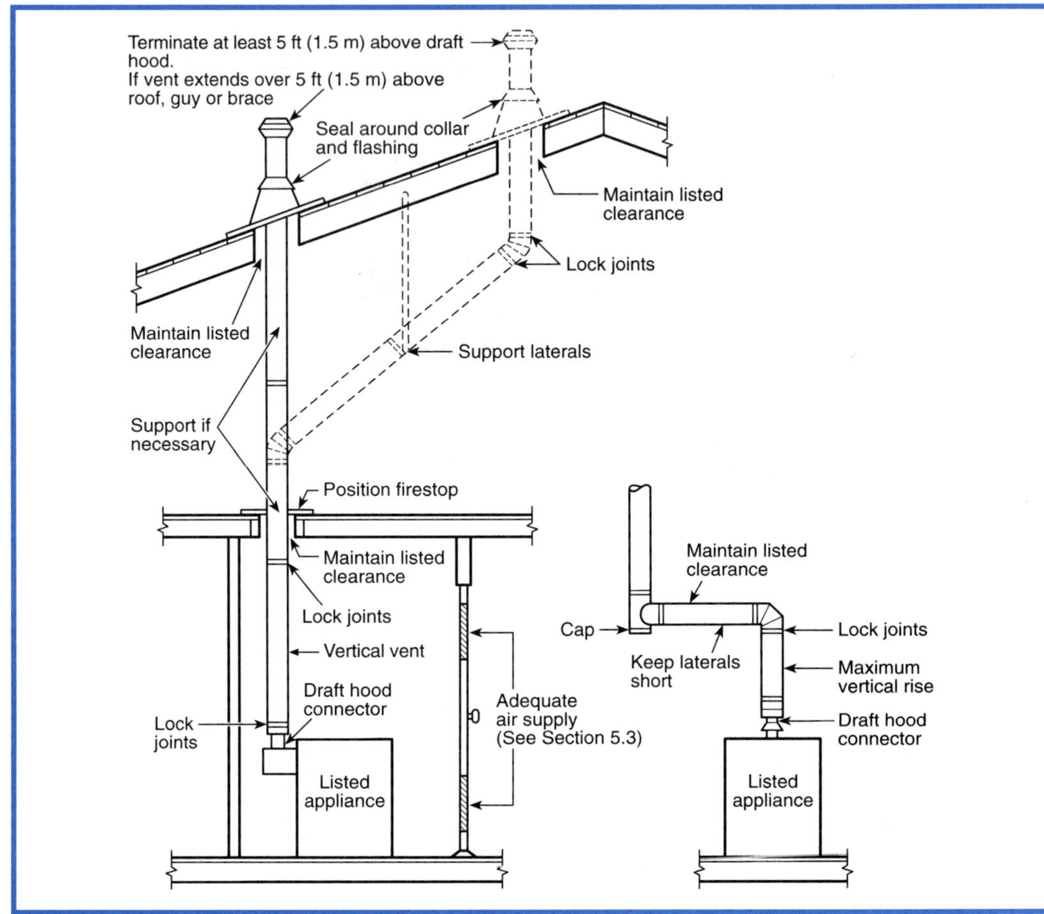

Exhibit 7.14 *Typical Type B gas vent — single appliance installation. (Courtesy of Gas Vent Institute.)*

If the system is designed in accordance with the tables in Chapter 10, a Type B gas vent can include two 90-degree turns and a horizontal run greater than 75 percent of the vertical portion of the vent. For systems with more that two 90-degree turns, 10.1.3 and 10.2.6 permit the use of Tables 10-1 through 10-10 with a derating of the maximum capacity values of 10 percent for each 90-degree turn over two 90-degree turns.

Exception: Systems designed and sized as provided in Chapter 10 or in accordance with other approved engineering methods.

Vents serving Category I fan-assisted appliances shall be installed in accordance with the appliance manufacturer's instructions and Chapter 10 or other approved engineering methods.

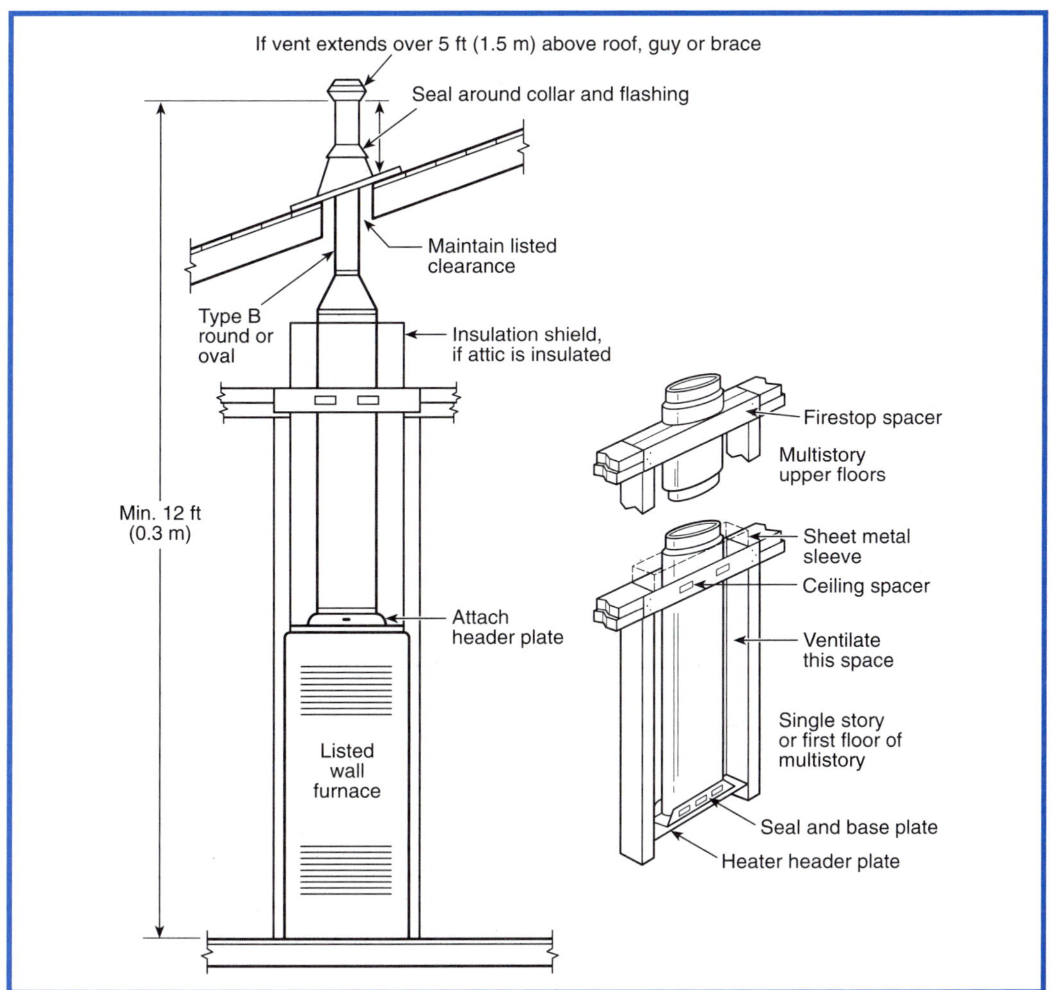

Exhibit 7.15 *Typical B-W gas vent installation. (Courtesy of Gas Vent Institute.)*

(e) Gas vents installed within masonry chimneys shall be installed in accordance with the terms of their listing and the manufacturers' installation instructions. Gas vents installed within masonry chimneys shall be identified with a permanent label installed at the point where the vent enters the chimney. The label shall contain the following language: "This gas vent is for appliances that burn gas. Do not connect to solid or liquid-fuel-burning appliances or incinerators."

This provision was amended in 1999 to require that gas vents installed in masonry chimneys be supplied with manufacturer's instructions specifically for that application. The installation of gas vents in existing masonry chimneys requires specialized techniques and hardware that are different from those used in the roof applications. The special considerations include inspection and cleaning of the chimney, hardware for supporting the pipe, sealing of the top of the chimney, and termination. Note the requirement to permanently mark the chimney, identifying the intended use of the vent. (See Exhibit 7.16.)

7.6.2 Gas Vent Termination.

Listed vent caps for Type B and Type B-W gas vents are designed and tested to serve the following four purposes:

(1) To reduce the chance of downdraft caused by wind effect
(2) To provide proper venting under wind conditions
(3) To permit vent termination at minimum height above the roof surface
(4) To prevent rain and debris from entering the vent

Notice that a listed vent cap allows the vent termination to be closer to the nearby roofline, because the terminations are tested to operate properly using the clearances shown in Figure 7.6.2(a) of the code.

(a) A gas vent shall terminate above the roof surface with a listed cap or listed roof assembly. Gas vents 12 in. (300 mm) in size or smaller with listed caps shall be permitted to be terminated in accordance with Figure 7.6.2(a), provided they are at least 8 ft (2.4 m) from a vertical wall or similar obstruction. All other gas vents shall terminate not less than 2 ft (0.6 m) above the highest point where they pass through the roof and at least 2 ft (0.6 m) higher than any portion of a building within 10 ft (3.1 m).

Exception No. 1: Industrial gas utilization equipment as provided in 7.2.4.

Exception No. 2: Direct-vent systems as provided in 7.2.5.

Exception No. 3: Equipment with integral vents as provided in 7.2.6.

Exception No. 4: Mechanical draft systems as provided in 7.3.4.

Exception No. 5: Ventilating hoods and exhaust systems as provided in 7.3.5.

Exception No. 1, which references 7.2.4, Well-Ventilated Spaces, is included here to clarify that vent termination does not apply to such spaces. Providing specific references within exceptions allows the reader to easily determine whether the exceptions apply to vent termination.

(b) A Type B or a Type L gas vent shall terminate at least 5 ft (1.5 m) in vertical height above the highest connected equipment draft hood or flue collar.

(c) A Type B-W gas vent shall terminate at least 12 ft (3.7 m) in vertical height above the bottom of the wall furnace.

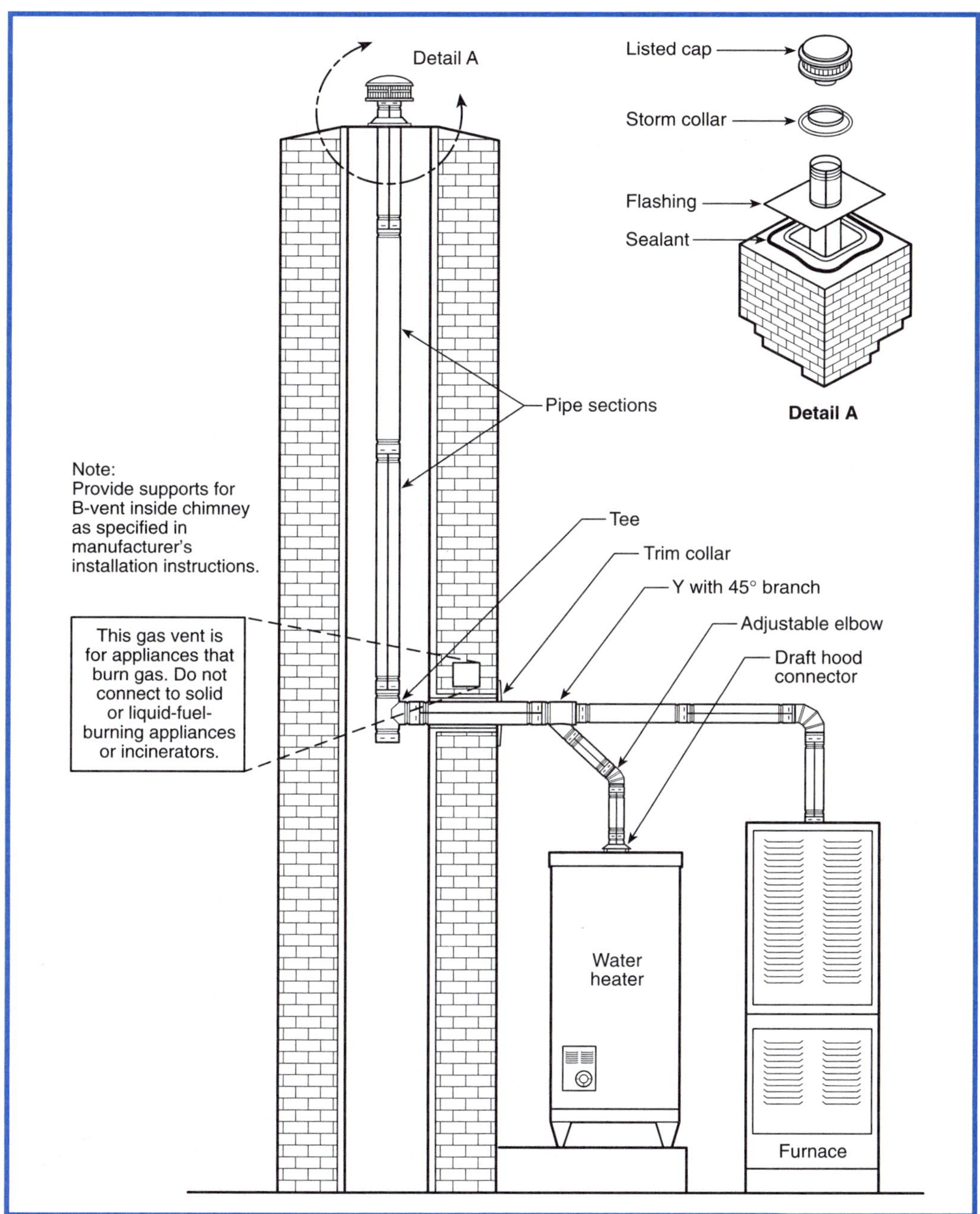

Exhibit 7.16 *Typical installation using Type B vent to line chimney. (Courtesy of Simpson Duravent.)*

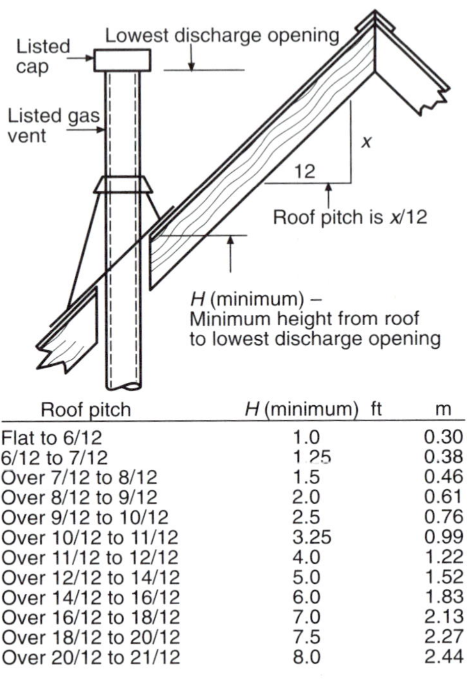

Roof pitch	H (minimum) ft	m
Flat to 6/12	1.0	0.30
6/12 to 7/12	1.25	0.38
Over 7/12 to 8/12	1.5	0.46
Over 8/12 to 9/12	2.0	0.61
Over 9/12 to 10/12	2.5	0.76
Over 10/12 to 11/12	3.25	0.99
Over 11/12 to 12/12	4.0	1.22
Over 12/12 to 14/12	5.0	1.52
Over 14/12 to 16/12	6.0	1.83
Over 16/12 to 18/12	7.0	2.13
Over 18/12 to 20/12	7.5	2.27
Over 20/12 to 21/12	8.0	2.44

Figure 7.6.2(a) *Gas vent termination locations for listed caps 12 in. (300 mm) or less in size at least 8 ft (2.4 m) from a vertical wall.*

(d) A gas vent extending through an exterior wall shall not terminate adjacent to the wall or below eaves or parapets, except as provided in 7.2.5 and 7.3.4.

7.6.3 Size of Gas Vents.

Venting systems shall be sized and constructed in accordance with Chapter 10 or other approved engineering methods and the gas vent and gas equipment manufacturers' instructions.

(a)* *Category I Appliances.* The sizing of natural draft venting systems serving one or more listed appliances equipped with a draft hood or appliances listed for use with Type B gas vent, installed in a single story of a building, shall be in accordance with Chapter 10 or in accordance with sound engineering practice.

Exception No. 1: As an alternate method for sizing an individual gas vent for a single, draft hood-equipped appliance, the effective area of the vent connector and the gas vent shall be not less than the area of the appliance draft hood outlet or greater than seven times the draft hood outlet area. Vents serving fan-assisted combustion system appliances shall be sized in accordance with Chapter 10 or other approved engineering methods.

Exception No. 2: As an alternate method for sizing a gas vent connected to two appliances, with draft hoods, the effective area of the vent shall be not less than the area of the larger draft hood outlet plus 50 percent of the area of the smaller draft hood outlet or greater than seven times the smaller draft hood outlet area. Vents serving fan-assisted combustion system appliances or combinations of fan-assisted combustion system and draft hood-equipped appliances shall be sized in accordance with Chapter 10 or other approved engineering methods.

A.7.6.3(a) Additional information on sizing venting systems can be found in the following:

(1) Tables in Chapter 10
(2) The gas equipment manufacturer's instructions
(3) The venting equipment manufacturer's sizing instructions
(4) Drawings, calculations, and specifications provided by the venting equipment manufacturer
(5) Drawings, calculations, and specifications provided by a competent person
(6) The chapter on chimney, gas vent, and fireplace systems of the *ASHRAE Handbook — HVAC Systems and Equipment*

Category I appliances may be either draft hood-equipped or fan-assisted combustion system in design. Different vent design methods are required for draft hood-equipped and fan-assisted combustion system appliances.

(b) *Category II, Category III, and Category IV Appliances.* The sizing of gas vents for Category II, Category III, and Category IV gas utilization equipment shall be in accordance with the equipment manufacturers' instructions.

The tables in Chapter 10 provide a safe method for sizing vents for all Category I appliances. The alternate methods permitted here offer simple techniques for installers to follow for commonly encountered vent arrangements. They provide a maximum size of seven times the smallest draft hood outlet area and a minimum size based on the draft hood outlet area. This provision limits the maximum size of the vent connector and gas vent to minimize gas vent condensation caused by insufficient flow of hot vent gases to heat an excessively large gas vent. Excessive condensation can cause premature failure of a gas vent.

Note that these alternate methods cannot be used for fan-assisted combustion appliances. The tables in Chapter 10 (or other engineering methods) must be used for all installations of fan-assisted combustion appliances. Vents for Category II, III, and IV appliances must be in accordance with the equipment manufacturer's instructions. These appliances are listed by definition, and the manufacturer's instructions reflect the test results determined by the listing laboratory.

Exception No. 2 was amended in 1999 to limit the use of this exception to only two appliances. Venting systems serving more than two appliances must use the vent sizing tables in Chapter 10, or other engineering methods. Refer to the commentary following 7.5.3, Exception No. 2 for more information.

7.6.4 Gas Vents Serving Equipment on More Than One Floor.

A single or common gas vent shall be permitted in multistory installations to vent Category I gas utilization equipment located on more than one floor level, provided the venting system is designed and installed in accordance with approved engineering methods.

All gas utilization equipment connected to the common vent shall be located in rooms separated from habitable space. Each of these rooms shall have provisions for an adequate supply of combustion, ventilation, and dilution air that is not supplied from habitable space. *(See Figure 7.6.4.)*

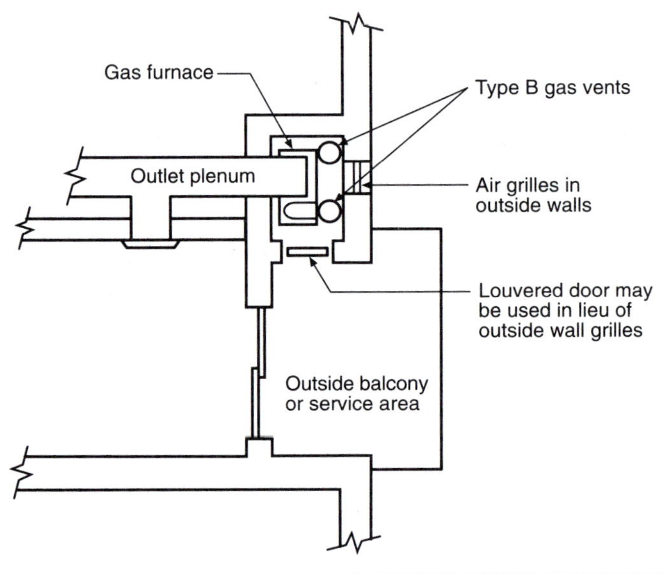

Figure 7.6.4 Plan view of practical separation method for multistory gas venting.

The size of the connectors and common segments of multistory venting systems for gas utilization equipment listed for use with Type B double-wall gas vent shall be in accordance with Table 10.6, provided:

(1) The available total height *(H)* for each segment of a multistory venting system is the vertical distance between the level of the highest draft hood outlet or flue collar on that floor and the centerline of the next highest interconnection tee. *(See Figure G.11.)*

(2) The size of the connector for a segment is determined from its gas utilization equipment heat input and available connector rise, and shall not be smaller than the draft hood outlet or flue collar size.

(3) The size of the common vertical vent segment, and of the interconnection tee at the base of that segment, shall be based on the total gas utilization equipment heat input entering that segment and its available total height.

This section provides requirements for the separation of appliances from the habitable space [shown in Figure 7.6.2(a) of the code]. Further information on sizing vents that serve appliances on more than one floor of a building can be found in 10.2.13 through 10.2.15 and Figures G.13 and G.14.

Safety can be achieved in high-rise, multilevel appliance installations by separating the appliance from any habitable or occupied space. The separation resolves the question of safety that arises when intercommunication of vents exists between various floors of a building. Separation ensures that no vent gases will enter the occupied space from the appliance room if the common vent becomes obstructed at any level or if the outlet is blocked. When such stoppage occurs, all vent gases from appliances operating below the obstruction will exit through the upper draft hood relief opening rather than through the vent outlet. Large quantities of vent gases will be dumped into the space containing those appliances immediately below the obstruction, while at the same time, the appliances located at lower levels will appear to be operating normally.

One practical plan to separate or isolate appliances, if installation can be made adjacent to an outside wall, is shown by Figure 7.6.2(a) of the code and Exhibit 7.17. Access to the appliance room is through a door that opens onto an outdoor balcony. A panel separates the appliance room from the inside of the building. If the appliance is a central furnace, the cold air return and outlet ducts are attached to the furnace.

Previous editions of this handbook suggested that a single combustion air duct could serve multiple floors of a building. In the 1999 edition, the technical committee prohibited this use by adding section 5.3.5, which requires that each confined space have its own combustion air duct and that there be separate ducts for the top and bottom combustion air opening into the space.

No requirements are given in the code for installation of Category II, III, or IV appliances using a common vent in multistory buildings. A Category II, III, or IV appliance must be installed in accordance with the terms of its listing and the manufacturer's instructions. Category I appliance common vents are not usually suitable for venting Category II, III, or IV appliances.

7.6.5 Support of Gas Vents.

Gas vents shall be supported and spaced in accordance with their listings and the manufacturers' instructions.

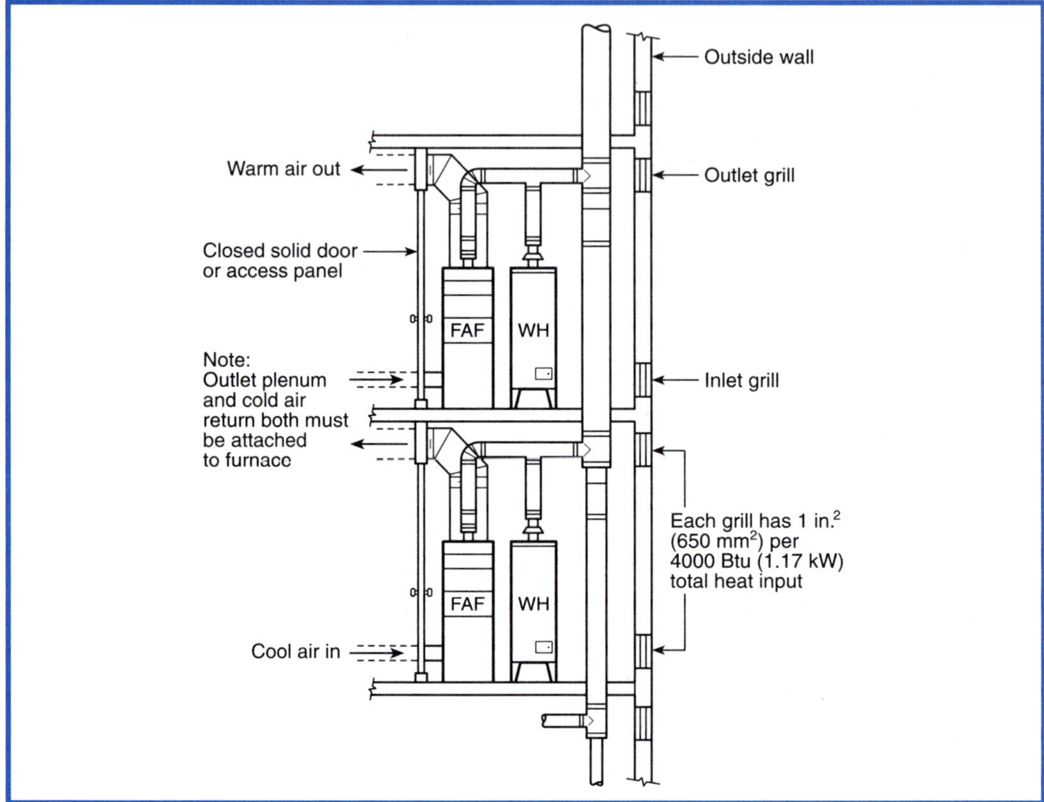

Exhibit 7.17 *Outside wall air supply in multistory building. (Courtesy of Gas Vent Institute.)*

7.6.6 Marking.

In those localities where solid and liquid fuels are used extensively, gas vents shall be permanently identified by a label attached to the wall or ceiling at a point where the vent connector enters the gas vent. The label shall read: "This gas vent is for appliances that burn gas. Do not connect to solid or liquid fuel-burning appliances or incinerators."

The authority having jurisdiction shall determine whether its area constitutes such a locality.

7.7 Single-Wall Metal Pipe

The use of single-wall metal pipe (e.g., stovepipe) as a vent for gas utilization equipment is limited by the code, especially in dwellings, for several reasons, including the following:

(1) The surface of the pipe is usually hot enough to burn a person who accidentally contacts the pipe.

(2) The high surface temperature can cause overheating of adjacent combustible material.

(3) The heat loss from a long length of pipe in a cool area can result in reduced draft and the condensation of water vapor in the flue gases.

The use of single-wall metal pipe as a vent, where permitted, is limited to installations such as those described in 7.6.3(a). Note that some building codes do not permit single-wall metal chimneys and unlisted metal chimneys inside one- and two-family dwellings.

7.7.1 Construction.

Single-wall metal pipe shall be constructed of galvanized sheet steel not less than 0.0304 in. (0.7 mm) thick or of other approved, noncombustible, corrosion-resistant material.

7.7.2 Cold Climate.

Uninsulated single-wall metal pipe shall not be used outdoors in cold climates for venting gas utilization equipment.

In the 1999 edition, this requirement was relocated here to place all requirements for the installation of gas vents in the same section. [It was formerly 7.8(d).] The outdoor use of single-wall metal pipe in cold climates is restricted to prevent premature corrosion of the vent. In cold climates, the low ambient temperature can cool vent gases to the point at which water vapor condenses in the vent and causes corrosion. The code does not attempt to define a "cold climate," because the particular installation and local conditions often determine the acceptability of using single-wall metal pipe. One simple guideline is checking to see how often the replacement of single-wall metal pipe is required in the area in question.

Tables 10-11, 10-12, and 10-13 were added in the 1996 edition to quantify the effects of cold ambient temperatures on masonry chimneys. They cannot be used for single-wall metal pipe connectors.

7.7.3 Termination.

Table 7.4.1 permits single-wall metal pipe to be used for venting low-heat and residential-type gas utilization equipment with draft hoods (with some restrictions); incinerators used outdoors; and gas-fired toilets. (See Exhibit 7.18.)

(a) Single-wall metal pipe shall terminate at least 5 ft (1.5 m) in vertical height above the highest connected equipment draft hood outlet or flue collar.

(b) Single-wall metal pipe shall extend at least 2 ft (0.6 m) above the highest point where it passes through a roof of a building and at least 2 ft (0.6 m) higher than any portion of a building within a horizontal distance of 10 ft (3.1 m). *[See Figure 7.5.2(a).]*

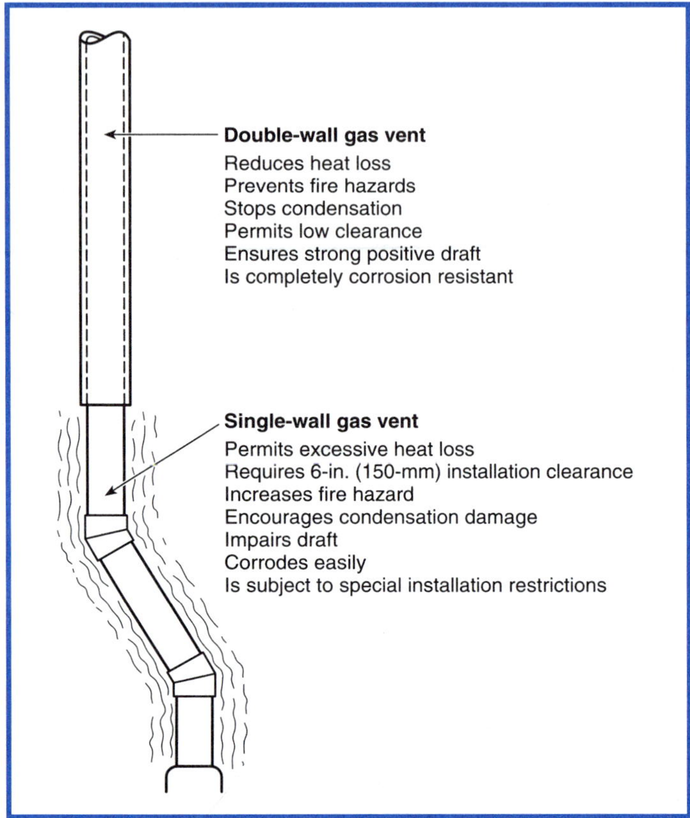

Exhibit 7.18 *Comparison of double-wall and single-wall gas vent charactcristics. (Courtesy of Gas Vent Institute)*

(c) An approved cap or roof assembly shall be attached to the terminus of a single-wall metal pipe. *[Also see 7.7.4(c).]*

> These paragraphs provide the limitations under which single-wall metal pipe can be installed. The limitations are based on the fact that single-wall metal pipe can operate at a high temperature that easily can ignite combustible material placed in contact with it. When this type of pipe passes through a combustible roof, protection of the roof is required, and a thimble is specified.

7.7.4 Installation with Equipment Permitted by 7.4.1.

(a) Single-wall metal pipe shall be used only for runs directly from the space in which the gas utilization equipment is located through the roof or exterior wall to the outer air. A pipe passing through a roof shall extend without interruption through the roof flashing, roof jacket, or roof thimble.

(b) Single-wall metal pipe shall not originate in any unoccupied attic or concealed space and shall not pass through any attic, inside wall, concealed space, or floor. For the installation of a single-wall metal pipe through an exterior combustible wall, see 7.10.15(b).

(c) Single-wall metal pipe used for venting an incinerator shall be exposed and readily examinable for its full length and shall have suitable clearances maintained.

(d) Minimum clearances from single-wall metal pipe to combustible material shall be in accordance with Table 7.7.4(d). The clearance from single-wall metal pipe to combustible material shall be permitted to be reduced where the combustible material is protected as specified for vent connectors in Table 6.2.3(b).

(e) Where a single-wall metal pipe passes through a roof constructed of combustible material, a noncombustible, nonventilating thimble shall be used at the point of passage. The thimble shall extend at least 18 in. (460 mm) above and 6 in. (150 mm) below the roof with the annular space open at the bottom and closed only at the top. The thimble shall be sized in accordance with 7.10.15(b).

Table 7.7.4(d) Clearance for Connectors

	Minimum Distance from Combustible Material			
Equipment	**Listed Type B Gas Vent Material**	**Listed Type L Vent Material**	**Single-Wall Metal Pipe**	**Factory-Built Chimney Sections**
Listed equipment with draft hoods and equipment listed for use with Type B gas vents	As listed	As listed	6 in.	As listed
Residential boilers and furnaces with listed gas conversion burner and with draft hood	6 in.	6 in.	9 in.	As listed
Residential appliances listed for use with Type L vents	Not permitted	As listed	9 in.	As listed
Residential incinerators	Not permitted	9 in.	18 in.	As listed
Listed gas-fired toilets	Not permitted	As listed	As listed	As listed
Unlisted residential appliances with draft hood	Not permitted	6 in.	9 in.	As listed
Residential and low-heat equipment other than those above	Not permitted	9 in.	18 in.	As listed
Medium-heat equipment	Not permitted	Not permitted	36 in.	As listed

For SI units, 1 in. = 25.4 mm.

Note: These clearances shall apply unless the listing of an appliance or connector specifies different clearances, in which case the listed clearances shall apply.

7.7.5 Size of Single-Wall Metal Pipe.

(a)* A venting system of a single-wall metal pipe shall be sized in accordance with one of the following methods and the gas equipment manufacturer's instructions:

(1) For a draft hood-equipped appliance, in accordance with Chapter 10
(2) For a venting system for a single appliance with a draft hood, the areas of the connector and the pipe each shall not be less than the area of the appliance flue collar or draft hood outlet, whichever is smaller. The vent area shall not be greater than seven times the draft hood outlet area.
(3) Other approved engineering methods

A.7.7.5(a) Reference can also be made to the chapter on chimney, gas vent, and fireplace systems of the *ASHRAE Handbook — HVAC Systems and Equipment.*

(b) Any shaped single-wall metal pipe shall be permitted to be used, provided its equivalent effective area is equal to the effective area of the round pipe for which it is substituted and provided the minimum internal dimension of the pipe is not less than 2 in. (50 mm).

(c) The vent cap or a roof assembly shall have a venting capacity not less than that of the pipe to which it is attached.

These alternate sizing methods are not permitted for vents serving Category I, fan-assisted combustion type appliances unless included in the manufacturer's instructions.

These methods provide a maximum size of seven times the smallest draft hood outlet area and a minimum size based on the draft hood outlet size. This provision limits the maximum size of the gas vent to minimize condensation caused by insufficient flow of hot vent gases to heat an excessively large gas vent. Excessive condensation can cause premature failure of single-wall metal pipe.

In 1999, this clause was revised to forbid any use of single wall galvanized pipe for vents serving more than one appliance (common venting).

7.7.6 Support of Single-Wall Metal Pipe.

All portions of single-wall metal pipe shall be supported for the design and weight of the material employed.

7.7.7 Marking.

Single-wall metal pipe shall comply with the marking provisions of 7.6.6.

7.8* Through the Wall Vent Termination

This section provides requirements for separation of the termination point of venting systems from the building openings for venting systems that terminate through the side of a building. In the 1999 edition, the title of the section was changed from "Venting System Location" to more accurately reflect its content. In addition, (d) was relocated to 7.7.2 because it does not cover termination of vents, but rather is a restriction on installation of gas vents and is better located in Section 7.7.

(a) A mechanical draft venting system shall terminate at least 3 ft (0.9 m) above any forced air inlet located within 10 ft (3.1 m).

Exception No. 1: This provision shall not apply to the combustion air intake of a direct-vent appliance.

The intent of 7.8(a) is to prevent gases from being drawn back into a building. It recognizes that vent gases are lighter than air. The exception regarding direct-vent appliance inlets recognizes that these appliances do not communicate with the air in a building.

Exception No. 2: This provision shall not apply to the separation of the integral outdoor air inlet and flue gas discharge of listed outdoor appliances.

Exception No. 2 prevents confusion in the installation of outdoor gas equipment. Some authorities have misinterpreted the code to prohibit such equipment or to require it to be modified in the field, which is not the intent of 7.8(a). An example of this type of equipment is a packaged rooftop air conditioner, which incorporates a gas vent and a circulation air inlet used for the building air supply.

(b) A mechanical draft venting system of other than direct-vent type shall terminate at least 4 ft (1.2 m) below, 4 ft (1.2 m) horizontally from, or 1 ft (300 mm) above any door, window, or gravity air inlet into any building. The bottom of the vent terminal shall be located at least 12 in. (300 mm) above grade.

(c) The vent terminal of a direct-vent appliance with an input of 10,000 Btu/hr (3 kW) or less shall be located at least 6 in. (150 mm) from any air opening into a building, and such an appliance with an input over 10,000 Btu/hr (3 kW) but not over 50,000 Btu/hr (14.7 kW) shall be installed with a 9-in. (230-mm) vent termination clearance, and an appliance with an input over 50,000 Btu/hr (14.7 kW) shall have at least a 12-in. (300-mm) vent termination clearance. The bottom of the vent terminal and the air intake shall be located at least 12 in. (300 mm) above grade.

Paragraphs (a) and (b) are concerned with preventing appliance combustion products from being drawn into a building through fresh air inlets. In that context, the "window" listed in 7.8(b) refers to openable windows rather than windows constructed only of glass with no movable parts, such as picture windows.

Paragraph (c) permits the vent terminals of direct-vent appliances to be located much closer to air inlets than permitted for nondirect-vent appliances in 7.8(a). In the 1996 edition, a new type of low-input (10,000 Btu/hr or less), direct-vent, space-heating appliance was recognized by the addition of a 6-in. (150-mm) clearance for such appliances. This change was based on a report prepared by the Canadian Gas Research Institute. Studies have shown that vent gases from direct-vent appliances disperse rapidly upon leaving the vent terminal, even when the terminal is located under an open window. However, a window is unlikely to be open when heat is needed.

Paragraph (c) also specifies the location of the exit terminal of direct-vent appliances. All of these locations are shown in Figure A.7.8.

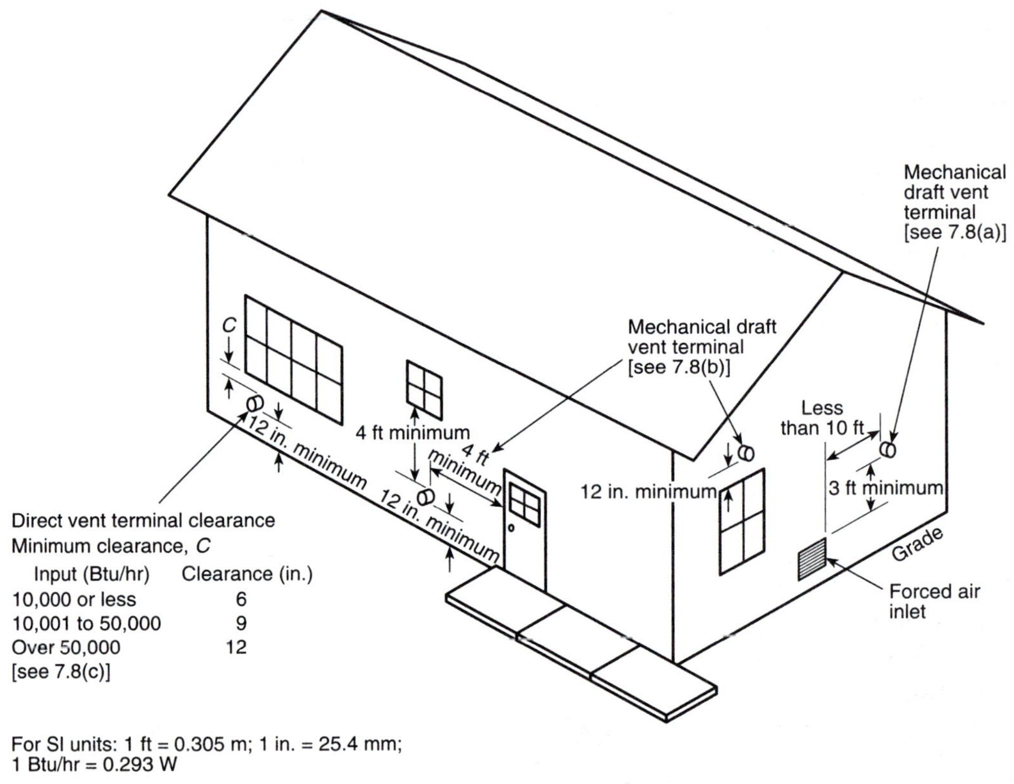

Figure A.7.8 *Exit terminals of mechanical draft and direct-vent venting systems.*

(d) Through-the-wall vents for Category II and Category IV appliances and noncategorized condensing appliances shall not terminate over public walkways or over an area where condensate or vapor could create a nuisance of hazard or could be detrimental to the operation of regulators, relief valves, or other equipment. Where local experience indicates that condensate is a problem with Category I and Category III appliances, this provision shall also apply.

Paragraph (d) provides for the protection of pedestrians and equipment, including gas meters. It places responsibility on the installer to locate vent terminations for

Category II and IV appliances away from walkways and gas equipment. These appliances are high efficiency with low vent temperatures. (See commentary following Section 7.4.1.) The last sentence of 7.8(d) recognizes that any appliance can present a condensation problem in a cold climate.

A.7.8 See Figure A.7.8.

7.9 Condensation Drain

(a) Provision shall be made to collect and dispose of condensate from venting systems serving Category II and Category IV gas utilization equipment and noncategorized condensing appliances in accordance with 7.8(d).

(b) Where local experience indicates that condensation is a problem, provision shall be made to drain off and dispose of condensate from venting systems serving Category I and Category III gas utilization equipment in accordance with 7.8(d).

Category II and IV vent systems operate at lower temperatures than do Category I and III systems, and condensation is expected. Other vent systems also are mandated to drain condensate where it is known to exist.

7.10 Vent Connectors for Category I Gas Utilization Equipment

Vent connectors are relatively short runs of single-wall metal pipe, Type B vent, or other vent materials that are used to connect an appliance to a chimney or vent. When two or more appliances are connected to a chimney or vent, at least one connector is used. When one appliance is connected to a vent with no change in size or venting material, there is no vent connector.

The vent connector has an important effect on the overall operation of the venting system. GRI-sponsored research has proven that a vent primes better and has a stronger draft if there is a connector rise directly over the appliance. It is also important that as much heat as possible is retained in the flue gases as they pass through the connector. For more information, refer to the GRI report by S. M. Ricci, et. al., *Vent Oversizing and the "Seven-Times" Rule: Analysis and Recommendations,* GRI-98/0285.

7.10.1 Where Required.

A vent connector shall be used to connect gas utilization equipment to a gas vent, chimney, or single-wall metal pipe, except where the gas vent, chimney, or single-wall metal pipe is directly connected to the equipment.

7.10.2 Materials.

(a) A vent connector shall be made of noncombustible, corrosion-resistant material capable of withstanding the vent gas temperature produced by the gas utilization equipment and of sufficient thickness to withstand physical damage.

(b) Where the vent connector used for gas utilization equipment having a draft hood or a Category I appliance is located in or passes through an attic space or other unconditioned area, that portion of the vent connector shall be listed Type B or Type L or listed vent material having equivalent insulation qualities.

Attics are commonly used for storage. Consequently, Type B and Type L vent material is specified for vent connectors of appliances installed in attics, because fires occur as a result of the ignition of combustible material that has been carelessly placed on or near single-wall metal vent connectors. In 1999, this was changed to apply to any area that is unconditioned. Type B or equivalent materials are needed to make sure that excessive condensation does not occur.

(c) Vent connectors for residential-type appliances shall comply with the following:

(1) *Vent Connectors Not Installed in Attics, Crawl Spaces, or Other Unconditioned Areas.* Vent connectors for listed gas appliances having draft hoods and for appliances having draft hoods and equipped with listed conversion burners that are not installed in attics, crawl spaces, or other unconditioned areas shall be one of the following:

   a. Type B or Type L vent material
   b. Galvanized sheet steel not less than 0.018 in. (0.46 mm) thick
   c. Aluminum (1100 or 3003 alloy or equivalent) sheet not less than 0.027 in. (0.69 mm) thick
   d. Stainless steel sheet not less than 0.012 in. (0.31 mm) thick
   e. Smooth interior wall metal pipe having resistance to heat and corrosion equal to or greater than that of b, c, or d above
   f. A listed vent connector

(2) Vent connectors shall not be covered with insulation.

Exception: Listed insulated vent connectors shall be installed in accordance with the terms of their listing.

(d) A vent connector for low-heat equipment shall be a factory-built chimney section or steel pipe having resistance to heat and corrosion equivalent to that for the appropriate galvanized pipe as specified in Table 7.10.2(d). Factory-built chimney sections shall be joined together in accordance with the chimney manufacturer's instructions.

(e) Vent connectors for medium-heat equipment and commercial and industrial incinerators shall be constructed of factory-built, medium-heat chimney sections or steel of a thickness not less than that specified in Table 7.10.2(e), and shall comply with the following:

(1) A steel vent connector for equipment with a vent gas temperature in excess of 1000°F (538°C) measured at the entrance to the connector shall be lined with medium-duty fire brick (ASTM C 64, *Specification for Refractories for Incinerators and Boilers, Type F*) or the equivalent.

(2) The lining shall be at least $2^1/_2$ in. (64 mm) thick for a vent connector having a diameter or greatest cross-sectional dimension of 18 in. (460 mm) or less.

(3) The lining shall be at least $4^1/_2$ in. (110 mm) thick laid on the $4^1/_2$-in. (110-mm) bed for a vent connector having a diameter or greatest cross-sectional dimension greater than 18 in. (460 mm).

(4) Factory-built chimney sections, if employed, shall be joined together in accordance with the chimney manufacturer's instructions.

Table 7.10.2(d) Minimum Thickness for Galvanized Steel Vent Connectors for Low-Heat Appliances

Diameter of Connector (in.)	Minimum Thickness (in.)	Diameter of Connector (in.)	Minimum Thickness (in.)
Less than 6	0.019	14 to 16 inclusive	0.034
6 to less than 10	0.023	Over 16	0.056
10 to 12 inclusive	0.029		

For SI units, 1 in. = 25.4 mm; 1 in.^{2} = 645 mm^2.

Table 7.10.2(e) Minimum Thickness for Steel Vent Connectors for Medium Heat Equipment and Commercial Industrial Incinerators

Vent Connector Size		Minimum Thickness (in.)
Diameter (in.)	Area (in.^{2})	
Up to 14	Up to 154	0.053
Over 14 to 16	154 to 201	0.067
Over 16 to 18	201 to 254	0.093
Over 18	Larger than 254	0.123

For SI units, 1 in. = 25.4 mm; 1 in.^{2} = 645 mm^2.

7.10.3* Size of Vent Connector.

The sizing method for vent connectors serving non-fan-assisted Category I appliances has not been changed by the revisions to the vent tables in Chapter 10. When sizing connectors serving fan-assisted Category I appliances, the user will note that the selection of single-wall metal pipe connectors is limited. This limit is imposed because of the lower vent-operating temperature of these appliances, which reduces the heat loss in the connector compared to non-fan-assisted combustion appliances. Single-wall connectors are not permitted when Tables 10-11 through 10-13 are being used to size connectors to exterior masonry chimneys.

(a) A vent connector for gas utilization equipment with a single draft hood or for a Category I fan-assisted combustion system appliance shall be sized and constructed in accordance with Chapter 10 and other approved engineering methods.

(b) For a single appliance having more than one draft hood outlet or flue collar, the manifold shall be constructed according to the instructions of the appliance manufacturer. If there are no instructions, the manifold shall be designed and constructed in accordance with approved engineering practices. As an alternate method, the effective area of the manifold shall equal the combined area of the flue collars or draft hood outlets and the vent connectors shall have a minimum 1 ft (0.3 m) rise.

In the absence of specific instructions, it is important to provide an adequate rise above the draft hoods. This rise is needed to ensure proper flue priming on startup. The minimum 1-ft (0.3-m) connector rise was added in the 1999 edition to limit the "rule of thumb" alternate to prevent spillage.

(c) Where two or more gas appliances are connected to a common vent or chimney, each vent connector shall be sized in accordance with Chapter 10 or other approved engineering methods.

As an alternative method applicable only when all of the appliances are draft hood–equipped, each vent connector shall have an effective area not less than the area of the draft hood outlet of the appliance to which it is connected.

(d) Where two or more gas appliances are vented through a common vent connector or vent manifold, the common vent connector or vent manifold shall be located at the highest level consistent with available headroom and clearance to combustible material and shall be sized in accordance with Chapter 10 or other approved engineering methods.

As an alternate method applicable only where there are two draft hood–equipped appliances, the effective area of the common vent connector or vent manifold and all junction fittings shall be not less than the area of the larger vent connector plus 50 percent of the areas of the smaller flue collar outlets.

This paragraph was revised in 1999 to limit the scope of the alternate "rule of thumb" method to the common venting of two appliances only. Common vents serving more than two appliances should be designed using the venting tables of Chapter 10 or other approved engineering methods.

(e) Where the size of a vent connector is increased to overcome installation limitations and obtain connector capacity equal to the equipment input, the size increase shall be made at the equipment draft hood outlet.

(f) The effective area of the vent connector, where connected to one or more appliances requiring draft for operation, shall be obtained by the application of approved engineering practices to perform as specified in 7.3.1 and 7.3.2.

A.7.10.3 Reference can also be made to the chapter on chimney, gas vent, and fireplace systems of the *ASHRAE Handbook — HVAC Systems and Equipment.*

7.10.4 Two or More Appliances Connected to a Single Vent.

(a) Where two or more vent connectors enter a common gas vent, chimney flue, or single-wall metal pipe, the smaller connector shall enter at the highest level consistent with the available headroom or clearance to combustible material.

(b) Vent connectors serving Category I appliances shall not be connected to any portion of a mechanical draft system operating under positive static pressure, such as those serving Category III or Category IV appliances.

7.10.5 Clearance.

Minimum clearances from vent connectors to combustible material shall be in accordance with Table 7.7.4(d).

Exception: The clearance between a vent connector and combustible material shall be permitted to be reduced where the combustible material is protected as specified for vent connectors in Table 6.2.3(b).

7.10.6 Avoid Unnecessary Bends.

A vent connector shall be installed so as to avoid turns or other construction features that create excessive resistance to flow of vent gases.

7.10.7 Joints.

Joints between sections of connector piping and connections to flue collars or draft hood outlets shall be fastened by sheet metal screws or other approved means.

Each bend or turn (elbow or tee) in a venting system reduces its capacity, especially in a natural draft vent. The capacities shown in the tables in Chapter 10 make an allowance for two 90-degree turns between the draft hood and the top of the venting system. If the venting system contains more than two 90-degree turns, a capacity reduction of 10 percent should be made for each additional turn; for example, the capacity of a venting system containing three 90-degree turns is 90 percent of that indicated in the tables, and with four 90-degree turns, 80 percent. (See 10.1.3 and 10.2.6.)

Exception: Vent connectors of listed vent material, which shall be assembled and connected to flue collars or draft hood outlets in accordance with the manufacturers' instructions.

7.10.8 Slope.

A vent connector shall be installed without any dips or sags and shall slope upward at least $^1/_4$ in./ft (20 mm/m).

7.10.9 Length of Vent Connector.

This subsection was revised in the 1999 edition to limit the horizontal length of a single wall vent or the Type B connector to 75 percent or 100 percent, respectively, of the total height of a chimney or vent.

(a) A vent connector shall be as short as practical and the gas utilization equipment located as close as practical to the chimney or vent.

(b)* Except as provided for in 7.10.3, the maximum horizontal length of a single-wall connector shall be 75 percent of the height of the chimney or vent. Except as provided for in 7.10.3, the maximum horizontal length of a Type B double-wall connector shall be 100 percent of the height of the chimney or vent. For a chimney or vent system serving multiple appliances, the maximum length of an individual connector, from the appliance outlet to the junction with the common vent or another connector, shall be 100 percent of the height of the chimney or vent.

A.7.10.9(b) See A.7.6.3(a).

7.10.10 Support.

A vent connector shall be supported for the design and weight of the material employed to maintain clearances and prevent physical damage and separation of joints.

This requirement was revised in the 1996 edition to emphasize using the tables in Chapter 10 for vent connector design rather than relying on an alternate method. Previously, the alternate method was the rule and the use of the tables in Chapter 10 was an exception.

7.10.11 Location.

Where the vent connector used for gas utilization equipment having a draft hood or for Category I appliances is located in or passes through an attic, crawl space, or other area that can be cold, that portion of the vent connector shall be of listed double-wall Type B, Type L vent material or listed material having equivalent insulation qualities.

7.10.12 Chimney Connection.

In entering a flue in a masonry or metal chimney, the vent connector shall be installed above the extreme bottom to avoid stoppage. A thimble or slip joint shall be permitted to be used to facilitate removal of the connector. The connector shall be firmly attached to or inserted into the thimble or slip joint to prevent the connector from falling out. Means shall be employed to prevent the connector from entering so far as to restrict the space between its end and the opposite wall of the chimney flue.

7.10.13 Inspection.

The entire length of a vent connector shall be readily accessible for inspection, cleaning, and replacement.

7.10.14 Fireplaces.

A vent connector shall not be connected to a chimney flue serving a fireplace unless the fireplace flue opening is permanently sealed.

7.10.15 Passage Through Ceilings, Floors, or Walls.

(a) A vent connector shall not pass through any ceiling, floor, fire wall, or fire partition. A single-wall metal pipe connector shall not pass through any interior wall.

In penetrating an interior wall, floor, or ceiling, a single-wall metal connector (e.g., stove pipe) would be entering a space or room of the building other than that in which the appliance is located. Such space could be a storeroom or other part of the building that is not normally occupied. In such cases, at least some portion of the connector would be out of sight. Therefore, any damage to the connector, such as separation of joints, or perforation by corrosion with consequent leakage of flue gases into the building, or placement of combustible material near or on the connector would escape early detection, creating a potentially hazardous situation.

Exception: Vent connectors made of listed Type B or Type L vent material and serving listed equipment with draft hoods and other equipment listed for Type B gas vents shall be permitted to pass through walls or partitions constructed of combustible material if the connectors are installed with not less than the listed clearance to combustible material.

(b) A vent connector made of a single-wall metal pipe shall not pass through a combustible exterior wall unless guarded at the point of passage by a ventilated metal thimble not smaller than the following:

(1) For listed gas utilization equipment equipped with draft hoods and equipment listed for use with Type B gas vents, 4 in. (100 mm) larger in diameter than the vent connector. Where there is a run of not less than 6 ft (1.8 m) of vent connector in the opening between the draft hood outlet and the thimble, the thimble shall be permitted to be 2 in. (50 mm) larger in diameter than the vent connector.

(2) For unlisted equipment having draft hoods, 6 in. (150 mm) larger in diameter than the vent connector

(3) For residential incinerators and all other residential and low-heat equipment, 12 in. (300 mm) larger in diameter than the vent connector

Exception: In lieu of thimble protection, all combustible material in the wall shall be removed from the vent connector a sufficient distance to provide the specified clearance from such vent connector to combustible material. Any material used to close up such opening shall be noncombustible.

(c) Vent connectors for medium-heat equipment shall not pass through walls or partitions constructed of combustible material.

Note that the term *combustible* is defined in the *National Fuel Gas Code,* Section 1.7, as follows:

Combustible Material. As pertaining to materials adjacent to or in contact with heat-producing appliances, vent connectors, gas vents, chimneys, steam and hot water pipes, and warm air ducts shall mean materials made of or surfaced with wood, compressed paper, plant fibers, or other materials that are capable of being ignited and burned. Such material shall be considered combustible even though flame-proofed, fire-retardant treated, or plastered.

The definition is specific and includes wood and other fibrous materials, even if they are treated to be flame-proof or covered with plaster.

7.11 Vent Connectors for Category II, Category III, and Category IV Gas Utilization Equipment

(*See Section 7.4.*)

The table in 7.4.1 requires that venting materials be furnished or specified by the equipment manufacturer for Category II, III, and IV gas utilization equipment.

7.12 Draft Hoods and Draft Controls

7.12.1 Equipment Requiring Draft Hoods.

Vented gas utilization equipment shall be installed with draft hoods.

Exception: Dual oven-type combination ranges, incinerators, direct-vent equipment, fan-assisted combustion system appliances, equipment requiring chimney draft for operation, single-firebox boilers equipped with conversion burners with inputs greater than 400,000 Btu/hr (117 kW), equipment equipped with blast, power, or pressure burners that are not listed for use with draft hoods, and equipment designed for forced venting.

7.12.2 Installation.

A draft hood supplied with or forming a part of listed vented gas utilization equipment shall be installed without alteration, exactly as furnished and specified by the equipment manufacturer. If a draft hood is not supplied by the equipment manufacturer where one is required, a draft hood shall be installed, be of a listed or approved type, and, in the absence of other instructions, be of the same size as the equipment flue collar. Where a draft hood is required with a conversion burner, it shall be of a listed or approved type.

Exception: If it is determined that a draft hood of special design is needed or preferable for a particular installation, the advice of the gas utilization equipment manufacturer and the approval of the authority having jurisdiction shall be secured.

7.12.3 Draft Control Devices.

Where a draft control device is part of the gas utilization equipment or is supplied by the equipment manufacturer, it shall be installed in accordance with the manufacturer's instructions. In the absence of manufacturer's instructions, the device shall be attached to the flue collar of the equipment or as near to the equipment as practical.

7.12.4* Additional Devices.

Gas utilization equipment (except incinerators) requiring controlled chimney draft shall be permitted to be equipped with a listed double-acting barometric draft regulator installed and adjusted in accordance with the manufacturers' instructions.

A.7.12.4 A device that will automatically shut off gas to the burner in the event of sustained backdraft is recommended if such backdraft might adversely affect burner operation or if flue gas spillage might introduce a hazard. Figure A.7.12.4 shows examples of correct and incorrect locations for barometric draft regulators.

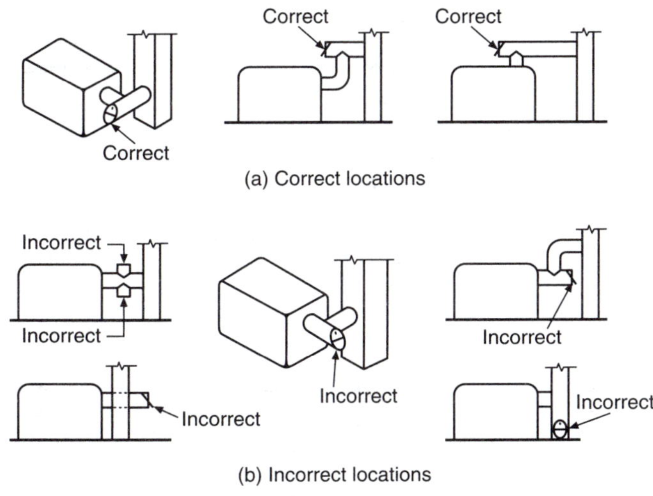

***Figure A.7.12.4** Locations for barometric draft regulators.*

7.12.5 Incinerator Draft Regulator.

A listed gas-fired incinerator shall be permitted to be equipped with a listed single acting barometric draft regulator where recommended by the incinerator manufacturer. This draft regulator shall be installed in accordance with the instructions accompanying the incinerator.

7.12.6 Install in Same Room.

A draft hood or a barometric draft regulator shall be installed in the same room or enclosure as the equipment in such a manner as to prevent any difference in pressure between the hood or regulator and the combustion air supply.

7.12.7 Positioning.

A draft hood or draft regulator shall be installed in the position for which it was designed with reference to the horizontal and vertical planes and shall be located so that the relief opening is not obstructed by any part of the gas utilization equipment or adjacent construction. The equipment and its draft hood shall be located so that the relief opening is accessible for checking vent operation.

7.12.8 Clearance.

A draft hood shall be located so that its relief opening is not less than 6 in. (150 mm) from any surface except that of the gas utilization equipment it serves and the venting system to which the draft hood is connected. Where a greater or lesser clearance is indicated on the equipment label, the clearance shall not be less than that specified on the label. These clearances shall not be reduced.

7.13 Manually Operated Dampers

A manually operated damper shall not be placed in the vent connector from any gas utilization equipment. Fixed baffles shall not be classified as manually operated dampers.

Exception: A connector serving a listed gas-fired incinerator where recommended by the incinerator manufacturer and installed in accordance with the instructions accompanying the incinerator.

7.14 Automatically Operated Vent Dampers

An automatically operated vent damper shall be of a listed type.

Preferably, automatic vent dampers should be included as part of the listed gas utilization equipment. If they are not, then Appendix I, J, or K should be carefully followed as applicable. Note that all of these appendixes also reference, as a first step, the use of Appendix H for a recommended safety inspection procedure for the existing appliance installation prior to the field installation of an automatic vent damper.

7.15 Obstructions

A device that retards the flow of vent gases shall not be installed in a vent connector, chimney, or vent.

Exception No. 1: Draft regulators and safety controls (1) specifically listed for installation in venting systems and installed in accordance with the terms of their listing and (2) designed and installed in accordance with approved engineering methods and approved by the authority having jurisdiction.

Exception No. 2: Listed heat reclaimers and automatically operated vent dampers installed in accordance with the terms of their listing.

Exception No. 3: Approved economizers, heat reclaimers, and recuperators installed in venting systems of equipment not required to be equipped with draft hoods, provided the gas utilization equipment manufacturer's instructions cover the installation of such a device in the venting system and performance in accordance with 7.3.1 and 7.3.2 is obtained.

> Exception No. 3 was revised in the 1996 edition of the code to add the qualification that the gas utilization equipment manufacturer anticipate the addition to the venting system.

Tables in Chapter 10 shall not apply where the equipment covered in Section 7.15, Exception No. 1, 2, or 3 is installed in the vent. Other approved engineering methods shall be used to size these vents.

References Cited in Commentary

The following publications are available from the National Fire Protection Association, 1 Batterymarch Park, P.O. Box 9101, Quincy, MA 02269-9101.

NFPA 91, *Standard for Exhaust Systems for Air Conveying of Vapors, Gases, Mists, and Noncombustible Particulate Solids,* 1999 edition.

NFPA 96, *Standard for Ventilation Control and Fire Protection of Commercial Cooking Operations,* 1998 edition.

NFPA 211, *Standard for Chimneys, Fireplaces, Vents, and Solid Fuel-Burning Appliances,* 2000 edition.

The following publications are available from the Gas Research Institute, 8600 West Bryn Mawr Avenue, Chicago, IL 60631-3562.

S. M. Ricci, et. al., *Vent Oversizing and the "Seven-Times" Rule: Analysis and Recommendations,* Topical Report, Gas Research Institute, GRI-98/0285, 1998.

The following publication is available from the U.S. Government Printing Office, Washington, DC 20402.

PL 100-12, *National Appliance Energy Conservation Act of 1987.*

The following publications are available from Underwriters Laboratories Inc., Publication Stock, 333 Pfingsten Road, Northbrook, IL 60062.

UL 103, *Factory-Built Chimneys for Residential Type and Building Heating Appliances,* 1995.

UL 1738, *Standard for Safe Venting Systems for Gas-Burning Appliances, Categories II, III and IV,* 1993.

Procedures to Be Followed to Place Equipment in Operation

Chapter 8 provides requirements for the final part of appliance installation. It describes the procedures that must be followed after an appliance is installed in place, piped, and connected to its venting system. The chapter covers the following:

- Adjusting the burner input, by pressure adjustment or orifice change, and adjustments needed due to high altitude [over 2000 ft (600 m)] (in Section 8.1)
- Air adjustments (in Section 8.2)
- Verifying operation of automatic ignition, safety shutoffs, and protective devices (in Sections 8.3 through 8.5)
- Checking draft (in Section 8.6)
- Providing operating instruction for the consumer (in Section 8.7)

8.1 Adjusting the Burner Input

Burner input for each gas appliance will be shown on the appliance rating plate. For units with multiple burners, divide the total input by the number of burners to determine the input rating per burner.

8.1.1* Adjusting Input.

The input shall be adjusted to the proper rate in accordance with the equipment manufacturers' instructions by changing the size of a fixed orifice, by changing the adjustment of an adjustable orifice, or by readjusting the gas pressure regulator outlet pressure (where a regulator is provided). Overfiring shall be prohibited. *(See Table 8.1.1.)*

Gas appliances are furnished with low-pressure gas regulators that are specific to the type of gas (i.e., LP-Gas or natural gas) to be used. Each burner will have an orifice that is rated for the proper input. While the code states, "Overfiring [of appliance] shall be prohibited," the intent is that the gas appliance must not be overfired at any rate in excess of its nameplate rating.

In accordance with instructions furnished by the manufacturer, minor adjustments to the appliance input rating can be made with the appliance regulator. Checking the manifold pressure, using either a "U Gauge" manometer or a pressure gauge, is necessary. Consult the appliance instructions for the location of the tap to be used for measuring manifold pressure; it will either be part of an appliance control or be located on the gas manifold itself.

A.8.1.1 Checking Burner Input.

(a) *Checking Burner Input Using a Meter.* To check the Btu input rate, the test hand on the meter should be timed for at least one revolution and the input determined from this timing. Test dials are generally marked $^1/_2$, 1, 2, or 5 ft³/revolution, depending on the size of the meter. Instructions for converting the test hand readings to cubic feet per hour are given in Table 8.1.1.

Use the data in Table 8.1.1, after timing the meter for at least one full revolution. The table shows the input in cubic feet per hour.

(b) *Checking Burner Input Not Using a Meter.* The fixed orifice size for each burner can be determined in accordance with Table F.1 for utility gases and Table F.2 for undiluted liquefied petroleum gases.

To determine the proper orifice size at the prescribed manifold pressure for the appliance, refer to Appendix F, Table F.1 for utility gases and Table F.2 for undiluted LP-Gases. For additional information on orifice sizing, refer to the Appendix F commentary.

Table 8.1.1 Gas Input to Burner in Cubic Feet per Hour

Seconds for One Revolution	Size of Test Meter Dial			
	½ ft³	1 ft³	2 ft³	5 ft³
10	180	360	720	1800
11	164	327	655	1636
12	150	300	600	1500
13	138	277	555	1385
14	129	257	514	1286
15	120	240	480	1200
16	112	225	450	1125

Table 8.1.1 Continued.

Seconds for One Revolution	Size of Test Meter Dial			
	$\frac{1}{2}$ ft^3	1 ft^3	2 ft^3	5 ft^3
17	106	212	424	1059
18	100	200	400	1000
19	95	189	379	947
20	90	180	360	900
21	86	171	343	857
22	82	164	327	818
23	78	157	313	783
24	75	150	300	750
25	72	144	288	720
26	69	138	277	692
27	67	133	267	667
28	64	129	257	643
29	62	124	248	621
30	60	120	240	600
31	58	116	232	581
32	56	113	225	563
33	55	109	218	545
34	53	106	212	529
35	51	103	206	514
36	50	100	200	500
37	49	97	195	486
38	47	95	189	474
39	46	92	185	462
40	45	90	180	450
41	44	88	176	440
42	43	86	172	430
43	42	84	167	420
44	41	82	164	410
45	40	80	160	400
46	39	78	157	391
47	38	77	153	383
48	37	75	150	375
49	37	73	147	367

(continues)

Table 8.1.1 Continued.

Seconds for One Revolution	Size of Test Meter Dial			
	$\frac{1}{2}$ ft^3	1 ft^3	2 ft^3	5 ft^3
50	36	72	144	360
51	35	71	141	353
52	35	69	138	346
53	34	68	136	340
54	33	67	133	333
55	33	65	131	327
56	32	64	129	321
57	32	63	126	316
58	31	62	124	310
59	30	61	122	305
60	30	60	120	300
62	29	58	116	290
64	29	56	112	281
66	29	54	109	273
68	28	53	106	265
70	26	51	103	257
72	25	50	100	250
74	24	48	97	243
76	24	47	95	237
78	23	46	92	231
80	22	45	90	225
82	22	44	88	220
84	21	43	86	214
86	21	42	84	209
88	20	41	82	205
90	20	40	80	200
94	19	38	76	192
98	18	37	74	184
100	18	36	72	180
104	17	35	69	173
108	17	33	67	167
112	16	32	64	161
116	15	31	62	155
120	15	30	60	150
130	14	28	55	138

Table 8.1.1 Continued.

Seconds for One Revolution	Size of Test Meter Dial			
	½ ft³	1 ft³	2 ft³	5 ft³
140	13	26	51	129
150	12	24	48	120
160	11	22	45	112
170	11	21	42	106
180	10	20	40	100

Note: To convert to Btu per hour, multiply by the Btu heating value of the gas used.

8.1.2 High Altitude.

Ratings of gas utilization equipment are based on sea level operation and shall not be changed for operation at elevations up to 2000 ft (600 m). For operation at elevations above 2000 ft (600 m), equipment ratings shall be reduced at the rate of 4 percent for each 1000 ft (300 m) above sea level before selecting appropriately sized equipment.

An example of the derating required for high altitudes [e.g., any installation above 2000 ft (600 m)] would be to multiply the altitude (in thousands of feet) by 4 percent and then deduct that rating from the sea level rating. For example, a 100,000-Btu (29-kW) furnace installed at an elevation of 5000 ft (1500 m) would have to be reduced by 20 percent to a rating of 80,000 Btu (23 kW) (5×4 percent = 20 percent reduction). For additional information on altitude derating, refer to the Appendix F commentary.

Exception No. 1: As permitted by the authority having jurisdiction.

Exception No. 2: Listed appliances shall be permitted to be derated in accordance with the terms of the listing.

The addition of the new Exception No. 2 allows the appliances to be derated according to the manufacturer's listing, which is more accurate, because it involves testing specific to the appliance. This exception is applicable primarily to fan-assisted combustion appliances. The induced draft blower, depending on its design and motor characteristics, can provide sufficient air to compensate for the lower air density at higher elevations.

8.2* Primary Air Adjustment

The primary air for injection (Bunsen)-type burners shall be adjusted for proper flame characteristics in accordance with the manufacturers' instructions. After setting the primary air, the adjustment means shall be secured in position.

A typical primary air adjustment for burners is an air shutter at the inlet of the burner, which can be rotated to open or close the openings, thus varying the amount of primary air. Other burners can have an interrupter device (or spoiler) in the venturi of the burner. This device can be turned in accordance with the manufacturer's instructions to change the flame characteristics. Always adjust the burner for a proper blue flame with no yellow tips. With LP-Gas, a slight yellow tip can be normal. (See Exhibit 8.1.)

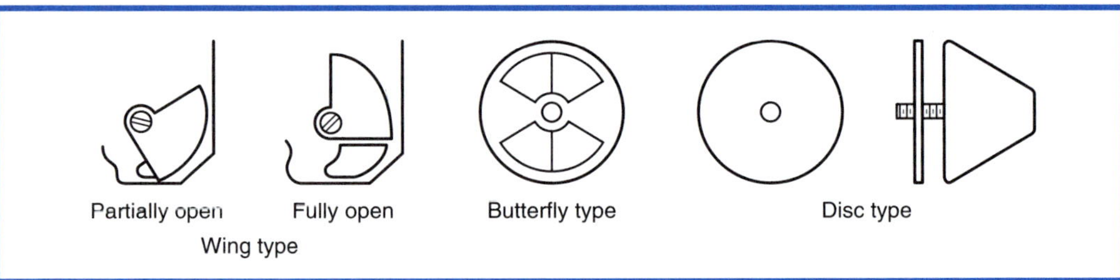

Exhibit 8.1 *Examples of burner primary air shutter adjustments. (Courtesy of American Gas Association.)*

A.8.2 Normally, the primary air adjustment should first be set to give a soft blue flame having luminous tips and then increased to a point where the yellow tips just disappear. If the burner cannot be so adjusted, the manufacturer or serving gas supplier should be contacted.

8.3 Safety Shutoff Devices

Where a safety shutoff device is provided, it shall be checked for proper operation and adjustment in accordance with the manufacturer's instructions. If the device does not function properly to turn off the gas supply in the event of pilot outage, it shall be properly serviced or replaced with new equipment.

The safety shutoff device will either shut off the supply of gas to the main burner or to both the main burner and the pilot burner. Consult the manufacturer's instructions to determine the type of device used. Test the equipment for proper operation of the shutoff device.

8.4 Automatic Ignition

Gas utilization equipment supplied with means for automatic ignition shall be checked for proper operation. If necessary, proper adjustments shall be made.

Most gas appliances are equipped with an automatic ignition system, incorporating either a constant burning pilot and safety shutoff device, or one of several different types of spark or glow coil ignition devices. Follow the manufacturer's instructions when checking these devices for proper operation.

8.5 Protective Devices

All protective devices furnished with the gas utilization equipment, such as a limit control, fan control to blower, temperature and pressure relief valve, low-water cutoff device, or manual operating features, shall be checked for proper operation.

8.6* Checking the Draft

Vent-connected gas utilization equipment shall be operated for several minutes and checked to see that the combustion products are going up the chimney or gas vent properly by passing a lighted match or taper around the edge of the relief opening of the draft hood. If the chimney or gas vent is drawing properly, the match flame will be drawn into the draft hood. If not, the combustion products will tend to extinguish this flame. If the combustion products are escaping from the relief opening of the draft hood, the equipment shall not be operated until proper adjustments or repairs are made to provide adequate draft through the chimney or gas vent.

If the appliance is installed in a closet with insufficient room for service personnel to be in the closet when the door is closed, then using a manometer or gauge with a probe inserted in the vent connector can be necessary to make sure that the burner, draft diverter, and vent are performing satisfactorily.

A.8.6 A procedure for checking draft can be found in Appendix H, steps 7, 8, and 10 through 14.

8.7 Operating Instructions

Operating instructions shall be furnished and shall be left in a prominent position near the equipment for the use of the consumer.

Operating instructions are required for each gas appliance and usually are included in the packet containing the installation instructions. The installer should be sure that the owner/user is aware of the proper operating procedures and, if necessary, should explain the function of the appliance. If the unit is equipped with a remote thermostat, show the owner how to operate the appliance from the thermostat location.

CHAPTER 9

Sizing Tables

This chapter was created first for the 1992 edition by relocating the pipe-sizing tables from Appendix C and by expanding the propane pipe-sizing tables between first- and second-stage regulators and the table covering piping of propane at appliance delivery pressure. In addition, new tables were added, providing sizing information for corrugated stainless steel tubing (CSST). In the 1996 edition, tables were added for CSST and Polyethylene, and the tables were reorganized for ease of use in the following order: (1) type of gas (e.g., Natural, Propane), (2) type of piping material (e.g., Steel, Copper, CSST. Polyethylene), and (3) pressure (lowest pressure first).

As appendix material in previous editions, the sizing tables were informative and were not adopted when the code was adopted by a government body. This sizing information was placed in the code to make it one of the mandatory design options. Other engineering methods that are approved by the authority having jurisdiction are also permitted as design options (see 2.4.3, A.2.4.3).

An example of the use of the pipe-sizing tables is found in Appendix C.

In the 1999 edition, several tables were revised for consistency and to correct minor errors.

9.1 Tables for Sizing Gas Piping Systems

Tables 9.1 through 9.34 can be used to size gas piping systems as required in Chapter 2. For SI units, 1 ft^3 = 0.028 m^3; 1 ft = 0.305 m; 1 in. water column = 0.249 kPa; 1 psi = 6.894 kPa; 1000 Btu/hr = 0.293 kW.

Table 9.1 Maximum Capacity of Pipe in Cubic Feet of Gas per Hour for Gas Pressures of 0.5 psi or Less and a Pressure Drop of 0.3 in. Water Column (Based on a 0.60 Specific Gravity Gas)

Nominal Iron Pipe Size (in.)	Internal Diameter (in.)	Length of Pipe (ft)													
		10	20	30	40	50	60	70	80	90	100	125	150	175	200
1/4	0.364	32	22	18	15	14	12	11	11	10	9	8	8	7	6
3/8	0.493	72	49	40	34	30	27	25	23	22	21	18	17	15	14
1/2	0.622	132	92	73	63	56	50	46	43	40	38	34	31	28	26
3/4	0.824	278	190	152	130	115	105	96	90	84	79	72	64	59	55
1	1.049	520	350	285	245	215	195	180	170	160	150	130	120	110	100
1 1/4	1.380	1050	730	590	500	440	400	370	350	320	305	275	250	225	210
1 1/2	1.610	1600	1100	890	760	670	610	560	530	490	460	410	380	350	320
2	2.067	3050	2100	1650	1450	1270	1150	1050	990	930	870	780	710	650	610
2 1/2	2.469	4800	3300	2700	2300	2000	1850	1700	1600	1500	1400	1250	1130	1050	980
3	3.068	8500	5900	4700	4100	3600	3250	3000	2800	2600	2500	2200	2000	1850	1700
4	4.026	17500	12000	9700	8300	7400	6800	6200	5800	5400	5100	4500	4100	3800	3500

Table 9.2 Maximum Capacity of Pipe in Cubic Feet of Gas per Hour for Gas Pressures of 0.5 psi or Less and a Pressure Drop of 0.5 in. Water Column (Based on a 0.60 Specific Gravity Gas)

Nominal Iron Pipe Size (in.)	Internal Diameter (in.)	Length of Pipe (ft)													
		10	20	30	40	50	60	70	80	90	100	125	150	175	200
$^1/_4$	0.364	43	29	24	20	18	16	15	14	13	12	11	10	9	8
$^3/_8$	0.493	95	65	52	45	40	36	33	31	29	27	24	22	20	19
$^1/_2$	0.622	175	120	97	82	73	66	61	57	53	50	44	40	37	35
$^3/_4$	0.824	360	250	200	170	151	138	125	118	110	103	93	84	77	72
1	1.049	680	465	375	320	285	260	240	220	205	195	175	160	145	135
$1^1/_4$	1.380	1400	950	770	660	580	530	490	460	430	400	360	325	300	280
$1^1/_2$	1.610	2100	1460	1180	990	900	810	750	690	650	620	550	500	460	430
2	2.067	3950	2750	2200	1900	1680	1520	1400	1300	1220	1150	1020	950	850	800
$2^1/_2$	2.469	6300	4350	3520	3000	2650	2400	2250	2050	1950	1850	1650	1500	1370	1280
3	3.068	11000	7700	6250	5300	4750	4300	3900	3700	3450	3250	2950	2650	2450	2280
4	4.026	23000	15800	12800	10900	9700	8800	8100	7500	7200	6700	6000	5500	5000	4600

Table 9.3 *Pipe Sizing Table for 2 psi Pressure Capacity of Pipes of Different Diameters and Lengths in Cubic Feet per Hour for an Initial Pressure of 2.0 psi with a 1.0 psi Pressure Drop and a Gas of 0.6 Specific Gravity*

Pipe Size of Schedule 40 Standard Pipe (in.)	Internal Diameter (in.)	Total Equivalent Length of Pipe (ft)													
		10	20	30	40	50	60	70	80	90	100	125	150	175	200
1/2	0.622	1506	1065	869	753	673	615	569	532	502	462	414	372	344	318
3/4	0.824	3041	2150	1756	1521	1360	1241	1150	1075	1014	934	836	751	695	642
1	1.049	5561	3932	3211	2781	2487	2270	2102	1966	1854	1708	1528	1373	1271	1174
1 1/4	1.380	11415	8072	6591	5708	5105	4660	4315	4036	3805	3508	3138	2817	2608	2413
1 1/2	1.610	17106	12096	9876	8553	7650	6983	6465	6048	5702	5257	4702	4222	3909	3613
2	2.067	32944	23295	19020	16472	14733	13449	12452	11647	10981	10125	9056	8130	7527	6959
2 1/2	2.469	52505	37127	30314	26253	23481	21435	19845	18563	17502	16138	14434	12960	11999	11093
3	3.068	92819	65633	53589	46410	41510	37893	35082	32817	30940	28530	25518	22911	21211	19608
4	4.026	189326	133873	109307	94663	84669	77292	71558	66937	63109	58194	52050	46732	43265	39997

Table 9.4 *Pipe Sizing Table for 5 psi Pressure Capacity of Pipes of Different Diameters and Lengths in Cubic Feet per Hour for an Initial Pressure of 5.0 psi with a 3.5 psi Pressure Drop and a Gas of 0.6 Specific Gravity*

Pipe Size of Schedule 40 Standard Pipe (in.)	Internal Diameter (in.)	Total Equivalent Length of Pipe (ft)													
		10	20	30	40	50	60	70	80	90	100	125	150	175	200
1/2	0.622	3185	2252	1839	1593	1425	1301	1204	1153	1062	979	876	786	728	673
3/4	0.824	6434	4550	3715	3217	2878	2627	2432	2330	2145	1978	1769	1589	1471	1360
1	1.049	11766	8320	6793	5883	5262	4804	4447	4260	3922	3617	3235	2905	2690	2487
1 1/4	1.380	24161	17084	13949	12080	10805	9864	9132	8542	8054	7427	6643	5964	5522	5104
1 1/2	1.610	36206	25602	20904	18103	16192	14781	13685	12801	12069	11128	9953	8937	8274	7649
2	2.067	69727	49305	40257	34864	31183	28466	26354	24652	23242	21433	19170	17211	15934	14729
2 1/2	2.469	111133	78583	64162	55566	49700	45370	42004	39291	37044	34159	30553	27431	25396	23478
3	3.068	196468	138924	113431	98234	87863	80208	74258	69462	65489	60387	54012	48494	44897	41504
4	4.026	400732	283361	231363	200366	179213	163598	151463	141680	133577	123173	110169	98911	91574	84656

Table 9.5 Pipe Sizing Table for Pressures Under 1 Pound Approximate Capacity of Pipes of Different Diameters and Lengths in Cubic Feet per Hour with Pressure Drop of 0.3 in. Water Column and 0.6 Specific Gravity

Pipe Size of Schedule 40 Standard Pipe (in.)	Internal Diameter (in.)	Total Equivalent Length of Pipe (ft)										
		50	100	150	200	250	300	400	500	1000	1500	2000
1.00	1.049	215	148	119	102	90	82	70	62	43	34	29
1.25	1.380	442	304	244	209	185	168	143	127	87	70	60
1.50	1.610	662	455	366	313	277	251	215	191	131	105	90
2.00	2.067	1275	877	704	602	534	484	414	367	252	203	173
2.50	2.469	2033	1397	1122	960	851	771	660	585	402	323	276
3.00	3.068	3594	2470	1983	1698	1505	1363	1167	1034	711	571	488
3.50	3.548	5262	3616	2904	2485	2203	1996	1708	1514	1041	836	715
4.00	4.026	7330	5038	4046	3462	3069	2780	2380	2109	1450	1164	996
5.00	5.047	13261	9114	7319	6264	5552	5030	4305	3816	2623	2106	1802
6.00	6.065	21472	14758	11851	10143	8990	8145	6971	6178	4246	3410	2919
8.00	7.981	44118	30322	24350	20840	18470	16735	14323	12694	8725	7006	5997
10.00	10.020	80130	55073	44225	37851	33547	30396	26015	23056	15847	12725	10891
12.00	11.938	126855	87187	70014	59923	53109	48120	41185	36501	25087	20146	17242

Table 9.6 Pipe Sizing Table for Pressures Under 1 Pound Approximate Capacity of Pipes of Different Diameters and Lengths in Cubic Feet per Hour with Pressure Drop of 0.5 in. Water Column and 0.6 Specific Gravity

Pipe Size of Schedule 40 Standard Pipe (in.)	Internal Diameter (in.)	Total Equivalent Length of Pipe (ft)										
		50	100	150	200	250	300	400	500	1000	1500	2000
1.00	1.049	284	195	157	134	119	108	92	82	56	45	39
1.25	1.380	583	400	322	275	244	221	189	168	115	93	79
1.50	1.610	873	600	482	412	366	331	283	251	173	139	119
2.00	2.067	1681	1156	928	794	704	638	546	484	333	267	229
2.50	2.469	2680	1842	1479	1266	1122	1017	870	771	530	426	364
3.00	3.068	4738	3256	2615	2238	1983	1797	1538	1363	937	752	644
3.50	3.548	6937	4767	3828	3277	2904	2631	2252	1996	1372	1102	943
4.00	4.026	9663	6641	5333	4565	4046	3666	3137	2780	1911	1535	1313
5.00	5.047	17482	12015	9649	8258	7319	6632	5676	5030	3457	2776	2376
6.00	6.065	28308	19456	15624	13372	11851	10738	9190	8145	5598	4496	3848
8.00	7.981	58161	39974	32100	27474	24350	22062	18383	16735	11502	9237	7905
10.00	10.020	105636	72603	58303	49900	44225	40071	34296	30396	20891	16776	14358
12.00	11.938	167236	114940	92301	78998	70014	63438	54295	48120	33073	26559	22731

Table 9.7 Pipe Sizing Table for 1 Pound Pressure Capacity of Pipes of Different Diameters and Lengths in Cubic Feet per Hour for an Initial Pressure of 1.0 psi with a 10 Percent Pressure Drop and a Gas of 0.6 Specific Gravity

Pipe Size of Schedule 40 Standard Pipe (in.)	Internal Diameter (in.)	Total Equivalent Length of Pipe (ft)										
		50	100	150	200	250	300	400	500	1000	1500	2000
1.00	1.049	717	493	396	338	300	272	233	206	142	114	97
1.25	1.380	1471	1011	812	695	616	558	478	423	291	234	200
1.50	1.610	2204	1515	1217	1041	923	836	716	634	436	350	300
2.00	2.067	4245	2918	2343	2005	1777	1610	1378	1222	840	674	577
2.50	2.469	6766	4651	3735	3196	2833	2567	2197	1947	1338	1075	920
3.00	3.068	11962	8221	6602	5650	5008	4538	3884	3442	2366	1900	1626
3.50	3.548	17514	12037	9666	8273	7332	6644	5686	5039	3464	2781	2381
4.00	4.026	24398	16769	13466	11525	10214	9255	7921	7020	4825	3875	3316
5.00	5.047	44140	30337	24362	20851	18479	16744	14330	12701	8729	7010	6000
6.00	6.065	71473	49123	39447	33762	29923	27112	23204	20566	14135	11351	9715
8.00	7.981	146849	100929	81049	69368	61479	55705	47676	42254	29041	23321	19960
10.00	10.020	266718	183314	147207	125990	111663	101175	86592	76745	52747	42357	36252
12.00	11.938	422248	290209	233048	199459	176777	160172	137087	121498	83505	67057	57392

Table 9.8 Pipe Sizing Table for 2 Pounds Pressure Capacity of Pipes of Different Diameters and Lengths in Cubic Feet per Hour for an Initial Pressure of 2.0 psi with a 10 Percent Pressure Drop and a Gas of 0.6 Specific Gravity

Pipe Size of Schedule 40 Standard Pipe (in.)	Internal Diameter (in.)	Total Equivalent Length of Pipe (ft)										
		50	100	150	200	250	300	400	500	1000	1500	2000
1.00	1.049	1112	764	614	525	466	422	361	320	220	177	151
1.25	1.380	2283	1569	1260	1079	956	866	741	657	452	363	310
1.50	1.610	3421	2351	1888	1616	1432	1298	1111	984	677	543	465
2.00	2.067	6589	4528	3636	3112	2758	2499	2139	1896	1303	1046	896
2.50	2.469	10501	7217	5796	4961	4396	3983	3409	3022	2077	1668	1427
3.00	3.068	18564	12759	10246	8769	7772	7042	6027	5342	3671	2948	2523
3.50	3.548	27181	18681	15002	12840	11379	10311	8825	7821	5375	4317	3694
4.00	4.026	37865	26025	20899	17887	15853	14364	12293	10895	7488	6013	5147
5.00	5.047	68504	47082	37809	32359	28680	25986	22240	19711	13547	10879	9311
6.00	6.065	110924	76237	61221	52397	46439	42077	36012	31917	21936	17616	15077
8.00	7.981	227906	156638	125786	107657	95414	86452	73992	65578	45071	36194	30977
10.00	10.020	413937	284497	228461	195533	173297	157020	134389	119106	81861	65737	56263
12.00	11.938	655315	450394	361682	309553	274351	248582	212754	188560	129596	104070	89071

Table 9.9 Pipe Sizing Table for 5 Pounds Pressure Capacity of Pipes of Different Diameters and Lengths in Cubic Feet per Hour for an Initial Pressure of 5.0 psi with a 10 Percent Pressure Drop and a Gas of 0.6 Specific Gravity

Pipe Size of Schedule 40 Standard Pipe (in.)	Internal Diameter (in.)	Total Equivalent Length of Pipe (ft)										
		50	100	150	200	250	300	400	500	1000	1500	2000
1.00	1.049	1989	1367	1098	940	833	755	646	572	393	316	270
1.25	1.380	4084	2807	2254	1929	1710	1549	1326	1175	808	649	555
1.50	1.610	6120	4206	3378	2891	2562	2321	1987	1761	1210	972	832
2.00	2.067	11786	8101	6505	5567	4934	4471	3827	3391	2331	1872	1602
2.50	2.469	18785	12911	10368	8874	7865	7126	6099	5405	3715	2983	2553
3.00	3.068	33209	22824	18329	15687	13903	12597	10782	9556	6568	5274	4514
3.50	3.548	48623	33418	26836	22968	20356	18444	15786	13991	9616	7722	6609
4.00	4.026	67736	46555	37385	31997	28358	25694	21991	19490	13396	10757	9207
5.00	5.047	122544	84224	67635	57887	51304	46485	39785	35261	24235	19461	16656
6.00	6.065	198427	136378	109516	93732	83073	75270	64421	57095	39241	31512	26970
8.00	7.981	407692	280204	225014	192583	170683	154651	132361	117309	80626	64745	55414
10.00	10.020	740477	508926	408686	349782	310005	280887	240403	213065	146438	117595	100646
12.00	11.938	1172269	805694	647001	553749	490777	444680	380588	337309	231830	186168	159336

Table 9.10 Pipe Sizing Table for 10 Pounds Pressure Capacity of Pipes of Different Diameters and Lengths in Cubic Feet per Hour for an Initial Pressure of 10.0 psi with a 10 Percent Pressure Drop and a Gas of 0.6 Specific Gravity

Pipe Size of Schedule 40 Standard Pipe (in.)	Internal Diameter (in.)	Total Equivalent Length of Pipe (ft)										
		50	100	150	200	250	300	400	500	1000	1500	2000
1.00	1.049	3259	2240	1798	1539	1364	1236	1058	938	644	517	443
1.25	1.380	6690	4598	3692	3160	2801	2538	2172	1925	1323	1062	909
1.50	1.610	10024	6889	5532	4735	4197	3802	3254	2884	1982	1592	1362
2.00	2.067	19305	13268	10655	9119	8082	7323	6268	5555	3818	3066	2624
2.50	2.469	30769	21148	16982	14535	12882	11672	9990	8854	6085	4886	4182
3.00	3.068	54395	37385	30022	25695	22773	20634	17660	15652	10757	8638	7393
3.50	3.548	79642	54737	43956	37621	33343	30211	25857	22916	15750	12648	10825
4.00	4.026	110948	76254	61235	52409	46449	42086	36020	31924	21941	17620	15080
5.00	5.047	200720	137954	110782	94815	84033	76140	65166	57755	39695	31876	27282
6.00	6.065	325013	223379	179382	153527	136068	123288	105518	93519	64275	51615	44176
8.00	7.981	667777	458959	358561	315440	279569	253310	216800	192146	132061	106050	90765
10.00	10.020	1212861	833593	659404	572924	507772	460078	393767	348988	239858	192614	164853
12.00	11.938	1920112	1319682	1059751	907010	803866	728361	623383	552493	379725	304933	260983

Table 9.11 Pipe Sizing Table for 20 Pounds Pressure Capacity of Pipes of Different Diameters and Lengths in Cubic Feet per Hour for an Initial Pressure of 20.0 psi with a 10 Percent Pressure Drop and a Gas of 0.6 Specific Gravity

Pipe Size of Schedule 40 Standard Pipe (in.)	Internal Diameter (in.)	Total Equivalent Length of Pipe (ft)										
		50	100	150	200	250	300	400	500	1000	1500	2000
1.00	1.049	5674	3900	3132	2680	2375	2152	1842	1633	1122	901	771
1.25	1.380	11649	8006	6429	5503	4877	4419	3782	3352	2304	1850	1583
1.50	1.610	17454	11996	9633	8245	7307	6621	5667	5022	3452	2772	2372
2.00	2.067	33615	23103	18553	15879	14073	12751	10913	9672	6648	5338	4569
2.50	2.469	53577	36823	29570	25308	22430	20323	17394	15416	10595	8509	7282
3.00	3.068	94714	65097	52275	44741	39653	35928	30750	27253	18731	15042	12874
3.50	3.548	138676	95311	76538	65507	58058	52604	45023	39903	27425	22023	18849
4.00	4.026	193187	132777	106624	91257	80879	73282	62720	55538	38205	30680	26258
5.00	5.047	349503	240211	192898	165096	146322	132578	113470	100566	69118	55505	47505
6.00	6.065	565926	388958	312347	267329	236928	214674	183733	162840	111919	89875	76921
8.00	7.981	1162762	799160	641754	549258	486797	441074	377502	334573	229950	184658	158043
10.00	10.020	2111887	1451488	1165596	997600	884154	801108	685645	607674	417651	335388	287049
12.00	11.938	3343383	2297888	1845285	1579326	1399727	1268254	1085462	962025	661194	530962	454435

Table 9.12 Pipe Sizing Table for 50 Pounds Pressure Capacity of Pipes of Different Diameters and Lengths in Cubic Feet per Hour for an Initial Pressure of 50.0 psi with a 10 Percent Pressure Drop and a Gas of 0.6 Specific Gravity

Pipe Size of Schedule 40 Standard Pipe (in.)	Internal Diameter (in.)	Total Equivalent Length of Pipe (ft)										
		50	100	150	200	250	300	400	500	1000	1500	2000
1.00	1.049	12993	8930	7171	6138	5440	4929	4218	3739	2570	2063	1766
1.25	1.380	26676	18335	14723	12601	11168	10119	8661	7676	5276	4236	3626
1.50	1.610	39970	27471	22060	18881	16733	15162	12976	11501	7904	6348	5433
2.00	2.067	76977	52906	42485	36362	32227	29200	24991	22149	15223	12225	10463
2.50	2.469	122690	84324	67715	57955	51365	46540	39832	35303	24263	19484	16676
3.00	3.068	216893	149070	119708	102455	90804	82275	70417	62409	42893	34445	29480
3.50	3.548	317564	218260	175271	150009	132950	120463	103100	91376	62802	50432	43164
4.00	4.026	442393	304054	244166	208975	185211	167814	143627	127294	87489	70256	60130
5.00	5.047	800352	550077	441732	378065	335072	303600	259842	230293	158279	127104	108784
6.00	6.065	1295955	890703	715266	612175	542559	491598	420744	372898	256291	205810	176147
8.00	7.981	2662693	1830054	1469598	1257785	1114752	1010046	864469	766163	526579	422862	361915
10.00	10.020	4836161	3323866	2669182	2284474	2024687	1834514	1570106	1391556	956409	768030	657334
12.00	11.938	7656252	5262099	4225651	3616611	3205335	2904266	2485676	2203009	1514115	1215888	1040643

Table 9.13 Maximum Capacity of Semi-Rigid Tubing in Cubic Feet of Gas per Hour for Gas Pressures of 0.5 psi or Less and a Pressure Drop of 0.3 in. Water Column (Based on a 0.60 Specific Gravity Gas)

Outside Diameter (in.)	Length of Tubing (ft)													
	10	20	30	40	50	60	70	80	90	100	125	150	175	200
$3/8$	20	14	11	10	9	8	7	7	6	6	5	5	4	4
$1/2$	42	29	23	20	18	16	15	14	13	12	11	10	9	8
$5/8$	86	59	47	40	36	33	30	28	26	25	22	20	18	17
$3/4$	150	103	83	71	63	57	52	49	46	43	38	35	32	30
$7/8$	212	146	117	100	89	81	74	69	65	61	54	49	45	42

Table 9.14 Maximum Capacity of Semi-Rigid Tubing in Cubic Feet of Gas per Hour for Gas Pressures of 0.5 psi or Less and a Pressure Drop of 0.5 in. Water Column (Based on a 0.60 Specific Gravity Gas)

Outside Diameter (in.)	Length of Tubing (ft)													
	10	20	30	40	50	60	70	80	90	100	125	150	175	200
$3/8$	27	18	15	13	11	10	9	9	8	8	7	6	6	5
$1/2$	56	38	31	26	23	21	19	18	17	16	14	13	12	11
$5/8$	113	78	62	53	47	43	39	37	34	33	29	26	24	22
$3/4$	197	136	109	93	83	75	69	64	60	57	50	46	42	39
$7/8$	280	193	155	132	117	106	98	91	85	81	71	65	60	55

Table 9.15 Maximum Capacity of Semi-Rigid Tubing in Cubic Feet of Gas per Hour for Gas Pressures of 0.5 psi or Less and a Pressure Drop of 1.0 in. Water Column (Based on a 0.6 Specific Gravity Gas)

Use this table to size tubing from house line regulator to the appliance.

Length (ft)	Diameter: Inside (Outside)						
	1/4 in. (0.315 in.)	3/8 in. (0.430 in.)	1/2 in. (0.545 in.)	5/8 in. (0.666 in.)	3/4 in. (0.785 in.)	1 in. (1.025 in.)	1 1/4 in. (1.265 in.)
10	42	95	177	300	461	928	1612
15	34	76	142	241	370	745	1294
20	29	65	122	206	317	638	1108
30	23	52	98	165	255	512	890
40	20	45	84	142	218	439	761
50	18	40	74	125	193	389	675
60	16	36	67	114	175	352	611
70	15	33	62	105	161	324	563
80	14	31	57	97	150	301	523
90	13	29	54	91	140	283	491
100	12	27	51	86	133	267	464
125	11	24	45	76	118	237	411
150	10	22	41	69	107	215	372
175	9	20	38	64	98	197	343
200	8	19	35	59	91	184	319
250	7	17	31	53	81	163	283
300	7	15	28	48	73	147	256

Table 9.16 Maximum Capacity of Semi-Rigid Tubing in Cubic Feet per Hour for a Gas Pressure of 2 psi or Less and Pressure Drop of 17 in. Water Column (Based on a 0.6 Specific Gravity Gas)

Length (ft)	Diameter: Inside (Outside)						
	$\frac{1}{4}$ in. (0.315 in.)	$\frac{3}{8}$ in. (0.430 in.)	$\frac{1}{2}$ in. (0.545 in.)	$\frac{5}{8}$ in. (0.666 in.)	$\frac{3}{4}$ in. (0.785 in.)	1 in. (1.025 in.)	$1\frac{1}{4}$ in. (1.265 in.)
10	201	454	845	1435	2200	4428	7690
15	161	364	678	1152	1766	3556	6175
20	138	312	581	986	1512	3044	5285
30	111	250	466	792	1214	2444	4244
40	95	214	399	678	1039	2092	3632
50	84	190	354	601	921	1854	3219
60	76	172	320	544	834	1680	2917
70	70	158	295	501	768	1545	2684
80	65	147	274	466	714	1438	2496
90	61	139	257	437	670	1349	2342
100	58	131	243	413	633	1274	2213
125	51	116	215	366	561	1129	1961
150	46	105	195	332	508	1023	1777
175	43	96	180	305	468	941	1635
200	40	90	167	284	435	876	1521
250	35	80	148	251	386	776	1348
300	32	72	134	228	349	703	1121

Table 9.17　Maximum Capacity of Semi-Rigid Tubing in Cubic Feet of Gas per Hour for a Gas Pressure of 2.0 psi or Less and a Pressure Drop of 1.0 psi (Based on a 0.6 Specific Gravity Gas)

Nominal Tubing Diameter Inside (in.)	Internal Diameter (in.)	Length of Tubing (ft)																	
		5	10	15	20	30	40	50	60	70	80	90	100	125	150	175	200	250	300
1/4	0.315	459	306	242	204	163	139	122	110	100	93	87	82	72	65	59	54	49	43
3/8	0.430	1071	722	569	484	382	323	285	255	234	217	204	191	168	151	139	124	119	102
1/2	0.545	2040	1385	1088	918	731	620	548	493	450	416	391	365	323	289	268	246	217	195
5/8	0.666	3527	2363	1827	1581	1258	1062	935	850	773	722	671	629	552	497	459	425	374	336
3/4	0.785	5524	3697	2932	2507	1955	1700	1487	1326	1215	1130	1045	986	871	782	718	671	586	527
1	1.025	8923	6459	5269	4589	3739	3229	2847	2592	2380	2252	2125	1997	1785	1615	1530	1445	1275	1147
1 1/4	1.265	17847	12748	10198	8923	7309	6374	5694	5184	459	4419	4164	3994	3627	3229	3017	2804	2507	2295
1 1/2	1.505	26345	18696	15297	12748	11048	9348	8328	7649	6969	6544	6119	5779	5184	4759	4419	4164	3654	3399
2	1.985	49291	34843	28894	24645	20396	16997	15297	14447	12748	11898	11473	10623	9603	8838	8243	7649	6884	6289
2 1/2	2.465	76485	54390	44192	38243	30594	27195	23795	22096	20396	18696	17847	16997	15297	13597	13172	11898	10623	9773

Table 9.18 Maximum Capacity of Semi-Rigid Tubing in Cubic Feet of Gas per Hour for a Gas Pressure of 5.0 psi or Less and a Pressure Drop of 3.5 psi (Based on a 0.6 Specific Gravity Gas)

Nominal Tubing Diameter Inside (in.)	Internal Diameter (in.)	Length of Tubing (ft)																	
		5	10	15	20	30	40	50	60	70	80	90	100	125	150	175	200	250	300
$1/4$	0.315	791	527	417	351	281	239	209	190	173	161	149	141	124	111	101	94	85	75
$3/8$	0.430	1845	1245	981	835	659	556	490	439	403	373	351	329	290	261	240	214	205	176
$1/2$	0.545	3514	2387	1874	1581	1259	1069	944	849	776	717	674	630	556	498	461	425	373	337
$5/8$	0.666	6076	4070	3148	2723	2167	1830	1611	1464	1332	1245	1157	1083	952	857	791	732	644	578
$3/4$	0.785	9517	6369	5051	4319	3368	2928	2562	2284	2094	1947	1801	1698	1501	1347	1237	1142	1010	906
1	1.025	15374	11127	9078	7906	6442	5564	4905	4466	4100	3880	3660	3441	3075	2782	2635	2489	2196	1977
$1^1/4$	1.265	30747	21962	17570	15374	12592	10981	9810	8931	8199	7614	7174	6881	6076	5564	5198	4832	4319	3853
$1^1/2$	1.505	45388	32211	26355	21962	19034	16106	14349	13177	12006	11274	10542	9956	8931	8199	7614	7174	6296	5857
2	1.985	84920	60030	49781	42460	35139	29283	26356	24890	21962	20498	19766	18302	16545	15227	14202	13177	11860	10835
$2^1/2$	2.465	131773	93705	76135	65886	52709	46853	40996	38068	35139	32211	30747	29283	26355	23426	22694	20498	18302	16838

Table 9.19 Maximum Capacity of CSST in Cubic Feet per Hour for Gas Pressure of 0.5 psi or Less and Pressure Drop of 0.5 in. Water Column (Based on a 0.60 Specific Gravity Gas)

EHD* Flow Designation	Tubing Length (ft)																
	5	10	15	20	25	30	40	50	60	70	80	90	100	150	200	250	300
13	46	32	25	22	19	18	15	13	12	11	10	10	9	7	6	5	5
15	63	44	35	31	27	25	21	19	17	16	15	14	13	10	9	8	7
18	115	82	66	58	52	47	41	37	34	31	29	28	26	20	18	16	15
19	134	95	77	67	60	55	47	42	38	36	33	32	30	23	21	19	17
23	225	161	132	116	104	96	83	75	68	63	60	57	54	42	38	34	32
25	270	192	157	137	122	112	97	87	80	74	69	65	62	48	44	39	36
30	471	330	267	231	206	188	162	144	131	121	113	107	101	78	71	63	57
31	546	383	310	269	240	218	188	168	153	141	132	125	118	91	82	74	67

Note: Table includes losses for four 90-degree bends and two end fittings. Tubing runs with larger numbers of bends or fittings shall be increased by an equivalent length of tubing to the following equation: $L = 1.3n$ where L is additional length (ft) of tubing and n is the number of additional fittings or bends.

*EHD — equivalent hydraulic diameter — a measure of the relative hydraulic efficiency between different tubing sizes. The greater the value of EHD, the greater the gas capacity of the tubing.

Table 9.20 Maximum Capacity of CSST in Cubic Feet per Hour for Gas Pressure of 0.5 psi or Less and a Pressure Drop of 3 in. Water Column (Based on a 0.60 Specific Gravity Gas)

EHD* Flow Designation	Tubing Length (ft)																
	5	10	15	20	25	30	40	50	60	70	80	90	100	150	200	250	300
13	120	83	67	57	51	46	39	35	32	29	27	26	24	19	17	15	13
15	160	112	90	78	69	63	54	48	44	41	38	36	34	27	23	21	19
18	277	197	161	140	125	115	100	89	82	76	71	67	63	52	45	40	37
19	327	231	189	164	147	134	116	104	95	88	82	77	73	60	52	46	42
23	529	380	313	273	245	225	196	176	161	150	141	133	126	104	91	82	75
25	649	462	379	329	295	270	234	210	192	178	167	157	149	122	106	95	87
30	1182	828	673	580	518	471	407	363	330	306	285	268	254	206	178	159	144
31	1365	958	778	672	599	546	471	421	383	355	331	311	295	240	207	184	168

Note: Table includes losses for four 90-degree bends and two end fittings. Tubing runs with larger numbers of bends or fittings shall be increased by an equivalent length of tubing to the following equation: $L = 1.3n$ where L is additional length (ft) of tubing and n is the number of additional fittings or bends.

*EHD — equivalent hydraulic diameter — a measure of the relative hydraulic efficiency between different tubing sizes. The greater the value of EHD, the greater the gas capacity of the tubing.

Table 9.21 Maximum Capacity of CSST in Cubic Feet per Hour for a Gas Pressure of 0.5 psi or Less and a Pressure Drop of 6 in. Water Column (Based on a 0.60 Specific Gravity Gas)

EHD* Flow Designation	Tubing Length (ft)																
	5	10	15	20	25	30	40	50	60	70	80	90	100	150	200	250	300
13	173	120	96	83	74	67	57	51	46	42	39	37	35	28	24	21	19
15	229	160	130	112	99	90	78	69	63	58	54	51	48	39	34	30	27
18	389	277	227	197	176	161	140	125	115	106	100	94	89	73	63	57	52
19	461	327	267	231	207	189	164	147	134	124	116	109	104	85	73	66	60
23	737	529	436	380	342	313	273	245	225	209	196	185	176	145	126	114	104
25	911	649	532	462	414	379	329	295	270	250	234	221	210	172	149	134	122
30	1687	1182	960	828	739	673	580	518	471	435	407	383	363	294	254	226	206
31	1946	1365	1110	958	855	778	672	599	546	505	471	444	421	342	295	263	240

Note: Table includes losses for four 90-degree bends and two end fittings. Tubing runs with larger numbers of bends or fittings shall be increased by an equivalent length of tubing to the following equation: $L = 1.3n$ where L is additional length (ft) of tubing and n is the number of additional fittings or bends.

*EHD — equivalent hydraulic diameter — a measure of the relative hydraulic efficiency between different tubing sizes. The greater the value of EHD, the greater the gas capacity of the tubing.

Table 9.22 Maximum Capacity of CSST in Cubic Feet per Hour for a Gas Pressure of 2 psi and a Pressure Drop of 1 psi (Based on a 0.60 Specific Gravity Gas)

EHD* Flow Designation	Tubing Length (ft)													
	10	25	30	40	50	75	80	100	150	200	250	300	400	500
13	270	166	151	129	115	93	89	79	64	55	49	44	38	34
15	353	220	200	172	154	124	120	107	87	75	67	61	52	46
18	587	374	342	297	266	218	211	189	155	135	121	110	96	86
19	700	444	405	351	314	257	249	222	182	157	141	129	111	100
23	1098	709	650	567	510	420	407	366	302	263	236	217	189	170
25	1372	876	801	696	624	512	496	445	364	317	284	260	225	202
30	2592	1620	1475	1273	1135	922	892	795	646	557	497	453	390	348
31	2986	1869	1703	1470	1311	1066	1031	920	748	645	576	525	453	404

Notes:

1. Table does not include effect of pressure drop across line regulator. If regulator loss exceeds $3/4$ psi, DO NOT USE THIS TABLE. Consult with regulator manufacturer for pressure drops and capacity factors. Pressure drops across regulator may vary with the flow rate.

2. CAUTION: Capacities shown in table may exceed maximum capacity for a selected regulator. Consult with regulator or tubing manufacturer for guidance.

3. Table includes losses for four 90-degree bends and two end fittings. Tubing runs with larger number of bends or fittings shall be increased by an equivalent length of tubing according to the following equation: $L = 1.3n$ where L is additional length (ft) of tubing and n is the number of additional fittings or bends.

*EHD — equivalent hydraulic diameter — a measure of the relative hydraulic efficiency between different tubing sizes. The greater the value of EHD, the greater the gas capacity of the tubing.

Table 9.23 Maximum Capacity of CSST in Cubic Feet per Hour for a Gas Pressure of 5 psi and a Pressure Drop of 3.5 psi (Based on 0.60 Specific Gravity Gas)

EHD* Flow Designation	Tubing Length (ft)													
	10	25	30	40	50	75	80	100	150	200	250	300	400	500
13	523	322	292	251	223	180	174	154	124	107	95	86	74	66
15	674	420	382	329	293	238	230	205	166	143	128	116	100	89
18	1084	691	632	549	492	403	391	350	287	249	223	204	177	159
19	1304	827	755	654	586	479	463	415	339	294	263	240	208	186
23	1995	1289	1181	1031	926	763	740	665	548	478	430	394	343	309
25	2530	1616	1478	1284	1151	944	915	820	672	584	524	479	416	373
30	4923	3077	2803	2418	2157	1752	1694	1511	1228	1060	945	860	742	662
31	5659	3543	3228	2786	2486	2021	1955	1744	1418	1224	1092	995	858	766

Notes:

1. Table does not include effect of pressure drop across line regulator. If regulator loss exceeds 1 psi, DO NOT USE THIS TABLE. Consult with regulator manufacturer for pressure drops and capacity factors. Pressure drop across regulator may vary with the flow rate.

2. CAUTION: Capacities shown in table may exceed maximum capacity of selected regulator. Consult with tubing manufacturer for guidance.

3. Table includes losses for four 90-degree bends and two end fittings. Tubing runs with larger numbers of bends or fittings shall be increased by an equivalent length of tubing to the following equation: $L = 1.3n$ where L is additional length (ft) of tubing and n is the number of additional fittings or bends.

*EHD — equivalent hydraulic diameter — a measure of the relative hydraulic efficiency between different tubing sizes. The greater the value of EHD, the greater the gas capacity of the tubing.

Table 9.24 Multipliers to Be Used with Tables 9.1 Through 9.12 When the Specific Gravity of the Gas Is Other Than 0.60.

Specific Gravity	Multiplier	Specific Gravity	Multiplier
0.35	1.31	1.00	0.78
0.40	1.23	1.10	0.74
0.45	1.16	1.20	0.71
0.50	1.10	1.30	0.68
0.55	1.04	1.40	0.66
0.60	1.00	1.50	0.63
0.65	0.96	1.60	0.61
0.70	0.93	1.70	0.59
0.75	0.90	1.80	0.58
0.80	0.87	1.90	0.56
0.85	0.84	2.00	0.55
0.90	0.82	2.10	0.54

Table 9.25 Pipe Sizing Between First Stage (High-Pressure Regulator) and Second Stage (Low-Pressure Regulator). Maximum undiluted propane capacities listed are based on a 10-psi first stage setting and 1 psi pressure drop. Capacities in 1000 Btu/hr.

Pipe Length (ft)	Schedule 40 Pipe Size, 1 psi drop								
	1/2 in. 0.622	3/4 in. 0.824	1 in. 1.049	1 1/4 in. 1.38	1 1/2 in. 1.61	2 in. 2.067	3 in. 3.068	3 1/2 in. 3.548	4 in. 4.026
30	1843	3854	7259	14904	22331	43008	121180	177425	247168
40	1577	3298	6213	12756	19113	36809	103714	151853	211544
50	1398	2923	5507	11306	16939	32623	91920	134585	187487
60	1267	2649	4989	10244	15348	29559	83286	121943	169877
70	1165	2437	4590	9424	14120	27194	76622	112186	156285
80	1084	2267	4270	8767	13136	25299	71282	104368	145393
90	1017	2127	4007	8226	12325	23737	66882	97925	136417
100	961	2009	3785	7770	11642	22422	63176	92499	128859
150	772	1613	3039	6240	9349	18005	50733	74280	103478
200	660	1381	2601	5340	8002	15410	43421	63574	88564
250	585	1224	2305	4733	7092	13658	38483	56345	78493
300	530	1109	2089	4289	6426	12375	34868	51052	71120
350	488	1020	1922	3945	5911	11385	32078	46967	65430
400	454	949	1788	3670	5499	10591	29843	43694	60870
450	426	890	1677	3444	5160	9938	28000	40997	57112
500	402	841	1584	3253	4874	9387	26449	38725	53948
600	364	762	1436	2948	4416	8505	23965	35088	48880
700	335	701	1321	2712	4063	7825	22047	32280	44969
800	312	652	1229	2523	3780	7279	20511	30031	41835
900	293	612	1153	2367	3546	6830	19245	28177	39253
1000	276	578	1089	2236	3350	6452	18178	26616	37078
1500	222	464	875	1795	2690	5181	14598	21373	29775
2000	190	397	748	1537	2302	4434	12494	18293	25483

Table 9.26 Pipe Sizing Between Single or Second Stage (Low-Pressure Regulator) and Appliance.

Pipe Length (ft)	Nominal Pipe Size, Schedule 40								
	$1/2$ in. 0.622	$3/4$ in. 0.824	1 in. 1.049	$1^1/4$ in. 1.38	$1^1/2$ in. 1.61	2 in. 2.067	3 in. 3.068	$3^1/2$ in. 3.548	4 in. 4.026
10	291	608	1146	2353	3525	6789	19130	28008	39018
20	200	418	788	1617	2423	4666	13148	19250	26817
30	161	336	632	1299	1946	3747	10558	15458	21535
40	137	287	541	1111	1665	3207	9036	13230	18431
50	122	255	480	985	1476	2842	8009	11726	16335
60	110	231	435	892	1337	2575	7256	10625	14801
80	94	198	372	764	1144	2204	6211	9093	12668
100	84	175	330	677	1014	1954	5504	8059	11227
125	74	155	292	600	899	1731	4878	7143	9950
150	67	141	265	544	815	1569	4420	6472	9016
200	58	120	227	465	697	1343	3783	5539	7716
250	51	107	201	412	618	1190	3353	4909	6839
300	46	97	182	374	560	1078	3038	4448	6196
350	43	89	167	344	515	992	2795	4092	5701
400	40	83	156	320	479	923	2600	3807	5303

Table 9.27 Copper Tube Sizing Between First Stage (High-Pressure Regulator) and Second Stage (Low-Pressure Regulator). Maximum undiluted propane capacities listed are based on a 10 psi first stage setting and 1 psi drop. Capacities in 1000 Btu/hr.

Tubing Length (ft)	Outside Diameter Copper Tubing, Type L				
	$3/_8$ in. 0.315	$1/_2$ in. 0.430	$5/_8$ in. 0.545	$3/_4$ in. 0.666	$7/_8$ in. 0.785
30	309	700	1303	2205	3394
40	265	599	1115	1887	2904
50	235	531	988	1672	2574
60	213	481	896	1515	2332
70	196	443	824	1394	2146
80	182	412	767	1297	1996
90	171	386	719	1217	1873
100	161	365	679	1149	1769
150	130	293	546	923	1421
200	111	251	467	790	1216
250	90	222	414	700	1078
300	89	201	375	634	976
350	82	185	345	584	898
400	76	172	321	543	836
450	71	162	301	509	784
500	68	153	284	481	741
600	61	138	258	436	671
700	56	127	237	401	617
800	52	118	221	373	574
900	49	111	207	350	539
1000	46	105	195	331	509
1500	37	84	157	266	409
2000	32	72	134	227	350

Table 9.28 Copper Tube Sizing Between Single or Second Stage (Low-Pressure Regulator) and Appliance. Maximum undiluted propane capacities are based on an 11-in. water column setting and a 0.5-in. water column pressure drop. Capacities in 1000 Btu/hr.

Tubing Length (ft)	Outside Diameter Copper Tubing, Type L				
	$3/8$ in. 0.315	$1/2$ in. 0.430	$5/8$ in. 0.545	$3/4$ in. 0.666	$7/8$ in. 0.785
10	49	110	206	348	536
20	34	76	141	239	368
30	27	61	114	192	296
40	23	52	97	164	253
50	20	46	86	146	224
60	19	42	78	132	203
80	16	36	67	113	174
100	14	32	59	100	154
125	12	28	52	89	137
150	11	26	48	80	124
200	10	22	41	69	106
250	9	19	36	61	94
300	8	18	33	55	85
350	7	16	30	51	78
400	7	15	28	47	73

Table 9.29 Maximum Capacity of CSST in Thousands of Btu per Hour of Undiluted Liquefied Petroleum Gases at a Pressure of 11 in. Water Column and a Pressure Drop of 0.5 in. Water Column (Based on a 1.52 Specific Gravity Gas)

EHD* Flow Designation	Tubing Length (ft)																
	5	10	15	20	25	30	40	50	60	70	80	90	100	150	200	250	300
13	72	50	39	34	30	28	23	20	19	17	15	15	14	11	9	8	8
15	99	69	55	49	42	39	33	30	26	25	23	22	20	15	14	12	11
18	181	129	104	91	82	74	64	58	53	49	45	44	41	31	28	25	23
19	211	150	121	106	94	87	74	66	60	57	52	50	47	36	33	30	26
23	355	254	208	183	164	151	131	118	107	99	94	90	85	66	60	53	50
25	426	303	248	216	192	177	153	137	126	117	109	102	98	75	69	61	57
30	744	521	422	365	325	297	256	227	207	191	178	169	159	123	112	99	90
31	863	605	490	425	379	344	297	265	241	222	208	197	186	143	129	117	107

Note: Table includes losses for four 90-degree bends and two end fittings. Tubing runs with larger numbers of bends or fittings shall be increased by an equivalent length of tubing to the following equation: $L = 1.3n$ where L is additional length (ft) of tubing and n is the number of additional fittings or bends.

*EHD — equivalent hydraulic diameter — a measure of the relative hydraulic efficiency between different tubing sizes. The greater the value of EHD, the greater the gas capacity of the tubing.

Table 9.30 Maximum Capacity of CSST in Thousands of Btu per Hour of Undiluted Liquefied Petroleum Gases at a Pressure of 2 psi and a Pressure Drop of 1 psi (Based on 1.52 Specific Gravity Gas)

EHD* Flow Designation	Tubing Length (ft)													
	10	25	30	40	50	75	80	110	150	200	250	300	400	500
13	426	262	238	203	181	147	140	124	101	86	77	69	60	53
15	558	347	316	271	243	196	189	169	137	118	105	96	82	72
18	927	591	540	469	420	344	333	298	245	213	191	173	151	135
19	1106	701	640	554	496	406	393	350	287	248	222	203	175	158
23	1735	1120	1027	896	806	663	643	578	477	415	373	343	298	268
25	2168	1384	1266	1100	986	809	768	703	575	501	448	411	355	319
30	4097	2560	2331	2012	1794	1457	1410	1256	1021	880	785	716	616	550
31	4720	2954	2692	2323	2072	1685	1629	1454	1182	1019	910	829	716	638

Notes:

1. Table does not include effect of pressure drop across the line regulator. If regulator loss exceeds $1/2$ psi (based on 13-in. water column outlet pressure), DO NOT USE THIS TABLE. Consult with regulator manufacturer for pressure drops and capacity factors. Pressure drops across a regulator can vary with flow rate.

2. CAUTION: Capacities shown in table can exceed maximum capacity for a selected regulator. Consult with regulator or tubing manufacturer for guidance.

3. Table includes losses for four 90-degree bends and two end fittings. Tubing runs with larger number of bends or fittings shall be increased by an equivalent length of tubing according to the following equation: $L = 1.3n$ where L is additional length (ft) of tubing and n is the number of additional fittings or bends.

*EHD — equivalent hydraulic diameter — a measure of the relative hydraulic efficiency between different tubing sizes. The greater the value of EHD, the greater the gas capacity of the tubing.

Table 9.31 *Maximum Capacity of CSST in Thousands of Btu per Hour of Undiluted Liquefied Petroleum Gases at a Pressure of 5 psi and a Pressure Drop of 3.5 psi (Based on a 1.52 Specific Gravity Gas)*

EHD* Flow Designation	Tubing Length (ft)													
	10	25	30	40	50	75	80	100	150	200	250	300	400	500
13	826	509	461	396	352	284	275	243	196	169	150	136	117	104
15	1065	664	603	520	463	376	363	324	262	226	202	183	158	140
18	1713	1092	999	867	777	637	618	553	453	393	352	322	279	251
19	2061	1307	1193	1033	926	757	731	656	535	464	415	379	328	294
23	3153	2037	1866	1629	1463	1206	1169	1051	866	755	679	622	542	488
25	3999	2554	2336	2029	1819	1492	1446	1296	1062	923	828	757	657	589
30	7829	4864	4430	3822	3409	2769	2677	2388	1941	1675	1493	1359	1173	1046
31	8945	5600	5102	4404	3929	3194	3090	2756	2241	1934	1726	1572	1356	1210

Notes:

1. Table does not include effect of pressure drop across line regulator. If regulator loss exceeds 1 psi water column, DO NOT USE THIS TABLE. Consult with regulator manufacturer for pressure drops and capacity factors. Pressure drop across regulator can vary with the flow rate.

2. CAUTION: Capacities shown in table can exceed maximum capacity of selected regulator. Consult with tubing manufacturer for guidance.

3. Table includes losses for four 90-degree bends and two end fittings. Tubing runs with larger numbers of bends or fittings shall be increased by an equivalent length of tubing to the following equation: $L = 1.3n$ where L is additional length (ft) of tubing and n is the number of additional fittings or bends.

*EHD — equivalent hydraulic diameter — a measure of the relative hydraulic efficiency between different tubing sizes. The greater the value of EHD, the greater the gas capacity of the tubing.

Table 9.32 Polyethylene Plastic Pipe Sizing Between First-Stage and Second-Stage Regulator

Pipe Length (ft)	Plastic Pipe Nominal Outside Diameter (IPS) (dimensions in parentheses are inside diameter)					
	½ in. SDR 9.33 (0.660)	¾ in. SDR 11.0 (0.860)	1 in. SDR 11.00 (1.077)	1¼ in. SDR 10.00 (1.328)	1½ in. SDR 11.00 (1.554)	2 in. SDR 11.00 (1.943)
30	2143	4292	7744	13416	20260	36402
40	1835	3673	6628	11482	17340	31155
50	1626	3256	5874	10176	15368	27612
60	1473	2950	5322	9220	13924	25019
70	1355	2714	4896	8483	12810	23017
80	1261	2525	4555	7891	11918	21413
90	1183	2369	4274	7404	11182	20091
100	1117	2238	4037	6994	10562	18978
125	990	1983	3578	6199	9361	16820
150	897	1797	3242	5616	8482	15240
175	826	1653	2983	5167	7803	14020
200	778	1539	2775	4807	7259	13043
225	721	1443	2603	4510	6811	12238
250	681	1363	2459	4260	6434	11560
275	646	1294	2336	4046	6111	10979
300	617	1235	2228	3860	5830	10474
350	567	1136	2050	3551	5363	9636
400	528	1057	1907	3304	4989	8965
450	495	992	1789	3100	4681	8411
500	468	937	1690	2928	4422	7945
600	424	849	1531	2653	4007	7199
700	390	781	1409	2441	3686	6623
800	363	726	1311	2271	3429	6161
900	340	682	1230	2131	3217	5781
1000	322	644	1162	2012	3039	5461
1500	258	517	933	1616	2441	4385
2000	221	443	798	1383	2089	3753

Note: Maximum undiluted propane capacities listed are based on 10 psi first stage setting 1 psi pressure drop. Capacities are in 1000 Btu/hr.

Table 9.33 Polyethylene Plastic Tube Sizing Between First-Stage Regulator and Second-Stage Regulator. Maximum undiluted propane capacities listed are based on 10-psi first-stage setting and 1-psi pressure drop. Capacities in 1000 Btu/hr.

	Plastic Tubing Size (CTS) (dimensions in parentheses are inside diameter)	
Plastic Tubing Length (ft)	**$^1/_2$ in. CTS SDR 7.00 (0.445)**	**1 in. CTS SDR 11.00 (0.927)**
30	762	5225
40	653	4472
50	578	3964
60	524	3591
70	482	3304
80	448	3074
90	421	2884
100	397	2724
125	352	2414
150	319	2188
175	294	2013
200	273	1872
225	256	1757
250	242	1659
275	230	1576
300	219	1503
350	202	1383
400	188	1287
450	176	1207
500	166	1140
600	151	1033
700	139	951
800	129	884
900	121	830
1000	114	784
1500	92	629
2000	79	539

Table 9.34 Polyethylene Plastic Tube Sizing Between Single- or Second-Stage Regulator and Building. Maximum undiluted propane capacities listed are based on 11-in. water column setting and a 0.5-in. water column pressure drop. Capacities in 1000 Btu/hr.

	Plastic Tubing Size (CTS) (dimensions in parentheses are inside diameter)	
Plastic Tubing Length (ft)	**$1/_2$ in. CTS SDR 7.00 (0.445)**	**1 in. CTS SDR 11.00 (0.927)**
10	121	829
20	83	569
30	67	457
40	57	391
50	51	347
60	46	314
70	42	289
80	39	269
90	37	252
100	35	238
125	31	211
150	28	191
175	26	176
200	24	164
225	22	154
250	21	145
275	20	138
300	19	132
350	18	121
400	16	113

Sizing of Category I Venting Systems

Chapter 10 contains thirteen tables and explanatory text for calculating the required vent size for a variety of Category I venting systems. This chapter was first created for the 1992 edition by relocating the vent sizing tables and related notes from Appendix G and extensively modifying them to cover fan-assisted combustion appliances. In compliance with United States energy laws, these new appliances provide higher efficiency, which is achieved by reducing vent temperature without jeopardizing vent operation. The chapter was revised in the 1999 edition by relocating definitions from this chapter to Section 1.7, Definitions.

The tables, which were revised in the 1992 edition, added two new columns, FAN Min and FAN Max, which provide minimum and maximum capacities for each vent size. These capacities are used for sizing vents that serve fan-assisted combustion appliances. The NAT Max column is unchanged from the tables in previous editions of the code and should be used for appliances other than the fan-assisted combustion type. Examples demonstrating the use of these tables appear in Appendix G.

Category I systems are defined in Section 1.7, Definitions, as follows:

Vented Appliance, Category I. An appliance that operates with a nonpositive vent static pressure and with a vent gas temperature that avoids excessive condensate production in the vent.

Category I venting systems are conventional venting systems, in which the heat of the flue gases is the force that operates the vent.

Chapter 10 includes the following:

- Additional requirements for a single-appliance vent, which are qualifications that must be considered when using single-appliance vent Tables 10.1 through 10.5 (in Section 10.1)
- Five vent tables for single-appliance installations:

Table No.	Vent	Connector
10.1	Type B	None
10.2	Type B	Single Wall
10.3	Masonry Chimney	Type B
10.4	Masonry Chimney	Single Wall
10.5	Single-Wall Metal Pipe	None

- Additional requirements for a single-appliance vent, which are qualifications that must be considered when using single-appliance vent Tables 10.6 through 10.13 (in Section 10.2)
- Eight vent tables for multiple-appliance venting installations, with the first five covering interior installations:

Table No.	Vent	Connector
10.6	Type B	Type B
10.7	Type B	Single Wall
10.8	Masonry Chimney	Type B
10.9	Masonry Chimney	Single Wall
10.10	Single Wall	Single Wall

Three tables covering exterior masonry chimneys, which are organized by geographical regions of the United States, based on the lowest anticipated winter temperature:

Table No.	Vent	Appliance
10.11	Exterior Masonry Chimney	Single NAT
10.12	Exterior Masonry Chimney	NAT + NAT
10.13	Exterior Masonry Chimney	FAN + NAT

Tables 10.11 through 10.13 were added in the 1996 edition to provide sizing information for exterior masonry chimneys serving heating appliances. Note that the

new tables are for space and central heating appliances only. Water heaters and other nonheating appliances are ignored, because they do not run long enough to produce continuous condensation during normal operating cycles. The tables provide sizing options based on local 99 percent winter design temperatures in six temperature ranges. A map showing the boundaries of these temperature ranges in the United States is included in Appendix G (see Figure G.19). This appendix also includes an example that illustrates the use of these new tables.

In the 1996 edition, Sections 11.2 and 11.3 (now Sections 10.1 and 10.2) were titled "Additional Requirements to Single Appliance Vent Tables 11.1 through 11.5" and "Additional Requirements to Multiple Appliance Vent Tables 11.6 through 11.13," respectively. The information in these sections originally appeared as notes to the single and multiple appliance tables in the 1992 code.

Further information on the development of these new tables is included in this handbook as Supplement 1. This supplement covers the development of the revised vent tables and the new exterior chimney tables in extensive technical detail and is recommended for those who desire a greater understanding of this important subject.

The tables in this chapter begin on p. 358.

10.1 Additional Requirements to Single Appliance Vent Tables 10.1 Through 10.5

Tables 10.1 through 10.5 (pp. 358 to 367) were developed with the assumption that there are no restrictions in the path of the vent gas flow (Exhibit 10.1a). An additional assumption is that there is no deliberate attempt to remove heat. Therefore, devices such as draft controls and heat economizers may not be used in conjunction with the tables. The 1999 code adds language that specifies how to interpret the tables when a vent damper is part of the installation. Basically, Paragraph 10.1.1 requires the user to think of the appliance as a natural draft (NAT) appliance for determining the maximum capacity. Then, the appliance is treated as a fan-assisted (FAN) appliance to determine the minimum capacity. This is because of the reduction of dilution air caused by the damper.

The simplified visual renditions of code requirements given in this section have been provided courtesy of the Gas Research Institute (GRI).

Exhibit 10.1a *No obstructions.*

10.1.1 These venting tables shall not be used where obstructions, as described in the exceptions to Section 7.15, are installed in the venting system. The installation of vents serving listed appliances with vent dampers shall be in accordance with the appliance manufacturer's instructions or in accordance with the following:

(1) The maximum capacity of the vent system shall be determined using the "NAT Max" column.
(2) The minimum capacity shall be determined as if the appliance were a fan-assisted appliance, using the "FAN Min" column to determine the minimum capacity of the vent system. Where the corresponding "Fan Min" is "NA," the vent configuration shall not be permitted and an alternative venting configuration shall be utilized.

Paragraphs (1) and (2) were added in the 1999 edition to provide guidance for using the venting tables with draft hood appliances equipped with a vent damper, such as boilers, which can use vent dampers to obtain higher efficiencies. The installer can use the tables as instructed here or use the manufacturer's installation instructions, even if the manufacturer specifies a venting configuration that is not permitted by this chapter.

10.1.2 If the vent size determined from the tables is smaller than the appliance draft hood outlet or flue collar, the smaller size shall be permitted to be used, provided the following requirements are met:

(1) The total vent height *(H)* is at least 10 ft (3 m).
(2) Vents for appliance draft hood outlets or flue collars 12 in. (300 mm) in diameter or smaller are not reduced more than one table size.
(3) Vents for appliance draft hood outlets or flue collars larger than 12 in. (300 mm) in diameter are not reduced more than two table sizes.
(4) The maximum capacity listed in the tables for a fan-assisted appliance is reduced by 10 percent (0.90 × maximum table capacity).
(5) The draft hood outlet is greater than 4 in. (100 mm) in diameter. Do not connect a 3-in. (80-mm) diameter vent to a 4-in. (100-mm) diameter draft hood outlet. This provision shall not apply to fan-assisted appliances.

There is an economic incentive to using smaller vent and connector diameters. However, experience shows that placing severe diameter reductions near the beginning of the vent can have a negative effect on the ability of the vent to establish draft. Therefore, this paragraph places limits on "downsizing" (Exhibit 10.1b). In particular, note that a 4 in. (10.16 mm) draft hood outlet may not be reduced to 3 in. (7.62 mm).

10.1.3 Single-appliance venting configurations with zero (0) lateral lengths in Tables 10.1, 10.2, and 10.5 shall have no elbows in the venting system. For vent configurations with lateral lengths, the venting tables include allowance for two 90 degree turns. For each additional 90 degree turn, or equivalent, the maximum capacity listed in the venting tables shall be

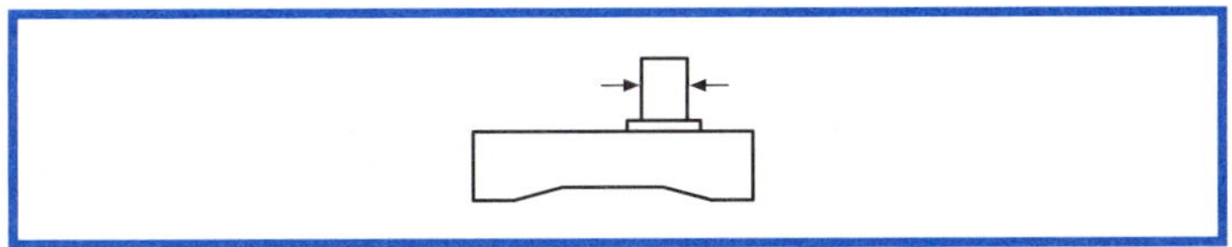

Exhibit 10.1b *Downsizing.*

reduced by 10 percent (0.90 × maximum table capacity). Two or more turns, the combined angles of which equal 90 degrees, shall be considered equivalent to one 90 degree turn.

The sizing tables were designed with an assumption that up to two 90 degree turns were part of the venting system — except for the zero lateral length case. The zero lateral case was assumed to extend straight up from the appliance outlet to the vent termination. It is possible to add additional elbows (Exhibit 10.1c). The Max capacity may be reduced by 10 percent for each 90 degree turn.

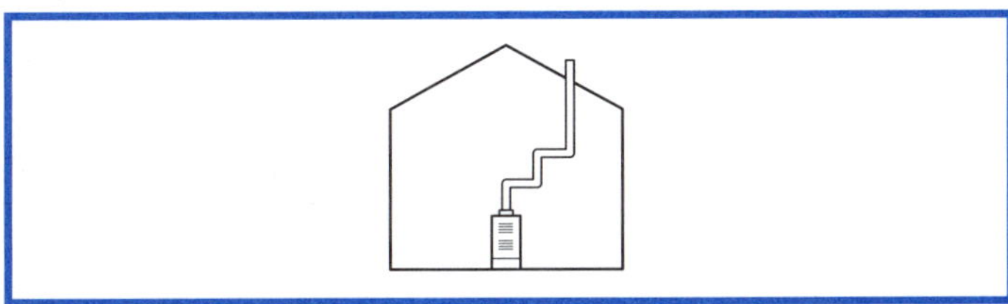

Exhibit 10.1c *Elbows.*

10.1.4 Zero (0) lateral *(L)* shall apply only to a straight vertical vent attached to a top outlet draft hood or flue collar.

It is not permitted to add elbows to a zero lateral length vent (Exhibit 10.1d). If this is necessary, use the table entries for 2 ft (0.6 m) lateral length.

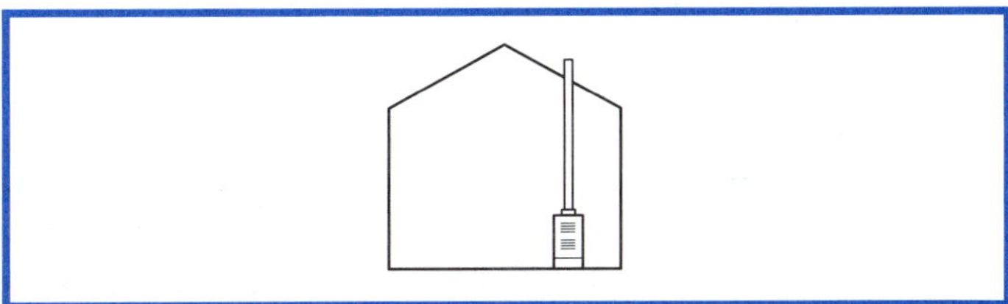

Exhibit 10.1d *No elbows.*

10.1.5 Sea level input ratings shall be used when determining maximum capacity for high-altitude installation. Actual input (derated for altitude) shall be used for determining minimum capacity for high-altitude installation.

Using the sea level input rating for the Max capacity is a conservative measure, since less draft is produced at high altitudes (Exhibit 10.1e). The derating process will also make condensation more likely. The reduced input rate should be used for the Min capacity.

Exhibit 10.1e High-altitude installations.

10.1.6 For appliances with more than one input rate, the minimum vent capacity (FAN Min) determined from the tables shall be less than the lowest appliance input rating, and the maximum vent capacity (FAN Max/NAT Max) determined from the tables shall be greater than the highest appliance rating input.

If the appliance has multiple input rates, the Min capacity is determined with the minimum input rate and the Max capacity is determined with the maximum input (Exhibit 10.1f).

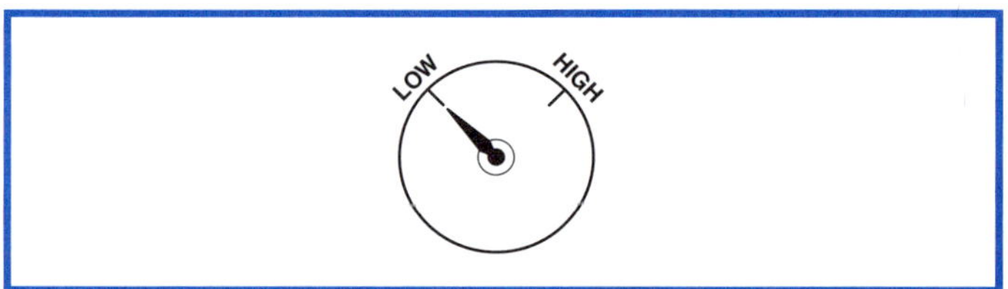

Exhibit 10.1f Two-stage or modulating furnaces.

10.1.7 Listed corrugated metallic chimney liner systems in masonry chimneys shall be sized by using Table 10.1 or 10.2 for Type B vents with the maximum capacity reduced by 20 percent (0.80 × maximum capacity) and the minimum capacity as shown in Table 10.1 or 10.2. Corrugated metallic liner systems installed with bends or offsets shall have their maximum capacity further reduced in accordance with 10.1.3.

Since properly installed corrugated chimney liners have heat loss similar to a Type B vent, they are sized using those tables. However, their corrugations and tendency to spiral in the chimney require a 20 percent Max capacity reduction. Many reliners begin at the breaching and then bend up vertically (Exhibit 10.1g). This bend must be counted as one of the two allowed elbows in the system.

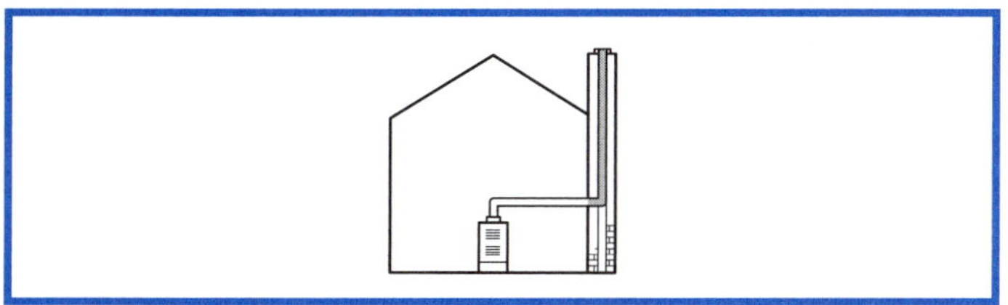

Exhibit 10.1g *Chimney reliners.*

10.1.8 If the vertical vent has a larger diameter than the vent connector, the vertical vent diameter shall be used to determine the minimum vent capacity, and the connector diameter shall be used to determine the maximum vent capacity. The flow area of the vertical vent shall not exceed seven times the flow area of the listed appliance categorized vent area, flue collar area, or draft hood outlet area unless designated in accordance with approved engineering methods.

In a vent system as described, the draft produced is limited by the small diameter, and condensation will begin in the larger diameter (Exhibit 10.1h). The Max and Min capacities must be determined accordingly. Experience has also shown that there are practical limits to how large the vertical vent may be relative to its source of vent gas flow. Therefore, the flow area of the vent may not be more than seven times the flow area of the outlet of the appliance or draft hood.

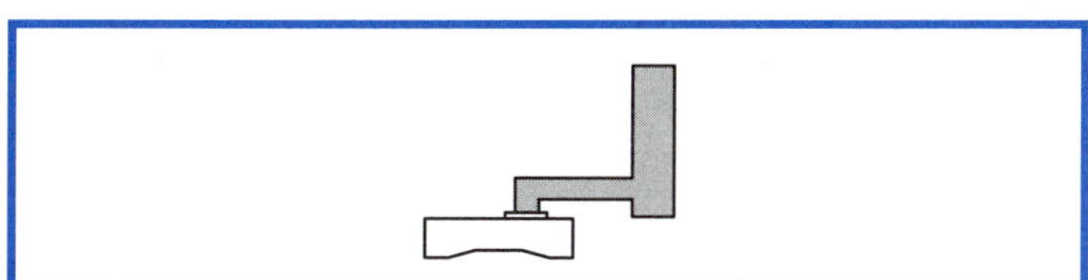

Exhibit 10.1h *Upsizing vertical vent.*

10.1.9 Tables 10.1 through 10.5 shall be used for chimneys and vents not exposed to the outdoors below the roof line. A Type B vent or listed chimney lining system passing through an unused masonry chimney flue shall not be considered to be exposed to the outdoors. Table 10.3 in combination with Table 10.11 shall be used for clay-tile-lined exterior masonry chimneys, provided all of the following requirements are met:

(1) The vent connector is Type B double wall.
(2) The vent connector length is limited to $1^1/_2$ ft for each inch (180 mm/mm) of vent connector diameter.
(3) The appliance is draft hood–equipped.
(4) The input rating is less than the maximum capacity given by in Table 10.3.
(5)   For a water heater, the outdoor design temperature shall not be less than 5°F (−15°C).

(6) For a space-heating appliance, the input rating is greater than the minimum capacity given by Table 10.11.

Exception: The installation of vents serving listed appliances shall be permitted to be in accordance with the appliance manufacturer's instructions and the terms of the listing.

In the 1999 edition, (5) was added, and the former (5) renumbered (6), to provide guidance for using masonry chimneys to vent waters. The minimum temperature of 5°F (−15°C) refers to Tables 10.11 and 10.12. Use these tables when sizing one or more heaters (and no other space or room heating appliances).

For exterior chimneys (Exhibit 10.1i), the Max capacity is determined by Table 10.2, while the Min capacity is determined by the by Table 10.11. Table 10.11, in turn, has different Min capacities based on the ambient temperatures expected. New requirement (5) sets the minimum design temperature where a conventional water heater may be vented alone into an exterior masonry chimney. The exception recognizes that manufacturers may produce products customized for exterior masonry chimneys and use their own instructions.

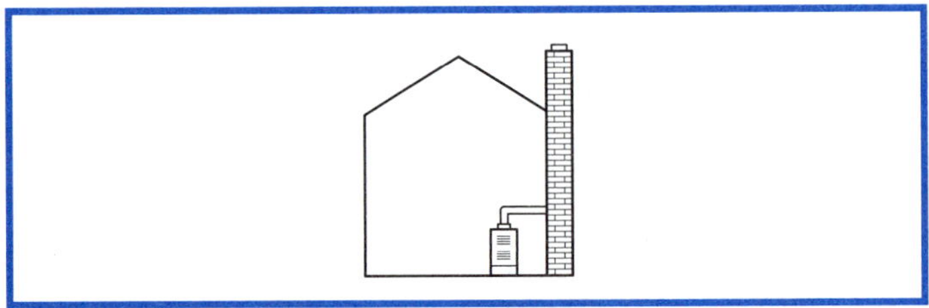

Exhibit 10.1i *Exterior chimneys.*

10.1.10 Vent connectors shall not be upsized more than two sizes greater than the listed appliance categorized vent diameter, flue collar diameter, or draft hood outlet diameter.

A sudden, large expansion of the vent connector diameter (Exhibit 10.1j) creates a pressure drop that may limit the draft and encourage condensation. Therefore, a limit is placed on it.

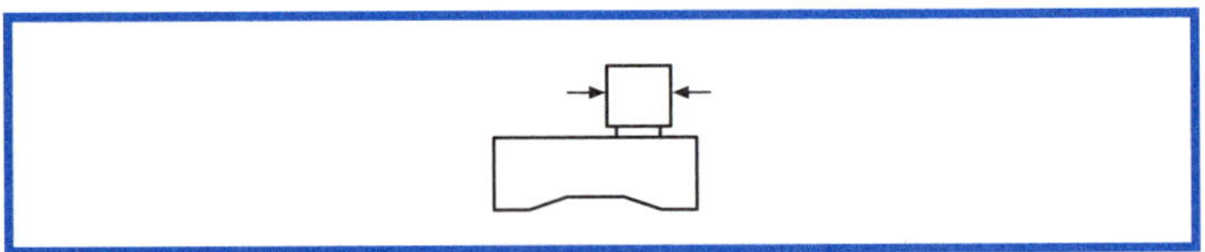

Exhibit 10.1j *Upsizing.*

10.1.11 In a single run of vent or vent connector, more than one diameter and type shall be permitted to be used, provided that all the sizes and types are permitted by the tables.

This paragraph effectively requires the installer to check the Min and Max capacities for the vent section for each vent type as if the entire vent section were made of it. See Exhibit 10.1k. For example, suppose a vent connector was half single wall and half Type B. The installer must show that a vent connector of that length would be allowed if it was all single wall and also if it was all Type B.

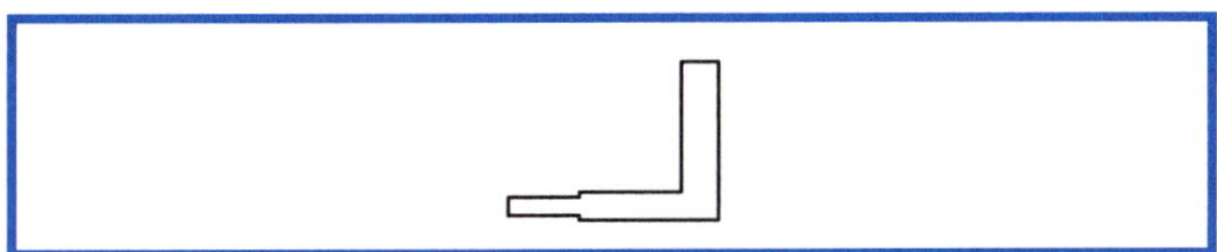

Exhibit 10.1k *Pot luck.*

10.1.12 Interpolation shall be permitted in calculating capacities for vent dimensions that fall between table entries. *(See Example 3, Appendix G.)*

10.1.13 Extrapolation beyond the table entries shall not be permitted.

If the installation dimensions fall between two table entries for which there are defined values, the installer may calculate the "in between" value, but it is not permitted to guess what the value might be outside of the table entries. See Exhibit 10.1l.

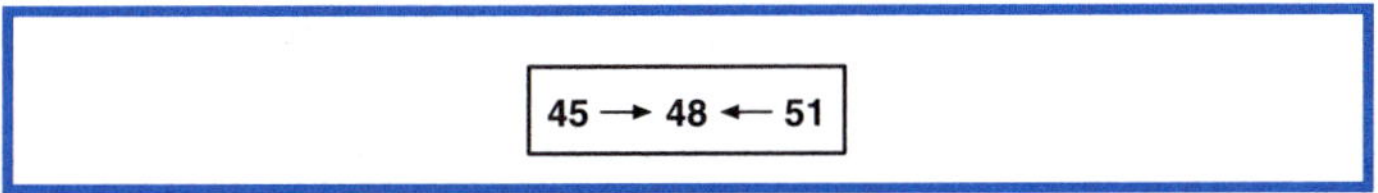

Exhibit 10.1l *Interpolation and extrapolation.*

10.1.14 For vent heights lower than 6 ft and higher than shown in the tables, engineering methods shall be used to calculate vent capacities.

New text in the 1999 edition reinforces the requirement that vent heights outside the tables must be calculated, and the tables cannot be used.

10.2 Additional Requirements to Multiple Appliance Vent Tables 10.6 Through 10.13(a) and (b)

Tables 10.6 through 10.13(a) and (b), on pages 368 through 383, were developed with the assumption that there are no restrictions in the path of the vent gas flow (Exhibit 10.2a). There is an additional assumption that there is no deliberate attempt

Exhibit 10.2a *No obstructions.*

to remove heat. Therefore, devices such as draft controls and heat economizers may not be used in conjunction with the tables. The 1999 code adds language that specifies how to interpret the tables when a vent damper is part of the installation. Basically, Paragraph 10.2.1 requires the user to think of the appliance as a natural draft (NAT) appliance for determining the maximum capacity. Then, the appliance is treated as a fan-assisted (FAN) appliance to determine the minimum capacity. This is because of the reduction of dilution air caused by the damper. The simplified visual renditions of code requirements given in this section have been provided courtesy of Gas Research Institute.

10.2.1 These venting tables shall not be used where obstructions, as described in the exceptions to Section 7.15, are installed in the venting system. The installation of vents serving listed appliances with vent dampers shall be in accordance with the appliance manufacturer's instructions, or in accordance with the following:

(1) The maximum capacity of the vent connector shall be determined using the NAT Max column.
(2) The maximum capacity of the vertical vent or chimney shall be determined using the FAN+NAT column when the second appliance is a fan-assisted appliance, or the NAT+NAT column when the second appliance is equipped with a draft hood.
(3) The minimum capacity shall be determined as if the appliance were a fan-assisted appliance.

   a. The minimum capacity of the vent connector shall be determined using the FAN Min column.
   b. The FAN+FAN column shall be used when the second appliance is a fan-assisted appliance, and the FAN+NAT column shall be used when the second appliance is equipped with a draft hood, to determine whether the vertical vent or chimney configuration is not permitted (NA). Where the vent configuration is NA, the vent configuration shall not be permitted and an alternative venting configuration shall be utilized.

Paragraphs (1), (2), and (3) were added in the 1999 edition to provide guidance for using the venting tables with draft hood appliances equipped with a vent damper, such as boilers, which can use vent dampers to obtain higher efficiencies. The installer can use the tables as instructed here, or use the manufacturers installation instructions, even if the manufacturer specifies a venting configuration that is not permitted by this chapter.

10.2.2 The maximum vent connector horizontal length shall be 18 in./in. (180 mm/mm) of connector diameter as follows:

Connector Diameter Maximum (in.)	Connector Horizontal Length (ft)
3	$4\frac{1}{2}$
4	6
5	$7\frac{1}{2}$
6	9
7	$10\frac{1}{2}$
8	12
9	$13\frac{1}{2}$
10	15
12	18
14	21
16	24
18	27
20	30
22	33
24	36

For SI units, 1 in. = 25.4 mm; 1 ft = 0.305 m.

10.2.3 The vent connector shall be routed to the vent utilizing the shortest possible route. Connectors with longer horizontal lengths than those listed in 10.2.2 are permitted under the following conditions:

(a) The maximum capacity (FAN Max or NAT Max) of the vent connector shall be reduced 10 percent for each additional multiple of the length listed in 10.2.2. For example, the maximum length listed for a 4-in. (100-mm) connector is 6 ft (1.8 m). With a connector length greater than 6 ft (1.8 m) but not exceeding 12 ft (3.7 m), the maximum capacity must be reduced by 10 percent (0.90 × maximum vent connector capacity). With a connector length greater than 12 ft (3.7 m) but not exceeding 18 ft (5.5 m), the maximum capacity must be reduced by 20 percent (0.80 × maximum vent capacity).

(b) For a connector serving a fan-assisted appliance, the minimum capacity (FAN Min) of the connector shall be determined by referring to the corresponding single appliance table. For Type B double-wall connectors, Table 10.1 shall be used. For single-wall connectors, Table 10.2 shall be used. The height *(H)* and lateral *(L)* shall be measured according to the procedures for a single appliance vent, as if the other appliances were not present.

> For brevity, the common venting tables are designed with the assumption that the vent connector is no more than 18 in. long for each in. of diameter (Exhibit 10.2b). Paragraphs 10.2.2 and 10.2.3 give guidance on how to handle connectors that are longer than this.

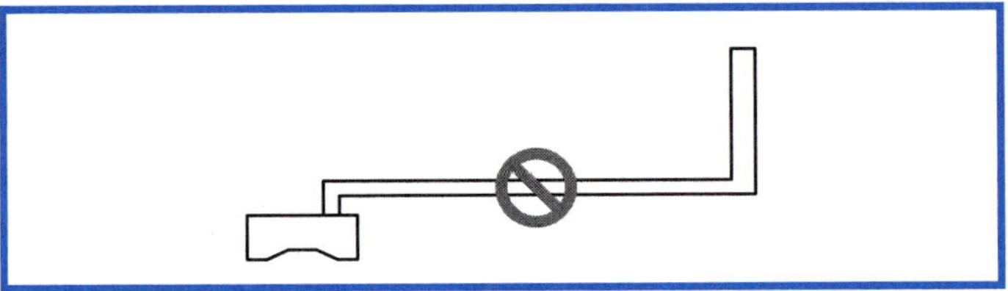

Exhibit 10.2b *Vent connector length.*

10.2.4 If the vent connectors are combined prior to entering the common vent, the maximum common vent capacity listed in the common venting tables shall be reduced by 10 percent (0.90 × maximum common vent capacity). The length of the common vent connector manifold *(L_M)* shall not exceed 18 in./in. (180 mm/mm) of common vent connector manifold diameter *(D)*. *(See Figure G.11.)*

> This paragraph considers the "common vent" to be the vertical vent or chimney. Note that a chimney relining system that comes out through the breaching, as many do, should be considered a manifold (Exhibit 10.2c).

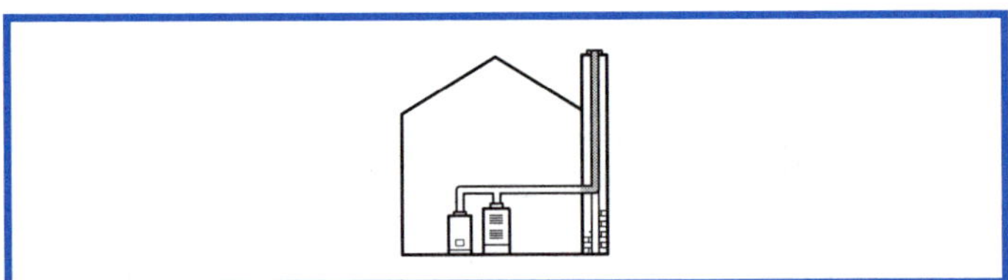

Exhibit 10.2c *Vent manifolds.*

10.2.5 If the common vertical vent is offset as shown in Figure G.12, the maximum common vent capacity listed in the common venting tables shall be reduced by 20 percent (0.80 × maximum common vent capacity), the equivalent of two 90 degree turns. The horizontal length of the common vent offset (L_O) shall not exceed 18 in./in. (180 mm/mm) of common vent diameter (D).

Vent offsets, which typically occur high in the venting system (Exhibit 10.2d), tend to reduce the draft produced. This is recognized by a 10 percent capacity reduction.

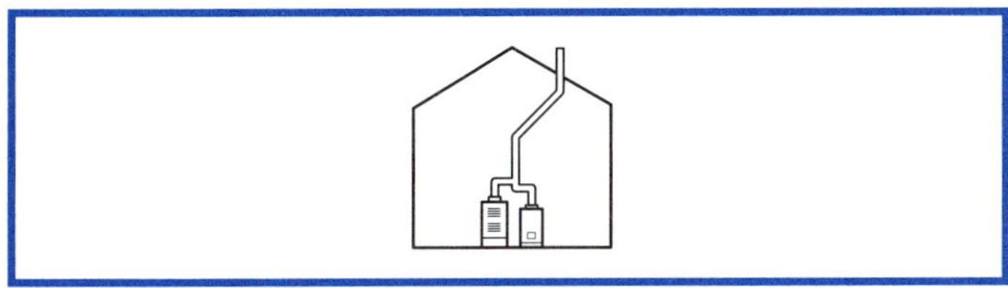

Exhibit 10.2d *Vent offsets.*

10.2.6 Excluding elbows counted in 10.2.5, for each additional 90 degree turn in excess of two, the maximum capacity of that portion of the venting system shall be reduced by 10 percent (0.90 × maximum common vent capacity). Two or more turns, the combined angles of which equal 90 degrees, shall be considered equivalent to one 90 degree turn.

The sizing tables were designed with an assumption that up to two 90 degree turns were part of the venting system — except for the zero lateral length case. The zero lateral case was assumed to extend straight up from the appliance outlet to the vent termination. It is possible to add additional elbows (Exhibit 10.2e). The Max capacity may be reduced by 10 percent for each 90 degree turn. Note that the capacity reduction is applied to the section of the vent where the extra elbows are found.

Exhibit 10.2e *Elbows.*

10.2.7 The cross-sectional area of the common vent shall be equal to or greater than the cross-sectional area of the largest connector.

10.2.8 Interconnection fittings shall be the same size as the common vent.

If pipe or fitting diameters suddenly decrease, the available draft is reduced. Paragraphs 10.2.7 and 10.2.8 require the installer to use a common vent and its tee-in connections that are at least the diameter of the largest vent connector (Exhibit 10.2f). On the other hand, experience has also shown that there are practical limits to how large the vertical vent may be relative to its source of vent gas flow. Therefore, the flow area of the common vent may not be more than seven times the flow area of the smallest of the outlets of any of the appliances or draft hoods in the common vent system.

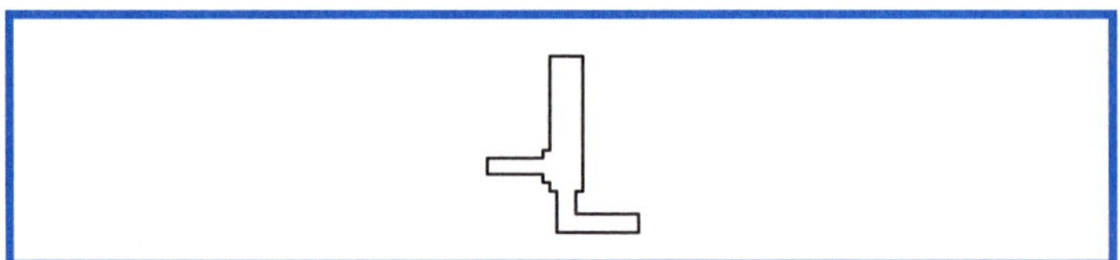

Exhibit 10.2f *Common vent diameter limitations.*

10.2.9 Sea level input ratings shall be used when determining maximum capacity for high-altitude installation. Actual input (derated for altitude) shall be used for determining minimum capacity for high-altitude installation.

Using the sea level input rating for the Max capacity is a conservative measure, since less draft is produced at high altitudes (Exhibit 10.2g). The derating process will also make condensation more likely. The reduced input rate should be used for the Min capacity.

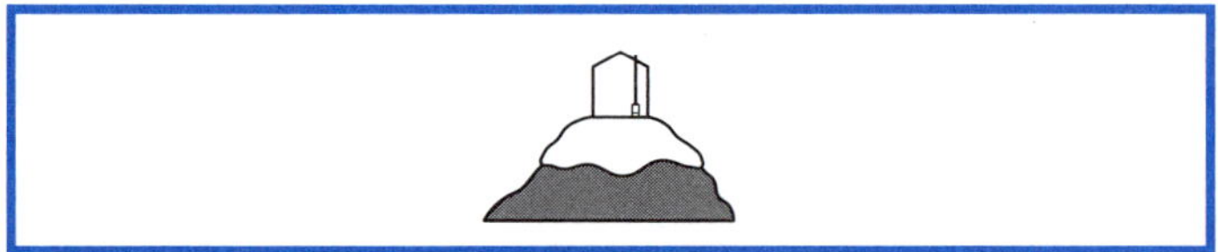

Exhibit 10.2g *High-altitude installations.*

10.2.10 The connector rise *(R)* for each appliance connector shall be measured from the draft hood outlet or flue collar to the centerline where the vent gas streams come together.

10.2.11 For multiple units of gas utilization equipment all located on one floor, available total height *(H)* shall be measured from the highest draft hood outlet or flue collar up to the level of the outlet of the common vent.

See Exhibit 10.2h and the figures in Appendix G.

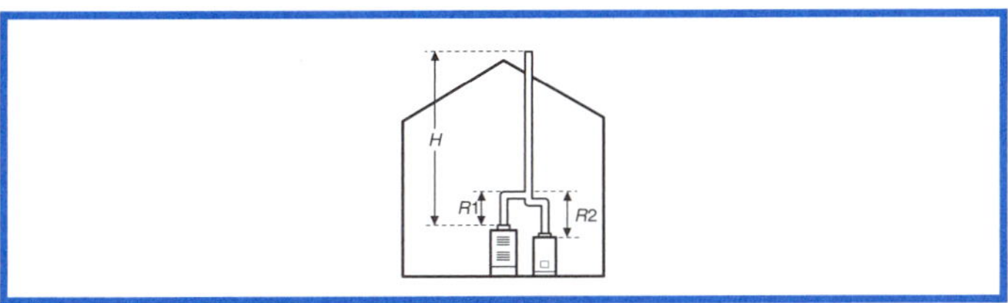

Exhibit 10.2h *Connector rise and vent height.*

10.2.12 For multistory installations, available total height *(H)* for each segment of the system shall be the vertical distance between the highest draft hood outlet or flue collar entering that segment and the centerline of the next higher interconnection tee. *(See Figure G.13.)*

10.2.13 The size of the lowest connector and of the vertical vent leading to the lowest interconnection of a multistory system shall be in accordance with Table 10.1 or 10.2 for available total height *(H)* up to the lowest interconnection. *(See Figure G.14.)*

10.2.14 Where used in multistory systems, vertical common vents shall be Type B double-wall and shall be installed with a listed vent cap. A multistory common vertical vent shall be permitted to have a single offset, provided all the following requirements are met:

(1) The offset angle does not exceed 45 degrees.
(2) The horizontal length of the offset does not exceed 18 in./in. (180 mm/mm) of common vent diameter of the segment in which the offset is located.
(3) For the segment of the common vertical vent containing the offset, the common vent capacity listed in the common venting tables is reduced by 20 percent (0.80 × maximum common vent capacity).
(4) A multistory common vent shall not be reduced in size above the offset.

See Exhibit 10.2i and the examples in Appendix G for more information on the multistory vents described in Paragraphs 10.2.12, 10.2.13, and 10.2.14. The vent on the first floor is treated like a normal venting system whose height ends at the next floor. The higher floors are treated as common vents whose height ends at the floor above. This tends to result in large diameter vertical vents. It may be more cost effective to use separate venting systems for groups of floors.

10.2.15 Where two or more appliances are connected to a vertical vent or chimney, the flow area of the largest section of vertical vent or chimney shall not exceed seven times the smallest listed appliance categorized vent areas, flue collar area, or draft hood outlet area unless designed in accordance with approved engineering methods.

See Exhibit 10.2i, Multistory vents.

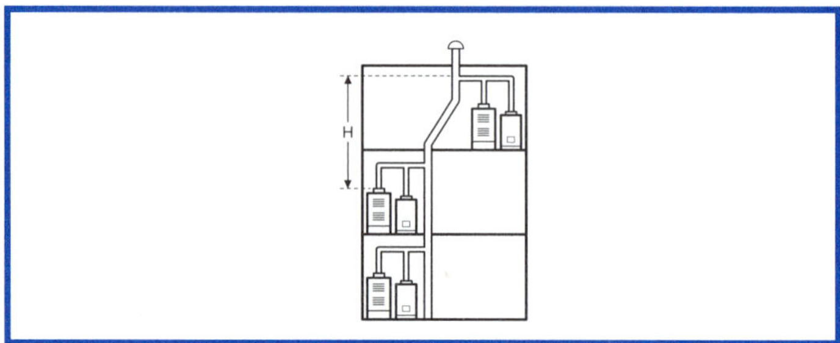

Exhibit 10.2i *Multistory vents.*

10.2.16 For appliances with more than one input rate, the minimum vent connector capacity (FAN Min) determined from the tables shall be less than the lowest appliance input rating, and the maximum vent connector capacity (FAN Max or NAT Max) determined from the tables shall be greater than the highest appliance input rating.

If the appliance has multiple input rates, the Min capacity is determined with the minimum input rate and the Max capacity is determined with the maximum input (Exhibit 10.2j).

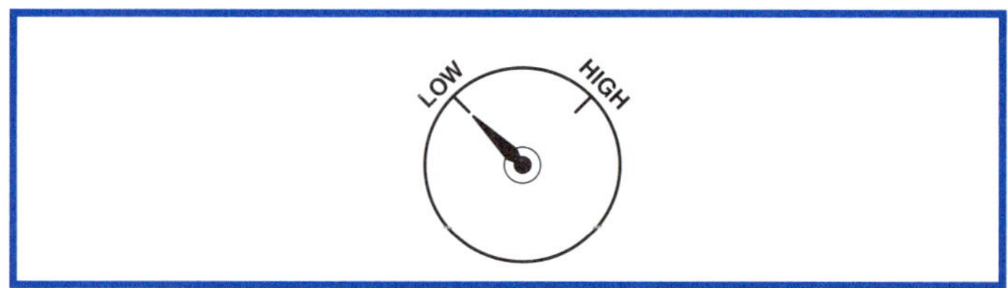

Exhibit 10.2j *Two-stage or modulating furnaces.*

10.2.17 Listed, corrugated metallic chimney liner systems in masonry chimneys shall be sized by using Table 10.6 or 10.7 for Type B vents, with the maximum capacity reduced by 20 percent (0.80 × maximum capacity) and the minimum capacity as shown in Table 10.6 or 10.7. Corrugated metallic liner systems installed with bends or offsets shall have their maximum capacity further reduced in accordance with 10.2.5 and 10.2.6.

Since properly installed corrugated chimney liners have heat loss similar to a Type B vent, they are sized using those tables. However, their corrugations and tendency to spiral in the chimney require a 20 percent Max capacity reduction. Many reliners begin at the breaching and then bend up vertically (Exhibit 10.2k). This bend must be counted as one of the two allowed elbows in the system.

Exhibit 10.2k *Chimney reliners.*

10.2.18 Tables 10.6 through 10.10 shall be used for chimneys and vents not exposed to the outdoors below the roof line. A Type B vent or listed chimney lining system passing through an unused masonry chimney flue shall not be considered to be exposed to the outdoors. Tables 10.12(a) and (b) and 10.13(a) and (b) shall be used for clay-tile-lined exterior masonry chimneys, provided all the following conditions are met:

(1) Vent connector is Type B double-wall.
(2) At least one appliance is draft hood–equipped.
(3) The combined appliance input rating is less than the maximum capacity given by Table 10.12(a) (for NAT+NAT) or Table 10.13(a) (for FAN+NAT).
(4) The input rating of each space-heating appliance is greater than the minimum input rating given by Table 10.12(b) (for NAT+NAT) or Table 10.13(b) (for FAN+NAT).
(5) The vent connector sizing is in accordance with Table 10.8.

If these conditions cannot be met, an alternative venting design shall be used, such as a listed chimney lining system.

Exception: The installation of vents serving listed appliances shall be permitted to be in accordance with the appliance manufacturer's instructions and the terms of the listing.

For exterior chimneys (Exhibit 10.2l), the Max capacity is determined by Table 10.2 while the Min capacity is determined by Table 10.12(a) or 10.13(a). Table 10.12(b) and 10.13(b), in turn, have different Min capacities based on the ambient temperatures expected. The vent connectors are sized with Table 10.8. The exception recognizes that manufacturers may produce products customized for exterior masonry chimneys and use their own instructions.

Exhibit 10.2l *Exterior chimneys.*

10.2.19 Vent connectors shall not be upsized more than two sizes greater than the listed appliance categorized vent diameter, flue collar diameter, or draft hood outlet diameter. Vent connectors for draft hood–equipped appliances shall not be smaller than the draft hood outlet diameter. If a vent connector size(s) determined from the tables for a fan-assisted appliance(s) is smaller than the flue collar diameter, the smaller size(s) shall be permitted to be used provided the following conditions are met:

(1) Vent connectors for fan-assisted appliance flue collars 12 in. (300 mm) in diameter or smaller are not reduced by more than one table size [e.g., 12 in. to 10 in. (300 mm to 250 mm) is a one-size reduction] and those larger than 12 in. (300 mm) in diameter are not reduced more than two table sizes [e.g., 24 in. to 20 in. (610 mm to 510 mm) is a two-size reduction].

(2) The fan-assisted appliance(s) is common vented with a draft hood–equipped appliance(s).

A sudden, large expansion of the vent connector diameter (Exhibit 10.2m) creates a pressure drop that may limit the draft and encourage condensation. Therefore, a limit is placed on it.

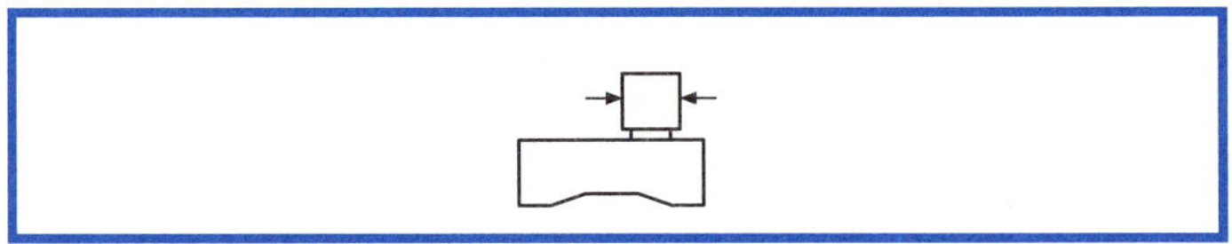

Exhibit 10.2m *Upsizing.*

10.2.20 All combination of pipe sizes, single-wall, and double-wall metal pipe shall be allowed within any connector run(s) or within the common vent, provided ALL of the appropriate tables permit ALL of the desired sizes and types of pipe, as if they were used for the entire length of the subject connector or vent. If single-wall and Type B double-wall metal pipes are used for vent connectors, the common vent must be sized using Table 10.7 or 10.9 as appropriate.

10.2.21 Where a table permits more than one diameter of pipe to be used for a connector or vent, all the permitted sizes shall be permitted to be used.

Paragraphs 10.2.20 and 10.2.21 effectively require the installer to check the Min and Max capacities for the vent section for each vent type as if the entire vent section were made of it. See Exhibit 10.2n. For example, suppose a vent connector was half single wall and half Type B. The installer must show that a vent connector of that length would be allowed if it was all single wall and also if it was all Type B.

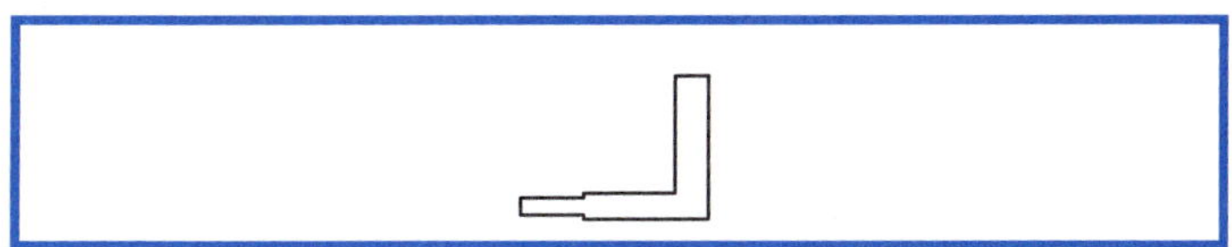

Exhibit 10.2n *Pot luck.*

10.2.22 Interpolation shall be permitted in calculating capacities for vent dimensions that fall between table entries. *(See Example 3, Appendix G.)*

10.2.23 Extrapolation beyond the table entries shall not be permitted.

If the installation dimensions fall between two table entries for which there are defined values, the installer may calculate the "in between" value, but it is not permitted to guess what the value might be outside of the table entries. See Exhibit 10.2o.

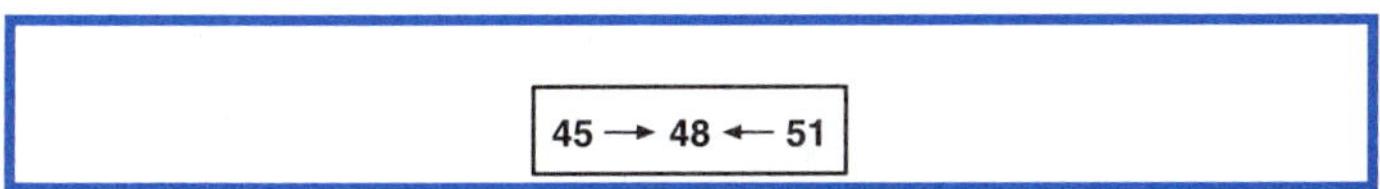

Exhibit 10.2o *Interpolation and extrapolation.*

10.2.24 For vent heights lower than 6 ft and higher than shown in the tables, engineering methods shall be used to calculate vent capacities.

For SI units, 1 in. = 25.4 mm; 1 in.^{2} = 645 mm^2; 1 ft = 0.305 m; 1000 Btu per hr = 0.293 kW.

New text in the 1999 edition reinforces the requirement that vent heights outside the tables must be calculated, and the tables cannot be used.

Table 10.1 Capacity of Type B Double-Wall Gas Vents When Connected Directly to a Single Category I Appliance

Height H (ft)	Lateral L (ft)	3 in. FAN Min	3 in. FAN Max	3 in. NAT Max	4 in. FAN Min	4 in. FAN Max	4 in. NAT Max	5 in. FAN Min	5 in. FAN Max	5 in. NAT Max	6 in. FAN Min	6 in. FAN Max	6 in. NAT Max	7 in. FAN Min	7 in. FAN Max	7 in. NAT Max	8 in. FAN Min	8 in. FAN Max	8 in. NAT Max	9 in. FAN Min	9 in. FAN Max	9 in. NAT Max
6	0	0	78	46	0	152	86	0	251	141	0	375	205	0	524	285	0	698	370	0	897	470
	2	13	51	36	18	97	67	27	157	105	32	232	157	44	321	217	53	425	285	63	543	370
	4	21	49	34	30	94	64	39	153	103	50	227	153	66	316	211	79	419	279	93	536	362
	6	25	46	32	36	91	61	47	149	100	59	223	149	78	310	205	93	413	273	110	530	354
8	0	0	84	50	0	165	94	0	276	155	0	415	235	0	583	320	0	780	415	0	1006	537
	2	12	57	40	16	109	75	25	178	120	28	263	180	42	365	247	50	483	322	60	619	418
	5	23	53	38	32	103	71	42	171	115	53	255	173	70	356	237	83	473	313	99	607	407
	8	28	49	35	39	98	66	51	164	109	64	247	165	84	347	227	99	463	303	117	596	396
10	0	0	88	53	0	175	100	0	295	166	0	447	255	0	631	345	0	847	450	0	1096	585
	2	12	61	42	17	118	81	23	194	129	26	289	195	40	402	273	48	533	355	57	684	457
	5	23	57	40	32	113	77	41	187	124	52	280	188	68	392	263	81	522	346	95	671	446
	10	30	51	36	41	104	70	54	176	115	67	267	175	88	376	245	104	504	330	122	651	427
15	0	0	94	58	0	191	112	0	327	187	0	502	285	0	716	390	0	970	525	0	1263	682
	2	11	69	48	15	136	93	20	226	150	22	339	225	33	475	316	45	633	414	53	815	544
	5	22	65	45	30	130	87	39	219	142	49	330	217	64	463	300	76	620	403	90	800	529
	10	29	59	41	40	121	82	51	206	135	64	315	208	84	445	288	99	600	386	116	777	507
	15	35	53	37	48	112	76	61	195	128	76	301	198	98	429	275	115	580	373	134	755	491
20	0	0	97	61	0	202	119	0	349	202	0	540	307	0	776	430	0	1057	575	0	1384	752
	2	10	75	51	14	149	100	18	250	166	20	377	249	33	531	346	41	711	470	50	917	612
	5	21	71	48	29	143	96	38	242	160	47	367	241	62	519	337	73	697	460	86	902	599
	10	28	64	44	38	133	89	50	229	150	62	351	228	81	499	321	95	675	443	112	877	576
	15	34	58	40	46	124	84	59	217	142	73	337	217	94	481	308	111	654	427	129	853	557
	20	48	52	35	55	116	78	69	206	134	84	322	206	107	464	295	125	634	410	145	830	537

30	0	0	100	64	0	213	128	0	374	220	0	587	336	0	853	475	0	1173	650	0	1548	855
	2	9	81	56	13	166	112	14	283	185	18	432	280	27	613	394	33	826	535	42	1072	700
	5	21	77	54	28	160	108	36	275	176	45	421	273	58	600	385	69	811	524	82	1055	688
	10	27	70	50	37	150	102	48	262	171	59	405	261	77	580	371	91	788	507	107	1028	668
	15	33	64	NA	44	141	96	57	249	163	70	389	249	90	560	357	105	765	490	124	1002	648
	20	56	58	NA	53	132	90	66	237	154	80	374	237	102	542	343	119	743	473	139	977	628
	30	NA	NA	NA	73	113	NA	88	214	NA	104	346	219	131	507	321	149	702	444	171	929	594
50	0	0	101	67	0	216	134	0	397	232	0	633	363	0	932	518	0	1297	708	0	1730	952
	2	8	86	61	11	183	122	14	320	206	15	497	314	22	715	445	26	975	615	33	1276	813
	5	20	82	NA	27	177	119	35	312	200	43	487	308	55	702	438	65	960	605	77	1259	798
	10	26	76	NA	35	168	114	45	299	190	56	471	298	73	681	426	86	935	589	101	1230	773
	15	59	70	NA	42	158	NA	54	287	180	66	455	288	85	662	413	100	911	572	117	1203	747
	20	NA	NA	NA	50	149	NA	63	275	169	76	440	278	97	642	401	113	888	556	131	1176	722
	30	NA	NA	NA	69	131	NA	84	250	NA	99	410	259	123	605	376	141	844	522	161	1125	670
100	0	NA	NA	NA	0	218	NA	0	407	NA	0	665	400	0	997	560	0	1411	770	0	1908	1040
	2	NA	NA	NA	10	194	NA	12	354	NA	13	566	375	18	831	510	21	1155	700	25	1536	935
	5	NA	NA	NA	26	189	NA	33	347	NA	40	557	369	52	820	504	60	1141	692	71	1519	926
	10	NA	NA	NA	33	182	NA	43	335	NA	53	542	361	68	801	493	80	1118	679	94	1492	910
	15	NA	NA	NA	40	174	NA	50	321	NA	62	528	353	80	782	482	93	1095	666	109	1465	895
	20	NA	NA	NA	47	166	NA	59	311	NA	71	513	344	90	763	471	105	1073	653	122	1438	880
	30	NA	NA	NA	NA	NA	NA	78	290	NA	92	483	NA	115	726	449	131	1029	627	149	1387	849
	50	NA	NA	NA	NA	NA	NA	NA	NA	NA	147	428	NA	180	651	405	197	944	575	217	1288	787

(continues)

Table 10.1 Continued.

Vent Diameter — D

Appliance Input Rating in Thousands of Btu per Hour

Height H (ft)	Lateral L (ft)	10 in. FAN Min	10 in. FAN Max	10 in. NAT Max	12 in. FAN Min	12 in. FAN Max	12 in. NAT Max	14 in. FAN Min	14 in. FAN Max	14 in. NAT Max	16 in. FAN Min	16 in. FAN Max	16 in. NAT Max	18 in. FAN Min	18 in. FAN Max	18 in. NAT Max	20 in. FAN Min	20 in. FAN Max	20 in. NAT Max	22 in. FAN Min	22 in. FAN Max	22 in. NAT Max	24 in. FAN Min	24 in. FAN Max	24 in. NAT Max
6	0	0	1121	570	0	1645	850	0	2267	1170	0	2983	1530	0	3802	1960	0	4721	2430	0	5737	2950	0	6853	3520
	2	75	675	455	103	982	650	138	1346	890	178	1769	1170	225	2250	1480	296	2782	1850	360	3377	2220	426	4030	2670
	4	110	668	445	147	975	640	191	1338	880	242	1761	1160	300	2242	1475	390	2774	1835	469	3370	2215	555	4023	2660
	6	128	661	435	171	967	630	219	1330	870	276	1753	1150	341	2235	1470	437	2767	1820	523	3363	2210	618	4017	2650
8	0	0	1261	660	0	1858	970	0	2571	1320	0	3399	1740	0	4333	2220	0	5387	2750	0	6555	3360	0	7838	4010
	2	71	770	515	98	1124	745	130	1543	1020	168	2030	1340	212	2584	1700	278	3196	2110	336	3882	2560	401	4634	3050
	5	115	758	503	154	1110	733	199	1528	1010	251	2013	1330	311	2563	1685	398	3180	2090	476	3863	2545	562	4612	3040
	8	137	746	490	180	1097	720	231	1514	1000	289	2000	1320	354	2552	1670	450	3163	2070	537	3850	2530	630	4602	3030
10	0	0	1377	720	0	2036	1060	0	2825	1450	0	3742	1925	0	4782	2450	0	5955	3050	0	7254	3710	0	8682	4450
	2	68	852	560	93	1244	850	124	1713	1130	161	2256	1480	202	2868	1890	264	3556	2340	319	4322	2840	378	5153	3390
	5	112	839	547	149	1229	829	192	1696	1105	243	2238	1461	300	2849	1871	382	3536	2318	458	4301	2818	540	5132	3371
	10	142	817	525	187	1204	795	238	1669	1080	298	2209	1430	364	2818	1840	459	3504	2280	546	4268	2780	641	5099	3340
15	0	0	1596	840	0	2380	1240	0	3323	1720	0	4423	2270	0	5678	2900	0	7099	3620	0	8665	4410	0	10393	5300
	2	63	1019	675	86	1495	985	114	2062	1350	147	2719	1770	186	3467	2260	239	4304	2800	290	5232	3410	346	6251	4080
	5	105	1003	660	140	1476	967	182	2041	1327	229	2696	1748	283	3442	2235	355	4278	2777	426	5204	3385	501	6222	4057
	10	135	977	635	177	1446	936	227	2009	1289	283	2659	1712	346	3402	2193	432	4234	2739	510	5159	3343	599	6175	4019
	15	155	953	610	202	1418	905	257	1976	1250	318	2623	1675	385	3363	2150	479	4192	2700	564	5115	3300	665	6129	3980
20	0	0	1756	930	0	2637	1350	0	3701	1900	0	4948	2520	0	6376	3250	0	7988	4060	0	9785	4980	0	11753	6000
	2	59	1150	755	81	1694	1100	107	2343	1520	139	3097	2000	175	3955	2570	220	4916	3200	269	5983	3910	321	7154	4700
	5	101	1133	738	135	1674	1079	174	2320	1498	219	3071	1978	270	3926	2544	337	4885	3174	403	5950	3880	475	7119	4662
	10	130	1105	710	172	1641	1045	220	2282	1460	273	3029	1940	334	3880	2500	413	4835	3130	489	5896	3830	573	7063	4600
	15	150	1078	688	195	1609	1018	248	2245	1425	306	2988	1910	372	3835	2465	459	4786	3090	541	5844	3795	631	7007	4575
	20	167	1052	665	217	1578	990	273	2210	1390	335	2948	1880	404	3791	2430	495	4737	3050	585	5792	3760	689	6953	4550

Height	Lateral																								
30	0	0	1977	1060	0	3004	1550	0	4252	2170	0	5725	2920	0	7420	3770	0	9341	4750	0	11483	5850	0	13848	7060
	2	54	1351	865	74	2004	1310	98	2786	1800	127	3696	2380	159	4734	3050	199	5900	3810	241	7194	4650	285	8617	5600
	5	96	1332	851	127	1981	1289	164	2759	1775	206	3666	2350	252	4701	3020	312	5863	3783	373	7155	4622	439	8574	5552
	10	125	1301	829	164	1944	1254	209	2716	1733	259	3617	2300	316	4647	2970	386	5803	3739	456	7090	4574	535	8505	5471
	15	143	1272	807	187	1908	1220	237	2674	1692	292	3570	2250	354	4594	2920	431	5744	3695	507	7026	4527	590	8437	5391
	20	160	1243	784	207	1873	1185	260	2633	1650	319	3523	2200	384	4542	2870	467	5686	3650	548	6964	4480	639	8370	5310
	30	195	1189	745	246	1807	1130	305	2555	1585	369	3433	2130	440	4442	2785	540	5574	3565	635	6842	4375	739	8239	5225
50	0	0	2231	1195	0	3441	1825	0	4934	2550	0	6711	3440	0	8774	4460	0	11129	5635	0	13767	6940	0	16694	8430
	2	41	1620	1010	66	2431	1513	86	3409	2125	113	4554	2840	141	5864	3670	171	7339	4630	209	8980	5695	251	10788	6860
	5	90	1600	996	118	2406	1495	151	3380	2102	191	4520	2813	234	5826	3639	283	7295	4597	336	8933	5654	394	10737	6818
	10	118	1567	972	154	2366	1466	196	3332	2064	243	4464	2767	295	5763	3585	355	7224	4542	419	8855	5585	491	10652	6749
	15	136	1536	948	177	2327	1437	222	3285	2026	274	4409	2721	330	5701	3534	396	7155	4511	465	8779	5546	542	10570	6710
	20	151	1505	924	195	2288	1408	244	3239	1987	300	4356	2675	361	5641	3481	433	7086	4479	506	8704	5506	586	10488	6670
	30	183	1446	876	232	2214	1349	287	3150	1910	347	4253	2631	412	5523	3431	494	6953	4421	577	8557	5444	672	10328	6603
100	0	0	2491	1310	0	3925	2050	0	5729	2950	0	7914	4050	0	10485	5300	0	13454	6700	0	16817	8600	0	20578	10300
	2	30	1975	1170	44	3027	1820	72	4313	2550	95	5834	3500	120	7591	4600	138	9577	5800	169	11803	7200	204	14264	8800
	5	82	1955	1159	107	3002	1803	136	4282	2531	172	5797	3475	208	7548	4566	245	9528	5769	293	11748	7162	341	14204	8756
	10	108	1923	1142	142	2961	1775	180	4231	2500	223	5737	3434	268	7478	4509	318	9447	5717	374	11658	7100	436	14105	8683
	15	126	1892	1124	163	2920	1747	206	4182	2469	252	5678	3392	304	7409	4451	358	9367	5665	418	11569	7037	487	14007	8610
	20	141	1861	1107	181	2880	1719	226	4133	2438	277	5619	3351	330	7341	4394	387	9289	5613	452	11482	6975	523	13910	8537
	30	170	1802	1071	215	2803	1663	265	4037	2375	319	5505	3267	378	7209	4279	446	9136	5509	514	11310	6850	592	13720	8391
	50	241	1688	1000	292	2657	1550	350	3856	2250	415	5289	3100	486	6956	4050	572	8841	5300	659	10979	6600	752	13354	8100

Table 10.2 Capacity of Type B Double-Wall Vents with Single-Wall Metal Connectors Serving a Single Category I Appliance

Height H (ft)	Lateral L (ft)	3 in. FAN Min	3 in. FAN Max	3 in. NAT Max	4 in. FAN Min	4 in. FAN Max	4 in. NAT Max	5 in. FAN Min	5 in. FAN Max	5 in. NAT Max	6 in. FAN Min	6 in. FAN Max	6 in. NAT Max	7 in. FAN Min	7 in. FAN Max	7 in. NAT Max	8 in. FAN Min	8 in. FAN Max	8 in. NAT Max	9 in. FAN Min	9 in. FAN Max	9 in. NAT Max	10 in. FAN Min	10 in. FAN Max	10 in. NAT Max	12 in. FAN Min	12 in. FAN Max	12 in. NAT Max
6	0	38	77	45	59	151	85	85	249	140	126	373	204	165	522	284	211	695	369	267	894	469	371	1118	569	537	1639	849
	2	39	51	36	60	96	66	85	156	104	123	231	156	159	320	213	201	423	284	251	541	368	347	673	453	498	979	648
	4	NA	NA	33	74	92	63	102	152	102	146	225	152	187	313	208	237	416	277	295	533	360	409	664	443	584	971	638
	6	NA	NA	31	83	89	60	114	147	99	163	220	148	207	307	203	263	409	271	327	526	352	449	656	433	638	962	627
8	0	37	83	50	58	164	93	83	273	154	123	412	234	161	580	319	206	777	414	258	1002	536	360	1257	658	521	1852	967
	2	39	56	39	59	108	75	83	176	119	121	261	179	155	363	246	197	482	321	246	617	417	339	768	513	486	1120	743
	5	NA	NA	37	77	102	69	107	168	114	151	252	171	193	352	235	245	470	311	305	604	404	418	754	500	598	1104	730
	8	NA	NA	33	90	95	64	122	161	107	175	243	163	223	342	225	280	458	300	344	591	392	470	740	486	665	1089	715
10	0	37	87	53	57	174	99	82	293	165	120	444	254	158	628	344	202	844	449	253	1093	584	351	1373	718	507	2031	1057
	2	39	61	41	59	117	80	82	193	128	119	287	194	153	400	272	193	531	354	242	681	456	332	849	559	475	1242	848
	5	52	56	39	76	111	76	105	185	122	148	277	186	190	388	261	241	518	344	299	667	443	409	834	544	584	1224	825
	10	NA	NA	34	97	100	68	132	171	112	188	261	171	237	369	241	296	497	325	363	643	423	492	808	520	688	1194	788
15	0	36	93	57	56	190	111	80	325	186	116	499	283	153	713	388	195	966	523	244	1259	681	336	1591	838	488	2374	1237
	2	38	69	47	57	136	93	80	225	149	115	337	224	148	473	314	187	631	413	232	812	543	319	1015	673	457	1491	983
	5	51	63	44	75	128	86	102	216	140	144	326	217	182	459	298	231	616	400	287	795	526	392	997	657	562	1469	963
	10	NA	NA	39	95	116	79	128	201	131	182	308	203	228	438	284	284	592	381	349	768	501	470	966	628	664	1433	928
	15	NA	NA	NA	NA	NA	72	158	186	124	220	290	192	272	418	269	334	568	367	404	742	484	540	937	601	750	1399	894
20	0	35	96	60	54	200	118	73	346	201	114	537	306	149	772	428	190	1053	573	238	1379	750	326	1751	927	473	2631	1346
	2	37	74	50	56	148	99	78	248	165	113	375	248	144	528	344	182	708	468	227	914	611	309	1146	754	443	1689	1098
	5	50	68	47	73	140	94	100	239	158	141	363	239	178	514	334	224	692	457	279	896	596	381	1126	734	547	1665	1074
	10	NA	NA	41	93	129	86	125	223	146	177	344	224	222	491	316	277	666	437	339	866	570	457	1092	702	646	1626	1037
	15	NA	NA	NA	NA	NA	80	155	208	136	216	325	210	264	469	301	325	640	419	393	838	549	526	1060	677	730	1587	1005
	20	NA	NA	NA	NA	NA	NA	136	192	126	254	306	196	309	448	285	374	616	400	448	810	526	592	1028	651	808	1550	973

Vent Diameter — D

Appliance Input Rating in Thousands of Btu per Hour

30	0	34	99	63	53	211	127	76	372	219	110	584	334	144	849	472	184	1168	647	229	1542	852	312	1971	1056	454	2996	1545
	2	37	80	56	55	164	111	76	281	183	109	429	279	139	610	392	175	823	533	219	1069	698	296	1346	863	424	1999	1308
	5	49	74	52	72	157	106	98	271	173	136	417	271	171	595	382	215	806	521	269	1049	684	366	1324	846	524	1971	1283
	10	NA	NA	NA	91	144	98	122	255	168	171	397	257	213	570	367	265	777	501	327	1017	662	440	1287	821	620	1927	1243
	15	NA	NA	NA	115	131	NA	151	239	157	208	377	242	255	547	349	312	750	481	379	985	638	507	1251	794	702	1884	1205
	20	NA	NA	NA	NA	NA	NA	181	223	NA	246	357	228	298	524	333	360	723	461	433	955	615	570	1216	768	780	1841	1166
	30	NA	NA	NA	NA	NA	NA	NA	NA	NA	NA	NA	NA	389	477	305	461	670	426	541	895	574	704	1147	720	937	1759	1101
50	0	33	99	66	51	213	133	73	394	230	105	629	361	138	928	515	176	1292	704	220	1724	948	295	2223	1189	428	3432	1818
	2	36	84	61	53	181	121	73	318	205	104	495	312	133	712	443	168	971	613	209	1273	811	280	1615	1007	401	2426	1509
	5	48	80	NA	70	174	117	94	308	198	131	482	305	164	696	435	204	953	602	257	1252	795	347	1591	991	496	2396	1490
	10	NA	NA	NA	89	160	NA	118	292	186	162	461	292	203	671	420	253	923	583	313	1217	765	418	1551	963	589	2347	1455
	15	NA	NA	NA	112	148	NA	145	275	174	199	441	280	244	646	405	299	894	562	363	1183	736	481	1512	934	668	2299	1421
	20	NA	NA	NA	NA	NA	NA	176	257	NA	236	420	267	285	622	389	345	866	543	415	1150	708	544	1473	906	741	2251	1387
	30	NA	NA	NA	NA	NA	NA	NA	NA	NA	315	376	NA	373	573	NA	442	809	502	521	1086	649	674	1399	848	892	2159	1318
100	0	NA	NA	NA	49	214	NA	69	403	NA	100	659	395	131	991	555	166	1404	765	207	1900	1033	273	2479	1300	395	3912	2042
	2	NA	NA	NA	51	192	NA	70	351	NA	98	563	373	125	828	508	158	1152	698	196	1532	933	259	1970	1168	371	3021	1817
	5	NA	NA	NA	67	186	NA	90	342	NA	125	551	366	156	813	501	194	1134	688	240	1511	921	322	1945	1153	460	2990	1796
	10	NA	NA	NA	85	175	NA	113	324	NA	153	532	354	191	789	486	238	1104	672	293	1477	902	389	1905	1133	547	2938	1763
	15	NA	NA	NA	132	162	NA	138	310	NA	188	511	343	230	764	473	281	1075	656	342	1443	884	447	1865	1110	618	2888	1730
	20	NA	NA	NA	NA	NA	NA	168	295	NA	224	487	NA	270	739	458	325	1046	639	391	1410	864	507	1825	1087	690	2838	1696
	30	NA	NA	NA	NA	NA	NA	231	264	NA	301	448	NA	355	685	NA	418	988	NA	491	1343	824	631	1747	1041	834	2739	1627
	50	NA	NA	NA	NA	NA	NA	NA	NA	NA	NA	NA	NA	540	584	NA	617	866	NA	711	1205	NA	895	1591	NA	1138	2547	1489

Table 10.3 Capacity of Masonry Chimney Flue with Type B Double-Wall Vent Connectors Serving a Single Category I Appliance

Type B Double-Wall Connector Diameter — D
To be used with chimney areas within the size limits at bottom

Appliance Input Rating in Thousands of Btu per Hour

Height H (ft)	Lateral L (ft)	3 in. FAN Min	3 in. FAN Max	3 in. NAT Max	4 in. FAN Min	4 in. FAN Max	4 in. NAT Max	5 in. FAN Min	5 in. FAN Max	5 in. NAT Max	6 in. FAN Min	6 in. FAN Max	6 in. NAT Max	7 in. FAN Min	7 in. FAN Max	7 in. NAT Max	8 in. FAN Min	8 in. FAN Max	8 in. NAT Max	9 in. FAN Min	9 in. FAN Max	9 in. NAT Max	10 in. FAN Min	10 in. FAN Max	10 in. NAT Max	12 in. FAN Min	12 in. FAN Max	12 in. NAT Max
6	2	NA	NA	28	NA	NA	52	NA	NA	86	NA	NA	130	NA	NA	180	NA	NA	247	NA	NA	320	NA	NA	401	NA	NA	581
	5	NA	NA	25	NA	NA	49	NA	NA	82	NA	NA	117	NA	NA	165	NA	NA	231	NA	NA	298	NA	NA	376	NA	NA	561
8	2	NA	NA	29	NA	NA	55	NA	NA	93	NA	NA	145	NA	NA	198	NA	NA	266	84	590	350	100	728	446	139	1024	651
	5	NA	NA	26	NA	NA	52	NA	NA	88	NA	NA	134	NA	NA	183	NA	NA	247	NA	NA	328	149	711	423	201	1007	640
	8	NA	NA	24	NA	NA	48	NA	NA	83	NA	NA	127	NA	NA	175	NA	NA	239	NA	NA	318	173	695	410	231	990	623
10	2	NA	NA	31	NA	NA	61	NA	NA	103	NA	NA	162	NA	NA	221	68	519	298	82	655	388	98	810	491	136	1144	724
	5	NA	NA	28	NA	NA	57	NA	NA	96	NA	NA	148	NA	NA	204	NA	NA	277	124	638	365	146	791	466	196	1124	712
	10	NA	NA	25	NA	NA	50	NA	NA	87	NA	NA	139	NA	NA	191	NA	NA	263	155	610	347	182	762	444	240	1093	668
15	2	NA	NA	35	NA	NA	67	NA	NA	114	NA	NA	179	53	475	250	64	613	336	77	779	441	92	968	562	127	1376	841
	5	NA	NA	35	NA	NA	62	NA	NA	107	NA	NA	164	NA	NA	231	99	594	313	118	759	416	139	946	533	186	1352	828
	10	NA	NA	28	NA	NA	55	NA	NA	97	NA	NA	153	NA	NA	216	126	565	296	148	727	394	173	912	567	229	1315	777
	15	NA	NA	NA	NA	NA	48	NA	NA	89	NA	NA	141	NA	NA	201	NA	NA	281	171	698	375	198	880	485	259	1280	742
20	2	NA	NA	38	NA	NA	74	NA	NA	124	NA	NA	201	51	522	274	61	678	375	73	867	491	87	1083	627	121	1548	953
	5	NA	NA	36	NA	NA	68	NA	NA	116	NA	NA	184	80	503	254	95	658	350	113	845	463	133	1059	597	179	1523	933
	10	NA	NA	NA	NA	NA	60	NA	NA	107	NA	NA	172	NA	NA	237	122	627	332	143	811	440	167	1022	566	221	1482	879
	15	NA	NA	NA	NA	NA	NA	NA	NA	97	NA	NA	159	NA	NA	220	NA	NA	314	165	780	418	191	987	541	251	1443	840
	20	NA	NA	NA	NA	NA	NA	NA	NA	83	NA	NA	148	NA	NA	206	NA	NA	296	186	750	397	214	955	513	277	1406	807
30	2	NA	NA	41	NA	NA	82	NA	NA	137	NA	NA	216	47	581	303	57	762	421	68	985	558	81	1240	717	111	1793	1112
	5	NA	NA	NA	NA	NA	76	NA	NA	128	NA	NA	198	75	561	281	90	741	393	106	962	526	125	1216	683	169	1766	1094
	10	NA	NA	NA	NA	NA	67	NA	NA	115	NA	NA	184	NA	NA	263	115	709	373	135	927	500	158	1176	648	210	1721	1025
	15	NA	NA	NA	NA	NA	NA	NA	NA	107	NA	NA	171	NA	NA	243	NA	NA	353	156	893	476	181	1139	621	239	1679	981
	20	NA	NA	NA	NA	NA	NA	NA	NA	91	NA	NA	159	NA	NA	227	NA	NA	332	176	860	450	203	1103	592	264	1638	940
	30	NA	NA	NA	NA	NA	NA	NA	NA	NA	NA	NA	NA	NA	NA	188	NA	NA	288	NA	NA	416	249	1035	555	318	1560	877

Height	Lateral L	3 in. FAN Min	3 in. FAN Max	3 in. NAT Max	4 in. FAN Min	4 in. FAN Max	4 in. NAT Max	5 in. FAN Min	5 in. FAN Max	5 in. NAT Max	6 in. FAN Min	6 in. FAN Max	6 in. NAT Max	7 in. FAN Min	7 in. FAN Max	7 in. NAT Max	8 in. FAN Min	8 in. FAN Max	8 in. NAT Max	9 in. FAN Min	9 in. FAN Max	9 in. NAT Max	10 in. FAN Min	10 in. FAN Max	10 in. NAT Max	12 in. FAN Min	12 in. FAN Max	12 in. NAT Max
50	2	NA	NA	NA	NA	NA	92	NA	NA	161	NA	NA	251	NA	NA	351	51	840	477	61	1106	633	72	1413	812	99	2080	1243
	5	NA	NA	NA	NA	NA	NA	NA	NA	151	NA	NA	230	NA	NA	323	83	819	445	98	1083	596	116	1387	774	155	2052	1225
	10	NA	NA	NA	NA	NA	NA	NA	NA	138	NA	NA	215	NA	NA	304	NA	NA	424	126	1047	567	147	1347	733	195	2006	1147
	15	NA	NA	NA	NA	NA	NA	NA	NA	127	NA	NA	199	NA	NA	282	NA	NA	400	146	1010	539	170	1307	702	222	1961	1099
	20	NA	NA	NA	NA	NA	NA	NA	NA	NA	NA	NA	185	NA	NA	264	NA	NA	376	165	977	511	190	1269	669	246	1916	1050
	30	NA	NA	NA	NA	NA	NA	NA	NA	NA	NA	NA	NA	NA	NA	NA	NA	NA	327	NA	NA	468	233	1196	623	295	1832	984
Minimum internal area of chimney (in.^{2})		12			19			28			38			50			63			78			95			132		
Maximum internal area of chimney (in.^{2})		49			88			137			198			269			352			445			550			792		

Table 10.4 Capacity of Masonry Chimney Flue with Single-Wall Vent Connectors Serving a Single Category I Appliance

Single-Wall Metal Connector Diameter — D
To be used with chimney areas within the size limits at bottom

Appliance Input Rating in Thousands of Btu per Hour

Height H (ft)	Lateral L (ft)	3 in. FAN Min	3 in. FAN Max	3 in. NAT Max	4 in. FAN Min	4 in. FAN Max	4 in. NAT Max	5 in. FAN Min	5 in. FAN Max	5 in. NAT Max	6 in. FAN Min	6 in. FAN Max	6 in. NAT Max	7 in. FAN Min	7 in. FAN Max	7 in. NAT Max	8 in. FAN Min	8 in. FAN Max	8 in. NAT Max	9 in. FAN Min	9 in. FAN Max	9 in. NAT Max	10 in. FAN Min	10 in. FAN Max	10 in. NAT Max	12 in. FAN Min	12 in. FAN Max	12 in. NAT Max
6	2	NA	NA	28	NA	NA	52	NA	NA	86	NA	NA	130	NA	NA	180	NA	NA	247	NA	NA	319	NA	NA	400	NA	NA	580
	5	NA	NA	25	NA	NA	48	NA	NA	81	NA	NA	116	NA	NA	164	NA	NA	230	NA	NA	297	NA	NA	375	NA	NA	560
8	2	NA	NA	29	NA	NA	55	NA	NA	93	NA	NA	145	NA	NA	197	NA	NA	265	NA	NA	349	382	725	445	549	1021	650
	5	NA	NA	26	NA	NA	51	NA	NA	87	NA	NA	133	NA	NA	182	NA	NA	246	NA	NA	327	NA	NA	422	673	1003	638
	8	NA	NA	23	NA	NA	47	NA	NA	82	NA	NA	126	NA	NA	174	NA	NA	237	NA	NA	317	NA	NA	408	747	985	621
10	2	NA	NA	31	NA	NA	61	NA	NA	102	NA	NA	161	NA	NA	220	216	518	297	271	654	387	373	808	490	536	1142	722
	5	NA	NA	28	NA	NA	56	NA	NA	95	NA	NA	147	NA	NA	203	NA	NA	276	334	635	364	459	789	465	657	1121	710
	10	NA	NA	24	NA	NA	49	NA	NA	86	NA	NA	137	NA	NA	189	NA	NA	261	NA	NA	345	547	758	441	771	1088	665

(continues)

Table 10.4 Continued.

		Single-Wall Metal Connector Diameter — *D* To be used with chimney areas within the size limits at bottom																							
		3 in.			4 in.			5 in.			6 in.			7 in.			8 in.			9 in.			10 in.			12 in.		
		Appliance Input Rating in Thousands of Btu per Hour																								
Height *H* (ft)	Lateral *L* (ft)	FAN		NAT	FAN		NAT	FAN		NAT	FAN		NAT	FAN		NAT	FAN		NAT	FAN		NAT	FAN		NAT	FAN		NAT
		Min	Max	Max	Min	Max	Max	Min	Max	Max	Min	Max	Max	Min	Max	Max	Min	Max	Max	Min	Max	Max	Min	Max	Max	Min	Max	Max
15	2	NA	NA	35	NA	NA	67	NA	NA	113	NA	NA	178	166	473	249	211	611	335	264	776	440	362	965	560	520	1373	840
	5	NA	NA	32	NA	NA	61	NA	NA	106	NA	NA	163	NA	NA	230	261	591	312	325	755	414	444	942	531	637	1348	825
	10	NA	NA	27	NA	NA	54	NA	NA	96	NA	NA	151	NA	NA	214	NA	NA	294	392	722	392	531	907	504	749	1309	774
	15	NA	NA	NA	NA	NA	46	NA	NA	87	NA	NA	138	NA	NA	198	NA	NA	278	452	692	372	606	873	481	841	1272	738
20	2	NA	NA	38	NA	NA	73	NA	NA	123	NA	NA	200	163	520	273	206	675	374	258	864	490	252	1079	625	508	1544	950
	5	NA	NA	35	NA	NA	67	NA	NA	115	NA	NA	183	NA	NA	252	255	655	348	317	842	461	433	1055	594	623	1518	930
	10	NA	NA	NA	NA	NA	59	NA	NA	105	NA	NA	170	NA	NA	235	312	622	330	382	806	437	517	1016	562	733	1475	875
	15	NA	NA	NA	NA	NA	NA	NA	NA	95	NA	NA	156	NA	NA	217	NA	NA	311	442	773	414	591	979	539	823	1434	835
	20	NA	NA	NA	NA	NA	NA	NA	NA	80	NA	NA	144	NA	NA	202	NA	NA	292	NA	NA	392	663	944	510	911	1394	800
30	2	NA	NA	41	NA	NA	81	NA	NA	136	NA	NA	215	158	578	302	200	759	420	249	982	556	340	1237	715	489	1789	1110
	5	NA	NA	NA	NA	NA	75	NA	NA	127	NA	NA	196	NA	NA	279	245	737	391	306	958	524	417	1210	680	600	1760	1090
	10	NA	NA	NA	NA	NA	66	NA	NA	113	NA	NA	182	NA	NA	260	300	703	370	370	920	496	500	1168	644	708	1713	1020
	15	NA	NA	NA	NA	NA	NA	NA	NA	105	NA	NA	168	NA	NA	240	NA	NA	349	428	884	471	572	1128	615	798	1668	975
	20	NA	NA	NA	NA	NA	NA	NA	NA	88	NA	NA	155	NA	NA	223	NA	NA	327	NA	NA	445	643	1089	585	883	1624	932
	30	NA	NA	NA	NA	NA	NA	NA	NA	NA	NA	NA	NA	NA	NA	182	NA	NA	281	NA	NA	408	NA	NA	544	1055	1539	865
50	2	NA	NA	NA	NA	NA	91	NA	NA	160	NA	NA	250	NA	NA	350	191	837	475	238	1103	631	323	1408	810	463	2076	1240
	5	NA	NA	NA	NA	NA	NA	NA	NA	149	NA	NA	228	NA	NA	321	NA	NA	442	293	1078	593	398	1381	770	571	2044	1220
	10	NA	NA	NA	NA	NA	NA	NA	NA	136	NA	NA	212	NA	NA	301	NA	NA	420	355	1038	562	447	1337	728	674	1994	1140
	15	NA	NA	NA	NA	NA	NA	NA	NA	124	NA	NA	195	NA	NA	278	NA	NA	395	NA	NA	533	546	1294	695	761	1945	1090
	20	NA	NA	NA	NA	NA	NA	NA	NA	NA	NA	NA	180	NA	NA	258	NA	NA	370	NA	NA	504	616	1251	660	844	1898	1040
	30	NA	NA	NA	NA	NA	NA	NA	NA	NA	NA	NA	NA	NA	NA	NA	NA	NA	318	NA	NA	458	NA	NA	610	1009	1805	970
Minimum internal area of chimney (in.^{2})		12			19			28			38			50			63			78			95			132		
Maximum internal area of chimney (in.^{2})		49			88			137			198			269			352			445			550			792		

Table 10.5 Capacity of Single-Wall Metal Pipe or Type B Asbestos Cement Vents Serving a Single Draft Hood–Equipped Appliance

Height *H* (ft)	Lateral *L* (ft)	Vent Diameter — *D*							
		3 in.	4 in.	5 in.	6 in.	7 in.	8 in.	10 in.	12 in.
		Maximum Appliance Input Rating in Thousands of Btu per Hour							
6	0	39	70	116	170	232	312	500	750
	2	31	55	94	141	194	260	415	620
	5	28	51	88	128	177	242	390	600
8	0	42	76	126	185	252	340	542	815
	2	32	61	102	154	210	284	451	680
	5	29	56	95	141	194	264	430	648
	10	24	49	86	131	180	250	406	625
10	0	45	84	138	202	279	372	606	912
	2	35	67	111	168	233	311	505	760
	5	32	61	104	153	215	289	480	724
	10	27	54	94	143	200	274	455	700
	15	NA	46	84	130	186	258	432	666
15	0	49	91	151	223	312	420	684	1040
	2	39	72	122	186	260	350	570	865
	5	35	67	110	170	240	325	540	825
	10	30	58	103	158	223	308	514	795
	15	NA	50	93	144	207	291	488	760
	20	NA	NA	82	132	195	273	466	726
20	0	53	101	163	252	342	470	770	1190
	2	42	80	136	210	286	392	641	990
	5	38	74	123	192	264	364	610	945
	10	32	65	115	178	246	345	571	910
	15	NA	55	104	163	228	326	550	870
	20	NA	NA	91	149	214	306	525	832
30	0	56	108	183	276	384	529	878	1370
	2	44	84	148	230	320	441	730	1140
	5	NA	78	137	210	296	410	694	1080
	10	NA	68	125	196	274	388	656	1050
	15	NA	NA	113	177	258	366	625	1000
	20	NA	NA	99	163	240	344	596	960
	30	NA	NA	NA	NA	192	295	540	890
50	0	NA	120	210	310	443	590	980	1550
	2	NA	95	171	260	370	492	820	1290
	5	NA	NA	159	234	342	474	780	1230
	10	NA	NA	146	221	318	456	730	1190
	15	NA	NA	NA	200	292	407	705	1130
	20	NA	NA	NA	185	276	384	670	1080
	30	NA	NA	NA	NA	222	330	605	1010

Table 10.6 Capacity of Type B Double-Wall Vents with Type B Double-Wall Connectors Serving Two or More Category I Appliances

Vent Connector Capacity

Type B Double-Wall Vent and Connector Diameter — D

Appliance Input Rating Limits in Thousands of Btu per Hour

Vent Height H (ft)	Connector Rise R (ft)	3 in. FAN Min	3 in. FAN Max	3 in. NAT Max	4 in. FAN Min	4 in. FAN Max	4 in. NAT Max	5 in. FAN Min	5 in. FAN Max	5 in. NAT Max	6 in. FAN Min	6 in. FAN Max	6 in. NAT Max	7 in. FAN Min	7 in. FAN Max	7 in. NAT Max	8 in. FAN Min	8 in. FAN Max	8 in. NAT Max	9 in. FAN Min	9 in. FAN Max	9 in. NAT Max	10 in. FAN Min	10 in. FAN Max	10 in. NAT Max
6	1	22	37	26	35	66	46	46	106	72	58	164	104	77	225	142	92	296	185	109	376	237	128	466	289
	2	23	41	31	37	75	55	48	121	86	60	183	124	79	253	168	95	333	220	112	424	282	131	526	345
	3	24	44	35	38	81	62	49	132	96	62	199	139	82	275	189	97	363	248	114	463	317	134	575	386
8	1	22	40	27	35	72	48	49	114	76	64	176	109	84	243	148	100	320	194	118	408	248	138	507	303
	2	23	44	32	36	80	57	51	128	90	66	195	129	86	269	175	103	356	230	121	454	294	141	564	358
	3	24	47	36	37	87	64	53	139	101	67	210	145	88	290	198	105	384	258	123	492	330	143	612	402
10	1	22	43	28	34	78	50	49	123	78	65	189	113	89	257	154	106	341	200	125	436	257	146	542	314
	2	23	47	33	36	86	59	51	136	93	67	206	134	91	282	182	109	374	238	128	479	305	149	596	372
	3	24	50	37	37	92	67	52	146	104	69	220	150	94	303	205	111	402	268	131	515	342	152	642	417
15	1	21	50	30	33	89	53	47	142	83	64	220	120	88	298	163	110	389	214	134	493	273	162	609	333
	2	22	53	35	35	96	63	49	153	99	66	235	142	91	320	193	112	419	253	137	532	323	165	658	394
	3	24	55	40	36	102	71	51	163	111	68	248	160	93	339	218	115	445	286	140	565	365	167	700	444
20	1	21	54	31	33	99	56	46	157	87	62	246	125	86	334	171	107	436	224	131	552	285	158	681	347
	2	22	57	37	34	105	66	48	167	104	64	259	149	89	354	202	110	463	265	134	587	339	161	725	414
	3	23	60	42	35	110	74	50	176	116	66	271	168	91	371	228	113	486	300	137	618	383	164	764	466
30	1	20	62	33	31	113	59	45	181	93	60	288	134	83	391	182	103	512	238	125	649	305	151	802	372
	2	21	64	39	33	118	70	47	190	110	62	299	158	85	408	215	105	535	282	129	679	360	155	840	439
	3	22	66	44	34	123	79	48	198	124	64	309	178	88	423	242	108	555	317	132	706	405	158	874	494
50	1	19	71	36	30	133	64	43	216	101	57	349	145	78	477	157	97	627	257	120	797	330	144	984	403
	2	21	73	43	32	137	76	45	223	119	59	358	172	81	490	234	100	645	306	123	820	392	148	1014	478
	3	22	75	48	33	141	86	46	229	134	61	366	194	83	502	263	103	661	343	126	842	441	151	1043	538
100	1	18	82	37	28	158	66	40	262	104	53	442	150	73	611	204	91	810	266	112	1038	341	135	1285	417
	2	19	83	44	30	161	79	42	267	123	55	447	178	75	619	242	94	822	316	115	1054	405	139	1306	494
	3	20	84	50	31	163	89	44	272	138	57	452	200	78	627	272	97	834	355	118	1069	455	142	1327	555

Vent Height H (ft)	Connector Rise R (ft)	12 in.			14 in.			16 in.			18 in.			20 in.			22 in.			24 in.		
		FAN		NAT	FAN		NAT	FAN		NAT	FAN		NAT	FAN		NAT	FAN		NAT	FAN		NAT
		Min	Max	Max	Min	Max	Max	Min	Max	Max	Min	Max	Max	Min	Max	Max	Min	Max	Max	Min	Max	Max
6	2	174	764	496	223	1046	653	281	1371	853	346	1772	1080	NA	NA	NA	NA	NA	NA	NA	NA	NA
	4	180	897	616	230	1231	827	287	1617	1081	352	2069	1370	NA	NA	NA	NA	NA	NA	NA	NA	NA
	6	NA	NA	NA	NA	NA	NA	NA	NA	NA	NA	NA	NA	NA	NA	NA	NA	NA	NA	NA	NA	NA
8	2	186	822	516	238	1126	696	298	1478	910	365	1920	1150	NA	NA	NA	NA	NA	NA	NA	NA	NA
	4	192	952	644	244	1307	884	305	1719	1150	372	2211	1460	471	2737	1800	560	3319	2180	662	3957	2590
	6	198	1050	772	252	1445	1072	313	1902	1390	380	2434	1770	478	3018	2180	568	3665	2640	669	4373	3130
10	2	196	870	536	249	1195	730	311	1570	955	379	2049	1205	NA	NA	NA	NA	NA	NA	NA	NA	NA
	4	201	997	664	256	1371	924	318	1804	1205	387	2332	1535	486	2887	1890	581	3502	2280	686	4175	2710
	6	207	1095	792	263	1509	1118	325	1989	1455	395	2556	1865	494	3169	2290	589	3849	2760	694	4593	3270
15	2	214	967	568	272	1334	790	336	1760	1030	408	2317	1305	NA	NA	NA	NA	NA	NA	NA	NA	NA
	4	221	1085	712	279	1499	1006	344	1978	1320	416	2579	1665	523	3197	2060	624	3881	2490	734	4631	2960
	6	228	1181	856	286	1632	1222	351	2157	1610	424	2796	2025	533	3470	2510	634	4216	3030	743	5035	3600
20	2	223	1051	596	291	1443	840	357	1911	1095	430	2533	1385	NA	NA	NA	NA	NA	NA	NA	NA	NA
	4	230	1162	748	298	1597	1064	365	2116	1395	438	2778	1765	554	3447	2180	661	4190	2630	772	5005	3130
	6	237	1253	900	307	1726	1288	373	2287	1695	450	2984	2145	567	3708	2650	671	4511	3190	785	5392	3790
30	2	216	1217	632	286	1664	910	367	2183	1190	461	2891	1540	NA	NA	NA	NA	NA	NA	NA	NA	NA
	4	223	1316	792	294	1802	1160	376	2366	1510	474	3110	1920	619	3840	2365	728	4861	2860	847	5606	3410
	6	231	1400	952	303	1920	1410	384	2524	1830	485	3299	2340	632	4080	2875	741	4976	3480	860	5961	4150
50	2	206	1479	689	273	2023	1007	350	2659	1315	435	3548	1665	NA	NA	NA	NA	NA	NA	NA	NA	NA
	4	213	1561	860	281	2139	1291	359	2814	1685	447	3730	2135	580	4601	2633	709	5569	3185	851	6633	3790
	6	221	1631	1031	290	2242	1575	369	2951	2055	461	3893	2605	594	4808	3208	724	5826	3885	867	6943	4620
100	2	192	1923	712	254	2644	1050	326	3490	1370	402	4707	1740	NA	NA	NA	NA	NA	NA	NA	NA	NA
	4	200	1984	888	263	2731	1346	336	3606	1760	414	4842	2220	523	5982	2750	639	7254	3330	769	8650	3950
	6	208	2035	1064	272	2811	1642	346	3714	2150	426	4968	2700	539	6143	3350	654	7453	4070	786	8892	4810

Appliance Input Rating Limits in Thousands of Btu per Hour

National Fuel Gas Code Handbook   1999

(continues)

Table 10.6 Continued.

Common Vent Capacity

Type B Double-Wall Common Vent Diameter — *D*

Combined Appliance Input Rating in Thousands of Btu per Hour

Vent Height *H* (ft)	4 in. FAN +FAN	4 in. FAN +NAT	4 in. NAT +NAT	5 in. FAN +FAN	5 in. FAN +NAT	5 in. NAT +NAT	6 in. FAN +FAN	6 in. FAN +NAT	6 in. NAT +NAT	7 in. FAN +FAN	7 in. FAN +NAT	7 in. NAT +NAT	8 in. FAN +FAN	8 in. FAN +NAT	8 in. NAT +NAT	9 in. FAN +FAN	9 in. FAN +NAT	9 in. NAT +NAT	10 in. FAN +FAN	10 in. FAN +NAT	10 in. NAT +NAT
6	92	81	65	140	116	103	204	161	147	309	248	200	404	514	260	547	434	335	672	520	410
8	101	90	73	155	129	114	224	178	163	339	275	223	444	548	290	602	480	378	740	577	465
10	110	97	79	169	141	124	243	194	178	367	299	242	477	577	315	649	522	405	800	627	495
15	125	112	91	195	164	144	283	228	206	427	352	280	556	444	365	753	612	465	924	733	565
20	136	123	102	215	183	160	314	255	229	475	394	310	621	499	405	842	688	523	1035	826	640
30	152	138	118	244	210	185	361	297	266	547	459	360	720	585	470	979	808	605	1209	975	740
50	167	153	134	279	244	214	421	353	310	641	547	423	854	706	550	1164	977	705	1451	1188	860
100	175	163	NA	311	277	NA	489	421	NA	751	658	479	1025	873	625	1408	1215	800	1784	1502	975

Combined Appliance Input Rating in Thousands of Btu per Hour

Vent Height *H* (ft)	12 in. FAN +FAN	12 in. FAN +NAT	12 in. NAT +NAT	14 in. FAN +FAN	14 in. FAN +NAT	14 in. NAT +NAT	16 in. FAN +FAN	16 in. FAN +NAT	16 in. NAT +NAT	18 in. FAN +FAN	18 in. FAN +NAT	18 in. NAT +NAT	20 in. FAN +FAN	20 in. FAN +NAT	20 in. NAT +NAT	22 in. FAN +FAN	22 in. FAN +NAT	22 in. NAT +NAT	24 in. FAN +FAN	24 in. FAN +NAT	24 in. NAT +NAT
6	900	696	588	1284	990	815	1735	1336	1065	2253	1732	1345	2838	2180	1660	3488	2677	1970	4206	3226	2390
8	994	773	652	1423	1103	912	1927	1491	1190	2507	1936	1510	3162	2439	1860	3890	2998	2200	4695	3616	2680
10	1076	841	712	1542	1200	995	2093	1625	1300	2727	2113	1645	3444	2665	2030	4241	3278	2400	5123	3957	2920
15	1247	986	825	1794	1410	1158	2440	1910	1510	3184	2484	1910	4026	3133	2360	4971	3862	2790	6016	4670	3400
20	1405	1116	916	2006	1588	1290	2722	2147	1690	3561	2798	2140	4548	3552	2640	5573	4352	3120	6749	5261	3800
30	1658	1327	1025	2373	1892	1525	3220	2558	1990	4197	3326	2520	5303	4193	3110	6539	5157	3680	7940	6247	4480
50	2024	1640	1280	2911	2347	1863	3964	3183	2430	5184	4149	3075	6567	5240	3800	8116	6458	4500	9837	7813	5475
100	2569	2131	1670	3732	3076	2450	5125	4202	3200	6749	5509	4050	8597	6986	5000	10681	8648	5920	13004	10499	7200

Table 10.7 Capacity of Type B Double-Wall Vent with Single-Wall Connectors Serving Two or More Category I Appliances

Vent Connector Capacity

Vent Height H (ft)	Connector Rise R (ft)	Single-Wall Metal Vent Connector Diameter — D																				
		3 in.			4 in.			5 in.			6 in.			7 in.			8 in.			9 in.			10 in.		
		Appliance Input Rating Limits in Thousands of Btu per Hour																				
		FAN		NAT	FAN		NAT	FAN		NAT	FAN		NAT	FAN		NAT	FAN		NAT	FAN		NAT	FAN		NAT
		Min	Max	Max	Min	Max	Max	Min	Max	Max	Min	Max	Max	Min	Max	Max	Min	Max	Max	Min	Max	Max	Min	Max	Max
6	1	NA	NA	26	NA	NA	46	NA	NA	71	NA	NA	102	207	223	140	262	293	183	325	373	234	447	463	286
	2	NA	NA	31	NA	NA	55	NA	NA	85	168	182	123	215	251	167	271	331	219	334	422	281	458	524	344
	3	NA	NA	34	NA	NA	62	121	131	95	175	198	138	222	273	188	279	361	247	344	462	316	468	574	385
8	1	NA	NA	27	NA	NA	48	NA	NA	75	NA	NA	106	226	240	145	285	316	191	352	403	244	481	502	299
	2	NA	NA	32	NA	NA	57	125	126	89	184	193	127	234	266	173	293	353	228	360	450	292	492	560	355
	3	NA	NA	35	NA	NA	64	130	138	100	191	208	144	241	287	197	302	381	256	370	489	328	501	609	400
10	1	NA	NA	28	NA	NA	50	119	121	77	182	186	110	240	253	150	302	335	196	372	429	252	506	534	308
	2	NA	NA	33	84	85	59	124	134	91	189	203	132	248	278	183	311	369	235	381	473	302	517	589	368
	3	NA	NA	36	89	91	67	129	144	102	197	217	148	257	299	203	320	398	265	391	511	339	528	637	413
15	1	NA	NA	29	79	87	52	116	138	81	177	214	116	238	291	158	312	380	208	397	482	266	556	596	324
	2	NA	NA	34	83	94	62	121	150	97	185	230	138	246	314	189	321	411	248	407	522	317	568	646	387
	3	NA	NA	39	87	100	70	127	160	109	193	243	157	255	333	215	331	438	281	418	557	360	579	690	437
20	1	49	56	30	78	97	54	115	152	84	175	238	120	233	325	165	306	425	217	390	538	276	546	664	336
	2	52	59	36	82	103	64	120	163	101	182	252	144	243	346	197	317	453	259	400	574	331	558	709	403
	3	55	62	40	87	107	72	125	172	113	190	264	164	252	363	223	326	476	294	412	607	375	570	750	457
30	1	47	60	31	77	110	57	112	175	89	169	278	129	226	380	175	296	497	230	378	630	294	528	779	358
	2	51	62	37	81	115	67	117	185	106	177	290	152	236	397	208	307	521	274	389	662	349	541	819	425
	3	54	64	42	85	119	76	122	193	120	185	300	172	244	412	235	316	542	309	400	690	394	555	855	482
50	1	46	69	34	75	128	60	109	207	96	162	336	137	217	460	188	284	604	245	364	768	314	507	951	384
	2	49	71	40	79	132	72	114	215	113	170	345	164	226	473	223	294	623	293	376	793	375	520	983	458
	3	52	72	45	83	136	82	119	221	123	178	353	186	235	486	252	304	640	331	387	816	423	535	1013	518
100	1	45	79	34	71	150	61	104	249	98	153	424	140	205	585	192	269	774	249	345	993	321	476	1236	393
	2	48	80	41	75	153	73	110	255	115	160	428	167	212	593	228	279	788	299	358	1011	383	490	1259	469
	3	51	81	46	79	157	85	114	260	129	168	433	190	222	603	256	289	801	339	368	1027	431	506	1280	527

(continues)

Table 10.7 Continued.

Common Vent Capacity

Vent Height H (ft)	Type B Double-Wall Vent Diameter — D																				
	4 in.			5 in.			6 in.			7 in.			8 in.			9 in.			10 in.		
	Combined Appliance Input Rating in Thousands of Btu per Hour																				
	FAN +FAN	FAN +NAT	NAT +NAT	FAN +FAN	FAN +NAT	NAT +NAT	FAN +FAN	FAN +NAT	NAT +NAT	FAN +FAN	FAN +NAT	NAT +NAT	FAN +FAN	FAN +NAT	NAT +NAT	FAN +FAN	FAN +NAT	NAT +NAT	FAN +FAN	FAN +NAT	NAT +NAT
6	NA	78	64	NA	113	99	200	158	144	304	244	196	398	310	257	541	429	332	665	515	407
8	NA	87	71	NA	126	111	218	173	159	331	269	218	436	342	285	592	473	373	730	569	460
10	NA	94	76	163	137	120	237	189	174	357	292	236	467	369	309	638	512	398	787	617	487
15	121	108	88	189	159	140	275	221	200	416	343	274	544	434	357	738	599	456	905	718	553
20	131	118	98	208	177	156	305	247	223	463	383	302	606	487	395	824	673	512	1013	808	626
30	145	132	113	236	202	180	350	286	257	533	446	349	703	570	459	958	790	593	1183	952	723
50	159	145	128	268	233	208	406	337	296	622	529	410	833	686	535	1139	954	689	1418	1157	838
100	166	153	NA	297	263	NA	469	398	NA	726	633	464	999	846	606	1378	1185	780	1741	1459	948

Table 10.8 Capacity of Masonry Chimney with Type B Double-Wall Connectors Serving Two or More Category I Appliances

Vent Connector Capacity

Type B Double-Wall Vent Connector Diameter — D

Appliance Input Rating Limits in Thousands of Btu per Hour

Vent Height H (ft)	Connector Rise R (ft)	3 in. FAN Min	3 in. FAN Max	3 in. NAT Max	4 in. FAN Min	4 in. FAN Max	4 in. NAT Max	5 in. FAN Min	5 in. FAN Max	5 in. NAT Max	6 in. FAN Min	6 in. FAN Max	6 in. NAT Max	7 in. FAN Min	7 in. FAN Max	7 in. NAT Max	8 in. FAN Min	8 in. FAN Max	8 in. NAT Max	9 in. FAN Min	9 in. FAN Max	9 in. NAT Max	10 in. FAN Min	10 in. FAN Max	10 in. NAT Max
6	1	24	33	21	39	62	40	52	106	67	65	194	101	87	274	141	104	370	201	124	479	253	145	599	319
	2	26	43	28	41	79	52	53	133	85	67	230	124	89	324	173	107	436	232	127	562	300	148	694	378
	3	27	49	34	42	92	61	55	155	97	69	262	143	91	369	203	109	491	270	129	633	349	151	795	439
8	1	24	39	22	39	72	41	55	117	69	71	213	105	94	304	148	113	414	210	134	539	267	156	682	335
	2	26	47	29	40	87	53	57	140	86	73	246	127	97	350	179	116	473	240	137	615	311	160	776	394
	3	27	52	34	42	97	62	59	159	98	75	269	145	99	383	206	119	517	276	139	672	358	163	848	452
10	1	24	42	22	38	80	42	55	130	71	74	232	108	101	324	153	120	444	216	142	582	277	165	739	348
	2	26	50	29	40	93	54	57	153	87	76	261	129	103	366	184	123	498	247	145	652	321	168	825	407
	3	27	55	35	41	105	63	58	170	100	78	284	148	106	397	209	126	540	281	147	705	366	171	893	463
15	1	24	48	23	38	93	44	54	154	74	72	277	114	100	384	164	125	511	229	153	658	297	184	824	375
	2	25	55	31	39	105	55	56	174	89	74	299	134	103	419	192	128	558	260	156	718	339	187	900	432
	3	26	59	35	41	115	64	57	189	102	76	319	153	105	448	215	131	597	292	159	760	382	190	960	486
20	1	24	52	24	37	102	46	53	172	77	71	313	119	98	437	173	123	584	239	150	752	312	180	943	397
	2	25	58	31	39	114	56	55	190	91	73	335	138	101	467	199	126	625	270	153	805	354	184	1011	452
	3	26	63	35	40	123	65	57	204	104	75	353	157	104	493	222	129	661	301	156	851	396	187	1067	505
30	1	24	54	25	37	111	48	52	192	82	69	357	127	96	504	187	119	680	255	145	883	337	175	1115	432
	2	25	60	32	38	122	58	54	208	95	72	376	145	99	531	209	122	715	287	149	928	378	179	1171	484
	3	26	64	36	40	131	66	56	221	107	74	392	163	101	554	233	125	746	317	152	968	418	182	1220	535
50	1	23	51	25	36	116	51	51	209	89	67	405	143	92	582	213	115	798	294	140	1049	392	168	1334	506
	2	24	59	32	37	127	61	53	225	102	70	421	161	95	604	235	118	827	326	143	1085	433	172	1379	558
	3	26	64	36	39	135	69	55	237	115	72	435	180	98	624	260	121	854	357	147	1118	474	176	1421	611
100	1	23	46	24	35	108	50	49	208	92	65	428	155	88	640	237	109	907	334	134	1222	454	161	1589	596
	2	24	53	31	37	120	60	51	224	105	67	444	174	92	660	260	113	933	368	138	1253	497	165	1626	651
	3	25	59	35	38	130	68	53	237	118	69	458	193	94	679	285	116	956	399	141	1282	540	169	1661	705

(continues)

Table 10.8 Continued.

Common Vent Capacity

Vent Height H (ft)	Minimum Internal Area of Masonry Chimney Flue (in.^{2})																						
	12			19			28			38			50			63			78			113		
	Combined Appliance Input Rating in Thousands of Btu per Hour																						
	FAN +FAN	FAN +NAT	NAT +NAT	FAN +FAN	FAN +NAT	NAT +NAT	FAN +FAN	FAN +NAT	NAT +NAT	FAN +FAN	FAN +NAT	NAT +NAT	FAN +FAN	FAN +NAT	NAT +NAT	FAN +FAN	FAN +NAT	NAT +NAT	FAN +FAN	FAN +NAT	NAT +NAT	FAN +FAN	FAN +FAN	NAT +NAT
6	NA	74	25	NA	119	46	NA	178	71	NA	257	103	NA	351	143	NA	458	188	NA	582	246	1041	853	NA
8	NA	80	28	NA	130	53	NA	193	82	NA	279	119	NA	384	163	NA	501	218	724	636	278	1144	937	408
10	NA	84	31	NA	138	56	NA	207	90	NA	299	131	NA	409	177	606	538	236	776	686	302	1226	1010	454
15	NA	NA	36	NA	152	67	NA	233	106	NA	334	152	523	467	212	682	611	283	874	781	365	1374	1156	546
20	NA	NA	41	NA	NA	75	NA	250	122	NA	368	172	565	508	243	742	668	325	955	858	419	1513	1286	648
30	NA	NA	NA	NA	NA	NA	NA	270	137	NA	404	198	615	564	278	816	747	381	1062	969	496	1702	1473	749
50	NA	NA	NA	NA	NA	NA	NA	NA	NA	NA	NA	NA	NA	620	328	879	831	461	1165	1089	606	1905	1692	922
100	NA	NA	NA	NA	NA	NA	NA	NA	NA	NA	NA	NA	NA	NA	348	NA	NA	499	NA	NA	669	2053	1921	1058

Table 10.9 Capacity of Masonry Chimney with Single-Wall Connectors Serving Two or More Category I Appliances

Vent Connector Capacity

Vent Height H (ft)	Connector Rise R (ft)	3 in. FAN Min	3 in. FAN Max	3 in. NAT Max	4 in. FAN Min	4 in. FAN Max	4 in. NAT Max	5 in. FAN Min	5 in. FAN Max	5 in. NAT Max	6 in. FAN Min	6 in. FAN Max	6 in. NAT Max	7 in. FAN Min	7 in. FAN Max	7 in. NAT Max	8 in. FAN Min	8 in. FAN Max	8 in. NAT Max	9 in. FAN Min	9 in. FAN Max	9 in. NAT Max	10 in. FAN Min	10 in. FAN Max	10 in. NAT Max
6	1	NA	NA	21	NA	NA	39	NA	NA	66	179	191	100	231	271	140	292	366	200	362	474	252	499	594	316
	2	NA	NA	28	NA	NA	52	NA	NA	84	186	227	123	239	321	172	301	432	231	373	557	299	509	696	376
	3	NA	NA	34	NA	NA	61	134	153	97	193	258	142	247	365	202	309	491	269	381	634	348	519	793	437
8	1	NA	NA	21	NA	NA	40	NA	NA	68	195	208	103	250	298	146	313	407	207	387	530	263	529	672	331
	2	NA	NA	28	NA	NA	52	137	139	85	202	240	125	258	343	177	323	465	238	397	607	309	540	766	391
	3	NA	NA	34	NA	NA	62	143	156	98	210	264	145	266	376	205	332	509	274	407	663	356	551	838	450
10	1	NA	NA	22	NA	NA	41	130	151	70	202	225	106	267	316	151	333	434	213	410	571	273	558	727	343
	2	NA	NA	29	NA	NA	53	136	150	86	210	255	128	276	358	181	343	489	244	420	640	317	569	813	403
	3	NA	NA	34	97	102	62	143	166	99	217	277	147	284	389	207	352	530	279	430	694	363	580	880	459
15	1	NA	NA	23	NA	NA	43	129	151	73	199	271	112	268	376	161	349	502	225	445	646	291	623	808	366
	2	NA	NA	30	92	103	54	135	170	88	207	295	132	277	411	189	359	548	256	456	706	334	634	884	424
	3	NA	NA	34	96	112	63	141	185	101	215	315	151	286	439	213	368	586	289	466	755	378	646	945	479
20	1	NA	NA	23	87	99	45	128	167	76	197	303	117	265	425	169	345	569	235	439	734	306	614	921	387
	2	NA	NA	30	91	111	55	134	185	90	205	325	136	274	455	195	355	610	266	450	787	348	627	986	443
	3	NA	NA	35	96	119	64	140	199	103	213	343	154	282	481	219	365	644	298	461	831	391	639	1042	496
30	1	NA	NA	24	86	108	47	126	187	80	193	347	124	259	492	183	338	665	250	430	864	330	600	1089	421
	2	NA	NA	31	91	119	57	132	203	93	201	366	142	269	518	205	348	699	282	442	908	372	613	1145	473
	3	NA	NA	35	95	127	65	138	216	105	209	381	160	277	540	229	358	729	312	452	946	412	626	1193	524
50	1	NA	NA	24	85	113	50	124	204	87	188	392	139	252	567	208	328	778	287	417	1022	383	582	1302	492
	2	NA	NA	31	89	123	60	130	218	100	196	408	158	262	588	230	339	806	320	429	1058	425	596	1346	545
	3	NA	NA	35	94	131	68	136	231	112	205	422	176	271	607	255	349	831	351	440	1090	466	610	1386	597
100	1	NA	NA	23	84	104	49	122	200	89	182	410	151	243	617	232	315	875	328	402	1181	444	560	1537	580
	2	NA	NA	30	88	115	59	127	215	102	190	425	169	253	636	254	326	899	361	415	1210	488	575	1570	634
	3	NA	NA	34	93	124	67	133	228	115	199	438	188	262	654	279	337	921	392	427	1238	529	589	1604	687

Single-Wall Metal Vent Connector Diameter — D

Appliance Input Rating Limits in Thousands of Btu per Hour

(continues)

Table 10.9 Continued.

Common Vent Capacity

Vent Height H (ft)	Minimum Internal Area of Masonry Chimney Flue (in.²)																					
	12			19			28			38			50			63			78			113		
	Combined Appliance Input Rating in Thousands of Btu per Hour																					
	FAN +FAN	FAN +NAT	NAT +NAT	FAN +FAN	FAN +NAT	NAT +NAT	FAN +FAN	FAN +NAT	NAT +NAT	FAN +FAN	FAN +NAT	NAT +NAT	FAN +FAN	FAN +NAT	NAT +NAT	FAN +FAN	FAN +NAT	NAT +NAT	FAN +FAN	FAN +NAT	NAT +NAT	FAN +FAN	FAN +FAN	NAT +NAT
6	NA	NA	25	NA	118	45	NA	176	71	NA	255	102	NA	348	142	NA	455	187	NA	579	245	NA	846	NA
8	NA	NA	28	NA	128	52	NA	190	81	NA	276	118	NA	380	162	NA	497	217	NA	633	277	1136	928	405
10	NA	NA	31	NA	136	56	NA	205	89	NA	295	129	NA	405	175	NA	532	234	771	680	300	1216	1000	450
15	NA	NA	36	NA	NA	66	NA	230	105	NA	335	150	NA	400	210	677	602	280	866	772	360	1359	1139	540
20	NA	NA	NA	NA	NA	74	NA	247	120	NA	362	170	NA	503	240	765	661	321	947	849	415	1495	1264	640
30	NA	NA	NA	NA	NA	NA	NA	NA	135	NA	398	195	NA	558	275	808	739	377	1052	957	490	1682	1447	740
50	NA	NA	NA	NA	NA	NA	NA	NA	NA	NA	NA	NA	NA	612	325	NA	821	456	1152	1076	600	1879	1672	910
100	NA	NA	NA	NA	NA	NA	NA	NA	NA	NA	NA	NA	NA	NA	NA	NA	NA	494	NA	NA	663	2006	1885	1046

Table 10.10 Capacity of Single-Wall Metal Pipe or Type B Asbestos Cement Venting Serving Two or More Draft Hood-Equipped Appliances

Vent Connector Capacity

Total Vent Height H (ft)	Connector Rise R (ft)	Vent Connector Diameter — D					
		3 in.	4 in.	5 in.	6 in.	7 in.	8 in.
		Maximum Appliance Input Rating in Thousands of Btu per Hour					
6–8	1	21	40	68	102	146	205
	2	28	53	86	124	178	235
	3	34	61	98	147	204	275
15	1	23	44	77	117	179	240
	2	30	56	92	134	194	265
	3	35	64	102	155	216	298
30 and up	1	25	49	84	129	190	270
	2	31	58	97	145	211	295
	3	36	68	107	164	232	321

Common Vent Capacity

Total Vent Height H (ft)	Common Vent Diameter						
	4 in.	5 in.	6 in.	7 in.	8 in.	10 in.	12 in.
	Combined Appliance Input Rating in Thousands of Btu per Hour						
6	48	78	111	155	205	320	NA
8	55	89	128	175	234	365	505
10	59	95	136	190	250	395	560
15	71	115	168	228	305	480	690
20	80	129	186	260	340	550	790
30	NA	147	215	300	400	650	940
50	NA	NA	NA	360	490	810	1190

Note: See Figure G.6 and Section 10.2.

Table 10.11 Exterior Masonry Chimney, Single NAT Installations with Type B Double-Wall Vent Connectors Minimum Allowable Input Rating of Space-Heating Appliance in Thousands of Btu per Hour

Vent Height H (ft)	Internal Area of Chimney (in.^{2})							
	12	19	28	38	50	63	78	113
37°F or greater	Local 99% winter design temperature: 37°F or greater							
6	0	0	0	0	0	0	0	0
8	0	0	0	0	0	0	0	0
10	0	0	0	0	0	0	0	0
15	NA	0	0	0	0	0	0	0
20	NA	NA	123	190	249	184	0	0
30	NA	NA	NA	NA	NA	393	334	0
50	NA	NA	NA	NA	NA	NA	NA	579
27°F to 36°F	Local 99% winter design temperature: 27°F to 36°F							
6	0	0	68	116	156	180	212	266
8	0	0	82	127	167	187	214	263
10	0	51	97	141	183	201	225	265
15	NA	NA	NA	NA	233	253	274	305
20	NA	NA	NA	NA	NA	307	330	362
30	NA	NA	NA	NA	NA	419	445	485
50	NA	NA	NA	NA	NA	NA	NA	763
17°F to 26°F	Local 99% winter design temperature: 17°F to 26°F							
6	NA	NA	NA	NA	NA	215	259	349
8	NA	NA	NA	NA	197	226	264	352
10	NA	NA	NA	NA	214	245	278	358
15	NA	NA	NA	NA	NA	296	331	398
20	NA	NA	NA	NA	NA	352	387	457
30	NA	NA	NA	NA	NA	NA	507	581
50	NA	NA	NA	NA	NA	NA	NA	NA
5°F to 16°F	Local 99% winter design temperature: 5°F to 16°F							
6	NA	NA	NA	NA	NA	NA	NA	416
8	NA	NA	NA	NA	NA	NA	312	423
10	NA	NA	NA	NA	NA	289	331	430
15	NA	NA	NA	NA	NA	NA	393	485
20	NA	NA	NA	NA	NA	NA	450	547
30	NA	NA	NA	NA	NA	NA	NA	682
50	NA	NA	NA	NA	NA	NA	NA	972

Table 10.11 Continued.

Vent Height *H* (ft)	Internal Area of Chimney (in.^{2})							
	12	19	28	38	50	63	78	113
−10°F to 4°F	Local 99% winter design temperature: −10°F to 4°F							
6	NA	NA	NA	NA	NA	NA	NA	484
8	NA	NA	NA	NA	NA	NA	NA	494
10	NA	NA	NA	NA	NA	NA	NA	513
15	NA	NA	NA	NA	NA	NA	NA	586
20	NA	NA	NA	NA	NA	NA	NA	650
30	NA	NA	NA	NA	NA	NA	NA	805
50	NA	NA	NA	NA	NA	NA	NA	1003
−11°F or lower	Local 99% winter design temperature: −11°F or lower. Not recommended for any vent configurations							

Note: See Figure G.19 for a map showing local 99 percent winter design temperatures in the United States.

Table 10.12(a) *Exterior Masonry Chimney, NAT+NAT Installations with Type B Double-Wall Vent Connectors: Combined Appliance Maximum Input Rating in Thousands of Btu per Hour*

Vent Height H (ft)	Internal Area of Chimney (in.^{2})							
	12	19	28	38	50	63	78	113
6	25	46	71	103	143	188	246	NA
8	28	53	82	119	163	218	278	408
10	31	56	90	131	177	236	302	454
15	NA	67	106	152	212	283	365	546
20	NA	NA	NA	NA	NA	325	419	648
30	NA	NA	NA	NA	NA	NA	496	749
50	NA	NA	NA	NA	NA	NA	NA	922
100	NA	NA	NA	NA	NA	NA	NA	NA

Table 10.12(b) Exterior Masonry Chimney, NAT+NAT Installations with Type B Double-Wall Vent Connections: Minimum Allowable Input Rating of Space-Heating Appliance in Thousands of Btu per Hour

Vent Height H (ft)	Internal Area of Chimney (in.^{2})							
	12	19	28	38	50	63	78	113
37°F or greater	Local 99% winter design temperature: 37°F or greater							
6	0	0	0	0	0	0	0	NA
8	0	0	0	0	0	0	0	0
10	0	0	0	0	0	0	0	0
15	NA	0	0	0	0	0	0	0
20	NA	NA	NA	NA	NA	184	0	0
30	NA	NA	NA	NA	NA	393	334	0
50	NA	NA	NA	NA	NA	NA	NA	579
100	NA	NA	NA	NA	NA	NA	NA	NA
27°F to 36°F	Local 99% winter design temperature: 27°F to 36°F							
6	0	0	68	NA	NA	180	212	NA
8	0	0	82	NA	NA	187	214	263
10	0	51	NA	NA	NA	201	225	265
15	NA	NA	NA	NA	NA	253	274	305
20	NA	NA	NA	NA	NA	307	330	362
30	NA	NA	NA	NA	NA	NA	445	485
50	NA	NA	NA	NA	NA	NA	NA	763
100	NA	NA	NA	NA	NA	NA	NA	NA
17°F to 26°F	Local 99% winter design temperature: 17°F to 26°F							
6	NA	NA	NA	NA	NA	NA	NA	NA
8	NA	NA	NA	NA	NA	NA	264	352
10	NA	NA	NA	NA	NA	NA	278	358
15	NA	NA	NA	NA	NA	NA	331	398
20	NA	NA	NA	NA	NA	NA	387	457
30	NA	NA	NA	NA	NA	NA	NA	581
50	NA	NA	NA	NA	NA	NA	NA	862
100	NA	NA	NA	NA	NA	NA	NA	NA
5°F to 16°F	Local 99% winter design temperature: 5°F to 16°F							
6	NA	NA	NA	NA	NA	NA	NA	NA
8	NA	NA	NA	NA	NA	NA	NA	NA
10	NA	NA	NA	NA	NA	NA	NA	430
15	NA	NA	NA	NA	NA	NA	NA	485
20	NA	NA	NA	NA	NA	NA	NA	547
30	NA	NA	NA	NA	NA	NA	NA	682
50	NA	NA	NA	NA	NA	NA	NA	NA
100	NA	NA	NA	NA	NA	NA	NA	NA
4°F or lower	Local 99% winter design temperature: 4°F or lower Not recommended for any vent configurations							

Note: See Figure G.19 for a map showing local 99 percent winter design temperatures in the United States.

National Fuel Gas Code Handbook   1999

Table 10.13(a) Exterior Masonry Chimney, FAN+NAT Installations with Type B Double-Wall Vent Connectors: Combined Appliance Maximum Input Rating in Thousands of Btu per Hour

Vent Height H (ft)	Internal Area of Chimney (in.^{2})							
	12	19	28	38	50	63	78	113
6	74	119	178	257	351	458	582	853
8	80	130	193	279	384	501	636	937
10	84	138	207	299	409	538	686	1010
15	NA	152	233	334	467	611	781	1156
20	NA	NA	250	368	508	668	858	1286
30	NA	NA	NA	404	564	747	969	1473
50	NA	NA	NA	NA	NA	831	1089	1692
100	NA	NA	NA	NA	NA	NA	NA	1921

Table 10.13(b) Exterior Masonry Chimney, FAN+NAT Installations with Type B Double-Wall Vent Connectors: Minimum Allowable Input Rating of Space-Heating Appliance in Thousands of Btu per Hour

Vent Height H (ft)	Internal Area of Chimney (in.^{2})							
	12	19	28	38	50	63	78	113
37°F or greater	Local 99% winter design temperature: 37°F or greater							
6	0	0	0	0	0	0	0	0
8	0	0	0	0	0	0	0	0
10	0	0	0	0	0	0	0	0
15	NA	0	0	0	0	0	0	0
20	NA	NA	123	190	249	184	0	0
30	NA	NA	NA	334	398	393	334	0
50	NA	NA	NA	NA	NA	714	707	579
100	NA	NA	NA	NA	NA	NA	NA	1600
27°F to 36°F	Local 99% winter design temperature: 27°F to 36°F							
6	0	0	68	116	156	180	212	266
8	0	0	82	127	167	187	214	263
10	0	51	97	141	183	210	225	265
15	NA	111	142	183	233	253	274	305
20	NA	NA	187	230	284	307	330	362
30	NA	NA	NA	330	319	419	445	485
50	NA	NA	NA	NA	NA	672	705	763
100	NA	NA	NA	NA	NA	NA	NA	1554

Table 10.13(b) Continued.

Vent Height *H* (ft)	Internal Area of Chimney (in.^{2})							
	12	19	28	38	50	63	78	113
17°F to 26°F	Local 99% winter design temperature: 17°F to 26°F							
6	0	55	99	141	182	215	259	349
8	52	74	111	154	197	226	264	352
10	NA	90	125	169	214	245	278	358
15	NA	NA	167	212	263	296	331	398
20	NA	NA	212	258	316	352	387	457
30	NA	NA	NA	362	429	470	507	581
50	NA	NA	NA	NA	NA	723	766	862
100	NA	NA	NA	NA	NA	NA	NA	1669
5°F to 16°F	Local 99% winter design temperature: 5°F to 16°F							
6	NA	78	121	166	214	252	301	416
8	NA	94	135	182	230	269	312	423
10	NA	111	149	198	250	289	331	430
15	NA	NA	193	247	305	346	393	485
20	NA	NA	NA	293	360	408	450	547
30	NA	NA	NA	377	450	531	580	682
50	NA	NA	NA	NA	NA	797	853	972
100	NA	NA	NA	NA	NA	NA	NA	1833
−10°F to 4°F	Local 99% winter design temperature: −10°F to 4°F							
6	NA	NA	145	196	249	296	349	484
8	NA	NA	159	213	269	320	371	494
10	NA	NA	175	231	292	339	397	513
15	NA	NA	NA	283	351	404	457	586
20	NA	NA	NA	333	408	468	528	650
30	NA	NA	NA	NA	NA	603	667	805
50	NA	NA	NA	NA	NA	NA	955	1003
100	NA	NA	NA	NA	NA	NA	NA	NA
−11°F or lower	Local 99% winter design temperature: −11°F or lower Not recommended for any vent configurations							

Note: See Figure G.19 for a map showing local 99 percent winter design temperatures in the United States.

CHAPTER 11

Referenced Publications

11.1

The following documents or portions thereof are referenced within this code as mandatory requirements and shall be considered part of the requirements of this code. The edition indicated for each referenced mandatory document is the current edition as of the date of the NFPA issuance of this code. Some of these mandatory documents might also be referenced in this code for specific informational purposes and, therefore, are also listed in Appendix M.

11.1.1 ASME Publications.

American Society of Mechanical Engineers, United Engineering Center, 345 East 47th Street, New York, NY 10017.

ANSI/ASME B1.20.1, *Standard for Pipe Threads, General Purpose (Inch),* 1983 (Reaffirmed 1992).

ANSI/ASME B16.1, *Standard for Cast Iron Pipe Flanges and Flanged Fittings, Class 25, 125, 250, and 800,* 1998.

ANSI/ASME B16.20, *Standard for Metallic Gaskets for Pipe Flanges, Ring-Joint, Spiral Wound, and Jacketed,* 1993.

ANSI/ASME B36.10, *Standard for Welded and Seamless Wrought-Steel Pipe,* 1996.

11.1.2 ASTM Publications.

American Society for Testing and Materials, 100 Barr Harbor Drive, West Conshohocken, PA 19428-2959.

ASTM A 53, *Standard Specification for Pipe, Steel, Black and Hot-Dipped, Zinc-Coated Welded and Seamless,* 1999.

ASTM A 106, *Standard Specification for Seamless Carbon Steel Pipe for High-Temperature Service,* 1999.

ASTM A 254, *Standard Specification for Copper Brazed Steel Tubing,* 1997.

ASTM A 539, *Standard Specification for Electric Resistance-Welded Coiled Steel Tubing for Gas and Fuel Oil Lines,* 1999.

ASTM B 88, *Specification for Seamless Copper Water Tube,* 1996.

ASTM B 210, *Specification for Aluminum-Alloy Drawn Seamless Tubes,* 1995.

ASTM B 241, *Specification for Aluminum-Alloy Seamless Pipe and Seamless Extruded Tube,* 1996.

ASTM B 280, *Specification for Seamless Copper Tube for Air Conditioning and Refrigeration Field Service,* 1997.

ASTM C 64-94, *Specification for Refractories for Incinerators and Boilers,* 1994.

ASTM D 2513, *Standard Specification for Thermoplastic Gas Pressure Pipe, Tubing, and Fittings,* 1999.

11.1.3 CSA International Publications.

CSA International, 8501 East Pleasant Valley Road, Cleveland, OH 44131.

ANSI Z21.8-1994, *Standard for Installation of Domestic Gas Conversion Burners.*

ANSI Z21.18/CGA 6.3-1995, Gas Appliance Pressure Regulators.

ANSI Z21.69/CGA 6.16, *Standard for Connectors for Movable Gas Appliances,* 1997.

ANSI Z21.80, *Standard for Line Pressure Regulators,* 1997.

ANSI Z83.18, *Standard for Direct Gas-Fired Industrial Air Heaters,* 1990 (1998).

ANSI LC 1/CSA 6.26, *Standard for Fuel Gas Piping Systems Using Corrugated Stainless Steel Tubing,* 1997.

11.1.4 MSS Publications.

Manufacturers Standardization Society of the Valve and Fittings Industry, 5203 Leesburg Pike, Suite 502, Falls Church, VA 22041.

MSS SP-6, *Standard Finishes for Contact Faces of Pipe Flanges and Connecting-End Flanges of Valves and Fittings,* 1999.

ANSI/MSS SP-58, *Pipe Hangers and Supports — Materials, Design and Manufacture,* 1993.

11.1.5 NFPA Publications.

National Fire Protection Association, 1 Batterymarch Park, P.O. Box 9101, Quincy, MA 02269-9101.

NFPA 30A, *Automotive and Marine Service Station Code,* 1996 edition.

NFPA 37, *Standard for the Installation and Use of Stationary Combustion Engines and Gas Turbines,* 1998 edition.

NFPA 51, *Standard for the Design and Installation of Oxygen-Fuel Gas Systems for Welding, Cutting, and Allied Processes,* 1997 edition.

NFPA 52, *Standard for Compressed Natural Gas (CNG) Vehicular Fuel Systems,* 1998 edition.

NFPA 58, *Liquefied Petroleum Gas Code,* 1998 edition.

NFPA 70, *National Electrical Code®,* 1999 edition.

NFPA 82, *Standard on Incinerators and Waste and Linen Handling Systems and Equipment,* 1999 edition.

NFPA 88A, *Standard for Parking Structures,* 1998 edition.

NFPA 90A, *Standard for the Installation of Air-Conditioning and Ventilating Systems,* 1999 edition.

NFPA 90B, *Standard for the Installation of Warm Air Heating and Air-Conditioning Systems,* 1999 edition.

NFPA 96, *Standard for Ventilation Control and Fire Protection of Commercial Cooking Operations,* 1998 edition.

NFPA 211, *Standard for Chimneys, Fireplaces, Vents, and Solid Fuel-Burning Appliances,* 1996 edition.

NFPA 409, *Standard on Aircraft Hangars,* 1995 edition.

NFPA 1192, *Standard on Recreational Vehicles,* 1999 edition.

11.1.6 U.S. Government Publication.

U.S. Government Printing Office, Washington, DC 20402.

Code of Federal Regulations, Title 49, Part 192.

Explanatory Material

The material contained in Appendix A of the 1999 edition of the *National Fuel Gas Code* is not a part of the requirements of the code but is included with the code for informational purposes only. For the convenience of readers, in this handbook, the Appendix A material immediately follows the corresponding code section or paragraph in Chapters 1 through 11 and, therefore, is not repeated here.

Coordination of Gas Utilization Equipment Design, Construction, and Maintenance

This appendix is not a part of the requirements of this code but is included for informational purposes only.

B.1 Coordination

B.1.1

Because industrial gas applications are so varied in nature, many agencies are jointly involved with their safe and satisfactory use. Prior to installation, the specific assignments should be agreed upon by the parties concerned. A typical, but not mandatory, delineation of assignments is given in B.1.2 through B.1.5, and a detailed checklist is given in B.2.

B.1.2

The person or agency planning an installation of gas equipment does the following:

(1) Verifies the adequacy of the gas supply, volume, pressure, and meter location
(2) Determines suitability of gas for the process
(3) Notifies gas suppliers of significant changes in requirements

B.1.3

Upon request, the gas supplier furnishes the user complete information on the following:

(1) Combustion characteristics and physical or chemical properties such as specific gravity, heating value, pressure, and the approximate analysis of the gas
(2) Conditions under which an adequate supply of gas at suitable pressure can be brought to the site
(3) Continuity of the gas supply

B.1.4

The gas equipment manufacturer or builder provides the following:

(1) Design and construction of all gas equipment or assemblies shipped from its plant
(2) Design and construction of all gas equipment fabricated, erected, or assembled by the gas equipment manufacturer or builder in the field
(3) A statement of the maximum hourly Btu input, type of gas, and design pressure range
(4) Written installation and operating instructions for the user

B.1.5

The person or agency installing the gas equipment and the person or agency authorizing the installation of gas equipment (purchaser) jointly should do the following:

(1) Select, erect, or assemble gas equipment, components, or designs purchased or developed by that person or agency
(2) Ensure conformance to codes, ordinances, or regulations applicable to the installation
(3) Provide adequate means of disposal of products of combustion
(4) Initially operate the gas equipment in a safe manner

B.2 Gas Equipment Design and Construction Checklist

B.2.1

The basic design and installation should consider the following:

(1) Suitability of equipment for process requirements
(2) Adequate structural strength and stability
(3) Reasonable life expectation
(4) Conformance to existing safety standards
(5) Adequate combustion space and venting
(6) Means for observation and inspection of combustion

B.2.2

Materials of construction used, other than pipe, fittings, and valves, should provide reasonable life expectancy for the service intended and should be capable of satisfactorily withstanding the following:

(1) Operating temperatures
(2) Chemical action
(3) Thermal shock
(4) Load stresses

B.2.3

Combustion systems should be selected for the characteristics of the available gas so that they will operate properly at the elevation at point of use and produce the following:

(1) Proper heat distribution
(2) Adequate operating temperature range
(3) Suitable flame geometry
(4) Flame stability
(5) Operating flexibility
(6) Desired heating chamber atmosphere

B.2.4

Pipe, fittings, and valves should conform to applicable American National Standards as indicated in Section 2.6. Piping, bushings, and material in fittings should not be selected nor used until the following factors have been considered:

(1) Correct size to handle required volume (consideration of pressure drop in controls and manifolds is particularly important in low pressure systems)
(2) Material specifications suitable for pressures and temperatures encountered
(3) Adequate supports and protection against physical damage
(4) Tight assembly and thorough leak inspection
(5) Use of sufficient unions and flanges, where permitted, for convenient field replacement or repair
(6) Arrangement of piping to provide accessibility for equipment adjustments and freedom from thermal damage

B.2.5

Information concerning the characteristics of the gas and electricity available at the point of utilization should be specific and complete. Gas controls and electrical equipment should be selected to conform to these characteristics, which include the following:

(1) Gas characteristics: Heat content, pressure, specific gravity, and approximate analysis
(2) Electrical characteristics: Voltages, number of phases, and frequencies for both control and power circuits

(3) Location of electrical equipment and wiring to avoid thermal damage and excessive concentrations of dust, dirt, or foreign material
(4) Requirements of applicable electrical codes and standards, with particular reference to NFPA 70, Article 500, of the *National Electrical Code*

Article 500 of the *National Electrical Code*®, "Hazardous (Classified) Locations," provides requirements for electrical equipment and wiring in locations where fire or explosion hazards can exist due to flammable gases or vapors, flammable liquids, combustible dust, or ignitable fibers or flyings.

B.2.6

Temperature controls, if used, should be carefully selected considering:

(1) Range and type of instruments and sensing elements
(2) Type of control action
(3) Suitability for service required
(4) Correlation of control instruments with operating equipment

B.2.7

In enclosed chambers, the accumulation of gas–air or solvent–air mixtures that can be accidentally ignited constitutes a potential hazard to life and property. For this reason, consideration should be given to the selection and installation of suitable protective equipment. The selection of a satisfactory protective system and components not otherwise covered by existing codes or standards should be based on the requirements of each individual installation after consultation with the various interested parties, including user, designer, insurance company, and local authorities having jurisdiction. Factors and considerations involved in the selection of protective equipment include the following:

(1) Feasibility of its installation
(2) Its adaptability to process and control requirements
(3) Conformance to existing standards, ordinances, requirements, and other regulations that apply *(See Appendix M for listing of standards and specifications.)*

Examples of equipment in which accumulations of gas–air mixtures can ignite are boilers, ovens, and furnaces. Consulting the following standards for additional information is recommended:

NFPA 86, *Standard for Ovens and Furnaces*

NFPA 86C, *Standard for Industrial Furnaces Using a Special Processing Atmosphere*

NFPA 86D, *Standard for Industrial Furnaces Using Vacuum as an Atmosphere*

NFPA 8501, *Standard for Single Burner Boiler Operation*

NFPA 8502, *Standard for the Prevention of Furnace Explosions/Implosions in Multiple Burner Boilers*

B.3 Maintenance of Gas Equipment

B.3.1

These recommendations are prepared for maintenance of gas equipment. Special types of equipment demand special attention.

B.3.2

Burners and pilots should be kept clean and in proper operating condition. Burner refractory parts should be examined at frequent regular intervals to ensure good condition.

> Refractory linings should be thoroughly inspected annually for spalling, cracking, and voids. Consult the equipment manufacturer for repair procedures if any defects are observed.

B.3.3

Where automatic flame safeguards are used, a complete shutdown and restart should be made at frequent intervals to check the components for proper operation.

B.3.4 Other Safeguard Equipment.

B.3.4.1   Accessory safeguard equipment, such as manual reset valves with pressure or vacuum switches, high-temperature limit switches, draft controls, shutoff valves, airflow switches, door switches, and gas valves, should be operated at frequent regular intervals to ensure proper functioning. If inoperative, they should be repaired or replaced promptly.

B.3.4.2   Where firechecks are installed in gas–air mixture piping to prevent flashbacks from traveling farther upstream, the pressure loss across the firechecks should be measured at regular intervals. When excessive pressure loss is found, screens should be removed and cleaned. Water-type backfire checks should be inspected at frequent regular intervals and liquid level maintained.

B.3.4.3   All safety shutoff valves should be checked for leakage and proper operation at frequent regular intervals.

B.3.5 Auxiliary Devices.

B.3.5.1   A necessary part of the gas equipment maintenance is the proper maintenance of auxiliary devices. Maintenance instructions as supplied by the manufacturers of these devices should be followed.

B.3.5.2   Gas combustion equipment, including blowers, mechanical mixers, control valves, temperature control instruments, air valves, and air filters, should be kept clean and should be examined at frequent regular intervals.

B.3.5.3   Necessary repairs and replacements should be made promptly.

B.3.6

Regulator and zero governor vents and impulse or control piping and tubing should be kept clear. Regulator valves that operate improperly should be cleaned, repaired, or replaced promptly.

B.3.7

A necessary part of the gas equipment maintenance is the proper maintenance of the gas piping system. It is recommended that gas piping be inspected and tested for leakage at regular intervals in accordance with the provisions of 4.1.5. Air piping should be kept internally clean to prevent accumulation of dust, lint, and grease in air jets and valves. Where conditions warrant, filters should be installed at the intake to the fans.

B.3.8

Standby or substitute fuel equipment and systems for gas equipment should be kept in good operating condition and tested periodically.

B.3.9

An adequate supply of repair parts should be maintained.

References Cited in Commentary

The following publications are available from the National Fire Protection Association, 1 Batterymarch Park, P.O. Box 9101, Quincy, MA 02269-9101.

NFPA 70, *National Electrical Code®,* 1999 edition.
NFPA 86, *Standard for Ovens and Furnaces,* 1999 edition.
NFPA 86C, *Standard for Industrial Furnaces Using a Special Processing Atmosphere,* 1999 edition.
NFPA 86D, *Standard for Industrial Furnaces Using Vacuum as an Atmosphere,* 1999 edition.
NFPA 8501, *Standard for Single Burner Boiler Operation,* 1997 edition.
NFPA 8502, *Standard for the Prevention of Furnace Explosions/Implosions in Multiple Burner Boilers,* 1999 edition.

Sizing and Capacities of Gas Piping

This appendix is not a part of the requirements of this code but is included for informational purposes only.

Appendix C provides additional information on the sizing of piping systems beyond what is provided in Chapters 2 and 9. An explanation, as well as an example, of the use of the sizing tables in Chapter 9 is included. A calculation method for sizing gas piping and a table of equivalent lengths of fittings that can be used with it is also provided. The commentary also includes a graphical method that can be used to size piping based on pipe length and flow of gas.

In addition to these methods of calculating the size of gas piping systems, the Polyflo® Computer, a circular slide rule, is referenced in a note in the code. This slide rule can be used to size gas piping and piping for other fluids. This excellent tool requires some practice for familiarity in its use. Besides these sizing methods, computer software is available for sizing pipe.

C.1 General

In determining the size of piping to be used in designing a gas piping system, the following factors must be considered:

(1) Allowable loss in pressure from point of delivery to equipment
(2) Maximum gas demand
(3) Length of piping and number of fittings

(4) Specific gravity of the gas

(5) Diversity factor

For any gas piping system, for special gas utilization equipment, or for conditions other than those covered by Tables 9.1, 9.2, 9.13, and 9.14, or Tables 9.26, 9.27, or 9.28 such as longer runs, greater gas demands, or greater pressure drops, the size of each gas piping system should be determined by standard engineering practices acceptable to the authority having jurisdiction.

C.2 Description of Tables

(a) The quantity of gas to be provided at each outlet should be determined, whenever possible, directly from the manufacturer's Btu input rating of the equipment to be installed. In case the ratings of the equipment to be installed are not known, Table C.2 shows the approximate consumption of average appliances of certain types in Btu per hour.

To obtain the cubic feet per hour of gas required, divide the total Btu input of all equipment by the average Btu heating value per cubic feet of gas. The average Btu per cubic feet of gas in the area of the installation can be obtained from the serving gas supplier.

(b) Capacities for gas at low pressure [0.5 psi (35 kPa) or less] in cubic feet per hour of 0.60 specific gravity gas for different sizes and lengths are shown in Tables 9.1 and 9.2 for iron pipe or equivalent rigid pipe, in Tables 9.13 and 9.14 for smooth wall semirigid tubing, and in Tables 9.19, 9.20, and 9.21 for corrugated stainless steel tubing. Tables 9.1 and 9.3 are based on a pressure drop of 0.3 in. (75 Pa) water column, whereas Tables 9.2, 9.14, and 9.17 are based upon a pressure drop of 0.5 in. (125 Pa) water column. Tables 9.20 and 9.21 are special low-pressure applications based upon pressure drops greater than 0.5 in. water column (125 Pa). In using these tables, no additional allowance is necessary for an ordinary number of fittings.

(c) Capacities in thousands of Btu per hour of undiluted liquefied petroleum gases based on a pressure drop of 0.5 in. (125 Pa) water column for different sizes and lengths are shown in Table 9.25 for iron pipe or equivalent rigid pipe, in Table 9.26 for smooth wall semi-rigid tubing, and in Table 9.29 for corrugated stainless steel tubing. Tables 9.30 and 9.31 for corrugated stainless steel tubing are based on pressure drops greater than 0.5 in. water column (125 Pa). In using these tables, no additional allowance is necessary for an ordinary number of fittings.

(d) Gas piping systems that are to be supplied with gas of a specific gravity of 0.70 or less can be sized directly from Tables 9.1, 9.2, 9.13, and 9.14, unless the authority having jurisdiction specifies that a gravity factor be applied. Where the specific gravity of the gas is greater than 0.70, the gravity factor should be applied.

Application of the gravity factor converts the figures given in Tables 9.1, 9.2, 9.13, and 9.14 to capacities with another gas of different specific gravity. Such application is accom-

Table C.2 Approximate Gas Input for Typical Appliances

Appliance	Input Btu/hr (Approx.)
Range, free-standing, domestic	65,000
Built-in oven or broiler unit, domestic	25,000
Built-in top unit, domestic	40,000
Water heater, automatic storage	
30- to 40-gal tank	45,000
Water heater, automatic storage	
50 gal tank	55,000
Water heater, automatic instantaneous	
Capacity at 2 gal/minute	142,800
Capacity at 4 gal/minute	285,000
Capacity at 6 gal/minute	428,400
Water heater, domestic, circulating, or side-arm	35,000
Refrigerator	3,000
Clothes dryer, Type 1 (domestic)	35,000
Gas light	2,500
Incinerator, domestic	35,000

For SI units, 1 Btu per hour = 0.293 W.

Note: For specific appliances or appliances not shown, the input should be determined from the manufacturer's rating.

plished by multiplying the capacities given in Tables 9.1, 9.2, 9.13, and 9.14 by the multipliers shown in Table 9.24. In case the exact specific gravity does not appear in the table, choose the next higher value specific gravity shown.

(e) Capacities for gas at pressures greater than 0.5 psi (3.5 kPa) in cubic feet per hour of 0.60 specific gravity gas for different sizes and lengths are shown in Tables 9.5 to 9.12 for iron pipe or equivalent rigid pipe and Tables 9.29 and 9.30 for corrugated stainless steel tubing.

The tables in Chapter 9 provide the designer with information for piping flow capacity for gas of 0.60 specific gravity (i.e., natural gas) and for propane. The tables were organized in the 1996 edition in the order of gas covered, piping material, and pressure; tables were added to provide additional sizing information on CSST, polyethylene, and additional pressures.

Tables 9.1 through 9.23 are for natural gas (0.60 specific gravity) and Table 9.24 provides conversions to other specific gravities. Tables 9.25 through 9.34 are for propane. For each gas, the tables for steel pipe are followed by tables for metallic tubing [i.e., Corrugated Stainless Steel Tubing (CSST) and polyethylene tubing].

Table C.2 shows approximate gas input for various appliances. Table C.3 shows equivalent lengths for fittings. Note, however, that if the system contains "typical" fittings, no allowance for fittings needs to be made to use the Appendix C tables.

The capacities of Tables 9.1 through 9.23 are given in standard cubic feet per hour, so the heating capacity can be obtained by multiplying the tabular value by the heating value of the gas. Since the heating value of natural gas is approximately 1000 Btu/ft^3 (37.5 MJ/m^3), the tabular values may be taken as the heating value in MBH (thousands of Btu per hour) for natural gas. Tables 9.25 through 9.34 for propane are in thousands of Btu per hour, which is commonly used for propane system calculations.

The tables include flow capacities for several initial pressures and for various pressure losses, so most smaller jobs will be covered by one of the tables. The example at the end of this appendix illustrates how to use the various tables.

C.3 Use of Capacity Tables

To determine the size of each section of gas piping in a system within the range of the capacity tables, proceed as follows. *(Also see sample calculation in Section C.4.)*

(a) Determine the gas demand of each appliance to be attached to the piping system. Where Tables 9.1, 9.2, 9.13, and 9.14 are to be used to select the piping size, calculate the gas demand in terms of cubic feet per hour for each piping system outlet. Where Tables 9.25 through 9.27 are to be used to select the piping size, calculate the gas demand in terms of thousands of Btu per hour for each piping system outlet.

(b) Where the piping system is for use with other than undiluted liquefied petroleum gases, determine the design system pressure, the allowable loss in pressure (pressure drop), and the specific gravity of the gas to be used in the piping system.

(c) Measure the length of piping from the point of delivery to the most remote outlet. Where a multipressure gas piping system is used, gas piping shall be sized for the maximum length of pipe measured from the gas pressure regulator to the most remote outlet of each similarly pressure section.

(d) In the appropriate capacity table, select the column showing the measured length, or the next longer length if the table does not give the exact length. This is the only length used in determining the size of any section of gas piping. If the gravity factor is to be applied, the values in the selected column of the table are multiplied by the appropriate multiplier from Table 9.24.

Capacities of smooth wall pipe or tubing can also be determined by using the following formulas:

High Pressure [1.5 psi (10.3 kPa) and above]:

$$Q = 181.6 \sqrt{\frac{D^5 \cdot (P_1^2 - P_2^2) \cdot Y}{Cr \cdot fba \cdot L}}$$

$$= 2237 D^{2.623} \left[\frac{(P_1^2 - P_2^2) \cdot Y}{Cr \cdot L}\right]^{0.541}$$

Low Pressure [Less than 1.5 psi (10.3 kPa)]:

$$Q = 187.3 \sqrt{\frac{D^5 \cdot \Delta H}{Cr \cdot fba \cdot L}}$$

$$= 2313 \, D^{2.623} \left(\frac{\Delta H}{Cr \cdot L}\right)^{0.541}$$

where:

Q = rate, cu ft per hr at 60°F and 30 in. mercury column

D = inside diameter of pipe, in.

P_1 = upstream pressure, psia

P_2 = downstream pressure, psia

Y = superexpansibility factor = 1/supercompressibility factor

Notes Applicable to Table C.3

For SI units, 1 foot = 0.305 m.

Note: Values for welded fittings are for conditions where bore is not obstructed by weld spatter or backing rings. If appreciably obstructed, use values for "Screwed Fittings."

[1] Flanged fittings have three-fourths the resistance of screwed elbows and tees.

[2] Tabular figures give the extra resistance due to curvature alone to which should be added the full length of travel.

[3] Small size socket-welding fittings are equivalent to miter elbows and miter tees.

[4] Equivalent resistance in number of diameters of straight pipe computed for a value of $f - 0.0075$ from the relation $n - k/4f$.

[5] For condition of minimum resistance where the centerline length of each miter is between d and $2^1/_2 d$.

[6] For pipe having other inside diameters, the equivalent resistance may be computed from the above n values.

Source: From *Piping Handbook*, by Sabin Crocker, 4th Ed., Copyright 1945 by McGraw-Hill, Inc., Table XIV, pp. 100–101. Used by permission of McGraw-Hill Book Company.

Table C.3 Equivalent Lengths of Pipe Fittings and Valves

Nominal pipe size, in.	Inside diam. d, in., Sched. 40[6]	Screwed Fittings[1]				90° Welding Elbows and Smooth Bends[2]					
		45°/Ell	90°/Ell	180° Close Return Bends	Tee	$R/d = 1$	$R/d = 1^1/_3$	$R/d = 2$	$R/d = 4$	$R/d = 6$	$R/d = 8$
k factor =		0.42	0.90	2.00	1.80	0.48	0.36	0.27	0.21	0.27	0.36
L/d' ratio[4] n =		14	30	67	60	16	12	9	7	9	12
		\multicolumn{10}{c}{L = Equivalent Length in Feet of Schedule 40 (standard weight) Straight Pipe[6]}
$1/_2$	0.622	0.73	1.55	3.47	3.10	0.83	0.62	0.47	0.36	0.47	0.62
$3/_4$	0.824	0.96	2.06	4.60	4.12	1.10	0.82	0.62	0.48	0.62	0.82
1	1.049	1.22	2.62	5.82	5.24	1.40	1.05	0.79	0.61	0.79	1.05
$1^1/_4$	1.380	1.61	3.45	7.66	6.90	1.84	1.38	1.03	0.81	1.03	1.38
$1^1/_2$	1.610	1.88	4.02	8.95	8.04	2.14	1.61	1.21	0.94	1.21	1.61
2	2.067	2.41	5.17	11.5	10.3	2.76	2.07	1.55	1.21	1.55	2.07
$2^1/_2$	2.469	2.88	6.16	13.7	12.3	3.29	2.47	1.85	1.44	1.85	2.47
3	3.068	3.58	7.67	17.1	15.3	4.09	3.07	2.30	1.79	2.30	3.07
4	4.026	4.70	10.1	22.4	20.2	5.37	4.03	3.02	2.35	3.02	4.03
5	5.047	5.88	12.6	28.0	25.2	6.72	5.05	3.78	2.94	3.78	5.05
6	6.065	7.07	15.2	33.8	30.4	8.09	6.07	4.55	3.54	4.55	6.07
8	7.981	9.31	20.0	44.6	40.0	10.6	7.98	5.98	4.65	5.98	7.98
10	10.02	11.7	25.0	55.7	50.0	13.3	10.0	7.51	5.85	7.51	10.0
12	11.94	13.9	29.8	66.3	59.6	15.9	11.9	8.95	6.96	8.95	11.9
14	13.13	15.3	32.8	73.0	65.6	17.5	13.1	9.85	7.65	9.85	13.1
16	15.00	17.5	37.5	83.5	75.0	20.0	15.0	11.2	8.75	11.2	15.0
18	16.88	19.7	42.1	93.8	84.2	22.5	16.9	12.7	9.85	12.7	16.9
20	18.81	22.0	47.0	105	94.0	25.1	18.8	14.1	11.0	14.1	18.8
24	22.63	26.4	56.6	126	113	30.2	22.6	17.0	13.2	17.0	22.6

Table C.3 Continued.

Miter Elbows[3] (No. of miters)					Welding Tees		Valves (screwed, flanged, or welded)			
1-45°	**1-60°**	**1-90°**	**2-90°**	**3-90°**	**Forged**	**Miter[3]**	**Gate**	**Globe**	**Angle**	**Swing Check**
0.45	**0.90**	**1.80**	**0.60**	**0.45**	**1.35**	**1.80**	**0.21**	**10**	**5.0**	**2.5**
15	**30**	**60**	**20**	**15**	**45**	**60**	**7**	**333**	**167**	**83**
L = Equivalent Length in Feet of Schedule 40 (standard weight) Straight Pipe[6]										
0.78	1.55	3.10	1.04	0.78	2.33	3.10	0.36	17.3	8.65	4.32
1.03	2.06	4.12	1.37	1.03	3.09	4.12	0.48	22.9	11.4	5.72
1.31	2.62	5.24	1.75	1.31	3.93	5.24	0.61	29.1	14.6	7.27
1.72	3.45	6.90	2.30	1.72	5.17	6.90	0.81	38.3	19.1	9.58
2.01	4.02	8.04	2.68	2.01	6.04	8.04	0.94	44.7	22.4	11.2
2.58	5.17	10.3	3.45	2.58	7.75	10.3	1.21	57.4	28.7	14.4
3.08	6.16	12.3	4.11	3.08	9.25	12.3	1.44	68.5	34.3	17.1
3.84	7.67	15.3	5.11	3.84	11.5	15.3	1.79	85.2	42.6	21.3
5.04	10.1	20.2	6.71	5.04	15.1	20.2	2.35	112	56.0	28.0
6.30	12.6	25.2	8.40	6.30	18.9	25.2	2.94	140	70.0	35.0
7.58	15.2	30.4	10.1	7.58	22.8	30.4	3.54	168	84.1	42.1
9.97	20.0	40.0	13.3	9.97	29.9	40.0	4.65	222	111	55.5
12.5	25.0	50.0	16.7	12.5	37.6	50.0	5.85	278	139	69.5
14.9	29.8	59.6	19.9	14.9	44.8	59.6	6.96	332	166	83.0
16.4	32.8	65.6	21.9	16.4	49.2	65.6	7.65	364	182	91.0
18.8	37.5	75.0	25.0	18.8	56.2	75.0	8.75	417	208	104
21.1	42.1	84.2	28.1	21.1	63.2	84.2	9.85	469	234	117
23.5	47.0	94.0	31.4	23.5	70.6	94.0	11.0	522	261	131
28.3	56.6	113	37.8	28.3	85.0	113	13.2	629	314	157

NOTE: For Y values for natural gas, refer to *Manual for Determination of Supercompressibility Factors for Natural Gas,* available from American Gas Association, 400 N. Capitol Street, N.W., Washington, DC 20001. For values for liquefied petroleum gases, refer to Engineering Data Book, available from Gas Processors Association, 1812 First Place, Tulsa, Oklahoma 74102.

Cr = factor for viscosity, density, and temperature

$$= 0.00354ST\left(\frac{Z}{S}\right)^{0.152}$$

S =   specific gravity of gas at 60°F and 30 in. mercury column

T =   absolute temperature, °F, or $t + 460$

t =   temperature, °F

Z =   viscosity of gas, centipoise (0.012 for natural gas, 0.008 for propane), or 1488μ

m =   viscosity, pounds per second ft

fba =   base friction factor for air at 60°F ($CF =1$)

ΔH =   pressure drop, in. water column (27.7 in. $H_2O =1$ psi)

L =   length of pipe, ft

CF =   factor $CF = \left(\frac{fb}{fba}\right)$

fb =   base friction factor for any fluid at a given temperature, °F

NOTE: For further details on the formulas, refer to "Polyflo Computer," available from Polyflo Company, 3412 High Bluff, Dallas, Texas 75234.

The Polyflo computer is shown in Exhibit C.1.

(e) Use this vertical column to locate ALL gas demand figures for this particular system of piping.

(f) Starting at the most remote outlet, find in the vertical column just selected the gas demand for that outlet. If the exact figure of demand is not shown, choose the next larger figure below in the column.

(g) Opposite this demand figure, in the first column at the left, will be found the correct size of gas piping.

(h) Proceed in a similar manner for each outlet and each section of gas piping. For each section of piping, determine the total gas demand supplied by that section.

The equivalent lengths in feet shown in Table C.3 have been computed on a basis that the inside diameter corresponds to that of Schedule 40 (standard-weight) steel pipe, which is close enough for most purposes involving other schedules of pipe. Where a more specific solution for equivalent length is desired, this can be made by multiplying the actual inside diameter of the pipe in inches by $n/12$, or the actual inside diameter in ft by n. N can be read from the table heading. The equivalent length values can be used with reasonable accuracy

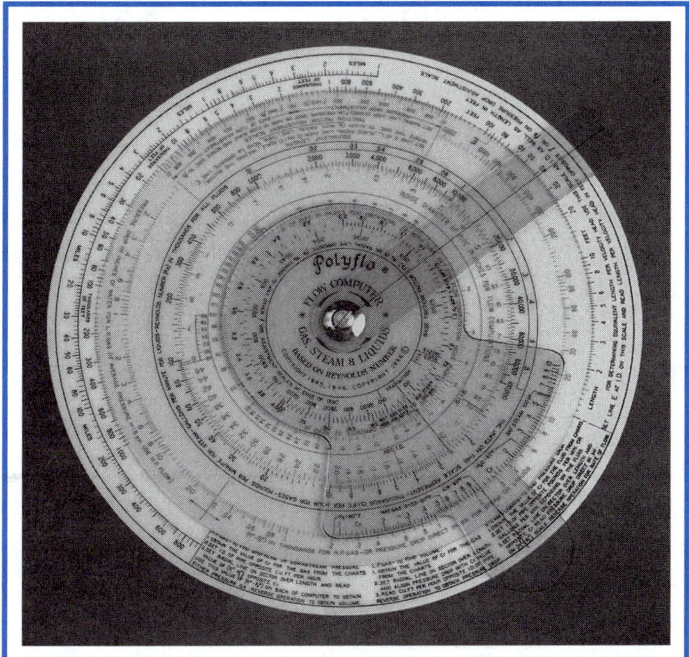

Exhibit C.1 *The Polyflo® computer. (Courtesy of Polyflo Company.)*

for copper or brass fittings and bends. For copper or brass valves, however, the equivalent length of pipe should be taken as 45 percent longer than the values in the table, which are for steel pipe. Resistance per foot of copper or brass pipe is less than that of steel.

C.4 Example of Piping System Design

Determine the required pipe size of each section and outlet of the piping system shown in Figure C.4, with a designated pressure drop of 0.50 in. water column (125 Pa). The gas to be used has 0.65 specific gravity and a heating value of 1000 Btu/ft^3 (37.5 MJ/m^3).

Solution

(a) Maximum gas demand for outlet A:

$$\frac{\text{Consumption (rating plate input, or Table C.2 if necessary)}}{\text{Btu of gas}} =$$

$$\frac{30{,}000 \text{ Btu/hr rating}}{1000 \text{ Btu/ft}^3} = 30 \text{ ft}^3/\text{hr}$$

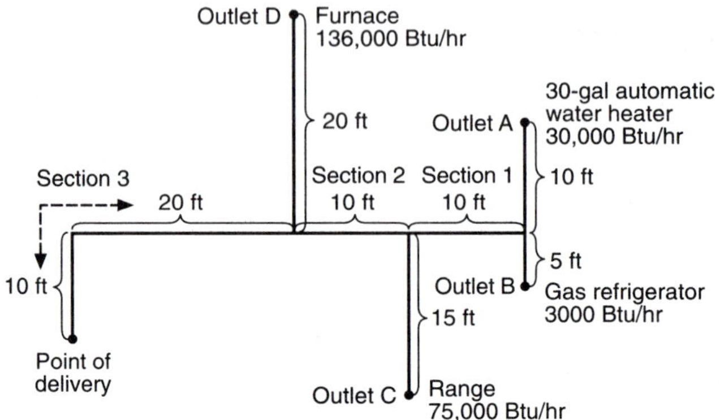

Figure C.4 *Piping plan.*

Maximum gas demand for outlet B:

$$\frac{\text{Consumption}}{\text{Btu of gas}} = \frac{3000}{1000} = 3 \text{ ft}^3/\text{hr}$$

Maximum gas demand for outlet C:

$$\frac{\text{Consumption}}{\text{Btu of gas}} = \frac{75,000}{1000} = 75 \text{ ft}^3/\text{hr}$$

Maximum gas demand for outlet D:

$$\frac{\text{Consumption}}{\text{Btu of gas}} = \frac{136,000}{1000} = 136 \text{ ft}^3/\text{hr}$$

(b) The length of pipe from the point of delivery to the most remote outlet (A) is 60 ft (18.3 m). This is the only distance used.

(c) Using the column marked 60 ft (18.3 m) in Table 9.2:

An excerpt from Table 9.2 is shown in Exhibit C.2.

(1) Outlet A, supplying 30 ft³/hr (0.8 m³/hr), requires ³/₈-in. pipe.
(2) Outlet B, supplying 3 ft³/hr (0.08 m³/hr), requires ¹/₄-in. pipe.
(3) Section 1, supplying outlets A and B, or 33 ft³/hr (0.9 m³/hr), requires ³/₈-in. pipe.

Note that these sizes selected from Table 9.2 are minimum sizes. If the installer elects to use ¹/₂-in. pipe as the smallest size, then the pipe supplying outlets A, B, and Section 1 could be increased to ¹/₂ inch. Also, if ³/₄-in. pipe was not available on the job site, the size of pipe supplying Outlets C and D could be increased.

Exhibit C.2 Excerpt from Table 9.2

Nominal Iron Pipe Size (in.)	Internal Diameter (in.)	Length of Pipe (Ft)		
		50	60	70
$1/4$	.364	18	16	15
$3/8$	.493	40	36	33
$1/2$	.622	73	66	61
$3/4$	.824	151	138	125
1	1.049	285	260	240
$1^1/4$	1.380	660	580	530
$1^1/2$	1.610	900	810	750

(4) Outlet C, supplying 75 ft³/hr (2.1 m³/hr), requires $3/4$-in. pipe.

A third method of sizing gas piping is detailed below as an option that is useful when large quantities of piping are involved in a job (e.g., an apartment house) and material costs are of concern. If the user is not completely familiar with this method, the resulting pipe sizing should be checked by a knowledgeable gas engineer.

(1) With the layout developed according to Section 2.1 of the code, indicate in each section the design gas flow under maximum operating conditions. For many layouts, the maximum design flow will be the sum of all connected loads. However, in some cases, certain combinations of utilization equipment will not occur simultaneously (e.g., gas heating and air conditioning). For these cases, the design flow is the greatest gas flow that can occur at any one time.
(2) Determine the inlet gas pressure for the system being designed. In most cases, the point of inlet will be the gas meter or service regulator, but in the case of a system addition, it could be the point of connection to the existing system.
(3) Determine the minimum pressure required at the inlet to the critical utilization equipment. Usually, the critical item will be the piece of equipment with the highest required pressure for satisfactory operation. If several items have the same required pressure, it will be the one with the greatest length of piping from the system inlet.
(4) The difference between the inlet pressure and critical item pressure is the allowable system pressure drop. Exhibits C.3 and C.4 show the relationship between gas flow, pipe size, and pipe length for natural gas with 0.60 specific gravity.

(5) Section 2, supplying outlets A, B, and C, or 108 ft³/hr (3.0 m³/hr), requires $3/4$-in. pipe.
(6) Outlet D, supplying 136 ft³/hr (3.8 m³/hr), requires $3/4$-in. pipe.
(7) Section 3, supplying outlets A, B, C, and D, or 244 ft³/hr (6.8 m³/hr), requires 1-in. pipe.

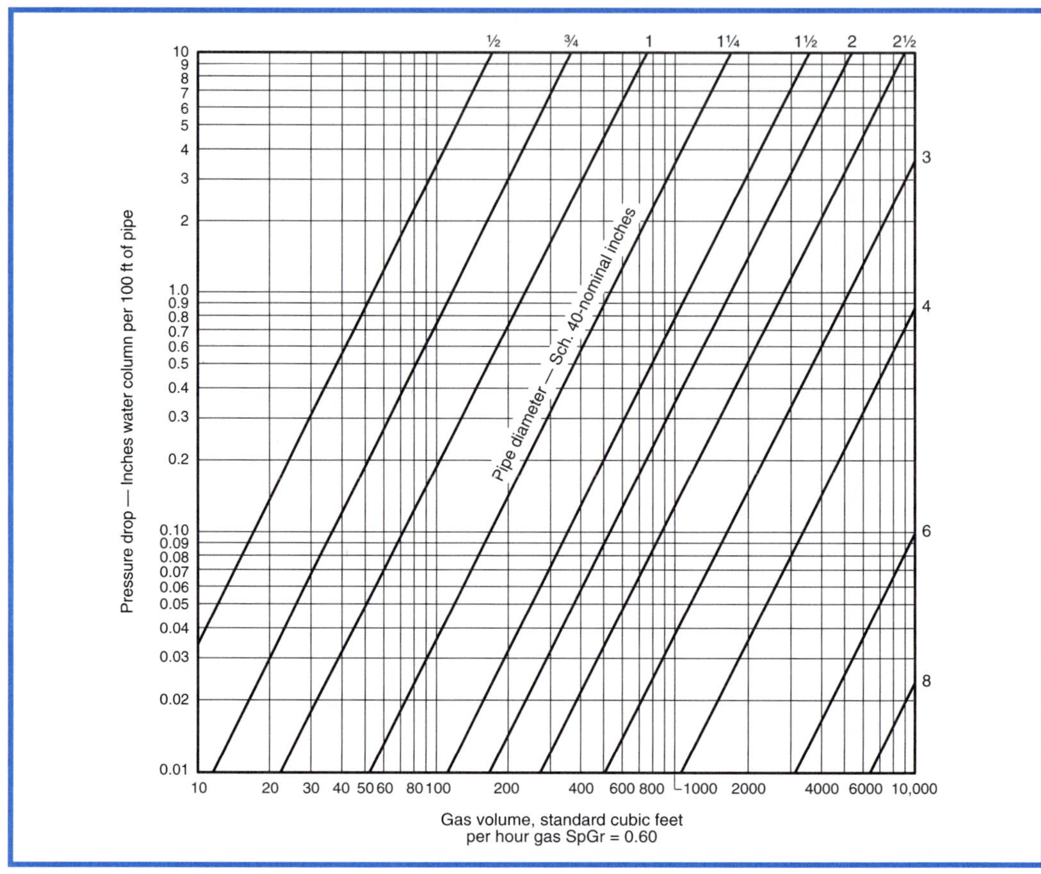

Exhibit C.3 *Capacity of natural gas piping, low pressure (0.60 in. water column at meter). (Courtesy of Richard E. White.)*

To use Exhibit C.3 (low-pressure applications), calculate the piping length from the inlet to the critical utilization equipment. Increase this length by 50 percent to allow for fittings. Divide the allowable pressure drop by the equivalent length (in hundreds of feet) to determine the allowable pressure drop per hundred feet. Select the pipe size from Exhibit C.3 for the required volume of flow.

To use Exhibit C.4 (high pressure applications), calculate the equivalent length as above. Calculate the index number for Exhibit C.4 by dividing the difference between the squares of the absolute values of inlet and outlet pressures by the equivalent length (in hundreds of feet). Select the pipe size from Exhibit C.4 for the gas volume required.

(d) If the gravity factor *[see C.2(d)]* is applied to this example, the values in the column marked 60 ft (18.3 m) of Table 9.2 would be multiplied by the multiplier (0.962) from Table 9.13 and the resulting cubic feet per hour values would be used to size the piping.

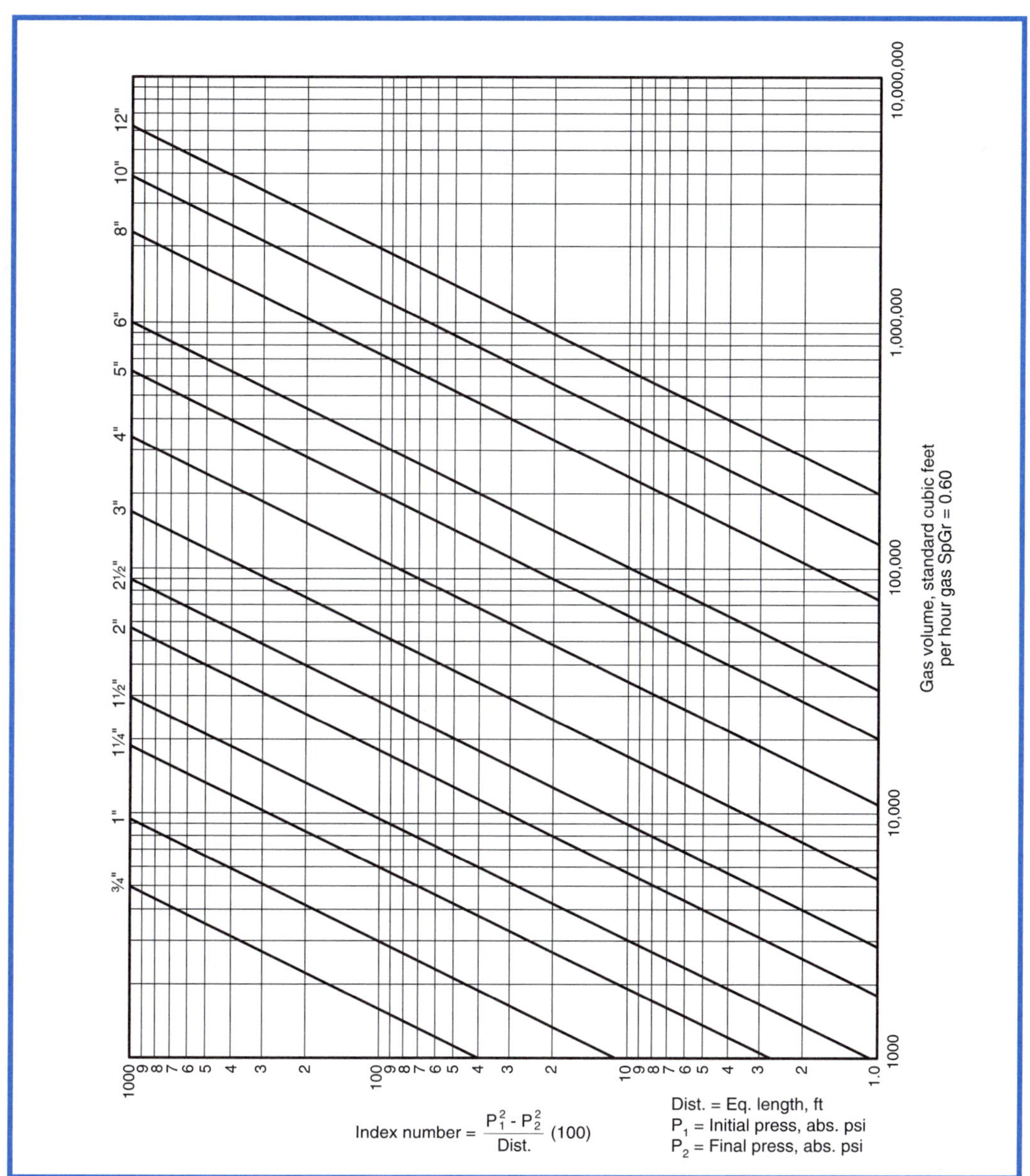

Exhibit C.4 *Capacity of natural gas piping, high pressure (1.5 psi and above). (Courtesy of Richard E. White.)*

Suggested Method for Checking for Leakage

This appendix is not a part of the requirements of this code but is included for informational purposes only.

D.1 Use of Lights

Artificial illumination used in connection with a search for gas leakage should be restricted to battery-operated flashlights (preferably of the safety type) or approved safety lamps. In searching for leaks, electric switches should not be operated. If electric lights are already turned on, they should not be turned off.

D.2 Testing for Leakage Using the Gas Meter

Immediately prior to the test, it should be determined that the meter is in operating condition and has not been bypassed.

Checking for leakage can be done by carefully watching the test dial of the meter to determine whether gas is passing through the meter. To assist in observing any movement of the test hand, wet a small piece of paper and paste its edge directly over the centerline of the hand as soon as the gas is turned on. This observation should be made with the test hand on the upstroke. Table D.2 can be used for determining the length of observation time.

Table D.2 Test Observation Times for Various Meter Dials

Dial Styles (ft^3)	Test Time (minutes)
$^1/_4$	5
$^1/_2$	5
1	7
2	10
5	20
10	30

For SI units, 1 ft^3 = 0.028 m^3.

In case carcful observation of the test hand for a sufficient length of time reveals no movement, the piping should be purged and a small gas burner turned on and lighted, and the hand of the test dial again observed. If the dial hand moves (as it should), it will show that the meter is operating properly. If the test hand does not move or register flow of gas through the meter to the small burner, the meter is defective and the gas should be shut off and the serving gas supplier notified.

D.3 Testing for Leakage Not Using a Meter

This test can be done by one of the following methods:

(a) *For Any Gas System.* To an appropriate checkpoint, attach a manometer or pressure gauge betwecn thc inlet to the piping system and the first regulator in the piping system, and momentarily turn on the gas supply and observe the gauging device for pressure drop with the gas supply shut off. No discernible drop in pressure should occur during a period of 3 minutes.

(b) *For Gas Systems Using Undiluted Liquefied Petroleum Gas. System Preparation for Propane.* A leak check performed on an LP-Gas system being placed back in service should include all regulators, including appliance regulators, and control valves in the system. Accordingly, each individual equipment shutoff valve should be supplying pressure to its appliance for the leak check. This check will prove the integrity of the 100 percent pilot shut-off of each gas valve so equipped, so the manual gas cock of each gas valve incorporating a 100 percent pilot shutoff should be in the on position. Pilots not incorporating a 100 percent pilot shutoff valve and all manual gas valves not incorporating safety shutoff systems are to be placed in the off position prior to leak checking, by using one of the following methods:

(1) By inserting a pressure gauge between the container gas shutoff valve and the first regu-lator in the system, admitting full container pressure to the system and then closing the

container shutoff valve. Enough gas should then be released from the system to lower the pressure gauge reading by 10 psi (69 kPa). The system should then be allowed to stand for 3 minutes without showing an increase or a decrease in the pressure gauge reading.

(2) For systems serving appliances that receive gas at pressures of $^1/_2$ psi (3.5 kPa) or less, by inserting a water manometer or pressure gauge into the system downstream of the final system regulator, pressurizing the system with either fuel gas or air to a test pressure of 9 in. $\pm$ $^1/_2$ in. (2.2 kPa $\pm$ 0.1 kPa) water column, and observing the device for a pressure change. If fuel gas is used as a pressure source, it is necessary to pressurize the system to full operating pressure, close the container service valve, and then release enough gas from the system through a range burner valve or other suitable means to drop the system pressure to 9 in. $\pm$ $^1/_2$ in. (2.2 kPa $\pm$ 0.1 kPa) water column. This ensures that all regulators in the system are unlocked and that a leak anywhere in the system is communicated to the gauging device. The gauging device should indicate no loss or gain of pressure for a period of 3 minutes.

Sections D.2 and D.3 describe two methods of checking for leakage. The first involves the use of a gas meter with a test hand. The tester must be sure that the gas meter is functioning and that all equipment valves are off. The system is then pressurized, and the test dial is observed carefully for the amount of time indicated in Table D.2. If there is a leak, the test dial will move.

The second method involves testing a system without a gas meter as follows:

(a) For gas systems supplied by a piped system, connect a manometer to the piping or appliance orifice and check for a drop in pressure while the main supply valve is shut off for three minutes.

(b) For LP-Gas systems without meters, two methods are provided. The first procedure is to charge the section between the first and second regulator to normal pressure, close off the supply, and then draw off enough gas to reduce the pressure to approximately 10 psi (69 kPa) below the initial pressure. The reason for reducing the pressure is to ensure that the supply valve is not leaking. If the pressure falls, then there is a leak in the piping system. If the pressure rises, then the supply valve is leaking.

The meter test hand or intermediate location gauge is then monitored for at least 10 minutes. Any movement of the test hand or drop in test gauge pressure indicates a leak, which must be located and repaired.

The second method was added in the 1992 edition and is useful for most residential and commercial systems that serve appliances operating at less than $^1/_2$ psi (3.5 kPa). It recognizes that some appliances incorporate regulators that can "lock up" and isolate the appliance from the piping system, which makes running a valid leak check impossible. Starting at system operating pressure and then lowering the pressure to 9 in. $\pm$ $^1/_2$ in. (2.2 kPa $\pm$ 0.1 kPa) water column ensures that regulator lockup does not interfere with the leak check.

D.4 When Leakage Is Indicated

If the meter test hand moves, or a pressure drop on the gauge is noted, all equipment or outlets supplied through the system should be examined to see whether they are shut off and do not leak. If they are found tight, there is a leak in the piping system.

Suggested Emergency Procedure for Gas Leaks

This appendix is not a part of the requirements of this code but is included for informational purposes only.

E.1

Where an investigation discloses a concentration of gas inside of a building, it is suggested the following immediate actions be taken:

(1) Clear the room, building, or area of all occupants. Do not re-enter the room, building, or area until the space has been determined to be safe.

(2) Use every practical means to eliminate sources of ignition. Take precautions to prevent smoking, striking matches, operating electrical switches or devices, opening furnace doors, and so on. If possible, cut off all electric circuits at a remote source to eliminate operation of automatic switches in the dangerous area. Safety flashlights designed for use in hazardous atmospheres are recommended for use in such emergencies.

(3) Notify all personnel in the area and the gas supplier from a telephone remote from the area of the leak.

(4) Ventilate the affected portion of the building by opening windows and doors.

(5) Shut off the supply of gas to the areas involved.

(6) Investigate other buildings in the immediate area to determine the presence of escaping gas therein.

If gas is suspected of having accumulated in a space, follow the procedures in this appendix to minimize the risk of fire or explosion. The most important point is item (1) — clear the building of all people. All possible ignition sources should be avoided, and the gas utility or supplier should be notified immediately to obtain all the assistance possible. It may also be advisable to notify the local fire department. The space(s) should then be ventilated. Safety flashlights listed for Class I, Group D locations should be the only source of illumination.

Be sure to check possible sources of gas outside the space where the gas has accumulated. A leak could be in an adjacent space or underground, in which case the gas could travel into and accumulate in the space where it was first noticed. Also, underground leaks can be deodorized by adsorption of the odorant by some soils; therefore, the lack of gas odor does not ensure that no gas is present. Use a gas detector to be sure.

Flow of Gas Through Fixed Orifices

This appendix is not a part of the requirements of this code but is included for informational purposes only.

This appendix provides orifice sizing tables and procedures for checking burner input. This knowledge enables field adjustment of input rate of appliances. Field adjustment can be applied on startup of large equipment, such as boilers and furnaces that require high-altitude adjustment.

F.1 Use of Orifice Tables

F.1.1 To Check Burner Input Not Using a Meter.

Gauge the size of the burner orifice and determine flow rate at sea level from Table F.1 — Utility Gases (cubic feet per hour) — or from Table F.2 — LP-Gases (Btu per hour). When the specific gravity of the utility gas is other than 0.60, select the multiplier from Table F.3 for the specific gravity of the utility gas served and apply to the flow rate as determined from Table F.1. When the altitude is above 2000 ft (600 m), first select the equivalent orifice size at sea level using Table F.4, then determine the flow rate from Table F.1 or Table F.2 as directed.

Having determined the flow rate (as adjusted for specific gravity and/or altitude where necessary), check the burner input at sea level with the manufacturer's rated input.

Table F.1 presents data for gas flow in standard cubic feet per hour at sea level, through orifices, as a function of orifice size and pressure at the orifice inlet. The data apply to gas with 0.60 specific gravity and an orifice coefficient of 0.90. The size is given in terms of numbered drill sizes, from #80 (the smallest) to #1 (the largest). For larger orifice sizes, refer to AGA Report No. 3, "Orifice Metering of Natural Gas and Other Related Hydrocarbon Fluids."

Table F.2 presents data for propane and butane in Btu per hour at sea level for orifices from 0.008 in. (0.2 mm) through #18 drill size [0.1695 in. (4.3 mm)]. Because the #18 drill size delivers 233 MBH (68 MW), the sizes in this table will cover all but the largest installations. The table covers gases with 11 in. (2.7 kPa) water column pressure at the orifice inlet and an orifice coefficient of 0.90 for the Btu per cubic foot and specific gravities given in the table.

Table F.3 presents the corrections (for Table F.1) for gases with specific gravities from 0.45 through 1.40. The volume taken from Table F.1 must be multiplied by the factor in Table F.3 to obtain the volume of gas of the new specific gravity that will flow through a given orifice size.

Table F.4 presents data for reducing the input rate of appliances installed at elevations of 2000 ft (600 m) and higher. As altitude increases, the atmosphere becomes less dense. Consequently, less oxygen is available (in the air) for the combustion of fuel. There may also be a slight reduction in the Btu per hour passing through the orifice, caused by a reduction in the density of the fuel gas (presuming the same orifice size and manifold pressure as at sea level), but not enough to compensate for the reduction in available oxygen.

The input rate specified on an appliance nameplate applies to elevations up to 2000 ft (600 m) above sea level. For elevations above 2000 ft (600 m), the standard cubic feet per hour (input rate) must be reduced by 4 percent for each 1000 ft (300 m) above sea level to provide sufficient oxygen for proper combustion of the fuel gas. Table F.4 also shows the proper burner orifice size to use at elevations of 2000 ft (600 m) and above, based on the orifice size for sea level operation from either Table F.1 (as corrected by Table F.3) or Table F.2.

Consider the following two examples:

(1) Burner elevation: 4000 ft (1200 m)

Sea level firing rate: 100 MBH (29 MW) total in two burners at 4 in. (1.0 kPa) water column pressure
Gas specific gravity: 0.80, 850 Btu/ft^3 (88 MW/m^3)

What burner orifices are needed?

Answer: Each burner is rated at 50,000 Btu/hr (15 MW)
(50,000)/(850) = 58.82 ft^3/hour
(58.82)/(0.866) = 67.92 equiv. ft^3/hour
(0.866 is obtained from Table F.3)

From Table F.1, 67.92 falls between a #26 and a #25 drill. (Select #26.) Table F.4 shows that a #28 drill should be used at 4000 ft (1200 m). Thus, two orifices are required with a #28 drill opening.

(2) Burner elevation: 5000 ft (1500 m)
     Inshot burner: 75 MBH (22 MW) sea level input rating
     Propane gas
     Table F.2 shows that a #41 drill is needed (74,924 Btu/hr).
     Table F.4 shows that a #43 drill should be used at a 5000-ft (1500-m) elevation.

If you are drilling on the job, drilling an orifice opening at least two sizes smaller than called for is good practice. If the orifice blank is held with pliers and the drill is powered by a handheld power tool, getting a straight, round hole is impossible. More gas will pass through the resulting opening than would be expected according to the tables. If, after test firing, increasing the opening is necessary, drilling one or two sizes larger will produce a hole much closer to the true size than the first hole drilled in the blank. Do not plan to compensate for an oversized or undersized orifice by reducing or increasing the appliance manifold pressure at the orifice by more than 10 percent of the manifold pressure specified for the appliance.

Some burners will operate satisfactorily only over a narrow range of orifice inlet pressures. If the pressure is too low, the flame may become soft and lazy. In extreme cases, sooting will occur. If the pressure is too high, the flame can become hard, noisy, and unstable. Also, lifting off the burner head is possible.

Table F.1 Utility Gases (cubic feet per hour at sea level)

Orifice or Drill Size	Pressure at Orifice — Inches Water Column								
	3	3.5	4	5	6	7	8	9	10
80	0.48	0.52	0.55	0.63	0.69	0.73	0.79	0.83	0.88
79	0.55	0.59	0.64	0.72	0.80	0.84	0.90	0.97	1.01
78	0.70	0.76	0.78	0.88	0.97	1.04	1.10	1.17	1.24
77	0.88	0.95	0.99	1.11	1.23	1.31	1.38	1.47	1.55
76	1.05	1.13	1.21	1.37	1.52	1.61	1.72	1.83	1.92
75	1.16	1.25	1.34	1.52	1.64	1.79	1.91	2.04	2.14
74	1.33	1.44	1.55	1.74	1.91	2.05	2.18	2.32	2.44
73	1.51	1.63	1.76	1.99	2.17	2.32	2.48	2.64	2.78
72	1.64	1.77	1.90	2.15	2.40	2.52	2.69	2.86	3.00
71	1.82	1.97	2.06	2.33	2.54	2.73	2.91	3.11	3.26
70	2.06	2.22	2.39	2.70	2.97	3.16	3.38	3.59	3.78
69	2.25	2.43	2.61	2.96	3.23	3.47	3.68	3.94	4.14
68	2.52	2.72	2.93	3.26	3.58	3.88	4.14	4.41	4.64
67	2.69	2.91	3.12	3.52	3.87	4.13	4.41	4.69	4.94
66	2.86	3.09	3.32	3.75	4.11	4.39	4.68	4.98	5.24

(continues)

Table F.1 Continued.

Orifice or Drill Size	Pressure at Orifice — Inches Water Column								
	3	3.5	4	5	6	7	8	9	10
65	3.14	3.39	3.72	4.28	4.62	4.84	5.16	5.50	5.78
64	3.41	3.68	4.14	4.48	4.91	5.23	5.59	5.95	6.26
63	3.63	3.92	4.19	4.75	5.19	5.55	5.92	6.30	6.63
62	3.78	4.08	4.39	4.96	5.42	5.81	6.20	6.59	6.94
61	4.02	4.34	4.66	5.27	5.77	6.15	6.57	7.00	7.37
60	4.21	4.55	4.89	5.52	5.95	6.47	6.91	7.35	7.74
59	4.41	4.76	5.11	5.78	6.35	6.78	7.25	7.71	8.11
58	4.66	5.03	5.39	6.10	6.68	7.13	7.62	8.11	8.53
57	4.84	5.23	5.63	6.36	6.96	7.44	7.94	8.46	8.90
56	5.68	6.13	6.58	7.35	8.03	8.73	9.32	9.92	10.44
55	7.11	7.68	8.22	9.30	10.18	10.85	11.59	12.34	12.98
54	7.95	8.59	9.23	10.45	11.39	12.25	13.08	13.93	14.65
53	9.30	10.04	10.80	12.20	13.32	14.29	15.27	16.25	17.09
52	10.61	11.46	12.31	13.86	15.26	16.34	17.44	18.57	19.53
51	11.82	12.77	13.69	15.47	16.97	18.16	19.40	20.64	21.71
50	12.89	13.92	14.94	16.86	18.48	19.77	21.12	22.48	23.65
49	14.07	15.20	16.28	18.37	20.20	21.60	23.06	24.56	25.83
48	15.15	16.36	17.62	19.88	21.81	23.31	24.90	26.51	27.89
47	16.22	17.52	18.80	21.27	23.21	24.93	26.62	28.34	29.81
46	17.19	18.57	19.98	22.57	24.72	26.43	28.23	30.05	31.61
45	17.73	19.15	20.52	23.10	25.36	27.18	29.03	30.90	32.51
44	19.45	21.01	22.57	25.57	27.93	29.87	31.89	33.96	35.72
43	20.73	22.39	24.18	27.29	29.87	32.02	34.19	36.41	38.30
42	23.10	24.95	26.50	29.50	32.50	35.24	37.63	40.07	42.14
41	24.06	25.98	28.15	31.69	34.81	37.17	39.70	42.27	44.46
40	25.03	27.03	29.23	33.09	36.20	38.79	41.42	44.10	46.38
39	26.11	28.20	30.20	34.05	37.38	39.97	42.68	45.44	47.80
38	27.08	29.25	31.38	35.46	38.89	41.58	44.40	47.27	49.73
37	28.36	30.63	32.99	37.07	40.83	43.62	46.59	49.60	52.17
36	29.76	32.14	34.59	39.11	42.76	45.77	48.88	52.04	54.74
35	32.36	34.95	36.86	41.68	45.66	48.78	52.10	55.46	58.34
34	32.45	35.05	37.50	42.44	46.52	49.75	53.12	56.55	59.49
33	33.41	36.08	38.79	43.83	48.03	51.46	54.96	58.62	61.55
32	35.46	38.30	40.94	46.52	50.82	54.26	57.95	61.70	64.89
31	37.82	40.85	43.83	49.64	54.36	58.01	61.96	65.97	69.39

(continues)

Table F.1 Continued.

Orifice or Drill Size	Pressure at Orifice — Inches Water Column								
	3	3.5	4	5	6	7	8	9	10
30	43.40	46.87	50.39	57.05	62.09	66.72	71.22	75.86	79.80
29	48.45	52.33	56.19	63.61	69.62	74.45	79.52	84.66	89.04
28	51.78	55.92	59.50	67.00	73.50	79.50	84.92	90.39	95.09
27	54.47	58.83	63.17	71.55	78.32	83.59	89.27	95.04	99.97
26	56.73	61.27	65.86	74.57	81.65	87.24	93.17	99.19	104.57
25	58.87	63.58	68.22	77.14	84.67	90.36	96.50	102.74	108.07
24	60.81	65.67	70.58	79.83	87.56	93.47	99.83	106.28	111.79
23	62.10	67.07	72.20	81.65	89.39	94.55	100.98	107.49	113.07
22	64.89	70.08	75.21	85.10	93.25	99.60	106.39	113.24	119.12
21	66.51	71.83	77.14	87.35	95.63	102.29	109.24	116.29	122.33
20	68.22	73.68	79.08	89.49	97.99	104.75	111.87	119.10	125.28
19	72.20	77.98	83.69	94.76	103.89	110.67	118.55	125.82	132.36
18	75.53	81.57	87.56	97.50	108.52	116.03	123.92	131.93	138.78
17	78.54	84.82	91.10	103.14	112.81	120.33	128.52	136.82	143.91
16	82.19	88.77	95.40	107.98	118.18	126.78	135.39	144.15	151.63
15	85.20	92.02	98.84	111.74	122.48	131.07	139.98	149.03	156.77
14	87.10	94.40	100.78	114.21	124.44	133.22	142.28	151.47	159.33
13	89.92	97.11	104.32	118.18	128.93	138.60	148.02	157.58	165.76
12	93.90	101.41	108.52	123.56	135.37	143.97	153.75	163.69	172.13
11	95.94	103.62	111.31	126.02	137.52	147.20	157.20	167.36	176.03
10	98.30	106.16	114.21	129.25	141.82	151.50	161.81	172.26	181.13
9	100.99	109.07	117.11	132.58	145.05	154.71	165.23	175.91	185.03
8	103.89	112.20	120.65	136.44	149.33	160.08	170.96	182.00	191.44
7	105.93	114.40	123.01	139.23	152.56	163.31	174.38	185.68	195.30
6	109.15	117.88	126.78	142.88	156.83	167.51	178.88	190.46	200.36
5	111.08	119.97	128.93	145.79	160.08	170.82	182.48	194.22	204.30
4	114.75	123.93	133.22	150.41	164.36	176.18	188.16	200.25	210.71
3	119.25	128.79	137.52	156.26	170.78	182.64	195.08	207.66	218.44
2	128.48	138.76	148.61	168.64	184.79	197.66	211.05	224.74	235.58
1	136.35	147.26	158.25	179.33	194.63	209.48	223.65	238.16	250.54

For SI units, 1 Btu/hr = 0.293 W; 1 ft3 = 0.028 m3; 1 ft = 0.305 m; 1 in. water column = 249 Pa.

Notes:

1. Specific Gravity = 0.60; orifice coefficient = 0.90.

2. For utility gases of another specific gravity, select multiplier from Table F.3. For altitudes above 2000 ft, first select the equivalent orifice size at sea level from Table F.4.

Table F.2 LP-Gases (Btu per hour at sea level)

	Propane	Butane
Btu per cubic foot =	2516	3280
Specific gravity =	1.52	2.01
Pressure at orifice, inches water column =	11	11
Orifice coefficient =	0.9	0.9

For altitudes above 2000 ft, first select the equivalent orifice size at sea level from Table F.4.

Orifice or Drill Size	Propane	Butane
0.008	519	589
0.009	656	744
0.010	812	921
0.011	981	1112
0.012	1169	1326
80	1480	1678
79	1708	1936
78	2080	2358
77	2629	2980
76	3249	3684
75	3581	4059
74	4119	4669
73	4678	5303
72	5081	5760
71	5495	6230
70	6375	7227
69	6934	7860
68	7813	8858
67	8320	9433
66	8848	10031
65	9955	11286
64	10535	11943
63	11125	12612
62	11735	13304
61	12367	14020
60	13008	14747
59	13660	15486

(continues)

Table F.2 Continued.

Orifice or Drill Size	Propane	Butane
58	14333	16249
57	15026	17035
56	17572	19921
55	21939	24872
54	24630	27922
53	28769	32615
52	32805	37190
51	36531	41414
50	39842	45168
49	43361	49157
48	46983	53263
47	50088	56783
46	53296	60420
45	54641	61944
44	60229	68280
43	64369	72973
42	71095	80599
41	74924	84940
40	78029	88459
39	80513	91215
38	83721	94912
37	87860	99605
36	92207	104532
35	98312	111454
34	100175	113566
33	103797	117672
32	109385	124007
31	117043	132689
30	134119	152046
29	150366	170466
28	160301	181728
27	168580	191114
26	175617	199092
25	181619	205896
24	187828	212935
23	192796	218567

(continues)

Table F.2 Continued.

Orifice or Drill Size	Propane	Butane
22	200350	227131
21	205525	232997
20	210699	238863
19	223945	253880
18	233466	264673

Table F.3 Multipliers for Utility Gases of Another Specific Gravity

Specific Gravity	Multiplier	Specific Gravity	Multiplier
0.45	1.155	0.95	0.795
0.50	1.095	1.00	0.775
0.55	1.045	1.05	0.756
0.60	1.000	1.10	0.739
0.65	0.961	1.15	0.722
0.70	0.926	1.20	0.707
0.75	0.894	1.25	0.693
0.80	0.866	1.30	0.679
0.85	0.840	1.35	0.667
0.90	0.817	1.40	0.655

Table F.4 Equivalent Orifice Sizes at High Altitudes (Includes 4% input reduction for each 1000 ft)

Orifice Size at Sea Level	Orifice Size Required at Other Elevations								
	2000	3000	4000	5000	6000	7000	8000	9000	10000
1	2	2	3	3	4	5	7	8	10
2	3	3	4	5	6	7	9	10	12
3	4	5	7	8	9	10	12	13	15
4	6	7	8	9	11	12	13	14	16
5	7	8	9	10	12	13	14	15	17
6	8	9	10	11	12	13	14	16	17
7	9	10	11	12	13	14	15	16	18
8	10	11	12	13	13	15	16	17	18

(continues)

Table F.4 Continued.

Orifice Size at Sea Level	Orifice Size Required at Other Elevations								
	2000	3000	4000	5000	6000	7000	8000	9000	10000
9	11	12	12	13	14	16	17	18	19
10	12	13	13	14	15	16	17	18	19
11	13	13	14	15	16	17	18	19	20
12	13	14	15	16	17	17	18	19	20
13	15	15	16	17	18	18	19	20	22
14	16	16	17	18	18	19	20	21	23
15	16	17	17	18	19	20	20	22	24
16	17	18	18	19	19	20	22	23	25
17	18	19	19	20	21	22	23	24	26
18	19	19	20	21	22	23	24	26	27
19	20	20	21	22	23	25	26	27	28
20	22	22	23	24	25	26	27	28	29
21	23	23	24	25	26	27	28	28	29
22	23	24	25	26	27	27	28	29	29
23	25	25	26	27	27	28	29	29	30
24	25	26	27	27	28	28	29	29	30
25	26	27	27	28	28	29	29	30	30
26	27	28	28	28	29	29	30	30	30
27	28	28	29	29	29	30	30	30	31
28	29	29	29	30	30	30	30	31	31
29	29	30	30	30	30	31	31	31	32
30	30	31	31	31	31	32	32	33	35
31	32	32	32	33	34	35	36	37	38
32	33	34	35	35	36	36	37	38	40
33	35	35	36	36	37	38	38	40	41
34	35	36	36	37	37	38	39	40	42
35	36	36	37	37	38	39	40	41	42
36	37	38	38	39	40	41	41	42	43
37	38	39	39	40	41	42	42	43	43
38	39	40	41	41	42	42	43	43	44
39	40	41	41	42	42	43	43	44	44
40	41	42	42	42	43	43	44	44	45

(continues)

Table F.4 Continued.

Orifice Size at Sea Level	Orifice Size Required at Other Elevations								
	2000	3000	4000	5000	6000	7000	8000	9000	10000
41	42	42	42	43	43	44	44	45	46
42	42	43	43	43	44	44	45	46	47
43	44	44	44	45	45	46	47	47	48
44	45	45	45	46	47	47	48	48	49
45	46	47	47	47	48	48	49	49	50
46	47	47	47	48	48	49	49	50	50
47	48	48	49	49	49	50	50	51	51
48	49	49	49	50	50	50	51	51	52
49	50	50	50	51	51	51	52	52	52
50	51	51	51	51	52	52	52	53	53
51	51	52	52	52	52	53	53	53	54
52	52	53	53	53	53	53	54	54	54
53	54	54	54	54	54	54	55	55	55
54	54	55	55	55	55	55	56	56	56
55	55	55	55	56	56	56	56	56	57
56	56	56	57	57	57	58	59	59	60
57	58	59	59	60	60	61	62	63	63
58	59	60	60	61	62	62	63	63	64
59	60	61	61	62	62	63	64	64	65
60	61	61	62	63	63	64	64	65	65
61	62	62	63	63	64	65	65	66	66
62	63	63	64	64	65	65	66	66	67
63	64	64	65	65	65	66	66	67	68
64	65	65	65	66	66	66	67	67	68
65	65	66	66	66	67	67	68	68	69
66	67	67	68	68	68	69	69	69	70
67	68	68	68	69	69	69	70	70	70
68	68	69	69	69	70	70	70	71	71
69	70	70	70	70	71	71	71	72	72
70	70	71	71	71	71	72	72	73	73
71	72	72	72	73	73	73	74	74	74
72	73	73	73	73	74	74	74	74	75

Table F.4 Continued.

Orifice Size at Sea Level	Orifice Size Required at Other Elevations								
	2000	**3000**	**4000**	**5000**	**6000**	**7000**	**8000**	**9000**	**10000**
73	73	74	74	74	74	75	75	75	76
74	74	75	75	75	75	76	76	76	76
75	75	76	76	76	76	77	77	77	77
76	76	76	77	77	77	77	77	77	77
77	77	77	77	78	78	78	78	78	78
78	78	78	78	79	79	79	79	80	80
79	79	80	80	80	80	.013	.012	.012	.01
80	80	.013	.013	.013	.012	.012	.012	.012	.011

F.4.2 To Select Correct Orifice Size for Rated Burner Input.

The selection of a fixed orifice size for any rated burner input is affected by many variables, including orifice coefficient, and it is recommended that the appliance manufacturer be consulted for that purpose. When the correct orifice size cannot be readily determined, the orifice flow rates, as stated in the tables in this appendix, can be used to select a fixed orifice size with a flow rate to approximately equal the required rated burner input.

For gases of the specific gravity and pressure conditions stipulated at elevations under 2000 ft (600 m), Table F.1 (in cu ft per hr) or Table F.2) (in Btu per hour) can be used directly.

Where the specific gravity of the gas is other than 0.60, select the multiplier from Table F.3 for the utility gas served and divide the rated burner input by the selected factor to determine equivalent input at a specific gravity of 0.60; then select orifice size as directed above.

Where the appliance is located at an altitude of 2000 ft (600 m) or above, first use the manufacturer's rated input at sea level to select the orifice size as directed, then use Table F.4 to select the equivalent orifice size for use at the higher altitude.

Reference Cited in Commentary

The following publication is available from the American Gas Association, 1515 Wilson Boulevard, Arlington, VA 22209.

ANSI/API 2530 (AGA Report No. 3), "Orifice Metering of Natural Gas and Other Related Hydrocarbon Fluids," 1985.

Sizing of Venting Systems Serving Appliances Equipped with Draft Hoods, Category I Appliances, and Appliances Listed for Use with Type B Vents

This appendix is not a part of the requirements of this code but is included for informational purposes only.

Appendix G provides additional information on the sizing of gas vents beyond what is provided in Chapter 7, "Venting of Equipment," and Chapter 10, "Sizing of Category I Venting Systems." The appendix is useful to the user of the code in the proper sizing of venting systems.

The figures in Appendix G illustrate common venting systems and are used in conjunction with the tables in Chapter 10. These figures illustrate the types of systems and the measurement of dimensions so that the tables in Chapter 10 can be used properly. The diagrams illustrate typical installations; other configurations that meet the code are possible. Therefore, the diagrams should not be interpreted as being the only permissible geometry.

A series of examples that show how to use the tables and other requirements of Chapter 10 are given. They also illustrate which dimensions on a real venting system correspond to the table entries. The examples present excerpts from the tables in Chapter 10 to clarify the calculations.

G.1 Examples Using Single Appliance Venting Tables

See Figures G.1 through G.14.

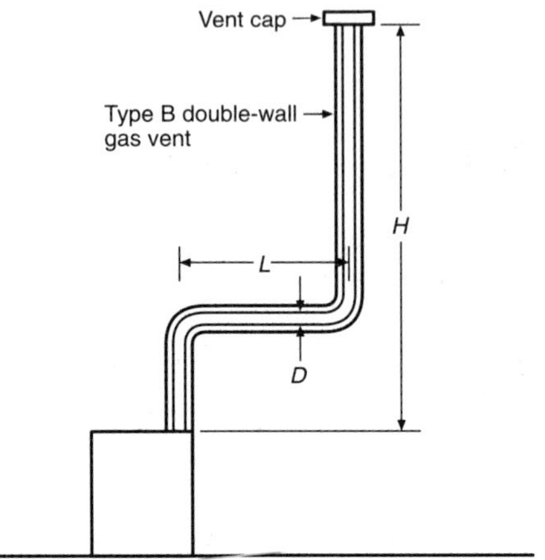

Table 10.1 is used when sizing Type B double-wall gas vent connected directly to the appliance.

Note: The appliance can be either Category I draft hood–equipped or fan-assisted type.

Figure G.1 *Type B double-wall vent system serving a single appliance with a Type B double-wall vent.*

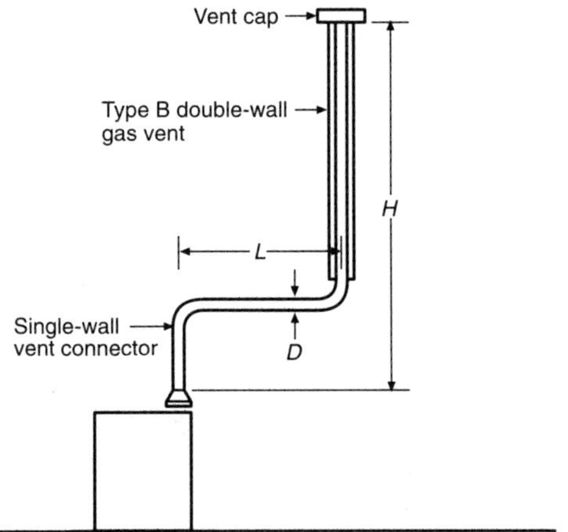

Table 10.2 is used when sizing a single-wall metal vent connector attached to a Type B double-wall gas vent.

Note: The appliance can be either Category I draft hood–equipped or fan-assisted type.

Figure G.2 *Type B double-wall vent system serving a single appliance with a single-wall metal vent connector.*

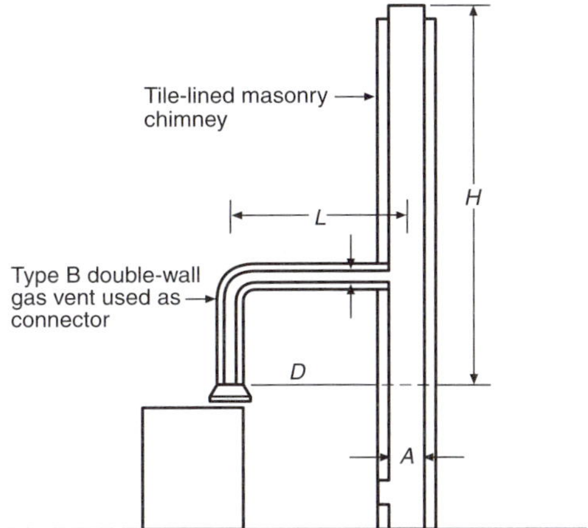

Table 10.3 is used when sizing a Type B double-wall
gas vent connector attached to a tile-lined masonry chimney.
Notes:
1. *A* is the equivalent cross-sectional area of the tile liner.
2. The appliance can be either Category I draft
   hood–equipped or fan-assisted type.

Figure G.3 *Vent system serving a single appliance with a masonry chimney and a Type B double-wall vent connector.*

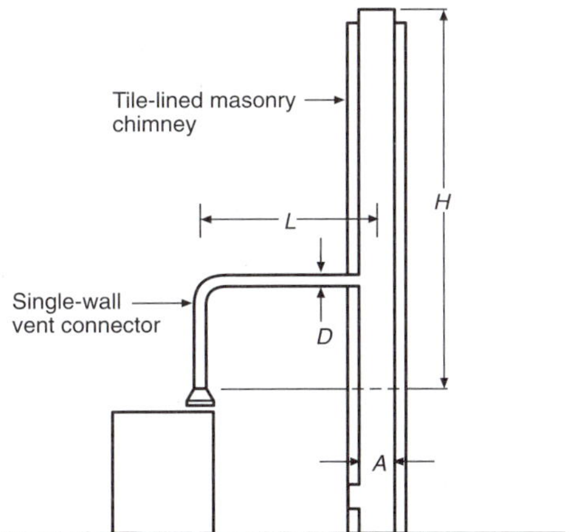

Table 10.4 is used when sizing a single-wall vent
connector attached to a tile-lined masonry chimney.
Notes:
1. *A* is the equivalent cross-sectional area of the tile liner.
2. The appliance can be either Category I draft
   hood–equipped or fan-assisted type.

Figure G.4 *Vent system serving a single appliance using a masonry chimney and a single-wall metal vent connector.*

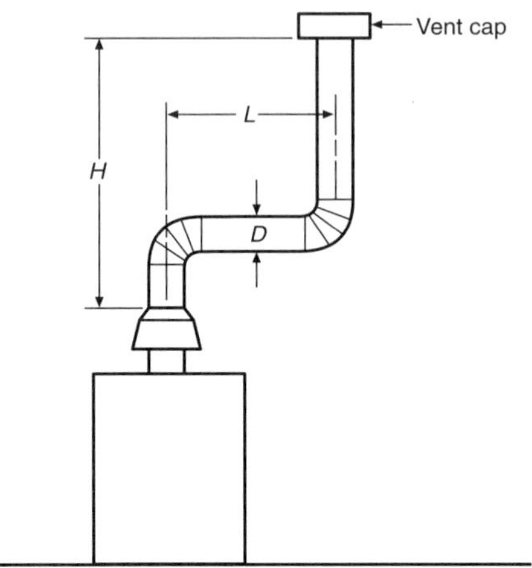

Figure G.5 *Asbestos cement Type B or single-wall metal vent system serving a single draft hood-equipped appliance.*

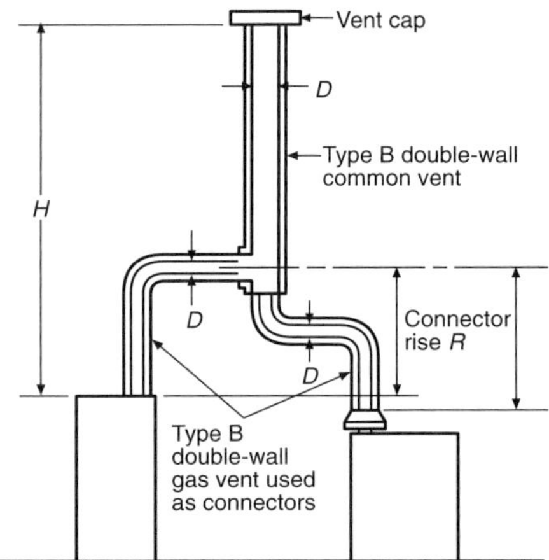

Figure G.6 *Vent system serving two or more appliances with Type B double-wall vent and Type B double-wall vent connectors.*

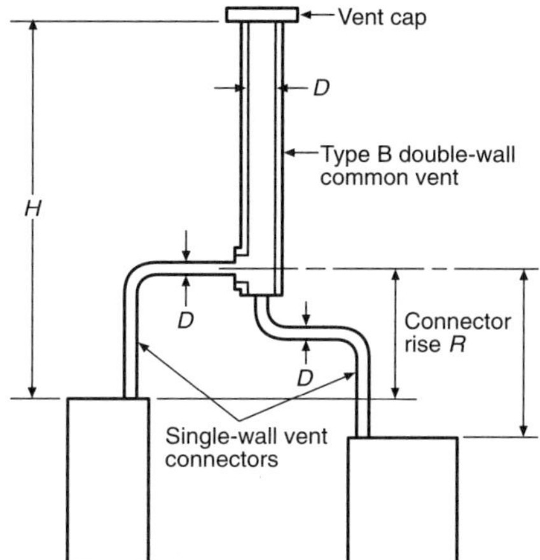

Table 10.7 is used when sizing single-wall vent
connectors attached to a Type B double-wall
common vent.
Note: Each appliance can be either Category I draft
hood–equipped or fan-assisted type.

Figure G.7 *Vent system serving two or more appliances with Type B double-wall vent and single-wall metal vent connectors.*

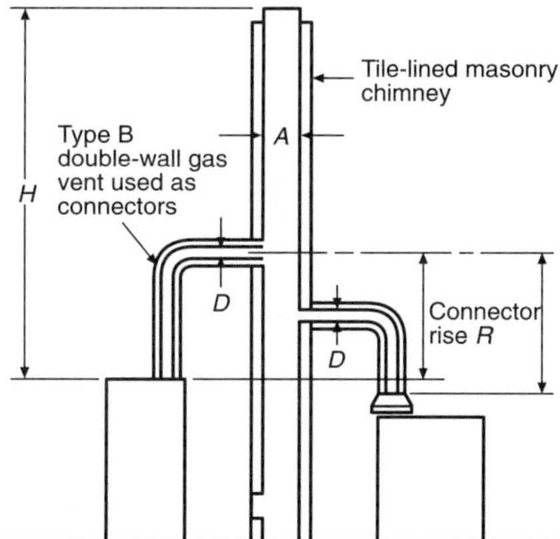

Table 10.8 is used when sizing Type B double-wall
vent connectors attached to a tile-lined masonry chimney.
Notes:
1. *A* is the equivalent cross-sectional area of the tile liner.
2. Each appliance can be either Category I draft
   hood–equipped or fan-assisted type.

Figure G.8 *Masonry chimney serving two or more appliances with Type B double-wall vent connectors.*

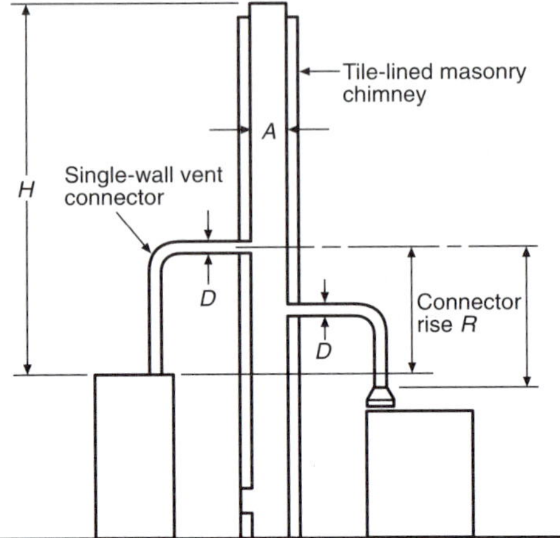

Table 10.9 is used when sizing single-wall metal vent connectors attached to a tile-lined masonry chimney.
Notes:
1. *A* is the equivalent cross-sectional area of the tile liner.
2. Each appliance can be either Category I draft hood–equipped or fan-assisted type.

Figure G.9 *Masonry chimney serving two or more appliances with single-wall metal vent connectors.*

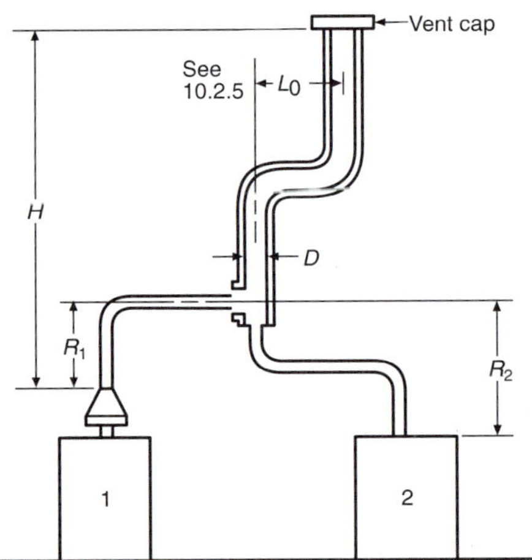

Asbestos cement Type B or single-wall metal pipe vent serving two or more draft hood–equipped appliances. *(See Table 10.10.)*

Figure G.10 *Asbestos cement Type B or single-wall metal vent system serving two or more draft hood-equipped appliances.*

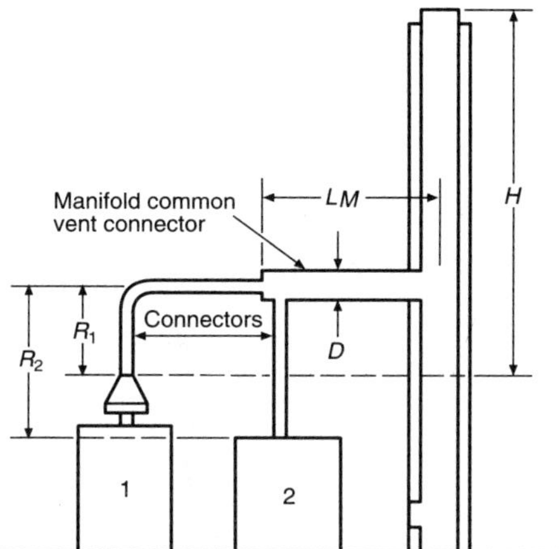

Example: Manifolded common vent connector L_M shall be no greater than 18 times the common vent connector manifold inside diameter; that is, a 4-in. (100-mm) inside diameter common vent connector manifold shall not exceed 72 in. (1800 mm) in length. *(See 10.2.4.)*
Note: This is an illustration of a typical manifolded vent connector. Different appliance, vent connector, or common vent types are possible. Consult Section 10.2.

Figure G.11 *Use of manifolded common vent connector.*

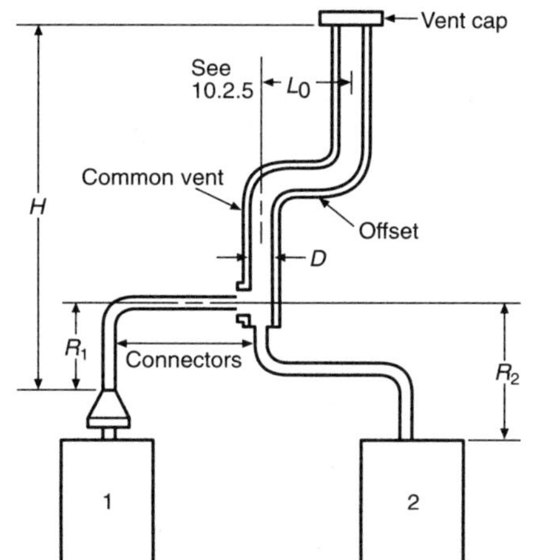

Example: Offset common vent
Note: This is an illustration of a typical offset vent. Different appliance, vent connector, or vent types are possible. *(Consult Sections 10.1. and 10.2.)*

Figure G.12 *Use of offset common vent.*

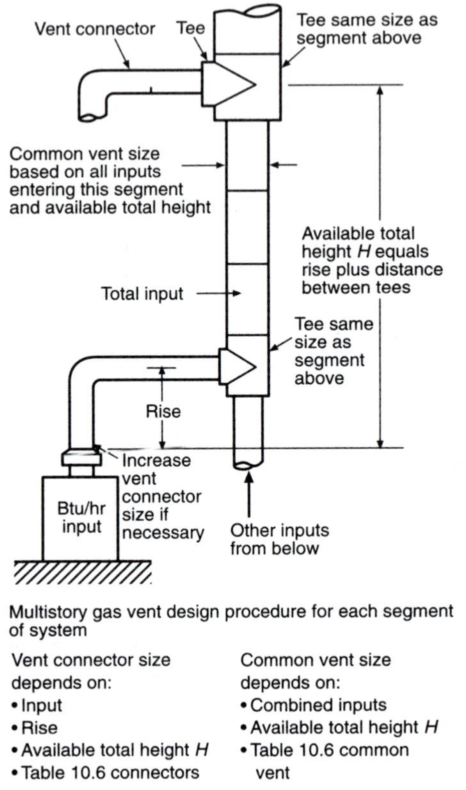

Figure G.13 *Multistory gas vent design procedure for each segment of system.*

Assume that Figure G.14 represents a four-story apartment building that has a listed fan-assisted combustion furnace installed on each floor. Each furnace has a 4-in. (100-mm) flue collar and an input of 80,000 Btu/hr (23 MW). All the furnaces are installed outside of the conditioned space (i.e., isolated combustion) and are to be vented into the common vent, which is located 5 ft (1.5 m) from the furnaces. For the purpose of calculation, the overall system is divided into smaller, simple vent systems for each level, as shown in Figure G.13.

The structure is such that the vent connector rise is to be 2 ft (0.6 m) and the available total height for each level is 10 ft (3.0 m), except for the top floor, which is 6 ft (1.8 m). The vent connector for the first floor or lowest furnace to the common vent is considered to be an individual vent terminating at the first tee or interconnection, and it is sized in accordance with Table 10.1. Every other vent connector

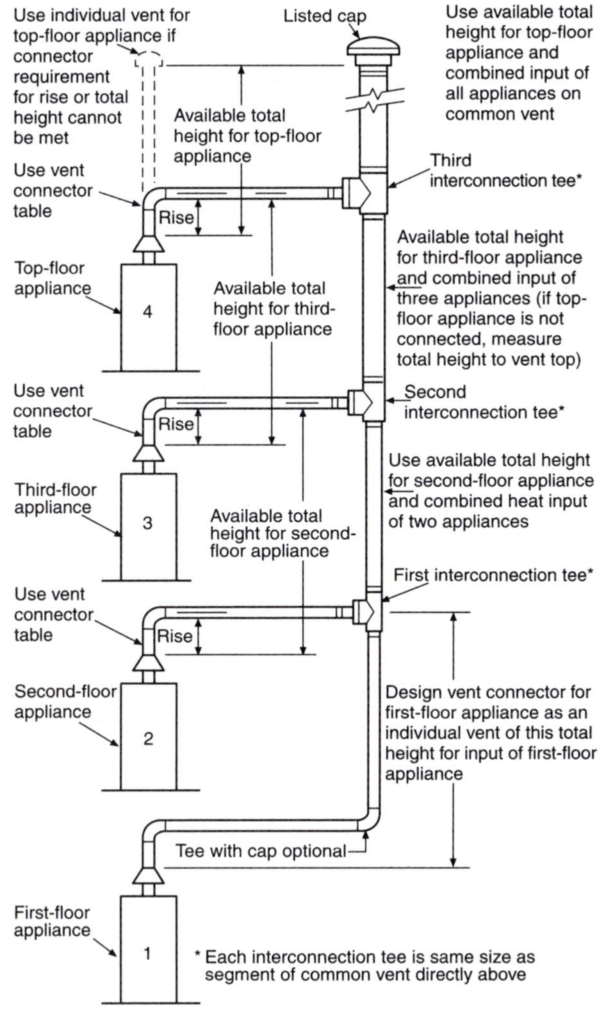

Figure G.14 *Principles of design of multistory vents using vent connector and common vent design tables. (See 10.2.10 through 10.2.15.)*

is sized in accordance with the Vent Connector Capacity section of Table 10.6. Each section of the common vent is sized in accordance with the Common Vent Capacity section of Table 10.6 to accommodate the accumulated total input of all appliances discharging into it, but sections are never smaller than the largest section below them.

Procedure for Sizing Connectors

First-Floor Connector (Furnace 1). The input is 80,000 Btu/hr, and the total height is 10 ft. This section of the vent is treated as a single-appliance vent with 5 ft of vent connector. Use Table 10.1, reading across the line for a 10-ft height (H) and a lateral (L) of 5 ft. Look under the FAN columns for a range that includes 80,000. The 4-in. column has a range of 32,000 Btu/hr to 113,000 Btu/hr, and the 5-in. column has a range of 41,000 Btu/hr to 187,000 Btu/hr. Both sizes are acceptable, and a 4-in. size is selected.

Paragraph 10.2.2 limits a 4-in. vent connector to a maximum length of 6 ft, which is not exceeded in this example.

Second- and Third-Floor Connectors (Furnaces 2 and 3). The input is 80,000 Btu/hr, the connector rise is 2 ft, and the vent height is 10 ft. Use the Vent Connector Capacity section of Table 10.6. Reading across the 10-ft height, 2-ft rise row in the FAN columns shows that a 4-in. connector has a range of 36,000 Btu/hr to 86,000 Btu/hr, a 5-in. connector has a range of 51,000 Btu/hr to 136,000 Btu/hr, and a 6-in. connector has a range of 67,000 Btu/hr to 206,000 Btu/hr. A connector with a diameter of 4 in., 5 in., or 6 in. will work, and a 4-in. connector is selected.

Fourth-Floor Connector (Furnace 4). The input is 80,000 Btu/hr, the connector rise is 2 ft, and the vent height is 6 ft. Again, use the Vent Connector Capacity section of Table 10.6. Reading across the 6-ft height, 2-ft rise row shows that a 5-in. connector has a range of 48,000 Btu/hr to 121,000 Btu/hr and a 6-in. connector has a range of 60,000 Btu/hr to 183,000 Btu/hr. Either a 5- or 6-in. connector will work, and a 5-in. connector is selected.

Paragraph 10.2.2 limits a 5-in. vent connector to a maximum length of $7^{1}/_{2}$ ft, which is not exceeded by the connector size selected.

Procedures for Sizing Common Vent

Common Vent for Furnaces 1 and 2. The input is 160,000 Btu/hr, and the vent height is 10 ft. Use the Common Vent Capacity section of Table 10.6, FAN+FAN columns only. Reading across the 10-ft line shows that a 5-in. vent has a capacity of 169,000 Btu/hr, which exceeds the 160,000 Btu/hr input. This size is used.

Common Vent for Third-Floor Furnace (3). The input is 240,000 Btu/hr, and the height is 10 ft. Use the Common Vent Capacity section of Table 10.6. Reading across the 10-ft line in the FAN+FAN columns shows that a 6-in. common vent has a capacity of 243,000 Btu/hr. This size is used.

Common Vent for Fourth-Floor Furnace (4). The input is 320,000 Btu/hr, and the height is 6 ft. Again, use the FAN+FAN columns in the Common Vent Capacity section of Table 10.6. Reading across the 6-ft line shows that an 8-in. vent has a capacity of 404,000 Btu/hr, and this size is selected.

These sizing procedures are summarized in Tables G.1 and G.2.

Table G.1 Common Vent Serving Appliances on Four Floors

Furnace	Total Input to Common Vent (Btu/hr)	Available Total Height (ft)	Vent Connector Size (in.)	Common Vent Size (in.)
1	80,000	10	4	4
2	160,000	10	4	5
3	240,000	10	4	6
4	320,000	6	5	8

Table G.2 Common Vent Serving Appliances on Three Floors; Independent Vent Serving Appliance on the Fourth Floor

Furnace	Total Input to Common Vent (Btu/hr)	Available Total Height (ft)	Vent Connector Size (in.)	Common Vent Size (in.)
1	80,000	10	4	4
2	160,000	10	4	5
3	240,000	10	4	6
4	80,000	6	4	None

Check for Excessive Vent Area. The vent connector and common vent has been sized using Table 10.6 and Section 10.2, Additional Requirements to Multiple Appliance Vent Tables. Refer to 10.2.15, which states:

10.2.15 Where two or more appliances are connected to a vertical vent or chimney, the flow area of the largest section of vertical vent or chimney shall not exceed seven times the smallest listed appliance categorized vent areas, flue collar area, or draft hood outlet area unless designed in accordance with approved engineering methods (**54:** 10.2.15).

In the example, the smallest vent connector diameter is 4 in., and the largest common vent diameter is 8 in.

For the 4-in. diameter connector,

$$\text{Area} = \Pi r^2 = 3.14(2)^2 = 12.56$$

For the 8-in. diameter common vent,

$$\text{Area} = \Pi r^2 = 3.14 (4)^2 = 50.25$$

$$\text{Ratio} = \text{area (smallest connector)/area (largest common vent section)}$$

$$\text{Ratio} = 50.25 / 12.56 = 4$$

As 4 is less than 7, the sizing is acceptable.

G.1.1 Example 1: Single Draft Hood-Equipped Appliance.

An installer has a 120,000-Btu/hr input appliance with a 5-in. diameter draft hood outlet that needs to be vented into a 10-ft high Type B vent system. What size vent should be used assuming (1) a 5-ft lateral single-wall metal vent connector is used with two 90 degree elbows or (2) a 5-ft lateral single-wall metal vent connector is used with three 90 degree elbows in the vent system? See Figure G.15.

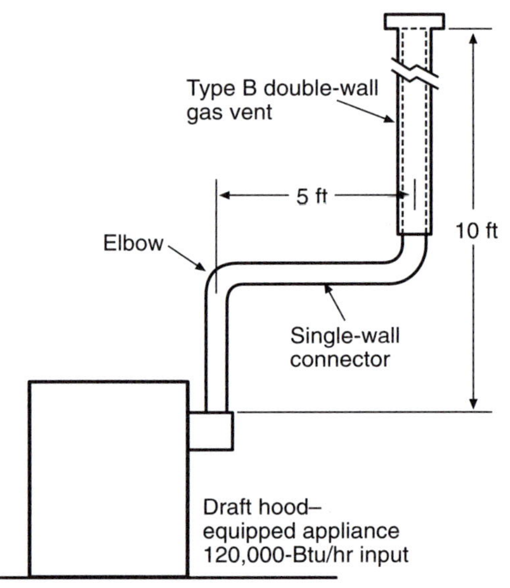

Figure G.15 *Single draft hood-equipped appliance (Example 1).*

Solution

Table 10.2 should be used to solve this problem, because single-wall metal vent connectors are being used with a Type B vent.

Refer to Table G.3, which is an extract from code Table 10.2.

(a) Read down the first column in Table 10.2 until the row associated with a 10-ft height and 5-ft lateral is found. Read across this row until a vent capacity greater than 120,000 Btu/hr is located in the shaded columns labeled NAT Max for draft hood–equipped appliances. In this case, a 5-in. diameter vent has a capacity of 122,000 Btu/hr and can be used for this application.

(b) If three 90-degree elbows are used in the vent system, then the maximum vent capacity listed in the tables must be reduced by 10 percent *(see 10.1.3)*. This implies that the 5-in.

Table G.3 Extract from Code Table 10.2

Height H (ft)	Lateral L (ft)	Vent Diameter — D						
		4 in.			5 in.			6 in.
		Appliance Input Rating in Thousands of Btu per Hour						
		FAN		NAT	FAN		NAT	FAN
		Min	Max	Max	Min	Max	Max	Min
10	0	57	174	99	82	293	165	120
	2	59	117	80	82	193	128	119
	5	76	111	76	105	185	122	148
	10	97	100	68	132	171	112	188

diameter vent has an adjusted capacity of only 110,000 Btu/hr. In this case, the vent system must be increased to 6 in. in diameter. See the following calculations:

$$122{,}000 \times 0.90 = 110{,}000 \text{ for 5-in. vent}$$

From Table 10.2, select 6-in. vent.

$$186{,}000 \times 0.90 = 167{,}000$$

This figure is greater than the required 120,000. Therefore, use a 6-in. vent and connector where three elbows are used.

G.1.2 Example 2: Single Fan-Assisted Appliance.

An installer has an 80,000-Btu/hr input fan-assisted appliance that must be installed using 10 ft of lateral connector attached to a 30 ft high Type B vent. Two 90 degree elbows are needed for the installation. Can a single-wall metal vent connector be used for this application? See Figure G.16.

Example 2 illustrates a very important point. The sizing tables for fan-assisted appliances are designed to reduce the allowable length of single-wall galvanized vent connector runs, as compared to those for type B connectors. The reason for this limitation is that single-wall connectors lose more heat in the vent gases than double-wall B vent connectors. This adversely affects the performance of the entire venting system. Type B connectors are of double-wall construction, with an air space. This design provides insulation and reduces heat loss. Remember that fan-assisted combustion appliances operate at a lower vent temperature than do draft hood–equipped appliances, so further heat loss in the venting system must be minimized to prevent condensation and corrosion of the vent.

There is no hard and fast rule of thumb for the maximum length of single-wall connectors. However, use of Table 10.2 in practical situations shows there are not many instances where a connector of more than 5 ft (1.5 m) long is allowed.

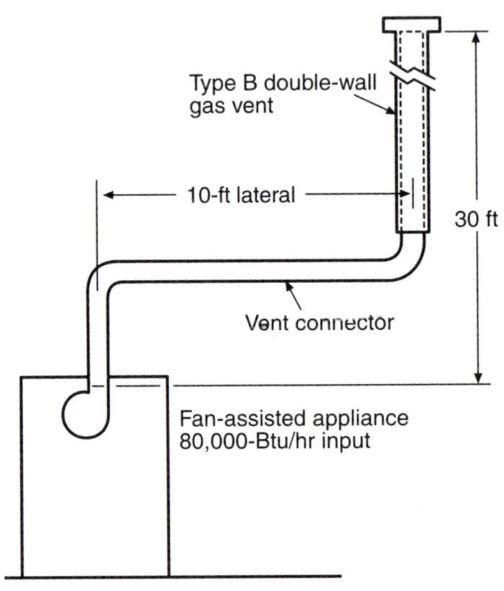

Figure G.16 *Single fan-assisted appliance (Example 2).*

Solution

Table 10.2 refers to the use of single-wall metal vent connectors with Type B vent. In the first column find the row associated with a 30-ft height and a 10-ft lateral. Read across this row, looking at the FAN Min and FAN Max columns, to find that a 3-in. diameter single-wall metal vent connector is not recommended. Moving to the next larger size single wall connector (4 in.), we find that a 4-in. diameter single-wall metal connector has a recommended minimum vent capacity of 91,000 Btu/hr and a recommended maximum vent capacity of 144,000 Btu/hr. The 80,000 Btu/hr fan-assisted appliance is outside this range, so the conclusion is that a single-wall metal vent connector cannot be used to vent this appliance using 10 ft of lateral for the connector.

Refer to Table G.4, which is an extract from code Table 10.2.

However, if the 80,000 Btu/hr input appliance could be moved to within 5 ft of the vertical vent, then a 4-in. single-wall metal connector could be used to vent the appliance. Table 10.2 shows the acceptable range of vent capacities for a 4-in. vent with 5 ft of lateral to be between 72,000 Btu/hr and 157,000 Btu/hr.

Table G.5 is an extract from code Table 10.2.

Table G.4 Extract from Code Table 10.2

Height H (ft)	Lateral L (ft)	3 in. FAN Min	3 in. FAN Max	3 in. NAT Max	4 in. FAN Min	4 in. FAN Max	4 in. NAT Max
30	0	34	99	63	53	211	127
	2	37	80	56	55	164	111
	5	49	74	52	72	157	106
	10	NA	NA	NA	91	144	98
	15	NA	NA	NA	115	131	NA
	20	NA	NA	NA	NA	NA	NA
	30	NA	NA	NA	NA	NA	NA

(Column groups: "Vent Diameter — D"; 3 in. and 4 in. each "Appliance Input Rating in Thousands of Btu per Hour", with FAN (Min, Max) and NAT (Max).)

Table G.5 Extract from Code Table 10.2

Height H (ft)	Lateral L (ft)	3 in. FAN Min	3 in. FAN Max	3 in. NAT Max	4 in. FAN Min	4 in. FAN Max	4 in. NAT Max
30	0	34	99	63	53	211	127
	2	37	80	56	55	164	111
	5	49	74	52	72	157	106
	10	NA	NA	NA	91	144	98
	15	NA	NA	NA	115	131	NA
	20	NA	NA	NA	NA	NA	NA
	30	NA	NA	NA	NA	NA	NA

(Column groups: "Vent Diameter — D"; 3 in. and 4 in. each "Appliance Input Rating in Thousands of Btu per Hour", with FAN (Min, Max) and NAT (Max).)

If the appliance cannot be moved closer to the vertical vent, then Type B vent could be used as the connector material. In this case, Table 10.1 shows that, for a 30-ft high vent with 10 ft of lateral, the acceptable range of vent capacities for a 4-in. diameter vent attached to a fan-assisted appliance is between 37,000 Btu/hr and 150,000 Btu/hr.

Table G.6 is an extract from code Table 10.1.

Table G.6 Extract from Code Table 10.1

Height H (ft)	Lateral L (ft)	Vent Diameter — D					
		3 in.			4 in.		
		Appliance Input Rating in Thousands of Btu per Hour					
		FAN		NAT	FAN		NAT
		Min	Max	Max	Min	Max	Max
30	0	0	100	64	0	213	128
	2	9	81	56	13	166	112
	5	21	77	54	28	160	108
	10	27	70	50	37	150	102
	15	33	64	NA	44	141	96
	20	56	58	NA	53	132	90
	30	NA	NA	NA	73	113	NA

G.1.3 Example 3: Interpolating Between Table Values.

An installer has an 80,000-Btu/hr input appliance with a 4-in. diameter draft hood outlet that needs to be vented into a 12 ft high Type B vent. The vent connector has a 5-ft lateral length and is also Type B. Can this appliance be vented using a 4-in. diameter vent?

Example 3 shows how to *interpolate* between existing table values (to find an intermediate value between two table values). However, it is not permissible to *extrapolate* out to values beyond the outer bounds of the tables (to infer a value beyond the values in the table). Some other approved engineering method must be used in this case.

Solution

Table 10.1 is used in the case of an all Type B vent system. However, since there is no entry in Table 10.1 for a height of 12 ft, interpolation must be used. Read down the 4-in. diameter NAT Max column to the row associated with 10-ft height and 5-ft lateral to find the capacity value of 77,000 Btu/hr. Read further down to the 15-ft height, 5-ft lateral row to find the capacity value of 87,000 Btu/hr. The difference between the 15-ft height capacity value and the 10-ft height capacity value is 10,000 Btu/hr. The capacity for a vent system with a 12-ft height is equal to the capacity for a 10-ft height plus $2/5$ of the difference between the 10-ft and 15-ft height values, or $77,000 + 2/5 \times 10,000 = 81,000$ Btu/hr. Therefore, a 4-in. diameter vent can be used in the installation.

Table G.7 is an extract from code Table 10.1.

Table G.7 Extract from Code Table 10.1

Height H (ft)	Lateral L (ft)	Vent Diameter — D 4 in. FAN Min	Vent Diameter — D 4 in. FAN Max	Vent Diameter — D 4 in. NAT Max	Height H (ft)	Lateral L (ft)	Vent Diameter — D 4 in. FAN Min	Vent Diameter — D 4 in. FAN Max	Vent Diameter — D 4 in. NAT Max
10	0	0	175	100	15	0	0	191	112
	2	17	118	81		2	15	136	93
	5	32	113	77		5	30	130	87
	10	41	104	70		10	40	121	82
						15	48	112	76

G.2 Examples Using Common Venting Tables

G.2.1 Example 4: Common Venting Two Draft Hood–Equipped Appliances.

A 35,000-Btu/hr water heater is to be common vented with a 150,000-Btu/hr furnace, using a common vent with a total height of 30 ft. The connector rise is 2 ft for the water heater with a horizontal length of 4 ft. The connector rise for the furnace is 3 ft with a horizontal length of 8 ft. Assume single-wall metal connectors will be used with Type B vent. What size connectors and combined vent should be used in this installation? See Figure G.17.

Solution

Table 10.7 should be used to size single-wall metal vent connectors attached to Type B vertical vents.

Table G.8 is an extract from code Table 10.7.

In the vent connector capacity portion of Table 10.7, find the row associated with a 30-ft vent height. For a 2-ft rise on the vent connector for the water heater, read the shaded columns for draft hood-equipped appliances to find that a 3-in. diameter vent connector has a capacity of 37,000 Btu/hr. Therefore, a 3-in. single-wall metal vent connector can be used with the water heater. For a draft hood–equipped furnace with a 3-ft rise, read across the appropriate row to find that a 5-in. diameter vent connector has a maximum capacity of 120,000 Btu/hr (which is too small for the furnace) and a 6-in. diameter vent connector has a maximum vent capacity of 172,000 Btu/hr. Therefore, a 6-in. diameter vent connector should be used with the 150,000 Btu per hr furnace. Since both vent connector horizontal lengths are less than the maximum lengths listed in 10.2.2, the table values can be used without adjustments.

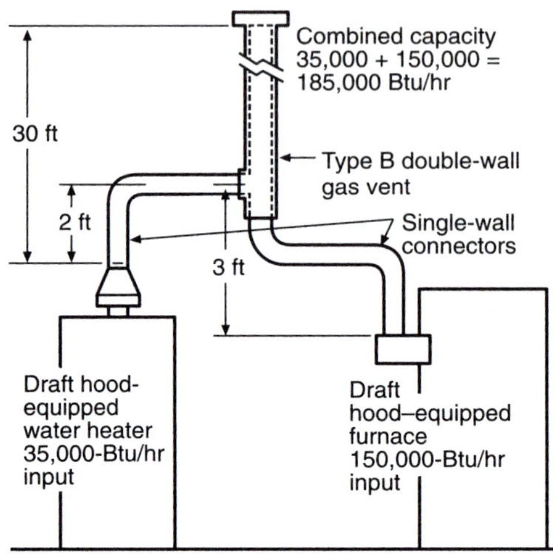

Figure G.17 *Common venting two draft hood–equipped appliances (Example 4).*

In the common vent capacity portion of Table 10.7, find the row associated with a 30-ft vent height and read over to the NAT + NAT portion of the 6-in. diameter column to find a maximum combined capacity of 257,000 Btu/hr. Since the two appliances total only 185,000 Btu/hr, a 6-in. common vent can be used.

Table G.8 is an extract from code Table 10.7. The common vent capacity is shown in the lower portion of the exhibit.

Table G.8 Extract from Code Table 10.7

Vent Connector Capacity

		Single-Wall Metal Vent Connector Diameter — D											
		3 in.			4 in.			5 in.			6 in.		
		Appliance Input Rating Limits in Thousands of Btu per Hour											
Height H	Rise R	FAN		NAT	FAN		NAT	FAN		NAT	FAN		NAT
(ft)	(ft)	Min	Max	Max	Min	Max	Max	Min	Max	Max	Min	Max	
30	1	47	60	31	77	110	57	112	175	89	169	278	129
	2	51	62	37	81	115	67	117	185	106	177	290	
	3	54	64	42	85	119	76	122	193	120	185	300	172

(continues)

Table G.8 Continued.

Common Vent Capacity

Vent Height H (ft)	Type B Double-Wall Vent Diameter — D								
	4 in.			5 in.			6 in.		
	Combined Appliance Input Rating in Thousands of Btu per Hour								
	FAN +FAN	FAN +NAT	NAT +NAT	FAN +FAN	FAN +NAT	NAT +NAT	FAN +FAN	FAN +NAT	NAT +NAT
15	121	108	88	189	159	140	275	221	200
20	131	118	98	208	177	156	305	247	223
30	145	132	113	236	202	180	350	286	257
50	159	145	128	268	233	208	406	337	296

G.2.2 Example 5(a): Common Venting of a Draft Hood–Equipped Water Heater with a Fan-Assisted Furnace into a Type B Vent.

In this case, a 35,000-Btu/hr input draft hood–equipped water heater with a 4-in. diameter draft hood outlet, 2 ft of connector rise, and 4 ft of horizontal length is to be common vented with a 100,000 Btu/hr fan-assisted furnace with a 4-in. diameter flue collar, 3 ft of connector rise, and 6 ft of horizontal length. The common vent consists of a 30-ft height of Type B vent. What are the recommended vent diameters for each connector and the common vent? The installer would like to use a single-wall metal vent connector. See Figure G.18.

Solution

(Table 10.7)

Water Heater Vent Connector Diameter. Since the water heater vent connector horizontal length of 4 ft is less than the maximum value listed in 10.7, the venting table values can be used without adjustments. Using the Vent Connector Capacity portion of Table 10.7, read down the Total Vent Height *(H)* column to 30 ft and read across the 2-ft Connector Rise *(R)* row to the first Btu/hr rating in the NAT Max column that is equal to or greater than the water heater input rating. The table shows that a 3-in. vent connector has a maximum input rating of 37,000 Btu/hr. Although this rating is greater than the water heater input rating, a 3-in. vent connector is prohibited by 10.2.19. A 4-in. vent connector has a maximum input rating of 67,000 Btu/hr and is equal to the draft hood outlet diameter. A 4-in. vent connector is selected. Since the water heater is equipped with a draft hood, there are no minimum input rating restrictions.

See Table G.9 for calculation of the water heater vent connector for Example 5 (a).

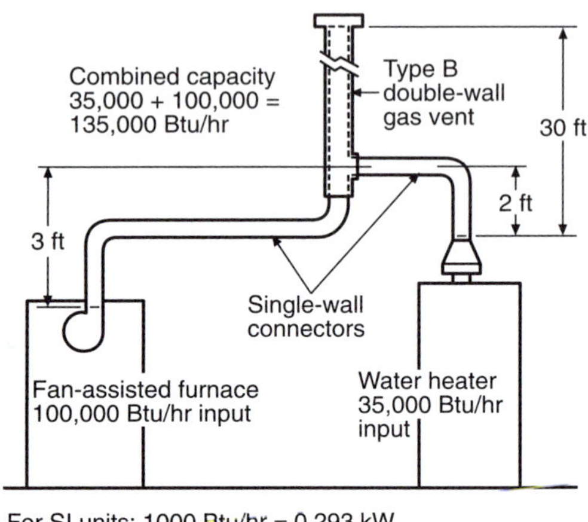

For SI units: 1000 Btu/hr = 0.293 kW

Figure G.18 *Common venting of a draft hood–equipped water heater with a fan-assisted furnace into a Type B double-wall common vent [Example 5(a)].*

Table G.9 Extract from Code Table 10.7

Vent Connector Capacity

Vent Height H (ft)	Connector Rise R (ft)	Single-Wall Metal Vent Connector Diameter — D					
		3 in.			4 in.		
		Appliance Input Rating Limits in Thousands of Btu per Hour					
		FAN		NAT	FAN		NAT
		Min	Max	Max	Min	Max	Max
30	1	47	60	31	77	110	57
	2	51	62	37	81	115	67
	3	54	64	42	85	119	76

Furnace Vent Connector Diameter. Using the Vent Connector Capacity portion of Table 10.7, read down the Total Vent Height *(H)* column to 30 ft and across the 3-ft Connector Rise *(R)* row. Since the furnace has a fan-assisted combustion system, find the first FAN Max column with a Btu/hr rating greater than the furnace input rating. The 4-in. vent connector has a maximum input rating of 119,000 Btu/hr and a minimum input rating of 85,000 Btu/hr.

See Table G.10.

Table G.10 Excerpt from Code Table 10.7

Vent Connector Capacity

Vent Height H (ft)	Connector Rise R (ft)	Single-Wall Metal Vent Connector Diameter — D							
		3 in.			4 in.			5 in.	
		Appliance Input Rating Limits in Thousands of Btu per Hour							
		FAN		NAT	FAN		NAT	FAN		NAT
		Min	Max	Max	Min	Max	Max	Min	Max	Max
30	1	47	60	31	77	110	57	112	175	89
	2	51	62	37	81	115	67	117	185	106
	3	54	64	42	85	119	76	122	193	120

The 100,000-Btu/hr furnace in this example falls within this range, so a 4-in. connector is adequate. Since the furnace vent connector horizontal length of 6 ft is less than the maximum value listed in Table 10.2.2, the venting table values can be used without adjustment. If the furnace had an input rating of 80,000 Btu/hr, then a Type B vent connector *[see Table 10.6]* would be needed in order to meet the minimum capacity limit.

Common Vent Diameter. The total input to the common vent is 135,000 Btu/hr. Using the Common Vent Capacity portion of Table 10.7, read down the Total Vent Height *(H)* column to 30 ft and across this row to find the smallest vent diameter in the FAN + NAT column that has a Btu/hr rating equal to or greater than 135,000 Btu/hr. The 4-in. common vent has a capacity of 132,000 Btu/hr and the 5-in. common vent has a capacity of 202,000 Btu/hr. Therefore, the 5-in. common vent should be used in this example.

See Table G.11.

Summary. In this example, the installer can use a 4-in. diameter, single-wall metal vent connector for the water heater and a 4-in. diameter, single-wall metal vent connector for the furnace. The common vent should be a 5-in. diameter Type B vent.

G.2.3 Example 5(b): Common Venting into an Interior Masonry Chimney.

In this case, the water heater and fan-assisted furnace of Example 5(a) are to be common-vented into a clay-tile-lined masonry chimney with a 30-ft height. The chimney is not exposed to the outdoors below the roof line. The internal dimensions of the clay tile liner are nominally 8 in. × 12 in. Assuming the same vent connector heights, laterals, and materials found in Example 5(a), what are the recommended vent connector diameters, and is this an acceptable installation?

Table G.11 Extract from Code Table 10.7

Common Vent Capacity

Vent Height *H* (ft)	Type B Double-Wall Vent Diameter — *D*								
	4 in.			5 in.			6 in.		
	Combined Appliance Input Rating in Thousands of Btu per Hour								
	FAN +FAN	FAN +NAT	NAT +NAT	FAN +FAN	FAN +NAT	NAT +NAT	FAN +FAN	FAN +NAT	NAT +NAT
15	121	108	88	189	159	140	275	221	200
20	131	118	98	208	177	156	305	247	223
30	145	132	113	236	202	180	350	286	257
50	159	145	128	268	233	208	406	337	296

Solution

Table 10.9 is used to size common venting installations involving single-wall connectors into masonry chimneys.

Water Heater Vent Connector Diameter. Using Table 10.9, Vent Connector Capacity, read down the Total Vent Height *(H)* column to 30 ft, and read across the 2-ft Connector Rise *(R)* row to the first Btu/hr rating in the NAT Max column that is equal to or greater than the water heater input rating. The table shows that a 3-in. vent connector has a maximum input of only 31,000 Btu/hr while a 4-in. vent connector has a maximum input of 57,000 Btu/hr. A 4-in. vent connector must therefore be used.

See Table G.12.

Furnace Vent Connector Diameter. Using the Vent Connector Capacity portion of Table 10.9, read down the Total Vent Height *(H)* column to 30 ft and across the 3-ft Connector Rise *(R)* row. Since the furnace has a fan-assisted combustion system, find the first FAN Max column with a Btu/hr rating greater than the furnace input rating. The 4-in. vent connector has a maximum input rating of 127,000 Btu/hr and a minimum input rating of 95,000 Btu/hr. The 100,000 Btu/hr furnace in this example falls within this range, so a 4-in. connector is adequate.

The portions of Table 10.9 used to calculate the diameter of the furnace vent connector are extracted in Table G.13.

Masonry Chimney. From Table G.2.4, the Equivalent Area for a Nominal Liner size of 8 in. × 12 in. is 63.6 in.^{2}. Using Table 10.9, Common Vent Capacity, read down the FAN + NAT column under the Minimum Internal Area of Chimney value of 63 to the row for 30-ft height

Table G.12 Extract from Code Table 10.9

Vent Connector Capacity

Vent Height *H* (ft)	Connector Rise *R* (ft)	Single-Wall Metal Vent Connector Diameter — *D*					
		3 in.			4 in.		
		Appliance Input Rating Limits in Thousands of Btu per Hour					
		FAN		NAT	FAN		NAT
		Min	Max	Max	Min	Max	Max
30	1	NA	NA	24	86	108	47
	2	NA	NA	31	91	119	57
	3	NA	NA	35	95	127	65

Table G.13 Extract from Code Table 10.9

Vent Connector Capacity

Vent Height *H* (ft)	Connector Rise *R* (ft)	Single-Wall Metal Vent Connector Diameter — *D*					
		3 in.			4 in.		
		Appliance Input Rating Limits in Thousands of Btu per Hour					
		FAN		NAT	FAN		NAT
		Min	Max	Max	Min	Max	Max
30	1	NA	NA	24	86	108	47
	2	NA	NA	31	91	119	57
	3	NA	NA	35	95	127	65

to find a capacity value of 739,000 Btu/hr. The combined input rating of the furnace and water heater, 135,000 Btu/hr, is less than the table value, so this is an acceptable installation.

The portions of Table G.2.4 and Table 10.9 used to answer the final portion of Example 5 (b) are extracted in Tables G.14 and G.15, respectively.

Section 10.2.15 requires the common vent area to be no greater than seven times the smallest listed appliance categorized vent area, flue collar area, or draft hood outlet area. Both appliances in this installation have 4-in. diameter outlets. From Table G.2.4, the equivalent area for an inside diameter of 4 in. is 12.2 in.^{2}. Seven times 12.2 equals 85.4, which is greater than 63.6, so this configuration is acceptable.

Table G.14 Extract from Code Table G.2.4

Nominal Liner Size (in.)	Inside Dimensions of Liner (in.)	Inside Diameter or Equivalent Diameter (in.)	Equivalent Area (in.^{2})
8 × 12	6$^1/_2$ × 10$^1/_2$	9	63.6

Table G.15 Extract from Code Table 10.9

Common Vent Capacity

Vent Height H (ft)	Minimum Internal Area of Masonry Chimney Flue (in.^{2})								
	50 in.			63 in.			78 in.		
	Combined Appliance Input Rating in Thousands of Btu per Hour								
	FAN +FAN	FAN +NAT	NAT +NAT	FAN +FAN	FAN +NAT	NAT +NAT	FAN +FAN	FAN +NAT	NAT +NAT
20	NA	503	240	765	661	321	947	849	415
30	NA	558	275	808	739	377	1052	957	490
50	NA	612	325	NA	821	456	1152	1076	600

G.2.4 Example 5(c): Common Venting into an Exterior Masonry Chimney.

In this case, the water heater and fan-assisted furnace of Examples 5(a) and 5(b) are to be common-vented into an exterior masonry chimney. The chimney height, clay-tile-liner dimensions, and vent connector heights and laterals are the same as in Example 5(b). This system is being installed in Charlotte, North Carolina. Does this exterior masonry chimney need to be relined? If so, what corrugated metallic liner size is recommended? What vent connector diameters are recommended? See Table G.2.4 and Figure G.19.

Table G.2.4 is included as a convenience. The Equivalent Diameter and Equivalent Area columns tabulate conversion between diameter and area of a circle for each whole inch diameter (8 in., 9 in., etc.). The table also offers suggestions for appropriate free areas to use for common clay tile sizes.

In practice, the dimensions of clay tile vary because of different manufacturing practices and tolerances. In addition, the internal corners of tile are not square. For this reason, the equivalent areas listed for the tiles are somewhat smaller than what would be calculated by the Inside Dimensions column.

Table G.2.4 Masonry Chimney Liner Dimensions with Circular Equivalents

Nominal Liner Size (in.)	Inside Dimensions of Liner (in.)	Inside Diameter or Equivalent Diameter (in.)	Equivalent Area (in.^{2})
4×8	$2^1/_2 \times 6^1/_2$	4	12.2
		5	19.6
		6	28.3
		7	38.3
8×8	$6^3/_4 \times 6^3/_4$	7.4	42.7
		8	50.3
8×12	$6^1/_2 \times 10^1/_2$	9	63.6
		10	78.5
12×12	$9^3/_4 \times 9^3/_4$	10.4	83.3
		11	95
12×16	$9^1/_2 \times 13^1/_2$	11.8	107.5
		12	113.0
		14	153.9
16×16	$13^1/_4 \times 13^1/_4$	14.5	162.9
		15	176.7
16×20	13×17	16.2	206.1
		18	254.4
20×20	$16^1/_2 \times 16^3/_4$	18.2	260.2
		20	314.1
20×24	$16^1/_2 \times 20^1/_2$	20.1	314.2
		22	380.1
24×24	$20^1/_4 \times 20^1/_4$	22.1	380.1
		24	452.3
24×28	$20^1/_4 \times 24^1/_4$	24.1	456.2
28×28	$24^1/_4 \times 24^1/_4$	26.4	543.3
		27	572.5
30×30	$25^1/_2 \times 25^1/_2$	27.9	607
		30	706.8
30×36	$25^1/_2 \times 31^1/_2$	30.9	749.9
		33	855.3
36×36	$31^1/_2 \times 31^1/_2$	34.4	929.4
		36	1017.9

For SI units, 1 in. = 25.4 mm; 1 in.^{2} = 645 mm^2.

Note: When liner sizes differ dimensionally from those shown in Table G.2.4, equivalent diameters can be determined from published tables for square and rectangular ducts of equivalent carrying capacity or by other engineering methods.

99% Winter Design Temperatures for the Contiguous United States

This map is a necessarily generalized guide to temperatures in the contiguous United States. Temperatures shown for areas such as mountainous regions and large urban centers may not be accurate. The data used to develop this map are from the 1993 *ASHRAE Fundamentals Handbook* (Chapter 24, Table 1: Climate Conditions for the United States).
For 99% winter design temperatures in Alaska, consult the *ASHRAE Fundamentals Handbook*.
99% winter design temperatures for Hawaii are greater than 37°F.

Figure G.19 *Range of winter design temperatures used in analyzing exterior masonry chimneys in the United States.*

Note that Table G.2.4 is in an informational appendix and is not mandated. In the field, it may be advantageous to carefully measure the actual free area of the tile.

Solution

According to 10.2.18, Type B vent connectors are required to be used with exterior masonry chimneys. Use Tables 10.13(a) and (b) to size FAN+NAT common venting installations involving Type-B double wall connectors into exterior masonry chimneys.

The local 99-percent winter design temperature needed to use Tables 10.13(a) and (b) can be found in *ASHRAE Handbook — Fundamentals*. For Charlotte, North Carolina, this design temperature is 19°F.

Chimney Liner Requirement. As in Example 5(b), use the 63 in.2 Internal Area columns for this size clay tile liner. Read down the 63 in.2 column of Table 10.13(a) to the 30-ft height row to find that the Combined Appliance Maximum Input is 747,000 Btu/hr.

See Table G.16.

Table G.16 Extract from Code Table 10.13 (a)

Vent Height H (ft)	Internal Area of Chimney (in.^{2})			
	38	50	63	78
15	334	467	611	781
20	368	508	668	858
30	404	564	747	969
50	NA	NA	831	1089

The combined input rating of the appliances in this installation, 135,000 Btu/hr, is less than the maximum value, so this criterion is satisfied. Table 10.13(b), at a 19°F Design Temperature, and at the same Vent Height and Internal Area used earlier, shows that the minimum allowable input rating of a space-heating appliance is 470,000 Btu/hr. The furnace input rating of 100,000 Btu/hr is less than this minimum value. So this criterion is not satisfied, and an alternative venting design needs to be used, such as a Type B vent shown in Example 5(a) or a listed chimney liner system shown in the remainder of the example.

According to 10.2.17, Tables 10.6 or 10.7 are used for sizing corrugated metallic liners in masonry chimneys, with the maximum common vent capacities reduced by 20 percent. This example will be continued assuming Type B vent connectors.

Water Heater Vent Connector Diameter. Using Table 10.6, Vent Connector Capacity, read down the Total Vent Height *(H)* column to 30 ft, and read across the 2-ft Connector Rise *(R)* row to the first Btu/hour rating in the NAT Max column that is equal to or greater than the water heater input rating. The table shows that a 3-in. vent connector has a maximum capacity of 39,000 Btu/hr. So the 35,000 Btu/hr water heater in this example can use a 3-in. connector.

The portions of Table 10.6 used to calculate the diameter of a single-wall vent connector and the furnace vent connector for Example 5 (c) are excerpted in Table G.17.

Furnace Vent Connector Diameter. Using Table 10.6, Vent Connector Capacity, read down the Total Vent Height *(H)* column to 30 ft, and read across the 3-ft Connector Rise *(R)* row to the first Btu/hr rating in the FAN Max column that is equal to or greater than the furnace input rating. The 100,000-Btu/hr furnace in this example falls within this range, so a 4-in. connector is adequate.

The portions of Table 10.6 used to calculate the diameter of the chimney liner for Example 5 (c) are excerpted in Table G.18.

Chimney Liner Diameter. The total input to the common vent is 135,000 Btu/hr. Using the Common Vent Capacity Portion of Table 10.6, read down the Total Vent Height *(H)* column to 30 ft and across this row to find the smallest vent diameter in the FAN+NAT column that has a Btu/hr rating greater than 135,000 Btu/hr. The 4-in. common vent has a capacity of 138,000 Btu/hr. Reducing the maximum capacity by 20 percent *(10.2.17)* results in a maximum capacity for a 4-in. corrugated liner of 110,000 Btu/hr, less than the total input of

Table G.17 Extract from Code Table 10.6

Vent Connector Capacity

Vent Height H (ft)	Connector Rise R (ft)	Single-Wall Metal Vent Connector Diameter — D					
		3 in.			4 in.		
		Appliance Input Rating Limits in Thousands of Btu per Hour					
		FAN		NAT	FAN		NAT
		Min	Max	Max	Min	Max	Max
30	1	20	62	33	31	113	59
	2	21	64	39	33	118	70
	3	22	66	44	34	123	79

Table G.18 Extract from Code Table 10.6

Common Vent Capacity

Vent Height H (ft)	Type B Double-Wall Vent Diameter — D								
	4 in.			5 in.			6 in.		
	Combined Appliance Input Rating in Thousands of Btu per Hour								
	FAN +FAN	FAN +NAT	NAT +NAT	FAN +FAN	FAN +NAT	NAT +NAT	FAN +FAN	FAN +NAT	NAT +NAT
15	125	112	91	195	164	144	283	228	206
20	136	123	102	215	183	160	314	255	229
30	152	138	118	244	210	185	361	297	266
50	167	153	134	279	244	214	421	353	310

135,000 Btu/hr. So a larger liner is needed. The 5-in. common vent capacity listed in Table 10.6 is 210,000 Btu/hr, and after reducing by 20 percent is 168,000 Btu/hr. Therefore, a 5-in. corrugated metal liner should be used in this example.

Single Wall Connectors. Once it has been established that relining the chimney is necessary, Type B double wall vent connectors are not specifically required. This example could be redone using Table 10.7 for single-wall vent connectors. For this case, the vent connector and liner diameters would be the same as found for Type B double-wall connectors.

Recommended Procedure for Safety Inspection of an Existing Appliance Installation

This appendix is not a part of the requirements of this code but is included for informational purposes only.

H.1 General

The following procedure is intended as a guide to aid in determining that an appliance is properly installed and is in a safe condition for continuing use.

This procedure is predicated on central furnace and boiler installations, and it should be recognized that generalized procedures cannot anticipate all situations. Accordingly, in some cases, deviation from this procedure is necessary to determine safe operation of the equipment.

(1) This procedure should be performed prior to any attempt to modify the appliance or the installation.
(2) If it is determined a condition that could result in unsafe operation exists, the appliance should be shut off and the owner advised of the unsafe condition.

The following steps should be followed in making the safety inspection:

(1) Conduct a test for gas leakage. *(See 5.5.4.)*
(2) Visually inspect the venting system for proper size and horizontal pitch, and determine there is no blockage or restriction, leakage, corrosion, and other deficiencies that could cause an unsafe condition.

(3) Shut off all gas to the appliance, and shut off any other fuel-gas-burning appliance within the same room. **Use the shutoff valve in the supply line to each appliance.**

(4) Inspect burners and crossovers for blockage and corrosion.

(5) **Applicable only to furnaces:** Inspect the heat exchanger for cracks, openings, or excessive corrosion.

(6) **Applicable only to boilers:** Inspect for evidence of water or combustion product leaks.

(7) Insofar as is practical, close all building doors and windows and all doors between the space in which the appliance is located and other spaces of the building. Turn on clothes dryers. Turn on any exhaust fans, such as range hoods and bathroom exhausts, so they will operate at maximum speed. Do not operate a summer exhaust fan. Close fireplace dampers. If, after completing Steps 8 through 13, it is believed sufficient combustion air is not available, refer to Section 5.3 of this code for guidance.

(8) Place the appliance being inspected in operation. **Follow the lighting instructions.** Adjust the thermostat so appliance will operate continuously.

(9) Determine that the pilot(s), where provided, is burning properly and that the main burner ignition is satisfactory by interrupting and re-establishing the electrical supply to the appliance in any convenient manner. If the appliance is equipped with a continuous pilot(s), test the pilot safety device(s) to determine whether it is operating properly by extinguishing the pilot(s) when the main burner(s) is off and determining, after 3 minutes, that the main burner gas does not flow upon a call for heat. If the appliance is not provided with a pilot(s), test for proper operation of the ignition system in accordance with the appliance manufacturer's lighting and operating instructions.

(10) Visually determine that the main burner gas is burning properly (i.e., no floating, lifting, or flashback). Adjust the primary air shutter(s) as required.

(11) If the appliance is equipped with high and low flame controlling or flame modulation, check for proper main burner operation at low flame.

(12) Test for spillage at the draft hood relief opening after 5 minutes of main burner operation. Use a flame of a match or candle or smoke from a cigarette, cigar, or pipe.

(13) Turn on all other fuel-gas-burning appliances within the same room so they will operate at their full inputs. **Follow lighting instructions for each appliance.**

(14) Repeat Steps 10 and 11 on the appliance being inspected.

(15) Return doors, windows, exhaust fans, fireplace dampers, and any other fuel-gas-burning appliance to their previous conditions of use.

(16) **Applicable only to furnaces:** Check both the limit control and the fan control for proper operation. Limit control operation can be checked by blocking the circulating air inlet or temporarily disconnecting the electrical supply to the blower motor and determining that the limit control acts to shut off the main burner gas.

(17) **Applicable only to boilers:** Determine that the water pumps are in operating condition. Test low water cutoffs, automatic feed controls, pressure and temperature limit controls, and relief valves in accordance with the manufacturer's recommendations to determine that they are in operating condition.

Recommended Procedure for Installing Electrically Operated Automatic Vent Damper Devices in Existing Vents

This appendix is not a part of the requirements of this code but is included for informational purposes only.

I.1 General

This procedure is intended as a guide to aid in safely installing an electrically operated automatic vent damper device in the vent serving an existing appliance.

This procedure is based on the assumption that the history of the specific appliance has been one of safe and satisfactory operation.

This procedure is predicated on central furnace, boiler, and water heater installations, and it should be recognized that generalized procedures cannot anticipate all situations. Accordingly, in some cases, deviation from this procedure is necessary to determine safe operation of the equipment.

The following steps should be followed in making the modifications:

(1) Perform a safety inspection of the existing appliance installation. See Appendix H for a recommended procedure for such a safety inspection.
(2) Shut off all gas and electricity to the appliance. **To shut off the gas, use the shutoff valve in the supply line to the appliance.**
(3) Install the damper device in strict accordance with the manufacturer's installation instructions. Make certain the device is not located in that portion of the venting system that serves any appliance other than the one for which the damper is installed.

(4) Make certain wiring connections are tight and wires are positioned and secured so they will not be able to contact high-temperature locations.

(5) Where an additional automatic valve has been incorporated or an existing gas control replaced, conduct a gas leakage test of the appliance piping and control system downstream of the shutoff valve in the supply line to the appliance.

(6) Visually inspect the modified venting system for proper horizontal pitch.

(7) Check that the damper operates properly and is properly sequenced with the appliance operating controls so that it opens when there is a call for heat and closes when the appliance is in a standby condition.

NOTE: If a boiler gas valve is sequenced by the aquastat, determine that the damper opens prior to the opening of the gas valve.

(8) Determine the amperage draw of the gas control circuit and damper device.

    a. Check the appliance transformer for adequate capacity.

    b. Check the heat anticipator in the comfort thermostat to determine that it is properly adjusted.

(9) Sequence the appliance through at least three normal operating cycles.

(10) Insofar as is practical, close all building doors and windows and all doors between the space in which the appliance is located and other spaces of the building. Turn on clothes dryers. Turn on any exhaust fans, such as range hoods and bathroom exhausts, so they will operate at maximum speed. Do not operate a summer exhaust fan. Close fireplace dampers.

(11) Place the appliance in operation. **Follow the lighting instructions.** Adjust the thermostat so the appliance will operate continuously.

(12) Test for spillage at the draft hood relief opening after 5 minutes of main burner operation. Use a flame of a match or candle or smoke from a cigarette, cigar, or pipe.

(13) Visually determine that the main burner gas is burning properly — that is, no floating, lifting, or flashback. Adjust the primary air shutter(s) as required.

    If the appliance is equipped with high and low flame controlling or flame modulation, check for proper main burner operation at low flame.

(14) Determine that the pilot(s), where provided, is burning properly and that the main burner ignition is satisfactory by interrupting and re-establishing the electrical supply to the appliance in any convenient manner. If the appliance is equipped with a continuous pilot(s), test the pilot safety device(s) to determine whether it is operating properly by extinguishing the pilot(s) when the main burner(s) is off and determining, after 3 minutes, that the main burner gas does not flow upon a call for heat. If this appliance is not provided with a pilot(s), test for proper operation of the ignition system in accordance with the appliance manufacturer's lighting and operating instructions.

(15) **Applicable only to furnaces:** Check both the limit control and the fan control for proper operation. Limit control operation can be checked by blocking the circulating air inlet or temporarily disconnecting the electrical supply to the blower motor and determining that the limit control acts to shut off the main burner gas.

(16) **Applicable only to boilers**:

    a. Determine that the water pumps are in operating condition.

    b. Test low water cutoffs, automatic feed controls, pressure and temperature limit controls, and relief valves in accordance with the manufacturer's recommendations to determine they are in operating condition.

(17) Label the damper device with information on the following:

    a. Name of qualified agency responsible for damper installation

    b. Date of installation

Recommended Procedure for Installing Mechanically Actuated Automatic Vent Damper Devices in Existing Vents

This appendix is not a part of the requirements of this code but is included for informational purposes only.

J.1 General

This procedure is intended as a guide to aid in safely installing a mechanically actuated automatic vent damper device in the vent serving an existing appliance.

This procedure is based on the assumption that the history of the specific appliance has been one of safe and satisfactory operation.

This procedure is predicated on central furnace, boiler, and water heater installations, and it should be recognized that generalized procedures cannot anticipate all situations. Accordingly, in some cases, deviation from this procedure is necessary to determine safe operation of the equipment.

The following steps should be followed in making the modifications:

(1) Perform a safety inspection of the existing appliance installation. See Appendix H for a recommended procedure for such a safety inspection.
(2) Shut off all gas and electricity to the appliance. **To shut off the gas, use the shutoff valve in the supply line to the appliance.**
(3) Install the damper device in strict accordance with the manufacturer's installation instructions. Make certain the device is not located in that portion of the venting system that serves any appliance other than the one for which the damper is installed.

(4) Make certain wiring and mechanical connections are tight and wires are positioned and secured so they will not be able to contact high-temperature locations.

(5) Where an additional automatic valve has been incorporated or an existing gas control replaced, conduct a gas leakage test of the appliance piping and control system downstream of the shutoff valve in the supply line to the appliance.

(6) Visually inspect the modified venting system for proper horizontal pitch.

(7) Check that the damper operates properly and is properly sequenced with the appliance operating controls so that it opens when there is a call for heat and closes when the appliance is in a standby condition.

NOTE: If a boiler gas valve is sequenced by the aquastat, determine that the damper opens prior to the opening of the gas valve.

(8) Determine the amperage draw of the gas control circuit and damper device.

    a. Check the appliance transformer for adequate capacity.

    b. Check the heat anticipator in the comfort thermostat to determine it is properly adjusted.

(9) Sequence the appliance through at least three normal operating cycles.

(10) Insofar as is practical, close all building doors and windows and all doors between the space in which the appliance is located and other spaces of the building. Turn on clothes dryers. Turn on any exhaust fans, such as range hoods and bathroom exhausts, so they will operate at maximum speed. Do not operate a summer exhaust fan. Close fireplace dampers.

(11) Place appliance in operation. **Follow the lighting instructions.** Adjust the thermostat so the appliance will operate continuously.

(12) Test for spillage at the draft hood relief opening after 5 minutes of main burner operation. Use a flame of a match or candle or smoke from a cigarette, cigar, or pipe.

(13) Visually determine that the main burner gas is burning properly (i.e., no floating, lifting, or flashback). Adjust the primary air shutter(s) as required.

    If the appliance is equipped with high and low flame controlling or flame modulation, check for proper main burner operation at low flame.

(14) Determine that the pilot(s), where provided, is burning properly and that the main burner ignition is satisfactory by interrupting and re-establishing the electrical supply to the appliance in any convenient manner. If the appliance is equipped with a continuous pilot(s), test the pilot safety device(s) to determine whether it is operating properly by extinguishing the pilot(s) when the main burner(s) is off and determining, after 3 minutes, that the main burner gas does not flow upon a call for heat. If this appliance is not provided with a pilot(s), test for proper operation of the ignition system in accordance with the appliance manufacturer's lighting and operating instructions.

(15) **Applicable only to furnaces:** Check both the limit control and the fan control for proper operation. Limit control operation can be checked by blocking the circulating air inlet or temporarily disconnecting the electrical supply to the blower motor and determining that the limit control acts to shut off the main burner gas.

(16) **Applicable only to boilers**:

    a. Determine that the water pumps are in operating condition.

    b. Test low-water cutoffs, automatic feed controls, pressure and temperature limit controls, and relief valves in accordance with the manufacturer's recommendations to determine they are in operating condition.

(17) Label the damper device with information on the following:

    a. Name of the qualified agency responsible for the damper installation

    b. Date of installation

Recommended Procedure for Installing Thermally Actuated Automatic Vent Damper Devices in Existing Vents

This appendix is not a part of the requirements of this code but is included for informational purposes only.

K.1 General

This procedure is intended as a guide to aid in safely installing a thermally actuated automatic vent damper device in the vent serving an existing appliance.

This procedure is based on the assumption that the history of the specific appliance has been one of safe and satisfactory operation.

This procedure is predicated on central furnace, boiler, and water heater installations, and it should be recognized that generalized procedures cannot anticipate all situations. Accordingly, in some cases, deviation from this procedure is necessary to determine safe operation of the equipment.

The following steps should be followed in making the modifications:

(1) Perform a safety inspection of the existing appliance installation. See Appendix H for a recommended procedure for such a safety inspection.
(2) Shut off all gas and electricity to the appliance. **To shut off the gas, use the shutoff valve in the supply line to the appliance.**
(3) Install the damper device in strict accordance with the manufacturer's installation instructions. Make certain the device is not located in that portion of the venting system that serves any appliance other than the one for which the damper is installed.

(4) Make certain that wiring connections are tight and wires are positioned and secured so they will not be able to contact high-temperature locations.

(5) Where an additional automatic valve has been incorporated or an existing gas control replaced, conduct a gas leakage test of the appliance piping and control system downstream of the shutoff valve in the supply line to the appliance.

(6) Visually inspect the modified venting system for proper horizontal pitch.

(7) Check the installation to determine there is no physical obstruction or deformation that could impair damper operation.

(8) Determine the amperage draw of the gas control circuit and damper device.

    a. Check the appliance transformer for adequate capacity.

    b. Check the heat anticipator in the comfort thermostat to determine that it is properly adjusted.

(9) Sequence the appliance through at least three normal operating cycles.

(10) Insofar as is practical, close all building doors and windows and all doors between the space in which the appliance is located and other spaces of the building. Turn on clothes dryers. Turn on any exhaust fans, such as range hoods and bathroom exhausts, so they will operate at maximum speed. Do not operate a summer exhaust fan. Close fireplace dampers.

(11) Place the appliance in operation. **Follow the lighting instructions.** Adjust the thermostat so the appliance will operate continuously.

(12) Test for spillage at the draft hood relief opening after 5 minutes of main burner operation. Use a flame of a match or candle or smoke from a cigarette, cigar, or pipe.

(13) Visually determine that the main burner gas is burning properly (i.e., no floating, lifting, or flashback). Adjust the primary air shutter(s) as required. If the appliance is equipped with high and low flame controlling or flame modulation, check for proper main burner operation at low flame.

(14) Determine that the pilot(s), where provided, is burning properly and that the main burner ignition is satisfactory by interrupting and re-establishing the electrical supply to the appliance in any convenient manner. If the appliance is equipped with a continuous pilot(s), test the pilot safety device(s) to determine whether it is operating properly by extinguishing the pilot(s) when the main burner(s) is off and determining, after 3 minutes, that the main burner gas does not flow upon a call for heat. If this appliance is not provided with a pilot(s), test for proper operation of the ignition system in accordance with the appliance manufacturer's lighting and operating instructions.

(15) Label the damper device with information on the following:

    a. Name of the qualified agency responsible for the damper installation

    b. Date of installation

Example of Air Opening Design for Combustion and Ventilation

This appendix is not a part of the requirements of this code but is included for informational purposes only.

This appendix was added in the 1999 edition of the code to provide a calculation example for a new method of supplying air for combustion and ventilation to confined spaces, which was added to Section 5.3.

L.1 Example of Combination Indoor and Outdoor Combustion Air Opening Design [5.3.3 (c)]

Determine the required combination of indoor and outdoor combustion air opening sizes for the following equipment installation example.

Example Installation.

A furnace and a water heater with the following inputs are located in a 10 ft × 10 ft equipment room with an 8-ft ceiling. An adjacent room that measures 15 ft × 20 ft with an 8-ft ceiling can be used to help meet the equipment combustion air needs. The house construction is not unusually tight.

Furnace input: 100,000 Btu/hr

Water heater input: 40,000 Btu/hr

Solution

(1) Determine the total equipment input and the total available room volumes:

Total equipment input: 100,000 Btu/hr + 40,000 Btu/hr = 140,000 Btu/hr

Equipment room volume: 10 ft × 10 ft with 8-ft ceiling = 800 ft^3

Adjacent room volume: 10 ft × 15 ft with 8-ft ceiling = 2400 ft^3

Total indoor room volume: 800 ft^3 + 2400 ft^3 = 3200 ft^3

(2) Determine whether location is an unconfined space:

Total equipment input = 140,000 Btu/hr

Unconfined space determinations: 140,000 Btu × 50 ft^3/1000 Btu/hr = 7000 ft^3

Conclusion: Equipment is located in a confined space since the total of 3200 ft^3 does not meet the unconfined criteria of 7000 ft^3.

(3) Determine ratio of actual confined volume to required unconfined volume, actual confined/required unconfined ratio:

$$\frac{3200 \text{ ft}^3}{7000 \text{ ft}^3} = 0.46$$

(4) Determine indoor combustion air opening size for openings communicating with adjacent indoor spaces as though all combustion air is to come from the indoors, indoor opening (high and low) sizes:

$$\frac{140,000 \text{ Btu/hr}}{1000 \text{ Btu/in.}^2} = 140 \text{ in.}^2$$

(5) Determine the outdoor combustion air opening (upper and lower) size as if all combustion air is to come from the outdoors. In this example, the combustion air openings directly communicate with the outdoors.

$$\frac{140,000 \text{ Btu/hr}}{4000 \text{ Btu/in.}^2} = 35 \text{ in.}^2$$

(6) Determine outdoor combustion air opening area ratio:

Subsection 5.3.3(c)6 requires the sum of the indoor and outdoor ratios must be equal to or greater than 1.00, outdoor combustion opening ratio:

$$1.00 - 0.46 \text{ (from Step 3)} = 0.54$$

(7) Determine reduced outdoor combustion air opening area:

$$\text{Opening area} = 0.54 \text{ (from Step 6)} \times 35 \text{ in.}^2$$

Each opening area = 19 in.^{2}

(8) Final design summary:

One high and one low indoor opening each sized 140 in.^{2}

One high and one low outdoor opening each sized 19 in.^{2}

The minimum dimension of the air opening should not be less than 3 in. (76 mm).

Referenced Publications

M.1

This list of documents is included because they pertain to equipment, accessories, materials, and installations that can be encountered in the application of this code and thus contain useful information. The documents are not considered part of the requirements of this code even though some are referenced in conjunction with the requirements in the code.

M.1.1 API Publication.

American Petroleum Institute, 1220 L Street, N.W., Washington, DC 20005.

API 1104, *Standard for Welding Pipelines and Related Facilities*, 1999.

M.1.2 ASHRAE Publications.

American Society of Heating, Refrigerating and Air Conditioning Engineers, Inc., 1791 Tullie Circle, N.E., Atlanta, GA 30329-2305.

ASHRAE Handbook — Fundamentals, 1997.
ASHRAE Handbook — HVAC Systems and Equipment, 1999.

M.1.3 ASME Publications.

American Society of Mechanical Engineers, United Engineering Center, 345 East 47th Street, New York, NY 10017.

ASME *Boiler and Pressure Vessel Code,* Section III ("Nuclear Power Plant Components"), Division 1-Subsections NB, NC, and ND, Article 3600, 1998.

ASME *Boiler and Pressure Vessel Code,* Section IX and Section IV, 1998.

ASME B1.20.1, *Pipe Threads, General Purpose, Inch,* 1983 (Reaffirmed: 1992).

ASME B16.1, *Cast Iron Pipe Flanges and Flanged Fittings, Class 25, 125, 250, and 800,* 1998.

ASME B16.5, *Pipe Flanges and Flanged Fittings,* 1996.

ASME B31.1, *Power Piping,* 1998.

ASME B31.3, *Process Piping,* 1999.

ASME B31.4, *Liquid Transportation Systems for Liquid Hydrocarbons and Other Liquids,* 1998.

ASME B31.5, *Refrigeration Piping,* 1992.

ASME B31.8, *Gas Transmission and Distribution Piping Systems,* 1995.

ASME B36.10, *Welded and Seamless Wrought-Steel Pipe,* 1996.

M.1.4 ASTM Publications.

American Society for Testing and Materials, 100 Barr Harbor Drive, West Conshohocken, PA 19428-2959.

ASTM A 53, *Standard Specification for Pipe, Steel, Black and Hot-Dipped, Zinc Coated Welded and Seamless,* 1999.

ASTM A 106, *Standard Specification for Seamless Carbon Steel Pipe for High-Temperature Service,* 1999.

ASTM A 254, *Standard Specification for Copper Brazed Steel Tubing,* 1997.

ASTM A 539, *Standard Specification for Electric Resistance-Welded Coiled Steel Tubing for Gas and Fuel Oil Lines,* 1999.

ASTM B 88, *Specification for Seamless Copper Water Tube,* 1996.

ASTM B 210, *Specification for Aluminum-Alloy Drawn Seamless Tubes,* 1995.

ASTM B 241, *Specification for Aluminum-Alloy Seamless Pipe and Seamless Extruded Tube,* 1996.

ASTM B 280, *Specification for Seamless Copper Tube for Air Conditioning and Refrigeration Field Service,* 1997.

ASTM D 2513, *Standard Specification for Thermoplastic Gas Pressure Pipe, Tubing, and Fittings,* 1999.

ASTM D 2517, *Specification for Reinforced Epoxy Resin Gas Pressure Pipe and Fittings,* 1994.

ANSI/ASTM D 2385, *Method of Test for Hydrogen Sulfide and Mercaptan Sulfur in Natural Gas (Cadmium Sulfate — Iodometric Titration Method),* 1981.

ANSI/ASTM D 2420, *Method of Test for Hydrogen Sulfide in Liquefied Petroleum (LP) Gases (Lead Acetate Method),* 1991 (Reaffirmed 1996).

M.1.5 AWS Publications.

American Welding Society, 550 N. W. LeJeune Road, Miami, FL 33126.

AWS B2.1, *Standard for Welding Procedure and Performance Qualification,* 1998.

AWS B2.2, *Standard for Brazing Procedure and Performance Qualification,* 1991.

M.1.6 CSA International Publications.

CSA International, 8501 East Pleasant Valley Road, Cleveland, OH 44131.

AGA NGV2, *Basic Requirements for Compressed Natural Gas Vehicle (NGV) Containers,* 1992.

AGA/CGA NGVI, *Compressed Natural Gas Vehicle (NGV) Fueling Connection Devices,* 1994.

AGA/CGA NGV3.1, *Fuel System Components for Natural Gas Powered Vehicles,* 1995.

ANSI LC 1/CSA 6.26, *Gas Piping Systems Using Corrugated Stainless Steel Tubing,* 1997.

ANSI LC 2, *Agricultural Heaters,* 1996.

ANSI Z21.1, *Household Cooking Gas Appliances,* 1996.

ANSI Z21.5.1/CGA 7.1, *Gas Clothes Dryers — Volume I — Type 1 Clothes Dryers,* 1995.

ANSI Z21.5.2, 7.2, *Gas Clothes Dryers — Volume II — Type 2 Clothes Dryers,* 1998.

ANSI Z21.10.1/CSA 4.1, *Gas Water Heaters — Volume I — Storage, Water Heaters with Input Ratings of 75,000 Btu per Hour or Less,* 1998.

ANSI Z21.10.3/CSA 4.3, *Gas Water Heaters — Volume III — Storage, Water with Input Ratings above 75,000 Btu per Hour, Circulating and Instantaneous,* 1998.

ANSI Z21.11.1, *Gas-Fired Room Heaters — Volume I — Vented Room Heaters,* 1991. (Included in Z21.86/CSA 2.32 after 1/1/ 98.)

ANSI Z21.11.2, *Gas-Fired Room Heaters — Volume II — Unvented Room Heaters,* 1996.

ANSI Z21.12, *Draft Hoods,* 1990 (Reaffirmed 1998).

ANSI Z21.13, *Gas-Fired Low-Pressure Steam and Hot Water Boilers,* 1991 (Reaffirmed 1998).

ANSI Z21.15/CGA 9.1, *Manually Operated Gas Valves for Appliances, Appliance Connector Valves, and Hose End Valves,* 1997.

ANSI Z21.17/CSA 2.7, *Domestic Gas Conversion Burners,* 1998.

ANSI Z21.18/CGA 6.3, *Gas Appliance Pressure Regulators,* 1995.

ANSI Z21.19, *Refrigerators Using Gas Fuel,* 1990 (Reaffirmed 1999).

ANSI Z21.20, *Automatic Gas Ignition Systems and Components,* 1997.

ANSI Z21.21/CGA 6.5, *Automatic Valves for Gas Appliances,* 1997.

ANSI Z21.22, *Relief Valves and Automatic Gas Shutoff Devices for Hot Water Supply Systems,* 1986 (Reaffirmed 1998).

ANSI Z21.23, *Gas Appliance Thermostats,* 1993 (Reaffirmed 1998).

ANSI Z21.24/CGA 6.10, *Metal Connectors for Gas Appliances,* 1997.

ANSI Z21.35/CGA 6.8, *Pilot Gas Filters,* 1995.

ANSI Z21.40.1/CGA 2.91-M99, *Gas-Fired Absorption Summer Air Conditioning Appliances,* 1996.

ANSI Z21.40.2/CGA 2.92, *Gas-Fired Work Activated Air-Conditioning and Heat Pump Appliances (Internal Combustion),* 1996.

ANSI Z21.40.4/CGA 2.94, *Performance Testing and Rating of Gas Fired, Air-Conditioning and Heat Pump Appliances,* 1996.

ANSI Z21.41/CGA 6.9, *Quick-Disconnect Devices for Use With Gas Fuel,* 1998.

ANSI Z21.42, *Gas-Fired Illuminating Appliances,* 1993 (Reaffirmed 1998).

ANSI Z21.47/CSA 2.3, *Gas-Fired Central Furnaces,* 1999.

ANSI Z21.48, *Gas-Fired Gravity and Fan Type Floor Furnaces,* 1992 (Included in Z21.86/CSA 2.32 after 1/1/98).

ANSI Z21.49, *Gas-Fired Gravity and Fan Type Vented Wall Furnaces,* 1993 (Included in Z21.86/CSA 2.32 after 1/1/98).

ANSI Z21.50/CSA 2.22, *Vented Gas Fireplaces,* 1998.

ANSI Z21.54/CSA 8.4, *Gas Hose Connectors for Portable Outdoor Gas-Fired Appliances,* 1996.

ANSI Z21.56/CSA 4.7, *Gas-Fired Pool Heaters,* 1998.

ANSI Z21.57, *Recreational Vehicle Cooking Gas Appliances,* 1993 (Reaffirmed 1998).

ANSI Z21.58/CGA 1.6, *Outdoor Cooking Gas Appliances,* 1995.

ANSI Z21.60/CGA 2.26, *Decorative Gas Appliances for Installation in Solid-Fuel Burning Fireplaces,* 1996.

ANSI Z21.61, *Gas-Fired Toilets,* 1983 (Reaffirmed 1996).

ANSI Z21.66/CGA 6.14, *Automatic Vent Damper Devices for Use with Gas-Fired Appliances,* 1996.

ANSI Z21.69/CSA 6.16, *Connectors for Movable Gas Appliances,* 1997.

ANSI Z21.71, *Automatic Intermittent Pilot Ignition Systems for Field Installations,* 1993 (Reaffirmed 1998).

ANSI Z21.77/CGA 6.23, *Manually-Operated Piezo-Electric Spark Gas Ignition Systems and Components,* 1995.

ANSI Z21.78, *Combination Gas Controls for Gas Appliances,* 1995.

ANSI Z21.86/CSA 2.32, *Vented Gas-Fired Space Heating Equipment,* 1998.

ANSI Z21.88/CSA 2.33, *Vented Gas-Fireplace Heaters,* 1998.

ANSI Z83.3, *Gas Utilization Equipment in Large Boilers,* 1971 (Reaffirmed 1995).

ANSI Z83.4, *Direct Gas-Fired Make-Up Air Heaters,* 1991 (Reaffirmed 1998).

ANSI Z83.6, *Gas-Fired Infrared Heaters,* 1990 (Reaffirmed 1998).

ANSI Z83.8/CGA 2.6, *Gas Fired Duct Furnaces and Unit Heaters,* 1996.

ANSI Z83.11/CGA 1.8, *Food Service Equipment,* 1996.

IAS U.S. 7, *Requirements for Gas Convenience Outlets and Optional Enclosures,* 1990.

IAS U.S. 9, *Requirements for Gas-Fired, Desiccant Type Dehumidifiers and Central Air Conditioners,* 1990.

IAS U.S. 42, *Requirement for Gas Fired Commercial Dishwashers,* 1992.

IAS NGV2, *Basic Requirements for Compressed Natural Gas Vehicle (NGV) Containers,* 1998.

M.1.7 MSS Publications.

Manufacturers Standardization Society of the Valve and Fittings Industry, 5203 Leesburg Pike, Suite 502, Falls Church, VA 22041.

MSS SP-6, *Standard Finishes for Contact Faces of Pipe Flanges and Connecting-End Flanges of Valves and Fittings,* 1999.

ANSI/MSS SP-58, *Pipe Hangers and Supports — Materials, Design and Manufacture,* 1993.

M.1.8 NACE Publication.

National Association of Corrosion Engineers, 1440 South Creek Drive, Houston, TX 77084.

NACE RP 0169, *Control of External Corrosion on Underground or Submerged Metallic Piping Systems,* 1996.

M.1.9 NFPA Publications.

National Fire Protection Association, 1 Batterymarch Park, P.O. Box 9101, Quincy, MA 02269-9101.

NFPA 30, *Flammable and Combustible Liquids Code*, 1996 edition.

NFPA 59, *Standard for the Storage and Handling of Liquefied Petroleum Gases at Utility Gas Plants*, 1998 edition.

NFPA 59A, *Standard for the Production, Storage, and Handling of Liquefied Natural Gas (LNG)*, 1996 edition.

NFPA 61, *Standard for the Prevention of Fires and Dust Explosions in Agricultural and Food Products Facilities*, 1999 edition.

NFPA 68, *Guide for Venting of Deflagrations*, 1998 edition.

NFPA 70, *National Electrical Code®*, 1999 edition.

NFPA 86, *Standard for Ovens and Furnaces*, 1999 edition.

NFPA 88B, *Standard for Repair Garages*, 1997 edition.

NFPA 90A, *Standard for the Installation of Air-Conditioning and Ventilating Systems*, 1999 edition.

NFPA 90B, *Standard for the Installation of Warm Air Heating and Air-Conditioning Systems*, 1999 edition.

NFPA 96, *Standard for Ventilation Control and Fire Protection of Commercial Cooking Operations*, 1998 edition.

NFPA 211, *Standard for Chimneys, Fireplaces, Vents, and Solid Fuel-Burning Appliances*, 1996 edition.

NFPA 501A, *Standard for Fire Safety Criteria for Manufactured Home Installations, Sites, and Communities*, 1999 edition.

NFPA 8501, *Standard for Single Burner Boiler Operation*, 1997 edition.

NFPA 8502, *Standard for the Prevention of Furnace Explosions/ Implosions in Multiple Burner Boilers*, 1999 edition.

M.1.10 UL Publications.

Underwriters Laboratories Inc., Publication Stock, 333 Pfingsten Road, Northbrook, IL 60062.

UL 103, *Chimneys, Factory-Built, Residential Type and Building Heating Appliances*, 1995.

ANSI/UL 441, *Gas Vents*, 1996.

ANSI/UL 641, *Low-Temperature Venting Systems*, 1995.

UL 1738, *Venting Systems for Gas Burning Appliances, Categories II, III and IV*, 1993.

UL 1777, *Chimney Liners*, 1996.

M.1.11 U.S. Government Publication.

U.S. Government Printing Office, Washington, DC 20402.

Manufactured Home Construction and Safety Standard, Title 24 *CFR*, Part 3280.

PART TWO

Supplements

In addition to the 1999 edition of NFPA 54 and commentary presented in Part One, the *National Fuel Gas Code Handbook* includes supplements. Part Two contains five supplements that explore the background of selected topics related to NFPA 54 in more detail than the commentary. These supplements are not part of the code; they are included as additional information for handbook users.

The five supplements in Part Two are

1 Development of Revised Venting Guidelines
2 Corrugated Stainless Steel Tubing Gas Piping Systems
3 Technical Background for Residential Carbon Monoxide Responders
4 Revision to American National Standard for Gas Water Heaters — Volume 1 — Storage Water Heaters with Input Ratings of 75,000 BTU per Hour or Less
5 Procedure to Estimate Infiltration Rate for Residential Structures

Part One of this handbook includes the complete text and figures of the 1999 edition of NFPA 54, *National Fuel Gas Code*, as well as Formal Interpretations and commentary that provide the history and other background information for specific paragraphs in the standard.

Development of Revised Venting Guidelines

Editor's Note: The 1992 edition of the National Fuel Gas Code first included changes to the tables used to size vents serving Category I appliances. The 1996 edition added three new tables that provide a method of sizing masonry chimneys exposed to the outdoors below the roofline. These changes were significant and resulted from research conducted by Battelle Laboratories, with the American Gas Association Laboratories, and funded by the Gas Research Institute, a nonprofit institute that sponsors projects promoting the use of natural gas. The need for this work was identified after problems arose in the field with these newer appliances. The exterior masonry chimney tables provide specific, scientifically based guidance for these chimneys that was not included in the research that led to the 1992 tables. No tables were added in the 1999 edition.

This supplement, written by the researchers, describes in detail the problem that led to the research and the technical aspects of the work. Many users of the code will find this explanation helpful in understanding the reasons for the new, more complex tables.

The editor thanks the Gas Research Institute for its cooperation in making this supplement possible and the researchers who worked with the National Fuel Gas Code Committee to make the new tables possible. Specifically, special thanks are extended to Darrel D. Paul and Allen L. Rutz of Battelle Laboratories, Robert A. Borgeson of AGA Research, and Douglas W. DeWerth of the International Approval Services — U.S., Inc. This supplement was taken from a topical report of research conducted as part of GRI's venting program and was revised in 1996 by the principal investigator, Allen L. Rutz.

INTRODUCTION

Conventional, atmospheric gas heating appliances (e.g., furnaces, boilers, and water heaters) have had an enviable field service record. These systems have provided the consumer with safe, economical space or water heating that requires minimal maintenance over a long service life. Single-wall galvanized vent connectors and either a masonry chimney or Type B double-wall metal vent pipe have been used to vent flue gases from these appliances. Here, too, the design principles, which were developed over the years, have resulted in trouble-free venting. The recent introduction of higher-efficiency appliances, however, has made altering the recommended way of designing vent systems necessary.

The new recommendations appear in Chapter 7, Chapter 10, and Appendix G. These changes have the largest impact on Category I fan-assisted furnaces and boilers. A Category I appliance is one whose vent is expected to operate under negative static pressure with a limited amount of condensation occurring. They include most traditional, draft hood–equipped units, as well as many mid-efficiency models designed for vertical venting. A more complete description of the categories is found on the following pages.

These recommendations apply to both new construction and retrofitted applications. **When a new appliance is retrofit into an existing installation or an existing appliance is removed from a common vent, the entire venting system, which may include an existing masonry chimney, should conform to current codes.** The existing venting system may need to be modified to match the new or retrofitted appliance installation.

Background of the Vent Sizing Tables

The venting recommendations found in the code for many years were developed in the 1950s for atmospheric appliances that were equipped with a draft hood. The efficiency of these appliances was low compared to the level required by the National Appliance Energy Conservation Act of 1987 (NAECA). About 35 percent of the annual energy input to a conventional appliance could have been lost from the building through the vent. This loss resulted in appliances that operated with an annual fuel utilization efficiency (AFUE) of approximately 65 percent to 70 percent and with high flue gas temperatures, which made venting relatively simple. NAECA required these losses to be reduced to no more than approximately 20 percent.

Today's mid-efficiency, Category I appliance (with an AFUE of approximately 80 percent) has lower flue gas temperatures and reduced off-cycle losses. Elimination of the draft hood or diverter in some new furnace designs also alters the vent gas temperature, dew point, flow rate, and the amount of dilution air in the vent. Overall, the potential for condensation increases, and vent system design for the modern furnace requires more care.

Modern lifestyles and energy-efficient buildings that use tighter construction practices also have increased the potential for indoor combustion air contamination. This fact, coupled with increased condensate, can increase the potential for vent system corrosion. Updated venting practice must take these factors into account for both ventilation and venting of appliances.

The primary factors affecting modern venting practice are as follows:

(1) The reduced level of vent system dilution air increases the potential for more condensate to form in the vent.
(2) Higher appliance efficiencies and lower vent gas temperatures mean that oversized vents will not heat up as quickly. The tendency for condensation also will increase.
(3) Limiting condensate formation reduces the chance of vent corrosion.

(4) The reduction of dilution air level increases the maximum capacity of the vent. In other words, a given vent can handle a larger input rate.

The results were reviewed, and the researchers determined that new vent sizing tables were needed. The following changes were made to the existing tables:

(1) **Minimum** vent system capacities (i.e., appliance input rates) were added for fan-assisted appliances to limit condensation.
(2) **Maximum** vent system capacities, which were designed to avoid positive vent pressures for fan-assisted appliances and are higher than those for draft hood–equipped models, also were added.
(3) The maximum vent system capacities for draft hood–equipped appliances were left unchanged.

Conventional Venting

The traditional, gas-heating appliance is equipped with a draft hood or draft diverter and an atmospheric burner (see Exhibit S1.1). The products of combustion are driven by the heat in the flue gases through the heat exchanger and the vent. The tendency of hot gases to rise is called buoyancy. In the vent, buoyancy creates what is usually called draft or stack action.

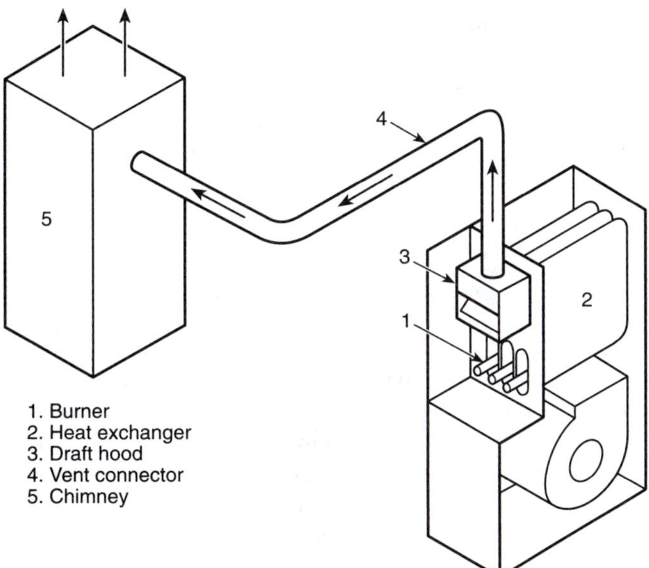

Exhibit S1.1 Draft hood–equipped furnace. (Courtesy of American Gas Association Laboratories.)

The purpose of the draft hood is to separate the appliance from the vent. The draft hood relief opening protects the appliance from wind or too much draft. It also allows additional air, called dilution air, to enter the vent and mix with the combustion products. This dilution air reduces the humidity or dew point of the vent gases, which in turn helps to reduce condensation.

Many people do not realize that products of combustion have water in them. In fact, a 100,000-Btu/hr (29-kW) appliance, operating continuously, will produce over 1 gallon (4 L) of water per hour in the flue gas. Of course, the water in the flue gas is vapor and cannot be seen unless it condenses into a liquid.

The amount of water vapor in the vent gases usually is measured as the dew point. The dew point of combustion products in a gas vent is generally about 90°F (32°C) to 130°F (54°C). If the vent gases contact a surface that has a temperature below the dew point, water will condense out of the vent gases onto the surface. As an analogy, this effect is seen on grass in the morning, when the grass temperature is below the dew point of the air. This type of condensation in the vent must be limited unless the vent has been designed for condensate handling.

Fan-Assisted Combustion Systems

The energy crisis of the 1970s caused consumers to demand appliances with higher AFUEs. Since then, appliances have been designed to reduce their off-cycle losses. In a few cases, a power burner or a combustion air or built-in flue damper is used for this purpose. However, most of these appliances use fan-assisted combustion systems with an induced draft blower to assist the combustion products through the heat exchanger (see Exhibit S1.2). Both mid-efficiency (AFUE = 78 percent to 83 percent) and high-efficiency (AFUE > 90 percent) condensing appliances can use fan-assisted systems.

A fan-assisted system does not use a draft hood, thus reducing the dilution air in the vent. The elimination of dilution air has the following important effects on vent performance:

(1) The vent gas dew point (or humidity) increases.
(2) The total amount of gases flowing in the vent decreases.
(3) When the appliance is off, there is much less air flow through the vent.

The higher dew point means that the vent must warm up more to stop condensation. At the same time, the lower flow rate makes warming up the vent more difficult. **All other things being equal, a mid-efficiency appliance will produce more condensate in the vent system than a conventional-efficiency draft hood–equipped model**.

Exhibits S1.3 and S1.4 show how condensation during startup lasts longer in a vent serving a fan-assisted appliance than in a traditional, draft hood–equipped appliance. In Exhibit S1.3, the average vent gas, vent wall, and dew points were measured at the same point in a Type B vent serving a conventional, draft hood–equipped appliance. The point when the wall temperature falls below the dew point is called the wet-time. At this point, condensation will be present on the vent wall surface. The dew point

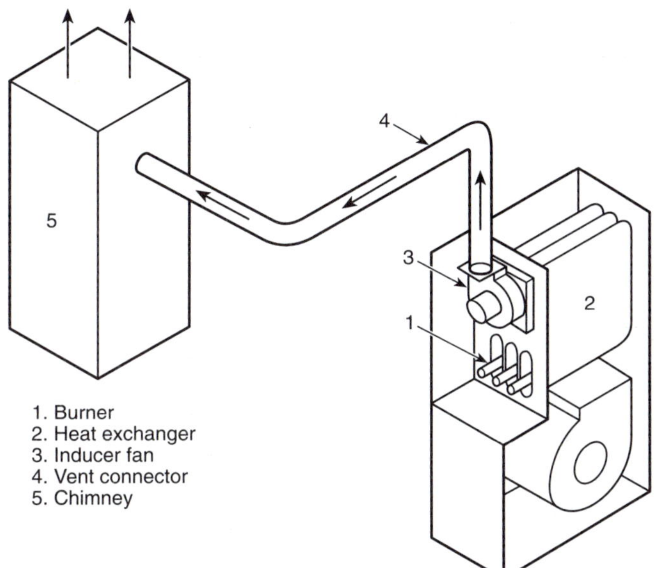

Exhibit S1.2 *Fan-assisted furnace. (Courtesy of American Gas Association Laboratories.)*

decreases with time, because the amount of dilution air increases as the draft becomes stronger.

Exhibit S1.4 presents the same temperature measurements found in Exhibit S1.3, but for a mid-efficiency, fan-assisted appliance instead of a draft hood–equipped appliance. Notice that the dew point no longer declines, because there is no source of dilution air. The higher dew point nearly doubles the wet-time. However, many mid-efficiency, fan-assisted appliances can successfully use traditional vents that are designed properly.

Appliance Categorization

Overall, the development of more efficient appliances resulted in many new product designs with different venting requirements. Space heating appliances were organized into four categories, based on the pressure produced in a special test rig and the difference between the actual temperature and dew point of the flue gas. Exhibit S1.5 shows the criteria for each appliance category, as well as the requirements that a particular vent system needs to meet.

Installers and code officials should understand the definitions of the different categories, because each space heating appliance has a rating plate containing its category number. This rating plate tells field personnel, in a generic sense, the venting requirements of each appliance. The category of an appliance is determined exclusively by the equipment design. Third-party testing laboratories verify that the appliance operates

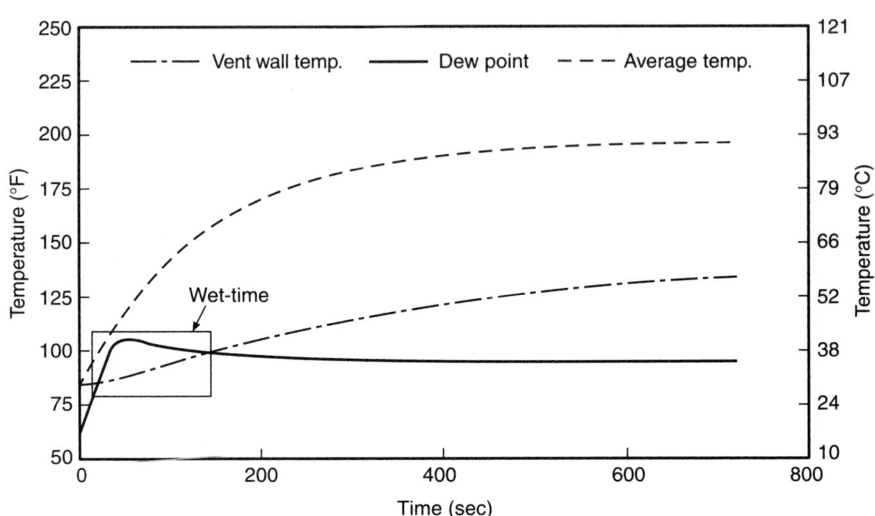

Exhibit S1.3 *Typical vent temperatures for a draft hood–equipped appliance. (Courtesy of American Gas Association Laboratories.)*

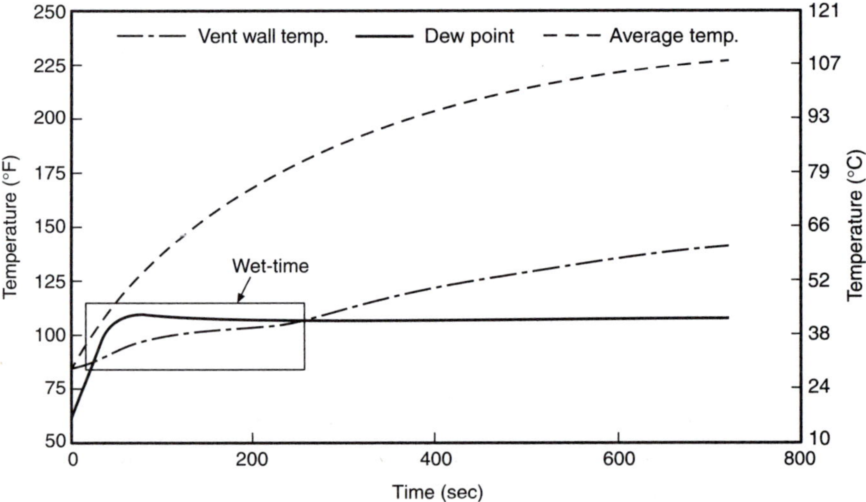

Exhibit S1.4 *Typical vent temperatures for a fan-assisted appliance. (Courtesy of American Gas Association Laboratories.)*

satisfactorily according to the category marked on the rating plate. There is no further need to check the category in the field.

Exhibit S1.5 illustrates the organization of the four categories. Category I appliances operate with a negative static vent pressure and have vent gases that are relatively

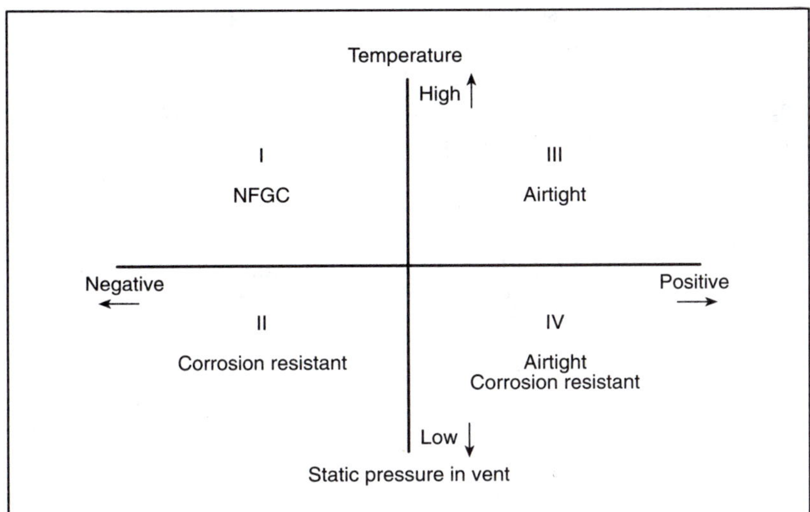

Exhibit S1.5 *ANSI categorization criteria. (Courtesy of American Gas Association Laboratories.)*

warm. They operate with only a limited amount of condensation. The traditional, draft hood–equipped appliance is considered Category I. However, many mid-efficiency, fan-assisted appliances are also Category I. How is this possible? First, the vent gas temperature of these appliances is in the same range as the traditional, draft hood appliances (after dilution). Second, the fan-assisted appliance is designed so that the vent system will remain at a negative pressure. If the vent is designed correctly, the vent pressure in a fan-assisted appliance is negative, because the draft action should be stronger than the fan pressure rise.

All Category I appliances should be vented using the manufacturer's installation instructions, which usually refer to the *National Fuel Gas Code*. Appliances in Categories II, III, or IV should also be vented according to the manufacturer's installation instructions. However, the *National Fuel Gas Code* does not contain any additional venting information for appliances in these categories.

If a special vent system is used, certain Category I appliances may also be vented in unconventional ways, including horizontal, through-the-wall arrangements. The special vent system must be supplied with the appliance by the manufacturer. The installer should follow the manufacturer's installation instructions exactly.

For residential installations, Category I appliances should never be sidewall vented through the wall using traditional venting products, such as Type B or single-wall galvanized vent pipe, because the pressure in the vent may be positive. These products should be used only for venting systems that are primarily vertical and sized according to the manufacturer's instructions.

Category II appliances also operate with negative vent pressure but with a vent gas temperature that could produce excessive condensation. These appliances require a corrosion-resistant vent. There are few, if any, Category II appliances on the market.

Category III appliances operate with a positive vent pressure and with a vent gas temperature that will produce limited condensation. These appliances require an airtight vent to avoid leakage. An example of a Category III appliance is a mid-efficiency furnace that is vented horizontally through the wall.

Category IV includes furnaces, boilers, and other appliances that operate with a positive vent pressure and with a low vent gas temperature. These appliances need an airtight, corrosion-resistant vent that includes a method of condensate disposal. Category IV appliances are usually high-efficiency, condensing models.

The category of an appliance determines the type, size, materials, and installation requirements of the venting system for that specific appliance. For example, a Category IV furnace requires a vent system built of corrosion-resistant materials, because condensate is corrosive. Gastight vent systems are essential for Category III or IV appliances, because the pressure within their vent systems exceeds the surrounding atmospheric pressure. On the other hand, Category I appliances can use traditional venting products, such as Type B vent pipe or masonry chimneys, within recommended limits.

Vent Sizing Tables

The pre-1992 vent sizing tables for Category I appliances in the *National Fuel Gas Code* were developed in the 1950s for draft hood–equipped models. Therefore, the tables assumed an appliance efficiency lower than that of our modern, mid-efficiency appliances. The tables also assumed that a large amount of dilution air would be in the vent. The research, sponsored by the Gas Research Institute, found that the tables indeed could be improved. The reasons for these changes, in summary, are as follows:

(1) Fan-assisted appliances are likely to produce more condensate in a vent than draft hood–equipped models. Therefore, a way to limit wet-time is needed.
(2) The reduction of dilution air increases the maximum capacity of the vent.
(3) The assumed efficiency for fan-assisted appliances used for determining the capacities in the tables should be raised to reflect modern appliances.

The potential to produce condensate must be controlled to limit corrosion. The corrosion can be accelerated by contamination of the combustion air by household chemicals. If contamination occurs, the condensate can become highly acidic.

Traditionally, the vent sizing tables have contained the maximum vent capacities that could be vented using a given vent diameter and length. These values were designed to prevent spillage.

The updated tables introduced the concept of minimum vent capacity. The minimum capacity is provided to prevent corrosion by limiting the wet-time. Minimum vent capacity is the smallest appliance input rating recommended for a given vent. A vent that is operating below its minimum capacity can experience excessive condensation and, possibly, corrosion.

The tables also contain separate, maximum vent capacities for fan-assisted appliances. These values are meant to prevent positive vent pressure. Therefore, a given vent for a fan-assisted appliance has both a minimum and a maximum capacity. Together, these capacity limitations form a range of acceptable appliance inputs and a range of acceptable vent sizes. The maximum capacity prevents vents from being too small; the minimum capacity prevents vents from being too large.

THE VENT-II COMPUTER PROGRAM

The venting tables for Category I fan-assisted appliances were generated using the VENT-II computer program to calculate the minimum and maximum vent capacities for each configuration. The VENT-II program uses a transient mathematical model to solve the mass, momentum, and energy balance equations for a user-specified appliance(s) connected to a user-specified vent system. The user specifies the number of appliances connected to the vent system (one or two), the appliance cycling times, and the number of consecutive appliance on/off-cycles in the transient analysis. VENT-II then calculates the temperatures, pressures, flow rates, and the amount of condensate formation for all of the user-specified sections of the venting system as a function of time.

The primary benefit of using a computer model is the laboratory time saved by not having to experimentally determine the minimum and maximum vent capacities at the precisely controlled, ambient temperatures specified for each of the over 3600 unique vent systems listed in the venting tables in Chapter 10. These vent systems consist of either single-wall metal vent connectors or Type B double-wall metal vent connectors attached to either a Type B double-wall vertical vent or to a tile-lined masonry chimney, meeting the requirements of NFPA 211, *Standard for Chimneys, Fireplaces, Vents, and Solid Fuel-Burning Appliances*. The analyzed vent systems range in vertical height from 6 ft to 100 ft (2 m to 30 m) and have horizontal, lateral lengths from 0 ft to 50 ft (0 m to 15 m).

Over the years, the original VENT-II computer program has been updated as new venting capabilities were needed. For instance, VENT-II, Version 3.0, allowed the program to calculate wet-times for each section of the vent system. VENT-II, Version 4.0, added the capability of simulating the vent system performance when two appliances were connected to the vent. VENT-II, Version 4.1, added additional features that allowed the program to accurately predict the performance of a masonry chimney when one or two appliances were vented. In all cases, the later versions of VENT-II retained the primary features of the earlier versions while improving the mathematical models.

The venting tables were calculated over a period of approximately five years, using the most up-to-date version of VENT-II available at the time that each table was generated. In chronological order, the single appliance venting tables for Type B double-wall vent systems were generated first. Next, the multiple appliance venting tables for Type B double-wall vent systems were generated. The single and multiple appliance venting tables for interior masonry chimneys followed. Finally, the single and multiple appliance venting tables for exterior chimneys were generated.

Validation of VENT-II

The origins of VENT-II date back to a 1983 GRI study in which Battelle developed a computer model that was experimentally validated by the American Gas Association Laboratories. In that study, the computer program predictions of vent static pressures and temperatures were compared to measurements on 20 unique combinations of appliance and vent configurations. In addition, the computer program predictions were compared to natural draft design data then contained in Table G.1 of the 1988 *National Fuel Gas Code*, Chapter 26 of the 1979 American Society of Heating, Refrigerating and Air-Conditioning Engineers (ASHRAE) *Equipment Handbook*, and an updated, hand-calculator, magnetic tape model developed by Richard Stone. Overall, there was close agreement between the vent simulation model predictions, the experimental data, and the existing information in the handbooks and codes.

Since 1983, VENT-II has been converted for use on an IBM® personal computer (Version 1.0), the user-interface and condensation models were refined (Version 2.0), a transient analysis capability was added (Version 3.0), a common vent model was included (Version 4.0), and an experimentally validated masonry chimney model was developed and refined (Version 4.1). At each step in the development of VENT-II, the new models have been compared against the previously developed sources of data, and additional validation experiments have been conducted.

In addition to experimental validation in the laboratory, VENT-II has undergone field validation. Battelle has used VENT-II to predict the venting performance of field installations in Newark, New Jersey, Hartford, Connecticut, and Columbus, Ohio. Battelle has also received numerous phone calls and letters from users of VENT-II who have successfully used the program to either diagnose problems or design vent systems. Chapter 31 of the 1992 *ASHRAE Handbook — HVAC Systems and Equipment* also recognizes VENT-II as a design tool for gas appliance venting systems.

BASELINE CONDITIONS FOR CATEGORY I FAN-ASSISTED APPLIANCE VENTING TABLES

Before venting tables could be generated, baseline conditions needed to be established to define a set of typical, appliance-operating parameters and ambient conditions. Where appropriate, the same conditions used to generate the existing *National Fuel Gas Code* tables were assumed for this analysis.

The appliance-operating parameters were set after consultations with and review by the research project's Technical Advisory Group, which included appliance manufacturers. These operating parameters were chosen so that the resulting venting tables would be applicable to the majority of Category I fan-assisted appliances being manufactured. However, ensuring that the venting guidelines are applicable to a particular appliance model is the responsibility of the appliance manufacturer, especially if the operation of that model deviates significantly from the baseline conditions assumed here. Table S1.1 summarizes the baseline appliance conditions used to generate the venting tables for Category I fan-assisted appliances.

Table S1.1   Baseline Appliance Conditions Used to Generate Venting Tables for Category I Fan-Assisted Appliances

Thermal Efficiency	83%
Excess Combustion Air for Natural Gas	65% (7% CO_2)
Off-Cycle Pressure Loss Coefficient	30

Appliance Thermal Efficiency

A very important appliance parameter affecting venting is the steady-state thermal efficiency of the appliance. In this case, steady-state thermal efficiency is defined as follows:

$$\text{Thermal efficiency (at steady state)} = \frac{\text{Appliance heat input rate} - \text{Appliance flue heat loss rate}}{\text{Appliance heat input rate}} \times 100$$

The thermal efficiency is the percentage of heat generated by the gas burner (i.e., appliance input rate) that is used for space or water heating. For natural draft venting systems (i.e., Category I appliances), a high appliance thermal efficiency means proportionately lower vent gas temperatures and, therefore, less buoyancy is available to exhaust the gases through the venting system. Lower vent gas temperatures can lead to an increased potential for condensate formation in the vent.

Category I fan-assisted appliances typically operate at thermal efficiencies between 80 percent and 83 percent, resulting in AFUEs greater than 78 percent. The venting tables for Category I fan-assisted appliances were generated assuming an appliance thermal efficiency of 83 percent. If all other parameters are held constant, the minimum and maximum vent capacity range, based on an efficiency of 83 percent, would be a subset of the range for lower appliance thermal efficiencies. Therefore, the venting tables generated at an 83 percent thermal efficiency are also applicable (and conservative) for Category I fan-assisted appliances with lower thermal efficiencies.

Appliance Excess Combustion Air

Appliance excess combustion air is another parameter that has a strong influence on venting. The excess combustion air level directly affects the quantity of vent gas to be vented; in turn, the maximum capacity of the vent is affected. At the same time, the amount of excess air also changes the dew point of the vent gas, influencing the amount of condensate formation in the vent and the minimum capacity of the vent.

Category I fan-assisted appliances typically operate with excess combustion air levels between 30 percent and 65 percent (9 percent to 7 percent CO_2). The venting tables for Category I fan-assisted appliances were generated assuming 65 percent excess combustion air (7 percent CO_2 level). If all other parameters are held constant, then a decrease in the excess combustion air generally will result in an increase in the minimum vent capacities and a decrease in maximum vent capacities.

Appliance Cycle Times

For space-conditioning equipment, appliance cycle times depend on the thermostat characteristics and setting, the outdoor temperature, the building heat loss characteristics, and the oversizing factor used at design load to determine the appliance input rating. Fortunately, the U.S. Department of Energy has done extensive research on typical appliance cycle times as part of their methodology for determining AFUEs of furnaces. The selection of cycle times that resulted in the new venting tables relies heavily on the results of the Department of Energy's analyses.

The generic, Category I fan-assisted appliance used to generate the venting tables was assumed to be installed in a typical midwestern residence with an average outdoor temperature during the heating season of 42°F (56°C). Under these circumstances, the appliance cycles on for 3.87 minutes and off for 13.3 minutes in accordance with the Department of Energy AFUE test.

Table S1.2 lists, as a function of time, the flue gas temperatures exiting the Category I fan-assisted appliance that were used to generate the venting tables. These temperatures are based on typical heatup and cooldown response times for a Category I fan-assisted appliance with a thermal efficiency of 83 percent and an excess combustion air level of 65 percent (7 percent CO_2).

During the off-cycle, the flow rate of air entering the vent system was calculated by assuming that the appliance had an off-cycle pressure loss coefficient of 30. The off-cycle pressure loss coefficient is the number of velocity heads ($^1/_2\rho rv^2$) of pressure that is lost between the appliance air intake and the appliance flue gas exit. A value of 30 was found to be representative of appliances with fan-assisted combustion systems as measured in the laboratory.

Gas Appliance Fuel Type

The four different fuel gases used in gas appliances in the United States were evaluated with respect to the products of combustion and the effect on venting performance. The four fuels and the reason for their inclusion in the comparison are as follows:

(1) *Natural gas.* The fuel assumed by VENT-II, Version 4.1, and used to develop the masonry chimney venting tables. The specific composition of natural gas (see Table S1.3) is derived from the *ASHRAE Handbook — Fundamentals* and has an equivalent Btu content of 1032 Btu/ft³ per cubic foot (38 kj/m³) of gas.

Table S1.2 Category I Fan-Assisted Appliance Outlet Temperatures Used to Generate Venting Tables

Time (min:sec)	Fan-Assisted Appliances (83% Eff., 7% CO_2)	
	(°F)	(°C)
00:00	131.8	55.4
0:23	181.6	83.1
0:46	215.4	101.9
1:10	238.0	111.4
1:33	254.0	123.3
1:56	264.6	129.2
2:19	271.8	133.2
2:43	276.7	135.9
3:06	280.0	137.8
3:29	282.2	139.0
3:52	283.8	139.9
4:46	244.4	116.0
5:38	215.1	101.7
6:32	193.4	84.7
7:25	177.1	80.6
8:18	165.1	73.9
9:11	156.1	68.9
10:05	149.4	65.2
10:58	144.5	62.5
11:51	140.8	60.4
12:44	138.0	58.9
13:37	136.0	57.8
14:31	134.4	56.9
15:24	133.3	56.3
16:17	132.5	55.8
17:15	131.8	55.4

(2) *Methane.* The fuel assumed by VENT-II prior to Version 4.1. (Natural gas is almost entirely methane.)

(3) *Propane.* A fuel used as a substitute for natural gas.

(4) *Butane.* A fuel used in the past in southern climates as a substitute for natural gas.

The important factors that affect venting performance with respect to the venting tables are the total amount of vent gas generated, the temperature of the vent gas, and the amount of moisture in the vent gas. The total volume and temperature of vent gas generated by an appliance per Btu/hr of energy are determining factors in calculating

Table S1.3   Natural Gas Composition (Mole Percents) Used in VENT-II, Version 4.1

Component	Mole Percent
N_2 nitrogen	0.362
CO_2 carbon dioxide	0.599
CH_4 methane	96.324
C_2H_6 ethane	2.061
C_3H_8 propane	0.367
C_4H_{10} isobutane	0.150
C_5H_{12} isopentane	0.056
C_6H_{14} N-hexane	0.081
Higher heating value, Btu/lb	23,207
Higher heating value, Btu/SCF	1,032
Specific gravity	0.5807

the maximum capacity of an appliance that is connected to a specified vent system. The amount of moisture in the vent gas affects the vent gas dew point and, in turn, the minimum capacity of a fan-assisted appliance.

Table S1.4 shows a comparison of the products of the stoichiometric combustion of the four gases, as derived from the *Gas Engineers Handbook*. Note that the volume of combustion products per kBtu fuel values in Table S1.4 are all within 1 percent of the value for natural gas. Also, note that the water vapor concentration in the combustion products is significantly less for both propane and butane than for natural gas, resulting in lower dew points in these fuels.

Although specific appliance designs will vary, certain conditions must be taken into consideration when characterizing the field conversion of an appliance from natural gas to propane. These conditions are the 1 percent to 2 percent increase in carbon dioxide that is measured at the appliance outlet and a slight increase in the efficiency of the appliance. Therefore, an appliance that operates at 65 percent excess air (7 percent CO_2) with natural gas may be expected to operate at approximately 48 percent excess air (9 percent CO_2) when converted to propane.

Table S1.5 outlines the change in vent gas based on the conversion from natural gas to propane. Note that the total volume of vent gas for propane and butane is less than the volume for natural gas. From laboratory test data of an appliance converted from natural gas to propane, an increase in the vent gas temperature was observed when burning propane. Also, note that the dew point of the vent gas is lower with both propane and butane than with natural gas. Therefore, based on the stoichiometric data presented in Table S1.4 and the example comparison in Table S1.5, the new venting tables are considered appropriate, without modification, for propane- and butane-fired appliances.

Table S1.4 Products of Stoichiometric Combustion of Four Different Gaseous Fuels

Combustion Products from Stoichiometric Combustion	Units	Methane CH_4	Propane C_3H_8	Butane C_4H_{10}	Natural Gas $C_{1.04}H_{3.88}$
Volume of combustion product per kBtu,	$\dfrac{ft^3 \text{ comb. product}}{kBtu \text{ fuel}}$	10.4	10.2	10.2	10.3
Water vapor concentration,	$\dfrac{ft^3 H_2O \text{ vapor}}{ft^3 \text{ comb. product}}$	0.190	0.155	0.149	0.184
Flue gas dew point,	°F	139	131	129	138
	(°C)	(59.4)	(55.0)	(53.9)	
Stoichiometric Combustion Parameters					
Combustion air required,	$\dfrac{ft^3 \text{ air}}{ft^3 \text{ fuel}}$	9.53	23.82	30.97	9.6
Ultimate CO_2,	(%)	11.73	13.75	14.05	12.1
Nitrogen in combustion products,	$\dfrac{ft^3 N_2}{ft^3 \text{ fuel}}$	7.53	18.82	24.47	7.58
Carbon dioxide in combustion products,	$\dfrac{ft^3 CO_2}{ft^3 \text{ fuel}}$	1.0	3.0	4.0	1.04
Water vapor in combustion products,	$\dfrac{ft^3 H_2O \text{ vapor}}{ft^3 \text{ gas}}$	2.0	4.0	5.0	1.94
Total combustion gas volume,	$\dfrac{ft^3 \text{ comb. product}}{ft^3 \text{ fuel}}$	10.53	25.82	33.47	10.56
Energy content,	$\dfrac{Btu}{ft^3 \text{ fuel}}$	1012	2524	3266	1030

Table S1.5   Rule-of-Thumb Vent Gas Comparison

Parameter	Units	Propane	Butane	Natural Gas
Carbon dioxide in vent gas	%	9.0	9.0	7.0
Flue product per Btu	$\dfrac{ft^3 \text{ vent gas}}{kBtu}$	14.79	15.11	16.28
Water vapor concentration	$\dfrac{ft^3 H_2O \text{ vapor}}{ft^3 \text{ vent gas}}$	0.1071	0.1011	0.1202
Vent gas dew point	°F	117.4	115.4	121.5

Ambient Conditions — Type B Vents and Interior Masonry Chimneys

In generating the venting tables for Type B vents and interior masonry chimneys, only interior venting configurations were considered, as shown in Exhibit S1.6. By definition, interior venting configurations are those that are not exposed to outdoor temperatures below the roofline. Above the roofline, only 3 ft (1 m) of vent or chimney is assumed to be exposed to outdoor temperatures.

Minimum vent capacities were calculated by assuming a 42°F (5.6°C) outdoor temperature and a 60°F (16°C) indoor temperature. Maximum vent capacities were calculated by assuming both the indoor and outdoor temperatures were 60°F (16°C). In both cases, zero house depressurization and no wind were assumed.

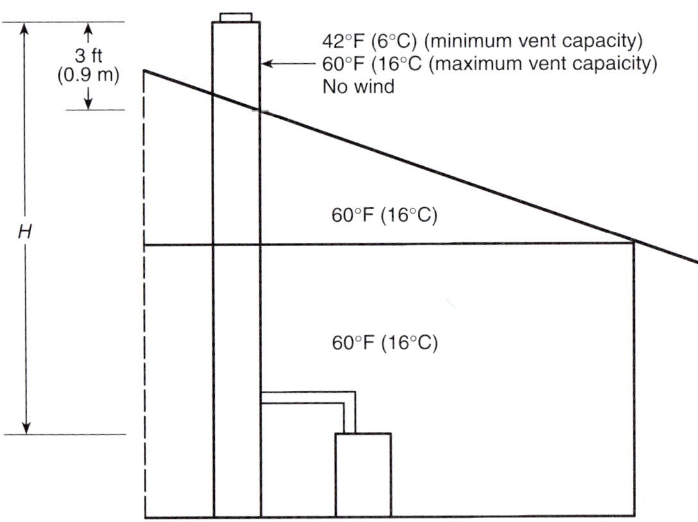

Exhibit S1.6 Ambient conditions assumed for generating venting tables for Type B vents and interior masonry chimneys. (Courtesy of Battelle.)

Maximum Vent Capacities

For Category I fan-assisted appliances, the maximum vent capacity was calculated by VENT-II, Version 4.1, at the appliance input rating that results in a zero static vent pressure 1 ft (0.3 m) from the appliance outlet and 1 minute after the start of the third appliance cycle. These criteria were selected because Category I appliance vent systems are not designed to operate under positive static pressures. However, when the appliance burner first ignites, the vent system undergoes a pressure spike until the vent primes (i.e., becomes filled with warm vent gases). Therefore, a time of 1 minute into the third burner cycle was selected to allow the vent to prime. Additional checks were put into place to ensure that, when fan-assisted appliances are common vented with a draft hood appliance, spillage of vent gas due to vent priming will remain below 1.5 percent of the total vent gas. The Type B double-wall venting tables that were generated using earlier versions of VENT-II (prior to Version 4.1) used a slightly different location and time for determining the maximum vent capacity that corresponded to a zero static pressure. However, essentially the same maximum capacities will be calculated if these criteria are used for Type B double-wall vent systems using VENT-II, Version 4.1.

Maximum vent capacities are relatively insensitive to which appliance cycle is chosen, because gas temperatures in the vent tend to stay approximately the same from one cycle to the next. Therefore, the third appliance cycle was chosen for calculating maximum vent capacities for both Type B double-wall vent systems and masonry chimneys.

Minimum Vent Capacities

The minimum vent capacities of Category I fan-assisted appliances assume that the vent materials do not exceed the specified wet-time limits. If these wet-time limits are exceeded, a high potential exists for the vent materials to experience excessive corrosion (in the case of metal vents) or for water damage to occur in surrounding structures (in the case of masonry chimneys).

For metal vent systems, the wet-time limits are based on conventional vent conditions that have existed for many years, as well as on the existing data available for the corrosion of condensing heat exchanger materials in the presence of vent gases. For vent connectors made of galvanized steel, aluminum, Type B vent, or 300 series stainless steels, the wet-time limits are set so that these materials can dry out within 3.8 minutes from the start of the third appliance cycle. For vertical Type B double-wall vents, the wet-time limits are set so that these materials can dry out within 12 minutes from the start of the third appliance cycle. The wet-time limits for vent connectors are lower than the wet-time limits for vertical vents, because the vent connectors usually experience a higher concentration of corrosive contaminants (e.g., chlorides, nitrates, sulfates, and fluorides) in the vent gases. By the time the vent gases enter the vertical vent, many of the corrosive contaminants have already been removed from the gas as it passed through the vent connector. Therefore, the vertical vent can accommodate a longer wet-time without necessarily experiencing excessive corrosion. In all cases, if the minimum vent capacity exceeds the maximum vent capacity, then the specific vent configuration is not permitted (NP) for use with Category I fan-assisted appliances.

Masonry Chimneys

For vent systems employing masonry chimneys, wet-time limits cannot be set based on the materials' degradation because of the lack of data available to make this determination. Therefore, the assumption was made that the increased wetness associated with Category I fan-assisted appliances did *not* increase the rate of degradation of the materials used to construct masonry chimneys. This assumption was confirmed through a series of spray chamber tests using masonry chimney material specimens exposed to synthetic vent gas condensate.

The criterion used to determine the adequacy of masonry chimneys is whether or not the chimney is wet continuously for long periods of time. Masonry chimneys built in accordance with NFPA 211, *Standard for Chimneys, Fireplaces, Vents, and Solid Fuel-Burning Appliances*, are not watertight and are not required to have condensate drains. Clay tile liners are porous, and mortar joints between clay tile liners can crack, which can allow moisture and condensate to migrate into surrounding structures, causing water damage. A homeowner will notice this problem when plaster walls adjacent to the chimney become wet and discolored or when paint or wallpaper starts to peel. Condensate can also collect in the bottom of the chimney and seep out, potentially causing water damage to the floor or furnishings in the vicinity. All of these possibilities point to the need for venting guidelines that will limit the amount of condensate formation in existing masonry chimneys when Category I fan-assisted appliances are employed.

Interior Masonry Chimneys

For interior masonry chimneys, the wet-time limit is checked at the maximum vent capacity and at the maximum liner size. In the VENT-II simulation, the 3 ft (1 m) of chimney that extends through the roof is assumed to be exposed to a 42°F (5.6°C) ambient temperature. The remainder of the chimney is assumed to be located in the interior of the house and exposed to a 60°F (16°C) ambient temperature. The VENT-II simulation begins by allowing the Category I fan-assisted appliance to operate continuously at its input rating for 1 hour to heat up the chimney, followed by an off-cycle lasting 13 minutes. Next, the appliance is allowed to cycle ON and OFF (according to the criteria stated in Section 3.3) for 42 cycles (approximately 12 hours). At the end of this time period, if the clay tile in the chimney is still wet, the table entry is NP (not permitted). If the clay tile has dried out at the end of this time period, then the minimum vent capacity is determined based on the wet-time limit of the vent connector material.

Exterior Masonry Chimneys

The maximum capacity of an exterior chimney is identical to the maximum capacity of an interior chimney of the same dimensions. However, since exterior masonry chimneys are exposed to the outdoor environment over their entire length, the impact of climate on the potential for excessive condensation is greater than the impact of climate for interior chimneys. Therefore, the tables for exterior chimneys are indexed by outdoor temperature range, becoming increasingly restrictive with increasingly colder climates.

Explicit minimum capacities, based on the chimney dimensions and the climate, are listed for exterior masonry chimneys. These minimum capacities were determined from VENT-II analysis, assuming that the chimney was allowed to be continuously wet for up to 1000 hours during a single heating season and was used to vent an appliance operating at a steady-state thermal efficiency of 83 percent.

The appliance efficiency is the most critical parameter affecting exterior masonry chimney performance. GRI research shows that when the appliance efficiency is around 80.5 percent rather than 83 percent, the guidelines for exterior masonry chimneys can be relaxed significantly. Use Tables 10.11 through 10.13 unless the installation instructions provide an alternate method.

SINGLE APPLIANCE VENTING TABLES

Single appliance venting tables apply when only one appliance is connected to the vent system. This section describes the rationale used to develop single appliance venting tables for Category I fan-assisted appliances. Four separate venting tables have been generated that cover the use of either single-wall metal vent connectors or Type B double-wall metal vent connectors with either Type B double-wall metal vertical vents or masonry chimneys.

Vent Configuration for Single Appliance Vents

For single appliance vents, as shown in Exhibit S1.7, the basic configurations for interior vents have been modeled using VENT-II. The following guidelines have been used to model the venting systems:

(1) Section A1 is 1 ft (0.3 m) in vertical height.
(2) The total lateral length (L) is split evenly among Sections A2 and A3.
(3) Section C4 is 3 ft (0.9 m) in vertical height.
(4) The remaining vertical height [the total height less 4 ft (1.2 m)] is split evenly among Sections C1, C2, and C3.

The A sections are used to designate the vent connector. These sections are modeled as either single-wall metal vent or Type B double-wall metal vent. All A sections are assumed to be exposed to a 60°F (16°C) indoor ambient temperature.

The C sections are used to designate the vertical vent. These sections are modeled as either Type B double-wall metal vent or as a tile-lined masonry chimney meeting NFPA 211 design requirements. The last vertical section extending through the roofline, C4, is assumed to be exposed to the outdoor air temperature, which is 42°F

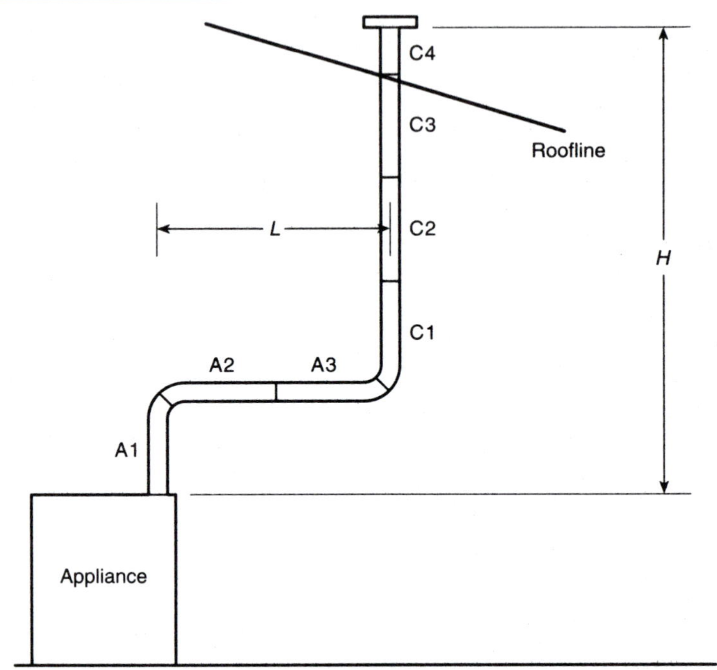

Exhibit S1.7 *VENT-II configuration for single appliance venting tables. (Courtesy of Battelle.)*

(5.6°C) when minimum vent capacities are calculated and 60°F (16°C) when maximum vent capacities are calculated. All other C sections (C1, C2, and C3) are assumed to be exposed to a 60°F (16°C) indoor ambient temperature.

For exterior chimneys, as shown in Exhibit S1.8, Sections C1, C2, and C3 are exposed on three sides to the outdoor air temperature while Section C4 is exposed on all sides. For exterior chimneys, calculations were made at a range of outdoor air temperatures corresponding to the major climates of the United States. (See Figure G.19 of the code.)

Although the exterior chimney calculations assumed three sides of the chimney were exposed, the potential for excessive condensation can exist even if only one side is exposed. The presence of any surface at a temperature below the dew point of the flue gases creates the potential for condensate formation. Therefore, the definition of an exterior chimney encompasses any chimney with one or more sides exposed to the outdoors below the roofline.

Format of Single Appliance Venting Tables

Venting tables for Category I draft hood appliances have been in place for many editions of the code and have withstood the test of time. They should be retained for vent-

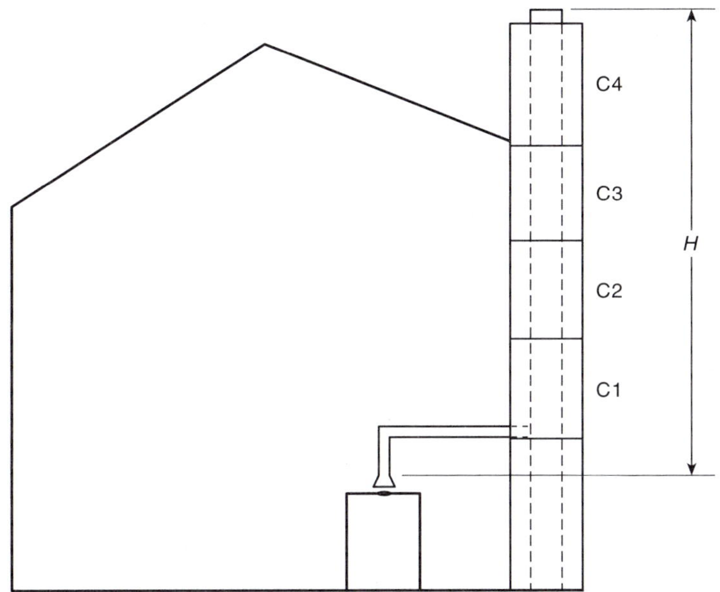

Exhibit S1.8 *VENT-II configuration for exterior masonry chimney venting tables.*

ing conventional draft hood appliances. For easy understanding, the format for the venting tables for Category I fan-assisted appliances was deliberately chosen to be similar to the format of the tables in prior editions of the code. At the suggestion of the Gas Research Institute's Technical Advisory Group, the venting tables for Category I fan-assisted appliances and Category I draft hood appliances were merged. However, the table entries for draft hood appliances were not changed from the existing code tables except where interpolation or extrapolation was required for consistency.

The following is an explanation of the terms and abbreviations used in the single appliance venting tables:

Fan-Assisted Combustion System. An appliance equipped with an integral, mechanical means to either draw or force products of combustion through the combustion chamber and heat exchangers.

FAN Min. The minimum input rating of a Category I appliance with a fan-assisted combustion system that could be attached to the vent.

FAN Max. The maximum input rating of a Category I appliance with a fan-assisted combustion system that could be attached to the vent.

NAT Max. The maximum input rating of a Category I appliance equipped with a draft hood that could be attached to the vent. There are no minimum appliance input ratings for draft hood–equipped appliances.

NA. Vent configuration is not permitted due to potential for condensate formation or pressurization of the venting system, or not applicable due to physical or geometric restraints.

*** (following number).** Indicates potential for continuous condensation.

In Appendix G of the code, Figure G.1 shows the configuration used to generate the venting table (Table 10.1) when the entire vent system is composed of Type B double-wall metal vent. Figure G.2 shows the configuration used to generate the venting table (Table 10.2) for Type B double-wall metal vents that are used with single-wall metal connectors. Figure G.3 shows the configuration used to generate the venting tables (Tables 10.3 and 10.11) for masonry chimneys that are used with Type B double-wall metal connectors. Figure G.4 shows the configuration used to generate the venting table (Table 10.4) for masonry chimneys that are used with single-wall metal connectors.

In all of these tables, the NAT Max values were derived directly from the tables for draft hood appliances in prior editions of the code. The FAN Min and FAN Max values for fan-assisted Category I appliances were derived using the VENT-II computer program with the criteria described in Section 3.0. In Tables 10.1 and 10.2, single appliance venting configurations with zero lateral lengths are assumed to have no elbows in the vent system. For vent configurations with lateral lengths, the venting tables include allowance for two 90° turns. For each additional 90° turn or equivalent, the maximum capacity listed in the venting tables should be reduced by 10 percent (0.90 @ maximum table capacity). Two 45° turns are equivalent to one 90° turn.

Sea level input ratings should be used when determining maximum capacity for high altitude installations. Actual input (derated for altitude) should be used to determine the minimum capacity for high altitude installations. For appliances with more than one input firing rate, the minimum vent capacity that is determined from the tables should be less than the lowest appliance input rating, and the maximum vent capacity that is determined from the tables should be greater than the highest appliance input rating.

In generating the masonry chimney venting tables, the minimum flow area of the chimney was assumed to be one size equivalent round larger than the vent connector flow area. The vent connector flow area was assumed to be equal to the appliance vent outlet flow area. The maximum vent capacities for fan-assisted appliances were calculated based on the minimum chimney flow area. This assumption also was used in the existing code table entries for draft hood appliances. For either Type B double-wall vertical vents or masonry chimneys, the vent flow area was assumed to be less than or equal to seven times the appliance outlet flow area.

For a single appliance venting into an exterior masonry chimney, the code specifies the following criteria:

(1) The appliance must be draft hood equipped.
(2) The vent connector must be Type B double-wall.

(3) The vent connector length must be less than $1^{1}/_{2}$ ft per inch of vent connector diameter.

(4) The appliance maximum input rate must be less than the rate listed in Tables 10.3.

(5) If the appliance is used for space heating, then its input rate must be greater than the rate listed in Table 10.11.

If the appliance is not used for space heating, then consult the appliance manufacturer, local gas supplier, or authority having jurisdiction. Table 10.11 lists space-heating appliance minimum input rates for exterior masonry chimneys for six climates (i.e., winter design temperatures).

COMMON VENTING TABLES FOR TWO APPLIANCES

Common venting tables apply when two appliances are connected to the same vent. This section describes the rationale used to develop common venting tables for Category I appliances. Venting tables have been generated that cover the use of either single-wall metal vent connectors or Type B double-wall vent connectors with either Type B double-wall metal vertical vents or masonry chimneys (interior and exterior).

Vent Size Combinations for Common Venting Tables

The common vent tables were generated assuming specific size combinations for the vent connectors on each appliance and for the common vent. For a Type B double-wall common vent, Table S1.6 lists the size combinations assumed for the vent connectors for each appliance and for the common vent. For a masonry chimney, Table S1.7 lists the size combinations used to generate common vent capacities, and Table S1.8 lists the size combinations used to generate vent connector capacities.

Format of Common Venting Tables

The common venting tables for fan-assisted appliances were combined with the common venting tables for draft hood appliances and put in the same format as the tables in prior editions of the *National Fuel Gas Code*. The table entries for draft hood appliances were not changed from the existing code tables, except where interpolation or extrapolation was required for consistency.

The following is an explanation of additional terms and abbreviations (see Section 1.7 of the code) used in the common venting tables:

FAN+FAN. The maximum combined appliance input rating of two or more Category I fan-assisted appliances that are attached to a common vent.

FAN+NAT. The maximum combined appliance input rating of one or more Category I fan-assisted appliances and one or more Category I draft hood–equipped appliances that are attached to a common vent.

Table S1.6 Size Combinations of Type B Vent Used to Generate Common Venting Tables

Appliance 1 Vent Connector Diameter		Appliance 2 Vent Connector Diameter		Type B Combined Vent Diameter	
in.	mm	in.	mm	in.	mm
3	80	3	80	4	100
3	80	4	100	5	130
3	80	5	130	6	150
4	100	7	178	8	200
5	130	8	200	9	230
5	130	9	230	10	250
6	150	10	250	12	310
7	180	12	310	14	360
8	200	14	360	16	410
9	230	16	410	18	457
10	250	18	460	20	510
11	280	20	510	22	560
12	310	22	566	24	610
14	360	26	660	28	710
16	410	30	760	32	810
18	450	32	810	36	910
20	510	36	910	40	1020
22	560	40	1010	44	1120
24	610	44	1020	48	1220

NAT+NAT. The maximum combined appliance input rating of two or more Category I draft hood–equipped appliances that are attached to a common vent.

NA. Vent configuration is not applicable due to physical or geometric constraints.

Figures G.6 and G.7 show the vent configuration used to generate the common venting tables for Type B double-wall vertical vents with either Type B double-wall connectors (Table 10.6) or single-wall metal connectors (Table 10.7). Figures G.8 and G.9 show the vent configuration used to generate the common venting tables for masonry chimneys with either Type B double-wall connectors (Tables 10.8, 10.12, and 10.13) or single-wall metal connectors (Table 10.9).

Vent Connector Capacity

In generating common venting tables, the vent connector capacities for each appliance were calculated by assuming that the appliance attached to the vent connector was on

Table S1.7   Size Combinations of Masonry Chimneys Used to Generate Common Vent Capacities in the Multiple Appliance Venting Tables

Appliance 1 Vent Connector Diameter		Appliance 2 Vent Connector Diameter		Minimum Masonry Chimney Flow Area		Maximum Masonry Chimney Flow Area	
in.	mm	in.	mm	in.^{2}	mm^2	in.^{2}	mm^2
3	80	3	80	12	7,740	49	31,600
3	80	4	100	19	12,300	88	55,800
3	80	5	130	28	18,100	137	88,400
3.5	89	6	150	38	24,500	198	128,000
4	100	7	180	50	33,300	269	173,000
4.5	110	8	200	63	40,600	352	550,000
5	130	9	230	78	50,300	445	286,000
6	150	10	250	113	72,900	550	355,000

Table S1.8   Size Combinations of Masonry Chimneys Used to Generate Vent Connector Capacity in the Multiple Appliance Venting Tables

Appliance 1 Vent Connector Diameter		Appliance 2 Vent Connector Diameter		Masonry Chimney Flow Area	
in.	mm	in.	mm	in.^{2}	mm^2
3	76	5	130	28	18,100
4	100	7	180	50	32,300
5	130	9	230	78	50,300
6	150	10	250	113	73,000
7	180	12	300	153	98,700
8	200	14	360	201	130,000
9	230	16	410	254	164,000
10	250	18	460	314	203,000

and that the other appliance was off the entire time. The lateral length of each vent connector was assumed to be equal to 1.5 ft per inch of connector diameter (see Table S1.9). The connector rise for appliance 1 was assumed to be equal to the value listed in the appropriate venting table, with the connector rise for appliance 2 set equal to the maximum value (either 3 ft or 6 ft) shown in the venting table for a given vent connector diameter. The maximum and minimum vent connector capacities were calculated as described in Sections 3.6 and 3.7.

The FAN Max column in the vent connector capacity tables refers to the maximum vent connector capacity when a Category I appliance with a fan-assisted combustion system is attached to the vent. The column labeled NAT Max refers to the maximum vent connector capacity for a draft hood–equipped appliance. The FAN Min column refers to the minimum vent connector capacity for use with fan-assisted appliances.

Table S1.9   Vent Connector Horizontal Length Used to Generate Common Venting Tables

Connector Diameter		Maximum Connector Horizontal Length	
in.	mm	ft	m
3	76	4	1.2
4	100	6	1.8
5	130	7.5	2.3
6	150	9	2.7
7	180	10.5	3.2
8	200	12	3.7
9	230	13.5	4.1
10	250	15	4.6
12	300	18	5.5
14	360	21	6.4
16	410	24	7.3
18	460	27	8.2
20	510	30	9.1
22	560	33	10.1
24	610	36	11.0

Common Vent Capacities

When determining common vent capacities for the tables, the size combinations assumed for the individual vent connectors affect the calculated capacities. Tables S1.6 and S1.7 list the size combinations used to generate the combined vent capacities. The connector rise for the first appliance is fixed at 1 ft, and the rise for the second appliance is fixed at 3 ft.

The common vent capacities are calculated assuming two Category I appliances are connected to the combined vent with both appliances cycling simultaneously. Only maximum capacity values are required for the common vent portion of the Type B double-wall venting tables. The FAN Min restrictions of the vent connector portion of the tables ensure that condensation in the common vent does not reach unacceptable levels.

In the venting tables, the column labeled FAN+FAN refers to the maximum combined vent capacity when two appliances with fan-assisted combustion systems are attached to the same vent. Both fan-assisted appliances are required to have zero static pressure in the first section of their respective vent connectors. The capacities of the two appliances are totaled to produce the value shown in the table.

The FAN+NAT column refers to the maximum combined vent capacity when one fan-assisted appliance and one draft hood–equipped appliance are connected to the same vent. The fan-assisted appliance is required to have zero static pressure in the first section of its vent connector, while the draft hood–equipped appliance, which is assumed to operate at 78 percent thermal efficiency with an 8 percent CO_2 excess combustion air level, is required to have 29 percent dilution air entering through the draft hood. The capacities of the two appliances are added to produce the value shown in the table.

Finally, the column labeled NAT+NAT shows the maximum combined vent capacity when two appliances that are equipped with draft hoods are connected to the same vent. Common vent capacities for two draft hood–equipped appliances were not computed using VENT-II. Instead, the values from prior editions of the code were retained in the NAT+NAT column.

Exterior Masonry Chimneys

For two or more appliances, with common venting into an exterior masonry chimney, Tables 10.12 and 10.13 provide minimum and maximum appliance input rates. To use these tables for sizing an exterior masonry chimney, the vent connector must be Type B double-wall, at least one appliance must be draft hood equipped, and at least one appliance must be used for space heating.

Table 10.12 is used when all appliances are draft hood equipped. Table 10.12a provides combined appliance maximum input rates, and Table 10.12b provides minimum space-heating appliance input rates for five climates. Table 10.13 is used when at least one appliance (but not all) is fan assisted. Table 10.13a provides combined maximum input rates, and Table 10.13b provides minimum space-heating appliance input rates for five climates.

Common Vent Connector Manifolds and Offsets

Connector manifolds in common vent systems result when the vent connectors of the appliances are combined before entering the vertical vent, as depicted in Figure G.11. The vertical vent may also be offset when going through the roof, as depicted in Figure G.12. These two cases are addressed separately in the following paragraphs. In Figures G.10 and G.12 the horizontal length is labeled L_0.

Case 1: Manifolding Vent Connectors. Based on the connector rise, the connector diameter, and the vent height, the vent connectors of both appliances are sized as for any common vent system. However, the common vent capacities listed in Table 10.6 through Table 10.9 are reduced by 10 percent to account for the additional restriction caused by the connector manifold. The horizontal length of the connector manifold should not exceed 1.5 ft per inch of common vent connector manifold diameter.

Case 2: Offsets in the Vertical Vent. Again, the vent connectors of both appliances are sized as for any common vent system. However, the common vent capacity is reduced by 20 percent for each offset. The horizontal length of the offset should not exceed 1.5 ft per inch of combined vent diameter.

Additional Guidelines for Common Venting Tables

The common venting tables were generated by assuming a vent connector horizontal length of 1.5 ft (18 in.) (460 mm) for each inch of connector diameter, as shown in the table in 10.2.2. The vent connector should be routed to the vent using the shortest possible route. Longer connectors than those listed above are permitted under the following conditions:

(a) The maximum capacity (i.e., FAN Max or NAT Max) of the vent connector should be reduced 10 percent for each additional multiple of the length listed in the table in 10.2.2 for Tables 10.6 through 10.10. For example, the maximum length listed in this table for a 4-in. (100-mm) connector is 6 ft (1.8 m). With a connector length greater than 6 ft (1.8 m) but not exceeding 12 ft (3.7 m), the maximum capacity must be reduced by 10 percent (0.90 @ maximum vent connector capacity). With a connector length greater than 12 ft (3.7 m) but not exceeding 18 ft (5.5 m), the maximum capacity must be reduced by 20 percent (0.80 @ maximum vent capacity).

(b) The minimum capacity (i.e., FAN Min) should be determined by referring to the corresponding single appliance tables (Table 10.1 to Table 10.4). In this case, for each appliance the entire vent connector and common vent from the appliance to the vent termination should be treated as a single appliance vent — that is, as if the other appliance were not present.

When two or more appliances are connected to a vertical vent or chimney, the flow area of the largest section of vertical vent or chimney is assumed to be less than or equal to seven times the vent outlet flow area of the smallest appliance. For appliances with more than one input rate, the minimum vent connector capacity (i.e., FAN Min) determined from the tables should be less than the lowest appliance input rating, and the maximum vent connector capacity (i.e., FAN Max or NAT Max) determined from the tables should be greater than the highest appliance input rating.

Vent connectors should not be increased more than two table sizes greater than the listed appliance categorized vent diameter, flue collar diameter, or draft hood outlet diameter. Vent connectors should not be smaller than the listed appliance outlet diameter.

SUMMARY

In 1987 the National Appliance Energy Conservation Act began to mandate higher minimum efficiency standards for nearly all gas appliances. Effective January 1, 1992, all fan-assisted central furnaces manufactured in the United States must have a minimum AFUE of at least 78 percent. (This percent corresponds to a steady-state thermal efficiency based on flue loss considerations of approximately 80.5 percent.) Also, the Department of Energy is evaluating minimum efficiency standards of gas-fired water heaters.

The furnace manufacturers have responded to these new regulations by developing higher-efficiency Category I gas furnaces with fan-assisted combustion systems. These new Category I furnaces have steady-state thermal efficiencies of up to 83 percent based on flue loss considerations. New venting guidelines and tables have been developed for these appliances because of the higher operating efficiencies and the presence of a fan-assisted combustion system.

The majority of gas appliances installed in homes prior to 1992 are draft hood–equipped models with steady-state thermal efficiencies around 75 percent. These appliances can be vented using recommendations found in the *National Fuel Gas Code*, which are also included with the tables in this report.

SUMMARY OF RESEARCH FINDINGS

Guidelines and tables found in editions of the *National Fuel Gas Code* prior to 1992 are still valid for lower-efficiency, Category I draft hood appliances and remain unchanged. The new guidelines and tables developed herein apply only to higher-efficiency, Category I fan-assisted appliances currently being produced to meet the new minimum efficiency standards.

The VENT-II computer program has been used to develop venting guidelines and tables for Category I fan-assisted appliances with steady-state thermal efficiencies that are based on flue loss considerations of up to 83 percent. Experimental validation of the VENT-II program was performed using typical venting systems composed of Type B double-wall gas vents meeting Underwriters Laboratories Standard UL 441 and also using an interior masonry chimney constructed to meet NFPA 211 requirements.

Venting guidelines and tables have been developed for venting either a single Category I fan-assisted appliance or for common venting Category I fan-assisted appliances with other Category I appliances. Vent systems covered by these guidelines and tables include Type B double-wall vents and interior and exterior masonry chimneys connected to Category I fan-assisted appliances by using either single-wall metal vent connectors or Type B double-wall vent connectors.

Category I fan-assisted appliances, compared to draft hood–equipped Category I appliances, have a greater potential for producing condensate in the vent system. This condensate can cause excessive corrosion in metal vent systems and water damage to surrounding structures if the condensate should leak from the vent system through joints or cracks. To limit condensate problems in vent systems with Category I fan-assisted appliances, wet-time limits have been established for various vent materials.

For a given vent size and configuration, wet-time limits are used to calculate minimum fan-assisted appliance input ratings, and vent static pressure limits are used to calculate maximum fan-assisted appliance input ratings. This information has been put into the same format as the venting guidelines and tables in previous editions of the code.

For the sake of completeness, the venting tables presented herein contain entries for sizing vents for both fan-assisted appliances and draft hood appliances. However, the table entries for draft hood appliances are identical (or interpolated) to those values in previous editions of the code. The scope of this supplement is limited to establishing venting guidelines and tables for Category I fan-assisted appliances.

References

Agrawal, A. K., R. A. Cudnik, B. Hindin, D. W. Locklin, M. J. Murphy, J. H. Payer, R. Razgaitis, G. H. Stickford, and S. G. Talbert. "Technology Development for Corrosion-Resistant Condensing Heat Exchangers." Gas Research Institute Technical Report, GRI-85/0282 (NTIS No. PB86-172038), October 1985.

American Gas Association (Secretariat). *Gas-Fired Central Furnaces (Except Direct Vent Central Furnaces)*. ANSI Z21.47, 1990 edition.

American National Standards Institute and American Society of Heating, Refrigerating and Air-Conditioning Engineers. *Method of Testing for Annual Fuel Utilization Efficiency of Residential Central Furnaces and Boilers*. ANSI/ASHRAE 103, 1982 edition.

American Society of Heating, Refrigerating and Air-Conditioning Engineers. "Chimney, Gas Vent, and Fireplace Systems." Chapter 26 in *Equipment Handbook*, 1979.

American Society of Heating, Refrigerating and Air-Conditioning Engineers. "Combustion and Fuels." Chapter 15 in *Fundamentals Handbook*, 1985.

Ball, D. A., and D. W. Locklin, eds. "Symposium on Condensing Heat Exchangers." Conference Proceedings, Battelle Laboratories, GRI-82/0009.3 (NTIS No. PB82-240078), March 1982.

Borgeson, R. A., D. W. DeWerth, and V. Kam (A.G.A. Laboratories); and J. J. Crisafulli, A. L. Rutz, and S. G. Talbert (Battelle). "Development of Venting Guidelines for Category I Fan-Assisted Gas Appliances with Fan-Assisted Combustion Systems." Topical report of research conducted as part of GRI's Venting Program, American Gas Association Laboratories and Battelle Laboratories for Gas Research Institute, June 25, 1992.

Connelly, S. M., and D. W. DeWerth (A.G.A. Laboratories); and D. W. Locklin, D. D. Paul, and S. G. Talbert (Battelle). "Venting Requirements for High-Efficiency Gas-Fired Heating Equipment." American Gas Association Laboratories and Battelle Laboratories for Gas Research Institute, GRI Report No.83/0062 (NTIS No. PB84-159912), September 1983.

Crisafulli, J. J., R. D. Fischer, D. D. Paul, A. L. Rutz, S. G. Talbert, and G. R. Whitacre. "User's Manual for VENT-II (Version 3.0) with Diskette: A Dynamic Microcomputer Program for Analyzing Gas Venting Systems." Gas Research Institute Topical Report, GRI-88/0304, November 1988.

Crisafulli, J. J., R. D. Fischer, D. D. Paul, A. L. Rutz, S. G. Talbert, and G. R. Whitacre. "User's Manual for VENT-II (Version 4.0) with Diskette: A Dynamic Microcomputer Program for Analyzing Common Venting Systems for Two Gas Appliances." Gas Research Institute Topical Report, GRI-89/0094, April 1989.

Crisafulli, J. J., R. D. Fischer, D. D. Paul, A. L. Rutz, S. G. Talbert, and G. R. Whitacre. "User's Manual for VENT-II (Version 4.1) with Diskette: An Interactive Personal Computer Program for Design and Analysis of Venting Systems for One or Two Gas Appliances." Gas Research Institute Topical Report, GRI-90/0178 (NTIS No. PB91-509950), July 1990.

Cudnik, R. A., C. A. Farnsworth, D. W. Locklin, J. H. Payer, R. Razgaitis, G. H. Stickford, S. G. Talbert, and E. L. White. "Research on Heat-Exchanger Corrosion." Gas Research Institute for Battelle Laboratories, GRI-84/0157 (NTIS No. PB85-126316), September 1984.

Gas Engineers Handbook. 1st ed. New York: Industrial Press, 1965.

National Fire Protection Association. *Standard for Chimneys, Fireplaces, Vents, and Solid Fuel-Burning Appliances*, NFPA 211, 1996 edition.

Paul, D. D., A. L. Rutz, and S. G. Talbert. "User's Manual for VENT-II: A Microcomputer Program for Designing Vent Systems for High-Efficiency Gas Appliances." Gas Research Institute Topical Report, GRI-85/0226 (NTIS No. PB86-20785), 1985.

Paul, D. D., A. L. Rutz, and S. G. Talbert. "User's Manual for VENT-II (Version 2.0) with Diskette: A Microcomputer Program for Designing Vent Systems for Gas Appliances." Gas Research Institute Topical Report, GRI-88/0027, February 1988.

Philips, D. B., A. L. Rutz, S. G. Talbert, and G. R. Whitacre. "Venting Gas Appliances into Exterior Masonry Chimneys: Analysis and Recommendations." GRI Topical Report, GRI-94/0193, June 1994.

Private communication with Richard Stone of Metalbestos Systems, Wallace Murray Corp., Belmont, CA, 1983.

Talbert, S. G. "Evaluation of Masonry Chimney Materials for Venting Category I Gas Appliances." GRI Topical Report, GRI-94/0338, June 1993.

Underwriters Laboratories Inc. *Gas Vents.* UL 441, 1991 edition.

U.S. Department of Energy. *Energy Conservation Program for Consumer Products: Final Rule Regarding Test Procedures and Energy Conservation Standards for Water Heaters.* 10 CFR Part 430, Federal Register, Part III, 55(201):42162-42177. October 17, 1990.

U.S. Department of Energy. *National Appliance Energy Conservation Act of 1987.* Public Law 100-12, March 17, 1987.

U.S. Department of Energy and Brookhaven National Laboratory for Gas Research Institute. "1987 International Symposium on Condensing Heat Exchangers." Abstracts (GRI-87/0091.1) and Proceedings (GRI-87/0091.2 and /0091.3, BNL-52086), April 1987.

Corrugated Stainless Steel Tubing Gas Piping Systems

Robert Torbin
Foster-Miller, Inc.

Editor's Note: Requirements for the corrugated stainless steel tubing (CSST) system of piping for fuel gases in buildings have been included in the National Fuel Gas Code since 1988, and this system is becoming very popular with plumbers and mechanical contractors. This technology provides another piping option in addition to steel pipe and copper tubing, and it is an economically favorable alternative to steel pipe in areas where the use of copper tubing is not practical or not allowed by local practice.

This supplement to the National Fuel Gas Code Handbook updates the supplement to the 1996 handbook and provides the user with more information on CSST. This information will enable both code users and code enforcers to better understand this material and will assist in its proper use.

Special thanks are extended to the Gas Research Institute, which funded the research program that identified CSST and sponsored demonstration projects.

Beginning in 1988, the *National Fuel Gas Code* has included coverage for listed corrugated stainless steel tubing (CSST) systems that are used for gas piping in homes. In the 1992 edition of the code, a new standard *Standard for Fuel Gas Piping Systems Using Corrugated Stainless Steel Tubing* (ANSI LC-1), for this product was referenced, including sizing tables for CSST piping systems. Coverage was added for using CSST fittings in concealed locations where the fittings are listed for concealed installation. The 1996 edition of the code updated the sizing tables and the ANSI LC-1 standard. Major changes in CSST technology have not occurred since 1996.

The Gas Research Institute (GRI) has actively sponsored research in alternative interior distribution technologies since 1983. GRI is a not-for-profit R&D organization that works on behalf of the natural gas industry and its customers. This research has resulted in several innovations, including new tubing technology, installation practices, and has provided input to standard writing groups. Extensive laboratory and field evaluations have been conducted by Foster-Miller, Inc. (under contract to GRI), in all types of building environments, including newly constructed and rehabilitated single family and multifamily houses, commercial buildings, and low rise and high rise structures. As part of this research, alternative approaches to the design and installation of interior gas piping systems have been developed, optimized, and field evaluated. The results from the research clearly indicate that CSST systems (operating at either low or elevated pressure) are both safe and reliable and, in many situations, can be installed for less cost than conventional, low-pressure, steel pipe systems.

CSST is available in the following sizes: $3/_8$ in., $1/_2$ in., $3/_4$ in., 1 in., $11/_4$ in., and $11/_2$ in. ID (0.9 mm, 13 mm, 19 mm, 25 mm, 32 mm, and 38 mm ID). The tubing is packaged in long coils on spools that facilitate the dispensing of the tubing. The coils are available in lengths varying from 100 ft to 500 ft (30 m to 150 m), depending on tubing diameter. Special mechanical fittings are provided to connect the tubing with other components within the overall gas distribution network. The tubing can be cut and the fittings field assembled using standard hand tools. All fittings have standard NPT threads for connection to conventional plumbing components and appliances.

Although the tubing and fittings are components, the ANSI LC-1 standard requires that CSST be sold as a system. Other necessary components to be used within the overall gas piping system (see Exhibit S2.1) include the following:

- Multiport manifolds
- Appliance termination outlets
- Special metallic striker plates
- A pressure regulator (for elevated pressure systems)

The CSST manufacturer is also responsible for preparing and distributing a set of design and installation instructions, as well as the training of all installers. The installation instructions must be revised and reprinted as needed and the trained installers kept abreast of all pertinent changes to the hardware, installation practices, or sizing methods.

CSST can be used in either conventional low-pressure or elevated-pressure [up to 5 psi (34 kPa)] systems. The choice of design pressure and allowable pressure drop depends on several factors, including the application, availability of elevated pressure from the gas supplier, local plumbing code restrictions, required gas load, delivery distance, and the locations of the appliances within the building. Regardless of the system pressure, the gas distribution network can be installed in one of three arrangements — either in series, in parallel, or a combination of the two. A series layout, like the design of a conventional steel pipe system, is best described as branch lines, supplying individual appliances and extending from one main feed line. Thus, the total required gas load is varied as the appliance lines branch off from the main feed line. In the parallel arrangement, a main line is installed from the meter to a multiport distribution manifold. The manifold can be located near the meter or inside the house near the major gas-burning appliances. From this manifold, independent, parallel lines are run to each appliance as shown in Exhibit S2.2.

Exhibit S2.1 *CSST system components. (Courtesy of Titeflex.)*

The use of CSST has several installation advantages over rigid steel pipe. In most installations, there will be no intermediate joints because the tubing can be installed in one continuous run. A continuous run not only significantly reduces the total number of joints, but it also minimizes the number of potential leak sites. The flexibility of CSST permits it to bend but not break with any building settlement or movement caused by wind or earthquake. Since most gas piping is field run, the flexibility of the corrugated tubing allows the installer to seek the path of least resistance around existing obstacles. This flexibility eliminates the repetitive measuring, cutting, threading, and joint assembly that are common with steel pipe systems.

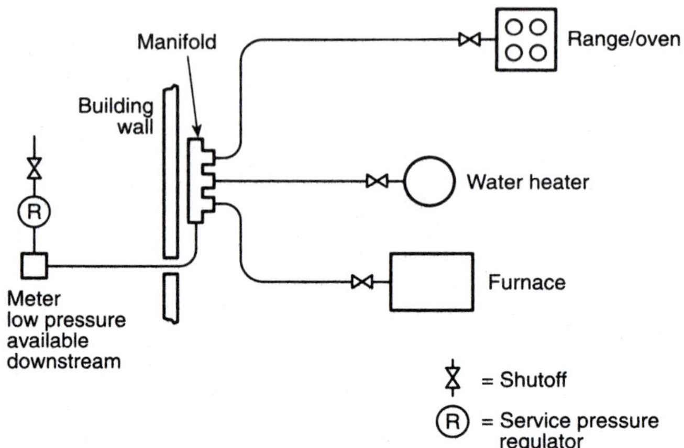

Exhibit S2.2 Typical parallel low-pressure system arrangement. (Courtesy of Foster-Miller, Inc.)

CSST has many applications, but it cannot be installed universally. CSST can only be used in above-ground installations, because at the time of this writing LC-1 does not include tests for underground applications. At some point in the future CSST may be recommended for buried installation. From a practical viewpoint, in areas where copper tubing is normally used for gas piping, it may offer economic advantages over CSST.

The dual-pressure system has become the standard approach to the design of the gas distribution network when using CSST. Typically, it is used when the gas load exceeds 100 ft^3/hr (2.8 m^3/hr), there is a significant distance between the gas meter and the gas appliances, and the pathways include complex turns and twists. Using Exhibit S2.3 as a reference, we can describe the system as follows:

(1) Natural gas at street pressure is delivered to the building, where a service regulator drops the pressure to 2 psi (14 kPa).
(2) The entire household gas load is then delivered through a single small-diameter line to a centrally located distribution point where a shutoff valve, pressure regulator, and distribution manifold are located (see Exhibit S2.4).
(3) The line pressure is reduced from 2 psi (14 kPa) to about $^1/_4$ psi (1.7 kPa), according to local requirements.
(4) The gas is then redistributed, at low pressure, through relatively short, independent, parallel lines to each appliance. Finally, the tubing is connected to the appliance, according to local practice.

The adoption and adaptation of the CSST system by the plumbing industry has necessitated the need for a change to the traditional system sizing method (Longest Length Method) provided in the code. CSST systems are usually sized using an

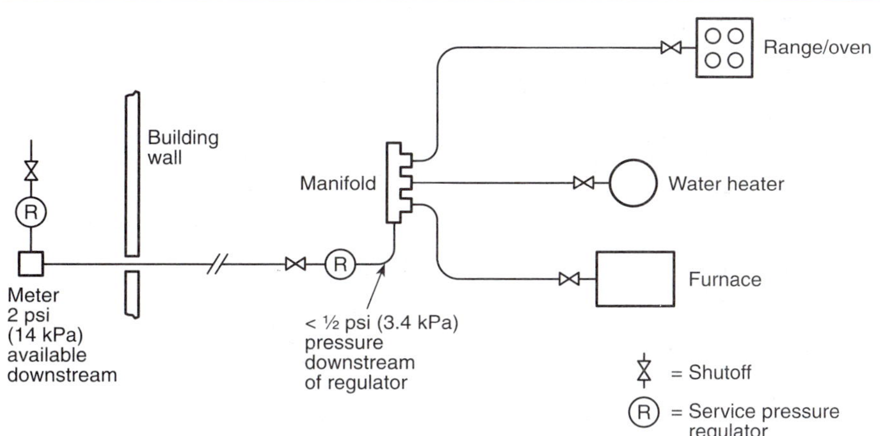

Exhibit S2.3 *Typical dual-pressure system arrangement. (Courtesy of Foster-Miller, Inc.)*

approved engineering approach as just described for the dual-pressure system. Associated with an engineered approach is the use of nontraditional pressure drops. Depending on the operating pressure, CSST systems are sized using pressure drops ranging from 0.5 in. to 6 in. (0.124 to 1.5 kPa) water column. Consultation with the local gas company and plumbing inspector is recommended prior to using one of the alternate sizing methods and/or pressure drops. New sizing examples have been developed that illustrate different approaches to sizing the pipe and tubing used in fuel gas networks, including examples of the dual-pressure approach and nontraditional system pressures and pressure drops. These examples are typically found in the manufacturer's installation instructions.

Over the course of many laboratory field trials and evaluations, a set of installation instructions and system sizing guidelines have been developed that represent practices that result in gas piping systems that have been proven to be reliable. These installation and sizing guidelines have been made available to all CSST manufacturers, which has led to the voluntary industry adoption of these practices. Industry-supported efforts have also led to standardized line sizing tables for CSST systems for a variety of line pressures and allowable pressure drops. These tables use a different line-sizing designation, called EHD (or Equivalent Hydraulic Diameter), which is assigned by the certifying organization in accordance with the ANSI LC-1 standard. These tables are constantly updated as new products come into the market.

Because CSST is constructed with a relatively thin wall thickness, it must be protected from puncture. This protection requirement extends to the installation of both concealed and exposed tubing where it is subject to damage. The tubing is most susceptible to punctures at points of support and at places where it passes through structural members such as studs, joists, plates, and so forth. Concealed tubing must be protected

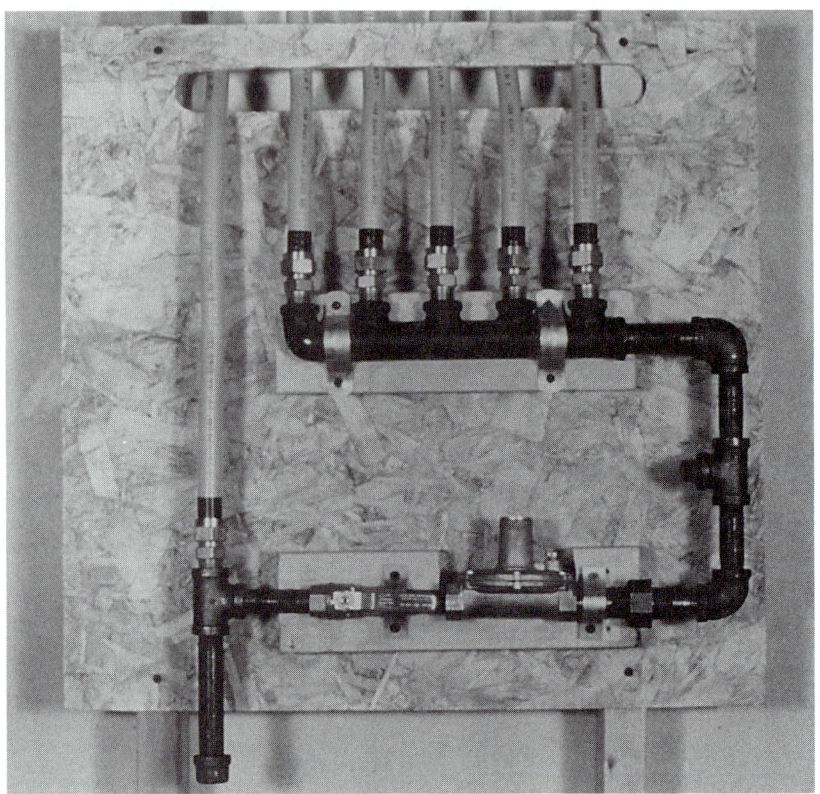

Exhibit S2.4 *Typical dual-pressure manifold center. (Courtesy of Foster-Miller, Inc.)*

with a listed metallic striker plate when it is located within $2^1/_2$ in. (64 mm) of the edge of any structural member. A variety of metallic striker plates that effectively deal with the variations of construction, common in both wood and metal frame buildings, have been developed. (See Exhibit S2.5.) Exhibits S2.6 and S2.7 represent typical situations requiring the use of striker plates. Minimum requirements for protection are included in the code, and extensive coverage can be found in the manufacturer's installation instructions.

As the use of CSST has grown, it has challenged both installers and enforcement officials with the need for understanding installation practices. Many of the traditional methods used to install gas piping are based on rigid steel pipe. This basis has led, in some cases, to different installation requirements for flexible tubing. For example, the gas meter can be installed without support if using steel piping. However, the meter cannot be installed using CSST without the use of supplemental mechanical support. Typical meter connections (with and without mechanical support) using CSST are

Exhibit S2.5 *Standard striker plate configurations. (Courtesy of Foster-Miller, Inc.)*

shown in Exhibit S2.8. CSST has also become very popular for retrofit applications, especially for the installation of additional appliances to an existing steel pipe system. This application has led to piping systems composed of two different materials, for which installation practices and line sizing for both the new tubing and the old piping must be checked.

One of the most innovative aspects of the CSST system is the qualification of the fittings for use in concealed locations. Each product must undergo rigorous testing (per the requirements of ANSI LC-1) that exposes the fitting/tubing to temperature extremes, static loads, vibration, and repeated cycling. The use of a concealed fitting is a last resort, according to the manufacturer's instructions. It should be employed only after all reasonable attempts to position the fitting in an accessible location have failed. This failure typically occurs when remodeling or repair activities require a change to the as-installed piping system. For example, when a permanent ceiling is installed in a previously unfinished basement, a once-accessible fitting becomes concealed. When a concealed tubing run is damaged accidentally, it can be repaired either by replacing the entire line or with a splice/coupling. However, replacing the entire line is not always a viable option, and a splice can end up behind the wallboard. The concealed joint qualification has been justified as a reasonable compromise to what would otherwise require an unnecessary and expensive replacement of an existing line to effect the desired repair/modification. Each manufacturer does provide a special termination fitting that, by its design, provides a means to support the tubing at the end of an appliance run. The

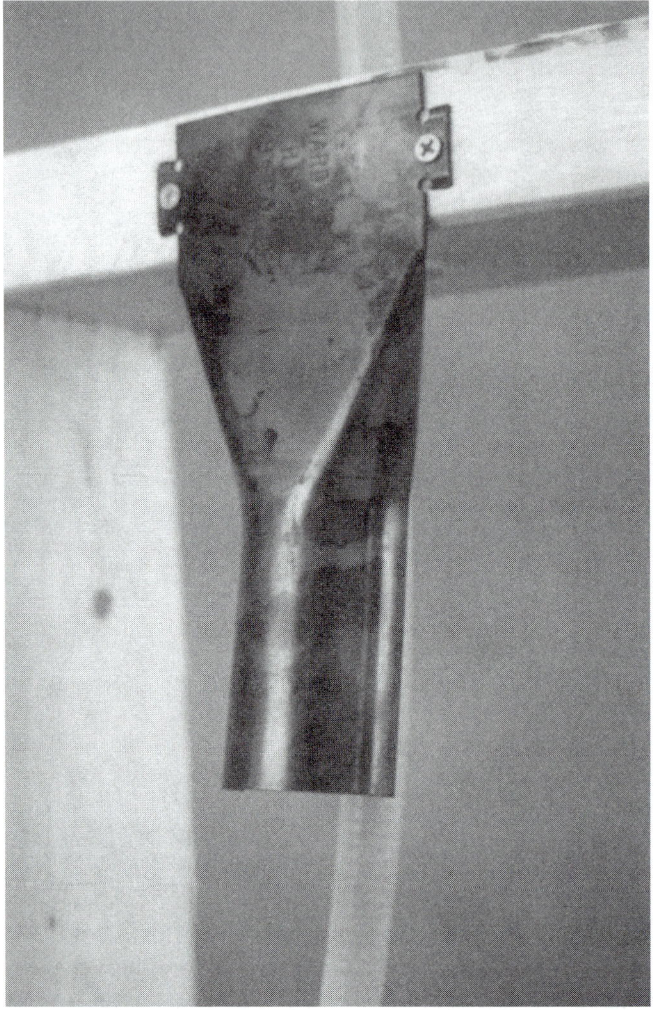

Exhibit S2.6 *Typical striker plate installation. (Courtesy of Foster-Miller, Inc.)*

termination fitting attaches to the wall stud behind the wallboard, eliminating the concealment of the fitting joint.

The introduction of CSST in 1989 initiated a decade-long process of code changes that reflect the adoption and adaptation of this new piping product by the plumbing and gas industries. Changes made to the code over this period have ensured that CSST systems are safe and reliable and that the industry's best engineering expertise has been applied to both the product and its installation practices.

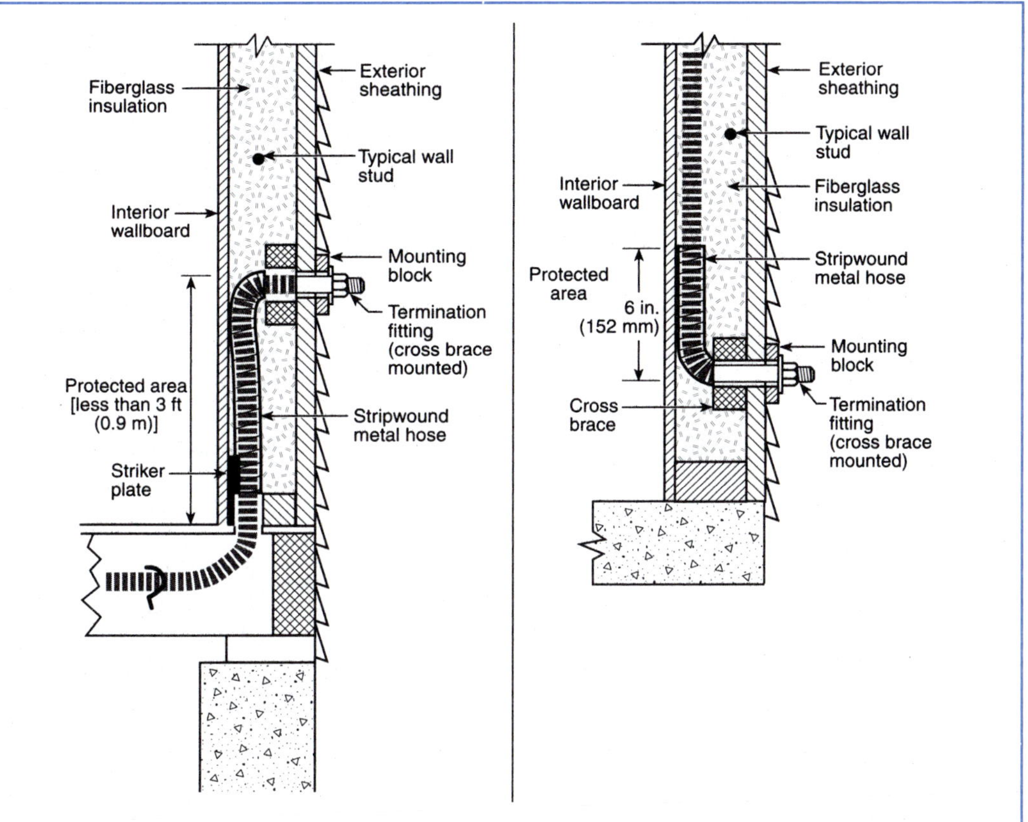

Exhibit S2.7 *Typical striker plate installation. (Courtesy of Foster-Miller, Inc.)*

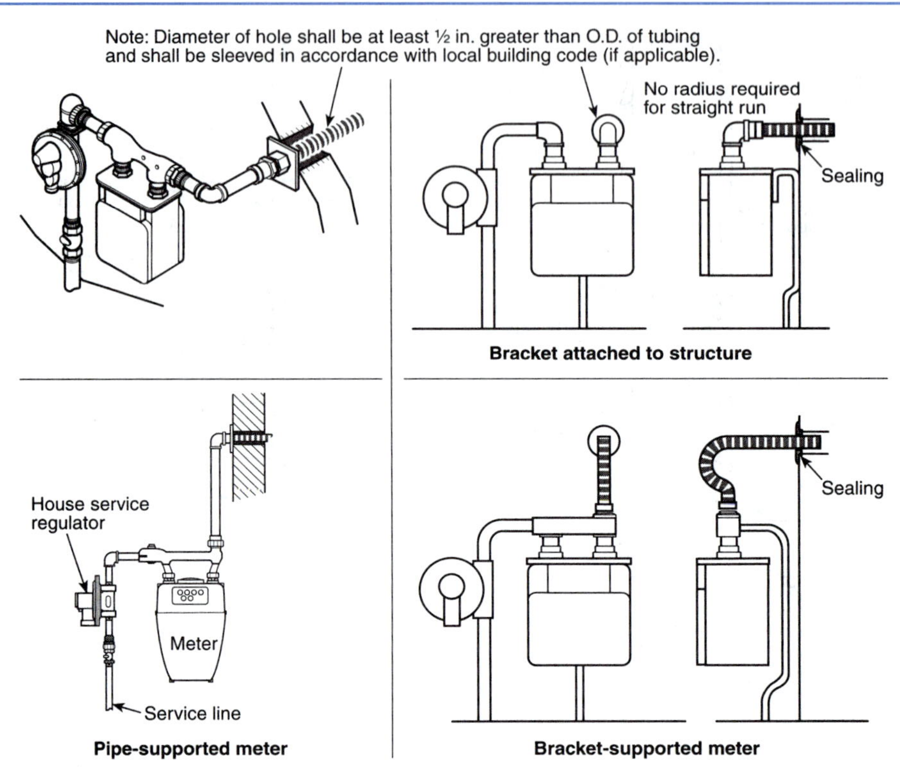

Exhibit S2.8 *Typical meter connections. (Courtesy of Foster-Miller, Inc.)*

Reference Cited

ANSI/IAS LC-1; CSA 6.26-M97, *Standard for Fuel Gas Piping Systems Using Corrugated Stainless Steel Tubing (CSST)*, 1997.

Technical Background for Residential Carbon Monoxide Responders

Irwin H. Billick
WEC Consulting Ltd.

Morrie Farbman
MJ Farbman Consulting

Editor's Note: This new supplement provides technical information to responders to carbon monoxide emergency calls that will enable them to establish a response policy.

The data on human response to carbon monoxide (CO) are variable, and the information presented in this supplement may vary from other published sources. However, these differences are not significant enough to affect the establishment of policy.

The editor thanks Messrs. Billick and Farbman for their contribution to addressing a problem that faces most fire departments, gas companies, and other emergency responders. The editor also expresses his appreciation to the Gas Research Institute for its support of the authors.

INTRODUCTION

In recent years there has been an increase in requests to both public and private organizations to provide assistance to residents in response to the real or suspected increased levels of potentially life-threatening carbon monoxide in their homes. To respond appropriately to these calls for help, responders need a well-conceived, technically sound plan or protocol that will ensure a rapid and efficient response.

The purpose of this supplement is to provide organizations that are the first responders to CO events with the technical background information needed to develop a protocol to meet their own unique local requirements. The protocol is a guideline for providing assistance to individuals who are, or suspect that they are, exposed to high levels of CO. Implicit in any protocol for responding to a CO call is resolving, or at least attempting to resolve, whether or not a suspected hazardous exposure actually occurred, or had the potential to occur, and whether or not an exposure, in fact, represented an acute exposure hazard threatening life and safety.

FUNCTIONS OF RESPONDERS

A distinction is made between first and secondary responders based on the major purpose for the call for assistance. The first responder's major purpose is to protect the lives and safety of the people from the acute toxic effects of high-level CO exposure. The major purpose of the secondary responder is to correct and eliminate the cause or source of the current or potentially hazardous exposure to CO. The distinction between primary and secondary responders is one of function and not organization. Indeed, the same organization may do both.

The basic function of fire departments, emergency medical services, and other similar organizations when they are the first responder to a CO call is to protect life and safety. It will be necessary for them to take action so that people will not be exposed to toxic levels of CO. These first responders generally do not have the training, tools, or resources to make detailed CO source identification and correction. In addition, these types of first responders may have other duties and manpower constraints that limit their capacity for detailed source identification and correction. In contrast, utility companies, frequently thought of as secondary responders, often are required to provide immediate action to protect life and safety. Their primary function, however, is to service potential sources of CO, such as combustion appliances. The material presented here will be useful to all those providing first-response capability, regardless of what their primary organization purpose may be.

OVERVIEW OF A CO PROTOCOL

Developing and implementing a CO protocol requires both good technical information and good public policy. It is not simply a matter of expert first responders reviewing conveniently available information and writing procedures based on measured CO concentrations as "action levels." Exhibit S3.1 is a flowchart of the elements that should be considered in the development of response protocols and procedures.

Several of the major elements that have a critical impact on the final protocol are beyond the scope of this supplement and will not be discussed here. These elements include legal, insurance, and organizational issues. The importance and impact of these issues cannot be overestimated and may be the ultimate determinant of the final protocol content. However, every effort should be made by first responders, who are ultimately charged with implementing the protocol, to ensure that these elements are consistent with basic scientific and technical facts and procedures and with the views of the public or private administrations within which they work. A detailed

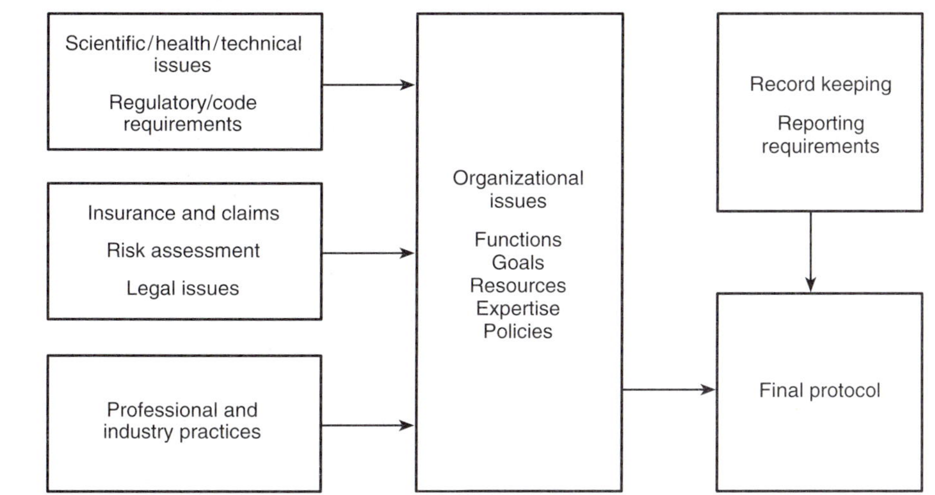

Exhibit S3.1 *Protocol development elements.*

review of current industry or professional practices is beyond the scope of this supplement. Many emergency response service and utility companies have produced CO response protocols. Some of these have been published on the Internet or are available on request.

A large number of these protocols were reviewed as preparation for this supplement. Several overall conclusions were derived from this review. First, one size does not fit all. Protocols and services provided are very locally driven and are highly affected by local organization or community demands, resources, and policies. Second, technical and factual information on which some of these protocols were developed may be inaccurate or misinterpreted, leading to inefficient overresponse. Finally, some sharing of protocols among organizations without critical review exists, which can lead to perpetuation of technical errors.

THE NATURE OF CARBON MONOXIDE POISONING

Carbon monoxide is an asphyxiant gas produced by the incomplete combustion of fossil fuels. In its pure form, it is odorless and colorless. In residential settings, it is often accompanied by other products of incomplete combustion, such as aldehydes, which have an acrid odor commonly identified by people as "fumes." In the absence of a properly working CO alarm or in the presence of illness, this smell may be an early warning of the incomplete combustion and CO. Exposure to elevated levels of CO may result in a variety of adverse health effects deriving from CO's affinity for hemoglobin to form carboxyhemoglobin (COHb), which disrupts oxygen transport in the body. Tissues with the highest oxygen needs — myocardium, brain, and exercising muscle — are the first affected. The percentage of COHb in the blood is a function of the level of CO exposure, the duration of exposure, and various physiological parameters such as respiration rate. Mathematically, various forms of the equation relate CO exposure to its impact on the body. Equation (S3.1) gives one such relationship between these parameters[1].

$$\%COHb_t = \%COHb_0^{(t/2398B)} + 218\,[1^{(t/2398B)}]\,[0.0003 + CO/1316] \tag{S3.1}$$

where:

$\%COHb_t$ = the percentage of COHb in the blood at time, t

$\%COHb_0$ = the percentage of COHb in the blood at time, 0

    CO = the level of CO exposure in ppm

      t = the time (min.)

        B = a parameter related to work or exercise level during exposure. B ranges from 0.1522 for sedentary work to 0.0404 for heavy work.

The importance of this equation is that it illustrates that both the level and time of CO exposure are associated with the COHb levels in the blood. As a result, interpreting CO exposure hazards from single- or short-term measurements can be misleading in interpreting acute safety hazards.

Exhibit S3.2 illustrates the relationship between CO concentrations and COHb levels in blood for a heavily working or exercising male adult. The curves for individuals with other characteristics, such as age, sex, physical activity, or health status will have similarly shaped curves.

Of vital importance for response protocol development is knowledge of the possible or expected adverse health effects or symptoms associated with various carboxyhemoglobin levels. The possible health outcomes for different levels of COHb that were used by Underwriters Laboratories in its standard for CO alarms[1] are given in Table S3.1. Other organizations, such as the Consumer Product Safety Commission (CPSC) and the Environmental Protection Agency (EPA), have presented similar guidelines[6]. The relationship between health effects and the measured or observed CO concentration levels are also shown in Exhibit S3.2 for that particular example.

It should be emphasized that the health effects/COHb/CO relationships discussed in this supplement are for general information only. The final levels adopted for actual action levels to be included in a response protocol should be based on actual recommendations by the appropriate public health authorities and on legal advice. United States government agencies such as the Environmental Protection Agency (EPA) and the Occupational Safety and Health Administration (OSHA) promulgate exposure standards (in terms of weighted average ppm over one to eight hours). These standards have specific applications in terms of subjects, levels of activity, and the health effects they are designed to prevent. These factors underlying the standards may or may not apply to the general public and, therefore, should not be applied to interpreting hazardous conditions in residences without careful consid-

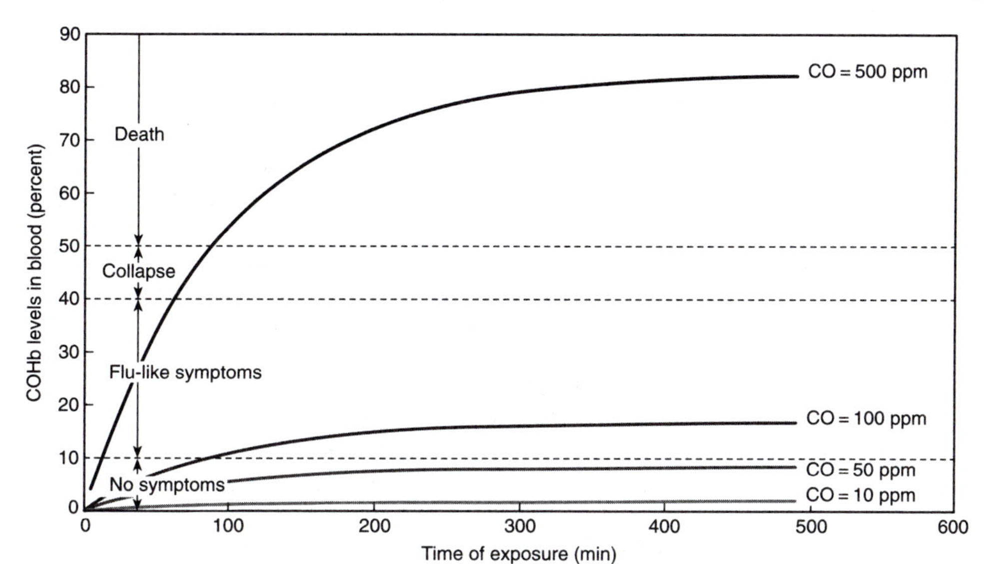

Exhibit S3.2 *Carboxyhemoglobin levels at constant carbon monoxide levels (high work effort).*

eration. Neither EPA nor OSHA currently supports use of such standard CO levels for use outside their immediate, statutory applications. Table S3.1 presents the current information on expected health outcomes provided by the American Lung Association, EPA, and CPSC[6].

Table S3.1   Expected Health Effects at Various Levels of Carboxyhemoglobin in Blood

Percent COHb	Expected Health Effects
50	Death or permanent brain damage
45	Coma and permanent brain damage
40	Collapse
35	Vomiting
30	Drowsiness
25	Headaches and nausea
20	Headache
15	Slight headache
10	None
5	None

OCCURRENCE OF UNINTENTIONAL CARBON MONOXIDE POISONING

An efficient and effective response protocol, which makes optimal use of the resources available, should be based on an accurate and realistic knowledge of adverse health effects from exposure to CO. The adverse health and safety impacts of CO of most concern to first responders are death and acute illness. It may be useful, in developing a protocol, to be aware of the incidence of CO poisoning, as well as individual fear of being poisoned. Current risk assessment practices usually consider two components. The first is factual or statistically based assessment of risk: how many people are affected, at what rate, are there any trends with time, what sources of CO are most likely, what levels and duration of exposure produce what effects, and so forth. Equally important is the perception of risk: What do people believe is the risk and how does this impact their demand for protection? The perception of risk is often referred to as the "outrage factor." Emphasis is given here to the factual assessment of risk. However, the perception of risk or outrage may have greater weight when considering legal, insurance, or policy issues.

Of primary concern, and the best documented, is the risk of unintentional death due to CO poisoning. Summary statistics on all the deaths specifically attributable to CO in the United States have been available from various federal agencies as far back as 1939. Exhibit S3.3 shows the percent distribution of CO-related deaths from all sources, including fire, suicides, unintentional accidents, and all other sources for the year 1989, using the CO data and cause classifications from the National Center for Health Statistics[7]. The general category "unintentional or accidental deaths," which is of concern for response protocols, accounts for less than 20 percent of the total.

In addition, starting in 1939, unintentional or accidental deaths were specifically attributed to CO and were further classified into three broad cause categories: illuminating or pipeline gas; motor vehicle exhaust gas; and other sources. With time, these general categories were broken down into several subclasses. Those subcategories of concern for residential response development are pipeline gas, LP-Gas, other domestic fuels, and motor vehicle exhaust. These historical data on CO deaths permit a long-term analysis of the changes that have taken place over sixty years. Detailed statistics on sex, age, location, time of year, and other demographic factors also are available. Several studies have been published that analyze the characteristics associated with unintentional CO deaths for varying time periods[7, 8, 9].

Exhibit S3.4 tracks the number of unintentional CO deaths since 1939, both in total and for the major underlying cause subcategories — motor vehicle exhaust, pipeline gas, and other causes. Total CO deaths peaked in 1945, as did CO deaths attributable to pipeline gas. Pipeline-related CO deaths were the major fraction of total deaths until 1959, when they were surpassed by motor vehicle deaths, which have been decreasing steadily since their peak in 1945. Motor vehicle deaths began to rise in the mid-1950s and did not begin to decrease until about 1980.

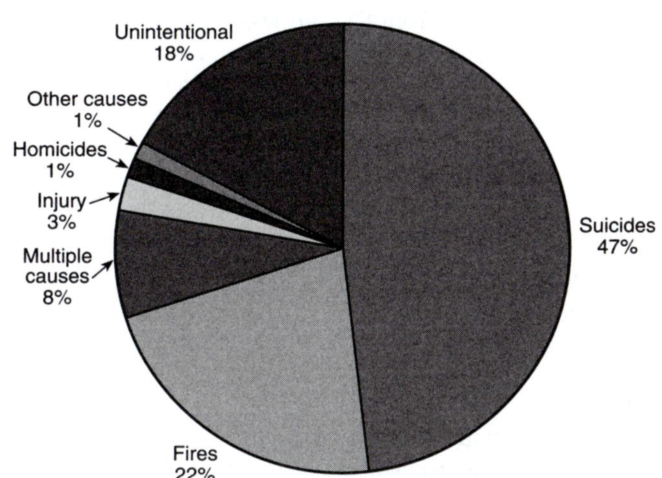

Exhibit S3.3 *Distribution of CO-related deaths (1989); 4630 total deaths. (Compiled from National Center for Health Statistics Data.)*

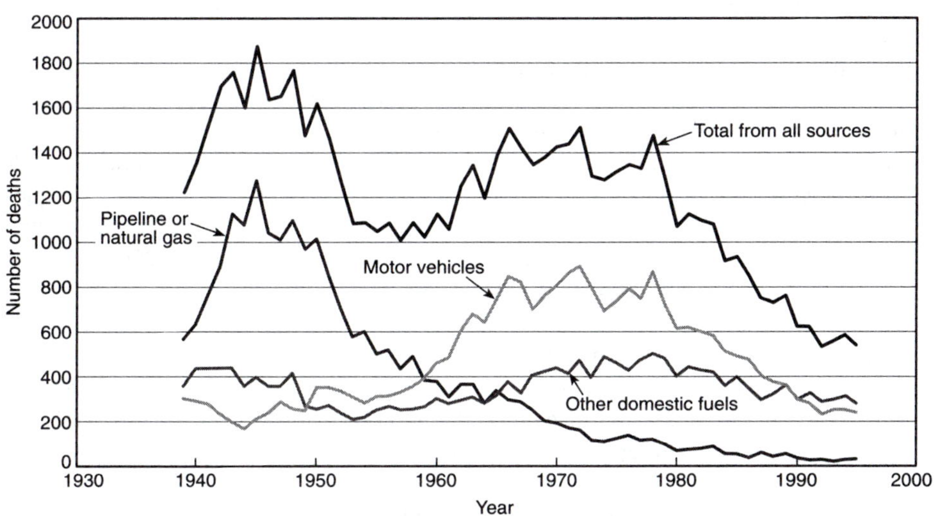

Exhibit S3.4 *Unintentional deaths, 1939 – 1995.*

Two observations are worth noting. First, the number of unintentional CO deaths for 1995 is 525, a decrease from a high of over 1800 in 1945. Second, CO deaths from all sources are currently decreasing at the rate of about 5 percent per year, a trend that shows no sign of stopping. This is a clear indication that the problem is diminishing.

Statistics for serious illness from CO are less well documented. The U.S. Consumer Product Safety Commission estimates about ten thousand persons per year require hospitalization or emergency treatment for CO poisoning.

REQUESTS FOR ASSISTANCE

The first point of contact for assistance related to a suspected carbon monoxide incident may be the result of the activation of a carbon monoxide alarm, illness, incapacitation, or suspected malfunction of appliances. The request may be made directly to a fire department, an emergency medical service, or a 911 center. In some cases, the request may be made to a local utility company. It would be advisable for all those organizations within a jurisdiction to develop a coordination plan defining their specific roles and responsibilities.

Dispatch personnel should be trained to elicit from the caller enough information to determine what level of response will be necessary. The proper response may be for emergency medical aid, investigation, or both. After determining what type of response will be needed, the dispatcher should advise the occupants to exit the building and await the arrival of the fire department. Remaining in the structure could continue to expose occupants to additional carbon monoxide, further elevating their COHb levels and accompanying symptoms.

Carbon monoxide (CO) alarms (sometimes referred to as "detectors") are intended to warn occupants of elevated CO levels well before those levels could cause a loss of ability to react to the dangers of carbon monoxide exposure. It is estimated that less than 30 percent of households have such alarms. There have been reports of CO alarm calls[2, 3] in which the responders have not found high levels of CO, and, therefore, questions have arisen about the reliability of alarm performance[4, 5]. However, the assumption of poor performance by the alarm in no way justifies a reduced or delayed response to an emergency call. All alarm calls should be treated as a call for assistance until on-scene investigation proves otherwise.

If the call is generated by an alarm activation, the call-taker should determine what kind of alarm (that is, a smoke alarm or carbon monoxide alarm) is activated. For both verified carbon monoxide alarm calls and non-alarm-generated calls, the call-taker must first determine whether any persons at the scene are exhibiting symptoms of carbon monoxide poisioning, specifically unconsciousness, convulsions, light-headedness, dizziness, nausea, headache, or other flu-like symptoms. In any incident where symptoms are reported, the response of emergency medical personnel (paramedic unit) as well as the other appropriate responders with a carbon monoxide metering device will be required. When the caller reports the activation of a carbon monoxide alarm or the suspected presence of carbon monoxide and there are no medical symptoms, only an investigative response may be required. In those situations where conditions are unknown or there is any doubt in the mind of the call-taker, a response by both an emergency medical unit and the investigative response organization is necessary.

INITIAL ARRIVAL ACTIONS

The first priority of the first responders is to take precautions to protect themselves from entering a potentially hazardous environment. Personnel should have self-contained breathing apparatus (SCBA) donned and ready for use. The decision to use SCBA can be made after the initial determination of CO levels using appropriate carbon monoxide meters.

Carbon monoxide meters are essential to the viability of a CO response program, and their reliability is directly related to their proper maintenance. Personnel must be trained in the proper operation and maintenance of carbon monoxide meters and procedures for conducting a carbon monoxide investigation. The manufacturer's instructions and recommendations for the maintenance of its meters should be followed. Maintenance normally includes the routine and consistent calibration of the meter, as well as periodic sensor and battery replacement.

Carbon monoxide meters with a digital readout are recommended for first responders. Single-gas, carbon monoxide–specific meters and multigas meters that monitor carbon monoxide as well as other gases are available, ranging in cost from approximately $500 to $2000. Regardless of the type of meter chosen, it is important that responders have a sufficient number of meters available for the level of service to be provided.

An initial CO meter reading is taken in ambient air prior to entering the structure. Some meters may need to be set to zero in ambient air prior to taking interior readings. Responders should operate meters according to manufacturer's instructions. A reading is first taken upon entering the structure. At this point, the determination must be made whether or not fire department personnel must use SCBA. Some fire departments require SCBA at CO levels as low as 35 ppm. The minimum level should be established by each fire department.

Upon entering the structure, the responder must first determine whether the occupants require medical attention and administer medical aid as needed. Once the need for medical attention has been addressed, the next action taken is to conduct a carbon monoxide investigation.

IDENTIFYING SOURCES AND LEVELS OF CO IN DWELLING UNITS

Any equipment that burns fossil fuels (such as vehicles or portable equipment fueled by gasoline; combustion appliances burning natural or LP-Gas; and fireplaces, stoves, and heaters using wood, charcoal, or kerosene) is a possible source of CO. Some of these devices may be permanently installed (such as ranges, space heaters, water heaters, furnaces, and fireplaces). Other devices may be mobile or portable (such as vehicles, barbecues, power washers, lawn mowers, and electric generators). Furthermore, possible sources of excessive CO may be outside the dwelling unit near a window or other air intake, in an attached garage or other dwelling unit, or in common areas of the structure. Therefore, a systematic search strategy to locate the source of excessive CO must be a basic part of a response.

Analysis of CO death data[7, 8, 9] provide useful information about the probable location and source of CO exposure. Exhibit S3.5 shows this information for 1988 through 1992. This analysis shows quite clearly that the home is the most likely location of deadly CO incidents. Within the home, the most likely sources are automobile exhaust and non-gasoline fuels.

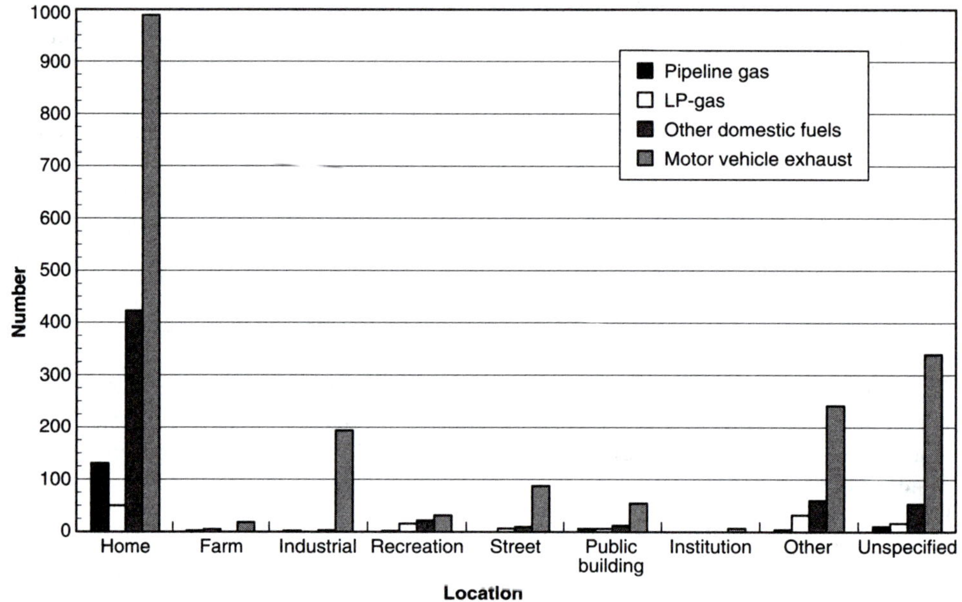

Exhibit S3.5 *CO deaths by source and location, 1988 – 1992.*

The responder must have the means to determine whether there are elevated levels of carbon monoxide present in air within a structure and must be able to measure those levels and trace the level of CO measured in the air back to the offending source. Appropriate action must be taken to eliminate current exposure to unacceptable levels of CO and to prevent future hazardous exposure.

One of the greatest challenges facing the first responder is interpreting the level of hazard from a limited number of CO measurements that must be taken in a short period of time. This interpretation is particularly relevant when the measured concentrations are relatively low compared to obviously hazardous environments such as those suggested by Exhibit S3.2. Single measurements tell very little about the time history of

CO concentrations in the structure, and it is the time history of CO concentrations that determine potential CO impact on occupants.

Measuring the levels of CO in the air is the first step in isolating specific high-CO-emitting sources, if any. Direct measurement of CO in the flue or vent of an appliance or at the tailpipe of a vehicle or other suspected source can be misleading. The levels of CO in the air are dependent on several other factors, such as house volume, air exchange or ventilation rate, time, and amount of fuel consumed, in addition to source strength. The relationships between these factors and the level of CO in the air have been studied and modeled extensively. An understanding of these relationships is needed to recognize the connection between levels of CO in the air and identification of a possible high-level source.

For the purpose of protocol development, the structure or building under investigation can be viewed as a single room with one or more sources emitting CO. It is also assumed that the CO coming from the sources is uniformly mixed throughout the entire structure. Equation (S3.2) describes an approximate relationship between the measured level of CO in the air at any given time and the sources. This approximation is based on the common practice that the CO metered was zeroed against the outside air before measurements where taken.

$$C(t) = \left(\frac{Sv}{Va}\right)(1^{(at)}) \tag{S3.2}$$

where:

$C(t)$ = the measured concentration of CO in air

  a = air exchange or ventilation rate per hour

  S = the concentration of CO in the source emission gas, in ppm

  v = volume of emission gases per hour, in ft^3 (depends on both the amount of fuel used and dilution due to combustion air)

  V = volume of the structure, in ft^3

The following important concepts are illustrated by Equation (S3.2).

(1) All possible sources should be running at the time of measurement. The occupants should be queried to determine where or not they turned off or removed any possible sources prior to the arrival of the responders. If so, these sources should be turned on so that the CO levels can be closely monitored for significant increase.

(2) Ventilation will have an impact on the observations as well as the ultimate CO levels reached. It should be determined whether the structure was ventilated prior to the arrival of the responders. If so, the ventilation should be returned to its original status.

(3) Measurements of CO should be taken in the air at a central location in the structure or room of interest.

(4) Measurements of concentration of CO taken in or near the vent are misleading. The level of CO in the flue gas is concentrated and is diluted by a factor related to the ratio of the volume of source emission gas (depends on volume of flue used) divided by the volume of the structure by the time it mixes throughout the house (v/V). A source with a given level of CO in the emission gas will give a different level in the air for a different size house or amount of fuel used. What may appear to be a high concentration of CO in the flue is diluted many times before reaching its final level in the air.

(5) CO levels, like carboxyhemoglobin levels in the blood, are time dependent, while decisions on what is the appropriate action are often made on a single or few measurements taken in a relatively short period of time. The short time measurement may or may not indicate what the ultimate CO level would have been if the source continued to operate. The maximum CO level that will be reached is Sv/Va.

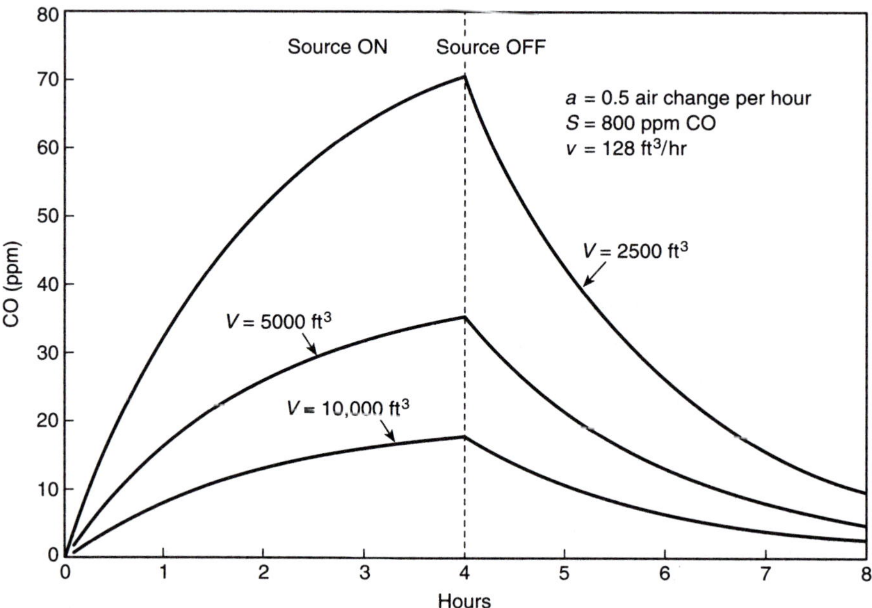

Exhibit S3.6 *CO levels for different house volumes at constant emission rate, using Equation (S3.2).*

The CO levels were modeled for three different house volumes, V — 2500, 5000, and 10,000 ft³ (see Exhibit S3.6). The other conditions were kept constant: a, air changes per hour, 0.5: S, source strength, 800 ppm CO in the flue; and v, volume of emission gas (fuel used), 128 ft^3/hr. The source was on for four hours and off for an additional four hours.

- The effect of volume of the house on dilution of CO from the source is readily apparent. A high CO emission in a small house is a more serious problem than the same emission in a large house.
- A single or a few closely spaced measurements are misleading. They give only the CO at the time taken and may not indicate what could ultimately take place if the source continued to operate.
- Readings taken with the source turned off are not indicative of what the levels may have been while the source was on.

If elevated levels of CO are found with all sources operating, it will be necessary to carry out the source identification procedure with only one possible source operating at a time to pinpoint the one that is the high producer of CO.

There are a number of potential sources that may be found in the home, which include but are not limited to the following:

- Emissions from an automobile left running. In colder climates, people may be tempted to warm up a vehicle while it is still in an attached garage. Even if the garage door is left open, high levels of CO may diffuse or infiltrate into the living areas of the structure.
- Blocked furnace, hot water heater, or fireplace flues and chimneys. Birds or other animals may nest in chimneys and partially or totally block the venting of combustion of gases, causing a backup of CO in the home. Poor design, construction, or installation may have the same effect.
- Improperly adjusted unvented appliances such as oven or cooktop, kitchen range, clothes dryer, space heaters, and gas refrigerators.
- Cracked furnace heat exchanger, which allows combustion products (including CO) to mix with heated air circulating in the home.
- Barbeque grills operated indoors. Both propane and charcoal-fueled barbeque grills can produce high levels of CO, and they are not intended for indoor use. [They pose a serious fire risk due to inadequate clearance for combustibles. NFPA 58, *Liquefied Petroleum Gas Code*, prohibits the storage and use of propane tanks indoors.]
- Portable generators, power washers, and lawnmowers, which contain internal combustion engines, have concerns similar to automobiles.

If the source that is causing high CO levels can be identified, it should be shut off and the structure should be ventilated. A second set of CO meter readings should be taken and the occupant should be informed to have the source of carbon monoxide repaired by a qualified service person before turning it back on. If a fuel-burning appliance has been shut off, the occupant must take corrective action before turning it back on (the information on levels of CO later in this article). If the natural gas supply has been shut off, the appropriate utility should be notified.

If there is no CO present or the occupant indicates that there have been previous activations, the possibility exists that the CO alarm is not working within specifications. In this case, the occupant should contact the alarm manufacturer or replace the alarm with a new one.

Depending on the level of CO found during the investigation, it is important to provide instructions to the occupant on what to do next. Studies have shown that there is a variation of CO levels in homes even under normal conditions. Exhibit S3.7 shows the distribution of maximum CO levels taken in 10-minute intervals during a 48-hour study period in 271 homes in California[10]. Over 95 percent of the homes had observations lower than 15 ppm. These observations are equivalent to those taken when the CO meter has been zeroed in the outside air.

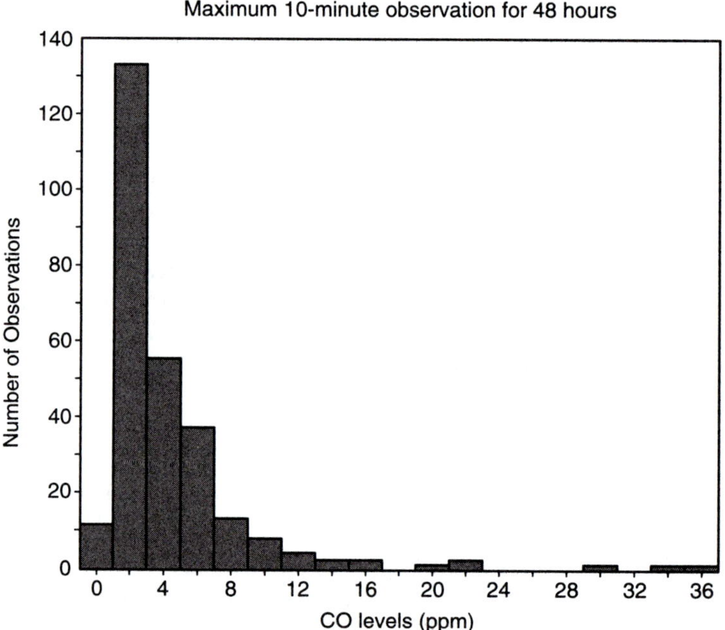

Exhibit S3.7 *Distribution of CO levels in residences.*

RECOMMENDATIONS FOR ACTION

Specific recommendations as to what actions should be taken at what CO level and what specific advice should be given to the occupant are beyond the scope of this supplement. This decision must be made locally, with input from all involved in the protocol development, as discussed earlier under "Overview of a CO Protocol." The relevant public health and legal experts should review guidance on the final choice of action levels and specific recommendations. The U.S. Consumer Product Safety Commission is preparing guidelines that will address these issues, and these guidelines should be consulted as part of the protocol development process.

Some of the possible actions and information that might be provided to the occupants for broad categories of CO levels are given in the following subsections as examples.

Low Level of CO

(1) Inform the resident(s) that your instruments did not detect significant levels of CO at this time.

(2) If the call was the result of a CO alarm, advise the resident(s) to take the following actions:

    a. Check the carbon monoxide alarm per the manufacturer recommendations.

    b. Reset the alarm.

    c. Call the manufacturer for additional information about the activation of the alarm at this level.

    d. Install a new CO alarm that meets the standard established by International Approval Services (IAS) 6-96 or Underwriters Laboratories (UL) 2034.

    e. If the alarm activates again, immediately call the responder back.

(3) If the call was the result of the complaint of a smell, determine if possible the source of smell. If the source is a gas leak, shut off the supply and inform the utility company. If the source is a spill, sewer backup, or other source, correct if possible or advise residents to have the problem corrected.

Mid Level of CO

(1) Inform the resident(s) that your instruments detected a higher than normal level of CO, and advise the resident(s) of the following:

    a. The observed levels are higher than normal and may indicate improper behavior (e.g., a vehicle left running in the garage, a closed fireplace damper, or possible malfunctioning equipment).

    b. This situation may lead to potentially dangerous levels of carbon monoxide in the future.

    c. A qualified contractor should be brought in as soon as possible to locate and repair the source of carbon monoxide.

    d. Remaining in the house until the source or action responsible for the increased levels are identified and corrective action is taken or repairs are made may not be safe.

High Level of CO

(1) Inform the resident(s) that instruments detected a hazardous and potentially lethal level of CO, and advise the resident(s) of the following:

    a. They must leave the building immediately; it is not safe to return until the source of CO is found and corrected.

    b. All possible sources of CO will be examined and repaired by a qualified contractor.

(2) Post the building "Not Suitable for Occupancy" until the sources of CO are found and eliminated.

(3) If a multifamily occupancy is involved and elevated levels of carbon monoxide are found in an individual living unit, perform the following:

    a. Check other units in the structure for CO.

    b. Consider the evacuation of all other occupants.

All Levels of CO

(1) Educational materials describing the dangers of CO poisoning and what should be done to avoid or protect against it should be given to the resident(s). Such materials are readily available from a number of public or private sources, including the Public Affairs Department of the National Fire Protection Association.

(2) A written Notice of Findings should be given to the resident(s) before leaving.

FORMS AND RECORD KEEPING

The use of standard forms or checklists is recommended. The first is a checklist for responding personnel to document vital information about the response call. An example of such a checklist is shown in Exhibit S3.8 (page 542).

A second form, Notice of Findings, should be given to the resident before leaving. Copies of these forms should be attached to the response report, and a second copy can be forwarded to cooperating agencies (such as the fire prevention bureau or building department) for follow-up. Standard incident reports should be filed with state or national data collection agencies. An example of a Notice of Findings form is shown in Exhibit S3.9 (page 544).

Both the Notice of Findings and the checklist should be reviewed and modified to suit local requirements and policies. To facilitate the tracking of the incidence of CO poisoning events, data should be submitted to the local fire department for inclusion in local and state incident databases and possibly the National Fire Incident Reporting System (NFIRS).

FOLLOW-UP

In those incidents where the source of CO was identified, the responding organization may want to follow up on the repair of the problem to ensure that the resident has followed instructions. Depending on the depth of resources of the jurisdiction, follow-up by the utility company, municipal building department, community development department, or the fire prevention bureau is desirable. In some communities, permits may be required or contractors may need to be approved prior to doing work. If the natural gas supply to the furnace or other appliances in the building is to be shut off, the gas utility should be notified.

References Cited

1. Underwriters Laboratories Inc., UL 2034, *Standard for Single- and Multiple-Station Carbon Monoxide Detectors,* August 8, 1995.
2. Tikalsky, S.M., and Dramer, J.M., *Carbon Monoxide Response Survey Analysis: Suburban Chicago Data-Interim Report.* GRI Report No. 95/0335. Gas Research Institute, October 1995.
3. Kramer, J.M., and Tikalsky, S.M., *Carbon Monoxide Response Survey Analysis: Utility Data-Supplemental Report,* GRI Report No. 97/0408. Gas Research Institute, April 1998.
4. Hedrick R.L., *Chamber Testing of Residential CO Alarms.* GRI Report No. 98/0140. Gas Research Institute, May 1998.
5. Clifford, P.K., and Siu, D.J., *Performance Testing of Residential CO Alarms.* GRI Report No. 98/0284. Gas Research Institute, December 1998.
6. American Lung Association, Environmental Protection Agency, Consumer Product Safety Commission, and the American Medical Association. *Indoor Air Pollution: An Introduction for Health Professionals.* 1995.
7. Koontz, M.D., and Niang, L.L., *Unintentional Carbon Monoxide-Related Deaths Between 1979 and 1993.* GRI Report No. 96/0038. Gas Research Institute, June 1997.
8. Cobb, N., and Etzel, R.A. *Unintentional Carbon Monoxide-Related Deaths in the United States, 1979 Through 1988.* JAMA, 1991; 266:659-663.
9. Ault, K.L. *Non-fire Carbon Monoxide Deaths and Injuries Associated with the Use of Consumer Products: Annual Estimates.* U.S. Consumer Product Safety Commission. September 1998.
10. Wilson, A.L., Colome, S.D., and Tan, Y. *California Indoor Air Quality Study.* Gas Research Institute, May 1993.

Acknowledgement

This supplement was prepared by WEC Consultants Ltd. with partial support from the Gas Research Institute under contract number GRI 6082. Neither GRI, members of GRI, the authors, nor any persons acting on their behalf:

(1) Makes any warranty or representation, express or implied, with respect to the accuracy, completeness, or usefulness of the information contained in this supplement.

(2) Assumes any liability with respect to the use of, or for any and all damages resulting from the use of the information contained in this report.

CARBON MONOXIDE INVESTIGATION
FIRST RESPONDER'S CHECKLIST

Location ____   Incident Number ______________

Date __________________   Time of Alarm __________   Time of Measurement __________

QUESTIONS TO ASK OCCUPANTS

Number of individuals involved: ________________

Are any members of the household feeling ill? ❑ Yes   ❑ No

Headache ❑ Yes ❑ No   Fatigue ❑ Yes ❑ No   Nausea ❑ Yes ❑ No

Dizziness ❑ Yes ❑ No   Shortness of breath ❑ Yes ❑ No   Confusion ❑ Yes ❑ No

Other ❑ Yes ❑ No ____

A "Yes" response to any item requires an EMS evaluation by paramedics.

Do you feel better when away from the house? ❑ Yes   ❑ No

What appliances were on at the time of activation? ________________________________

____

What appliances were in use for the 24 hours previous to activation? ________________

____

GAS DETECTION METER CHECKLIST

Area of	Room Location	PPM Reading	Area of	Room Location	PPM Reading
Outside reading	__________	________	Gas dryer	__________	________
Upon entering	__________	________	Garage	__________	________
Space heater	__________	________	Furnace	__________	________
Water heater	__________	________	Chimney	__________	________
Gas refrigerator	__________	________	Fireplace	__________	________
Stove/oven	__________	________	BBQ grill	__________	________
At CO detector	__________	________	Other	__________	________

National Fuel Gas Code Handbook — CO Checklist (1 of 2)

Exhibit S3.8 *Checklist for response call documentation.*

CO DETECTOR INFORMATION

Location of Detector: ____

Date Installed: _________________

Make: _________________   Model: _________________   Serial No.: _________________

Name of individual handling CO monitor: _______________________________________

Monitor type: ____

Individual completing checklist: ____

PREVENTION BUREAU FOLLOW-UP:

Follow-up comments: ____

____

____

____

____

____

____

____

____

____

____

____

____

____

FPB Signature: _______________________________________   Date: _________________

National Fuel Gas Code Handbook — CO Checklist (2 of 2)

Exhibit S3.8 *Continued.*

NOTICE OF FINDINGS

Carbon monoxide is an odorless, tasteless, colorless gas that is DEADLY. It is a by-product of a fuel-burning process. It can cause symptoms that can mimic the flu and lead to unconsciousness and even death. Many appliances around the home are capable of producing carbon monoxide when a fault or unusual condition exists. Since the source may be transient in nature, the source may not always be detectable.

The ________________________ responded to your building at ________________________
                                                                           Location

on ________________ . The incident number is ________________________ .
      Date   Number

Carbon monoxide at the highest level of ____________ ppm (parts per million) was found.

Reading*	Recommendation
Less than 30 ppm	Our instruments did not detect elevated levels at this time.
	Check your carbon monoxide detector per the manufacturer's recommendations. Call the manufacturer for additional information (number may be on the back of unit). If it activates again, call the fire department back.
30 to 99 ppm	We have detected potentially dangerous levels of carbon monoxide. We are recommending that you have a qualified contractor locate and repair the source of carbon monoxide immediately. It is not safe until repairs are made.
100 ppm or greater	We have detected a potentially lethal level of carbon monoxide in your home. **Leave your home immediately!** It is not safe to return until repairs are made or the source is found and corrected. The building will be posted "NOT SUITABLE FOR OCCUPANCY" and the building department will be notified.
	Have your sources of carbon monoxide examined and repaired by a qualified contractor. The contractor should contact the building department and fire prevention bureau for a re-inspection prior to re-occupying the building.

* The levels of CO and the recommendations for action given here are for illustration only.
  They should be modified according to local evaluation and policy.

National Fuel Gas Code Handbook — Notice of Findings (1 of 2)

Exhibit S3.9 *Notice of findings incident report.*

What did your investigating company find?______________________________________

____

____

What action did your company take?____

____

____

Comments: ____

____

____

____

____

____

____

____

____

Carbon monoxide affects individuals differently, depending on the size and medical history of the occupant. Therefore, families with young children or members with medical conditions should take extra precautions in the event that carbon monoxide was detected.

Issued by______________________ of the______________________ responder organization.

Received by: ____ Date:_________________

White: Responder   *Yellow: Other department*   *Pink: Occupant/Owner*

National Fuel Gas Code Handbook — Notice of Findings (2 of 2)

Exhibit S3.9 *Continued.*

Revision to American National Standard for Gas Water Heaters — Volume 1 — Storage Water Heaters with Input Ratings of 75,000 Btu per Hour or Less

Editor's Note: The following material was submitted to the ANSI Z21/83 committee in support of a proposal to add a requirement to ANSI Z21.10.1, Gas Water Heaters — Volume 1 — Storage Water Heaters with Input Ratings of 75,000 Btu per Hour or Less requiring that all newly manufactured water heaters pass a flammable vapor ignition resistance test. The ANSI Z21/83 committee accepted this proposal at its May 1999 meeting. At the time of this writing, an appeal has been filed on this new test, and it is pending. Following the hearing, the standard will be submitted to ANSI for a public comment period.

OBJECTIVE

This supplement presents the background, development, and rationale for a test using gasoline to determine the resistance of gas-fired water heaters to ignite flammable vapors that are outside the appliance. The research was done by reviewing studies sponsored by the Gas Appliance Manufacturers Association (GAMA) and the Gas Research Institute (GRI) that were aimed at identifying issues related to flammable vapors incidents.

BACKGROUND

The proposed gasoline-based tests to determine the resistance of a water heater to ignite flammable vapors both grew out of two related projects that were funded by GRI and GAMA. The original test room at AGA Research (AGAR) was built to support Task 2 of a GAMA project conducted by A. D. Little (ADL). It was later used in support of a GRI project that was intended to develop a test method that uses butane as a surrogate for a flammable liquid spill. The gasoline test method and test room have been improved continuously since then under funding from the Water Heater Consortium (WHC). The development of these test methods is a model of cooperative funding between GRI and the water heater industry.

The Gas Appliance Manufacturers Association (GAMA) began this overall research effort because of documented field problems involving property damage and injury to consumers. GAMA members then undertook a proactive program to improve consumers' awareness of the hazards associated with the misuse of gasoline with the consent of the U.S. Consumer Product Safety Commission (CPSC). CPSC continues to monitor the development of a test to verify the resistance of water heaters to ignite flammable vapors outside the appliance, as well as the development of new water heater conceptual designs that are expected to pass that test.

The first organized review of flammable vapors incidents involving gas-fired water heaters was sponsored by GAMA and conducted by ADL. The overall goal of this effort was to develop a comprehensive document detailing the extent of the hazard and the effectiveness of mitigating measures. In performing this task, the following data resources were reviewed:

(1) One hundred forty-two detailed incident reports from several sources (CPSC, NFPA, and NEISS)
(2) National and state fire incident databases
(3) Twenty-six interviews with persons knowledgeable about incidents
(4) Published reports on the subject

TYPICAL SCENARIOS IDENTIFIED

From analysis of the incidents in a population of 53 million water heaters, using the databases shown above, seven typical scenarios where flammable vapors incidents with water heaters occur were identified:

- One bathroom scenario
- Two utility room scenarios
- Three garage and basement scenarios
- One garage scenario

These scenarios occur in rooms ranging in size from small bathrooms up to larger rooms such as a garage. The casualty rate in small bathrooms was more than twice the average for all other gas-fired water heater flammable vapor ignition incidents.

The utility room scenarios involve the second-smallest room in which incidents were reported. The two utility room scenarios were characterized by:

- A 10-ft × 10-ft × 8-ft (3-m × 3-m × 2.4-m) room
- 1 gallon of gasoline for Scenario 1, spill outside room (3.8 L)
- 1-5 gallons of gasoline for Scenario 2, spill inside room (5.7 L)
- Movement involved in Scenario 2
- Gas-fired water heater located in corner

In the detailed description of one actual Scenario 1 incident, a person using gasoline to remove stains from trousers performed the cleaning operation outdoors near an open door to the utility room. The day was windy and the person thought that the wind would disperse the vapors. As the person lifted the trousers from the soak pot, a flame from the gas-fired water heater located behind a closed door ignited the flammable gasoline vapors. This incident highlights how air motion can amplify the danger from the flammable vapors.

Three garage and basement scenarios were characterized by the following:

- A 20-ft × 10-ft × 8-ft (6.1-m × 3-m × 2.4-m) room
- Spill of 1 qt to 5 gal of gasoline (0.9 L – 18.9 L)
- Activity or movement in the direct vicinity of the gas-fired water heater
- Water heater located in corner

One additional garage scenario was characterized by the following:

- A 20-ft × 10-ft × 8-ft (6.1-m × 3-m × 2.4-m) room
- Slow leakage of gasoline from a fuel tank
- No activity or movement
- Water heater located in corner

The incidents in the garage and basement scenarios typically involve use of gasoline as a cleaner, accidental spills, and leaking storage containers.

Conclusions

The GAMA Task 1 work identified the following factors as important in incidents involving ignition of flammable vapors by gas-fired water heaters:

(1) Gasoline is the major cause of incidents.
(2) Room size and spill size affect the time to ignition and severity.
(3) Motion in the room has a strong effect on the mixing of the vapors with the air.

It can be further concluded that the incidents occur in circumstances that include a wide range of unpredictable external variables that come together at the same time. To compound the problem, water heaters will operate on different cycles depending on their design and the owner's usage patterns. So, the water heater's propensity to draw in and ignite vapors varies in time in an unpredictable way as well. Overall, the final flammable vapor test method for gas-fired water heaters will need to be capable of incorporating as many of these factors as possible.

INITIAL GASOLINE TEST ROOM STUDIES

Gasoline and Lower Flammability Limit (LFL)

For a vapor mixture to be flammable, it must reach a composition (mixture of fuel and oxygen) that can ignite and sustain combustion. The range of flammable composition of vapors and air is bounded by a lower and an upper concentration of the vapor in air. The lower flammability limit (LFL) is the lowest concentration of vapor that will support a flame. Below this level the mixture is too lean. The upper flammability limit (UFL) is the maximum concentration of vapor that can support a flame. Above this level the mixture is too rich.

Gasoline is a mixture of hydrocarbon and other compounds. The component with the highest vapor pressure is butane. Therefore, it is likely that the initial vapor cloud over a gasoline spill will be rich in butane. The LFL of butane is about 1.8 percent and the HFL is about 8.4 percent. From a description of incidents, the LFL appears to be the point at which most incidents occur. This is the level that is first achieved following a spill. However, it is also possible that ignition could be avoided somehow immediately after the spill. In this case, the concentration could possibly rise above the UFL. If the UFL is exceeded, however, the possibility of ignition still exists as the gasoline concentration drops and becomes flammable again. It is also conceivable that the vapor concentration could remain within the flammable range for some period of time. In this case, ignition could occur at any time, until the concentration leaves the flammable range.

The time it takes to reach LFL after a spill is partially determined by the gasoline's volatility. Other factors include the size of the spill and room, temperature, air change rate, and room air motion. The term "volatility" refers to how quickly the liquid will evaporate, creating the flammable vapors. One measure of the volatility is the Reid Vapor Pressure (RVP). A higher value for the RVP indicates that evaporation will occur faster. The composition of gasoline varies by brand, time, and location, but the following two generic gasolines represent the extremes of volatility:

(1) "Summer blend" gasoline, with a Reid Vapor Pressure of about 9 psi (62 kPa)
(2) "Winter blend" gasoline, with a Reid Vapor Pressure of about 12–15 psi (82 – 103 Pa)

The winter blend gasoline is essentially summer blend gas that has had butane added to it to increase the volatility. This difference presents practical problems for developing a test method using a consistent blend of gasoline. It is well understood that the volatility of the winter blend makes it very difficult to store while preserving its Reid Vapor Pressure, because the butane is likely to be lost over time.

Large and Small Room Tests

Under the auspices of GAMA, the following gasoline spill tests were conducted in the gasoline test facility in the presence of operating gas-fired water heaters:

- Twenty-one tests in a 10-ft × 20-ft × 8-ft (3-m × 6.1-m × 2.4-m) room
- Ten tests in a 6-ft × 10-ft × 8-ft (1.8-m × 3-m × 2.4-m) room
- Six tests in an 8-ft × 8-ft × 8-ft (2.4-m × 2.4-m × 2.4-m) room

Results of these tests are presented in the GAMA Task 2 Report.

Conclusions

Task 2 of the GAMA study helped to understand the relative importance of the following factors that influence the potential ignition of flammable vapors by gas-fired water heaters:

(1) Spill surface
(2) Floor and room temperature
(3) Room size
(4) Flammable vapor liquid composition
(5) Ventilation rate

The room experiments resulted in several conclusions, including the following:

(1) A gasoline spill near a gas-fired water heater is likely to result in an ignition of the flammable vapors.
(2) Installation of a water heater on an 18-in. (0.46 m) stand may delay but cannot guarantee elimination of the ignition of flammable vapors.
(3) Rags soaked in gasoline can present ignition sources in small rooms.

In addition, there were several general observations that provide an insight to these experiments.

(1) Air motion is an important accelerator of ignition. Without forced convection in the room, the vapors will diffuse slowly away from the spill and be diluted by the room's ventilation. Therefore, without an induced air movement, a false sense of security can result.
(2) While elevation of the water heater may delay ignition of the vapors, the ignition may release more force than for floor-mounted water heaters because a larger volume of flammable vapors has accumulated by the time of ignition.
(3) Room size, spill size, and the ventilation rate have an important combined effect on the vapor profile over time.
(4) Room temperature is not as important as room size, motion, and size of the spill.

GASOLINE TEST ROOM

Construction History

The current flammable vapor test room at AGAR evolved over several years. The room was originally constructed to support the GAMA Task 2 work of ADL as described previously. The structure had two rooms that could be used to perform tests in different volumes. The structure was wood frame with drywall construction and was housed under a plastic film structure. The larger part of the facility was abandoned and sealed off because the smaller room was chosen as the preferred test condition.

During the summer of 1997, construction began on a new, improved, and automated structure reflecting the experience gained in the old room. This new room is sheltered under an improved plastic film structure and has proven to be much more reliable.

Description of Room

Exhibit S4.1 shows the floor plan for the test room. The room is described as follows:

- The room is constructed with metal studs covered with sheet metal.
- The foundation is a concrete pad embedded with hydronic heating coils.
- The floor is a single piece of stainless steel.
- The pressure relief opening is covered with plastic or foil with perforations to minimize pressure buildup within the chamber.
- The water heater is vented through the back wall. The vent pipe is terminated within the outer plastic structure underneath an 8-in. (0.2-m) diameter duct connected to the outside of the exterior structure. This duct acts to isolate the water heater vent from variable outdoor conditions such as temperature, rain, or wind.
- The mannequin is attached to a pneumatic cylinder with a 3-ft (0.9-m) stroke length. Cylinder movement is manually controlled from the control room.

The following data are taken during a test:

(1) Temperature

     a. Test chamber ambient
     b. Water heater flue before the draft hood
     c. Water within the heater at a level equal to the location of the T-P valve
     d. Test chamber floor
     e. Ambient at the combustion air inlet to the test room

(2) Pressure

     a. Differential between the test room and the exterior structure
     b. Differential between the water heater vent and the exterior structure

(3) Hydrocarbon concentration (measured as butane) sample points

     a. At the combustion air opening
     b. At half the height of the water heater on the front
     c. At half the height of the water heater on the back
     d. On the top of the water heater near the draft hood

(4) Miscellaneous

   a. Relative humidity in the exterior structure at the combustion air inlet to the test room
   b. Pilot millivoltage

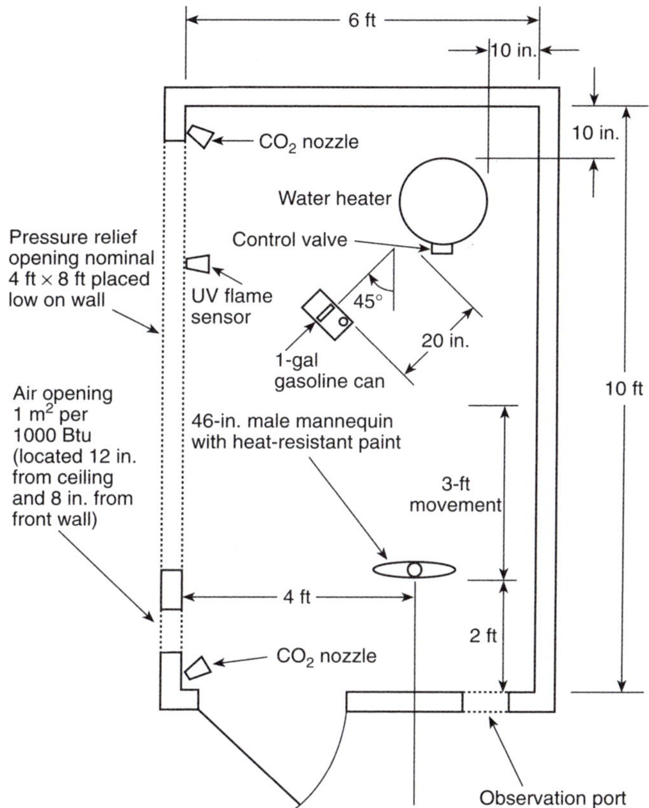

Exhibit S4.1 *Diagram of the current gasoline test room [housed within exterior structure (not shown) to control surrounding ambient conditions].*

Test Results

The new test facility provides a stable platform for performing a variety of tests with gasoline. Appropriately, the results obtained thus far are dependent on the specific water heater. In particular, very different vapor profiles have been observed as different gas water heaters react to the spill. Unfortunately, the test can damage the water heater.

   Tests of WHC prototypes in the new room, as the room design was being developed, show that the facility has been improved. Similar water heaters produce similar results. The tests are also instructive in showing how much the water heater's size or operation can change the resulting vapor profile. The following discussion highlights results obtained in the new room from December 1997 through January 1998. In the interest of protecting the confidentiality of the manufacturers, the specific design differences in each test will not be discussed. But, they are all different and the reader should not expect the results to be exactly the same.

Exhibit S4.2 shows the vapor profile results for one test of Prototype 34. The test presented is for summer blend gasoline, without movement. The burner was on at the start of the test.

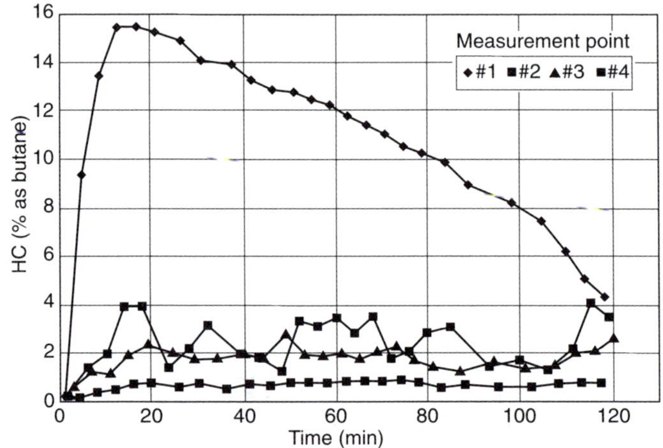

Exhibit S4.2 *Prototype 34, Test No.1, summer blend, with main burner on.*

In Exhibit S4.2, the four traces are hydrocarbon concentrations at different elevations in the room.

The hydrocarbon concentrations are measured as a butane equivalent. The sampling system read each measurement point once every 3 to 5 minutes. New hydrocarbon analyzers, which allow simultaneous measurements, have now been installed and were used in test runs presented in Exhibits S4.6 and S4.7. The long sampling runs, in use at the time of these tests, also introduce a lag time between what is happening in the test room and when the measurement is recorded. The most important measurement point is point 1, which is 3 in. above the floor and is closest to where the vapors are being pulled into the heater. Measurement points 2 to 4 are progressively higher in the room, as described in "description of room." Note that a large difference in concentration exists for an extended period between point 1 and the rest of the room. This difference is characteristic of a test with no movement. The floor concentration rises immediately and then gradually decreases as the heater consumes the vapors.

Exhibit S4.3 presents the same data from a repeated test of the same heater. Comparing Exhibits S4.2 and S4.3, we see a good correspondence between the vapor profiles. And, in the end, the water heater passed the test both times.

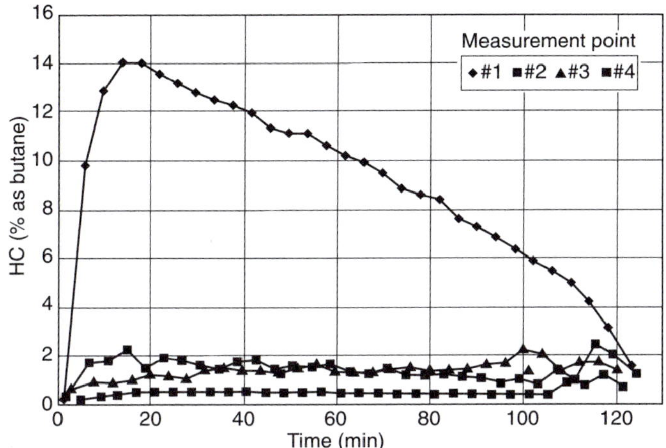

Exhibit S4.3 *Prototype 34, Test No. 2, summer blend, with main burner on.*

Exhibits S4.4 and S4.5 present continued tests of the same water heater. In this case, the two tests use winter blend gasoline, with room air motion caused by the flat mannequin. Early winter blend gasoline tests used a flat, wooden, silhouette shaped mannequin, which has since been replaced by a department-store mannequin. Both tests illustrate the different vapor profile caused by room air motion. In these tests, the vapor concentration rises quickly near floor level, represented by point 1. Note that the difference between the concentration near the floor and the concenration higher in the room diminishes more quickly compared to a test with no room air motion.

In these two tests, the vapor profiles are qualitatively quite similar. The difference in the absolute concentration is probably the result of the way the gasoline splashes at the start of the test. Despite the differences, the gasoline vapor at the level where the vapors can enter the heater was in the flammable range and the water heater passed the test both times.

It is certainly possible to design a standard way of spilling the gasoline. However, it is believed that spilling the gasoline by tipping over the container increases the realism of the test because of the splashing it creates.

Recent tests also show that the vapor profiles can be changed by relatively minor heater design changes. Exhibits S4.6 and S4.7 present the results of two summer blend tests. The same type of heater was used in each test, except they had slightly different burners that were being evaluated. In the first test, shown in Exhibit S4.6,

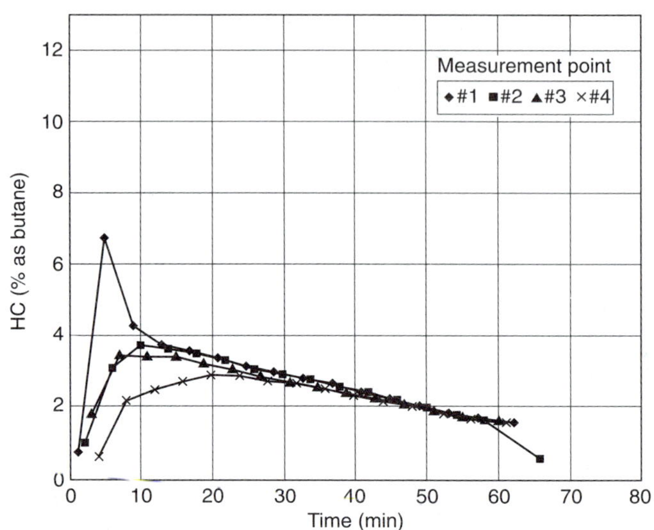

Exhibit S4.4 *Prototype 34, Test No. 1, winter blend.*

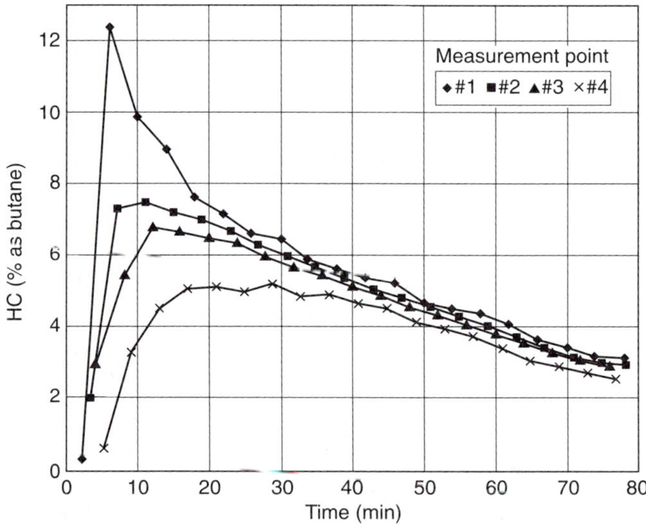

Exhibit S4.5 *Prototype 34, Test No. 3, winter blend.*

instrumentation indicated that gasoline vapors were burning in the combustion chamber. However, eventually all combustion was extinguished and the pilot dropped out. At this point, no subsequent ignition was possible and the test was terminated. As shown in Exhibit S4.7, the second test began much like the previous test. Gasoline vapor combustion occurred in the water heater but eventually ceased. However, in this case, the pilot did not drop out and the heater's main burner continued to cycle. This test was terminated after two hours with no ignition of the vapors in the room.

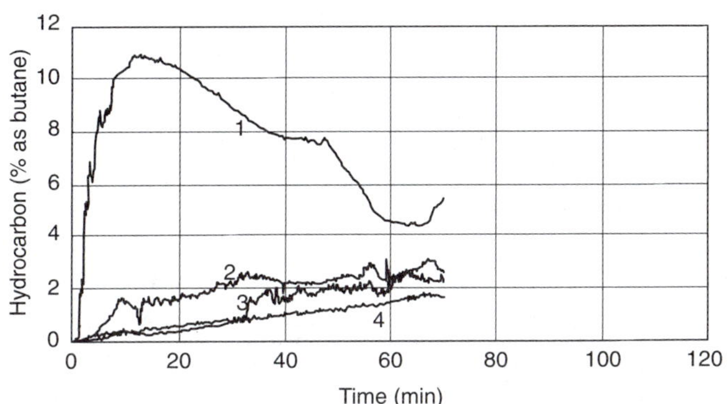

Exhibit S4.6 *Prototype 59b, summer blend, with main burner on.*

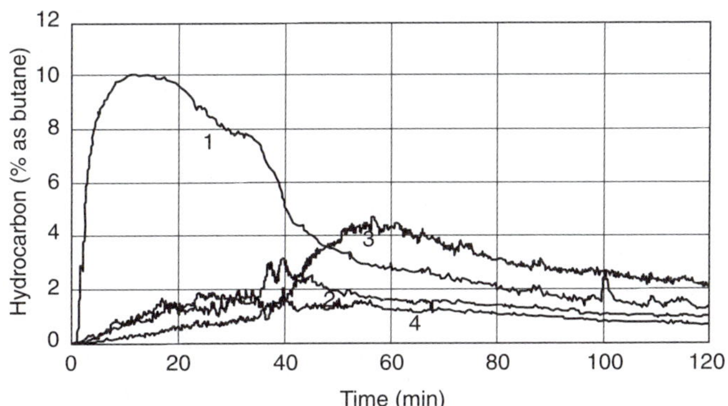

Exhibit S4.7 *Prototype 59c, summer blend, with main burner on.*

The tests in Exhibits S4.6 and S4.7 illustrate another important point. There are several different possible test outcomes that are all "passing." These possibilities include the following:

(1) All combustion in the heater is extinguished soon after the spill because the flammable vapors are above the higher flammable limit.
(2) Combustion of flammable vapors occurs in the heater for a period and then extinguishes itself and the normal burners.

(3) Combustion of flammable vapors occurs in the heater until the source of the vapors is depleted.

In each case, there is no ignition in the room. So, from a safety aspect, the outcomes are equivalent.

CONCLUSIONS

The gasoline test room, with a real gasoline spill, is more useful for the practical testing of water heaters in a way that includes the many design-specific parameters of the appliances. It is most appropriate for testing water heaters in the latter stages of development and for certification purposes.

Bibliography

Flammable Vapor Hazards Ignition Study, A.D. Little, June 16, 1993, GAMA Task 1 Report.

Flammable Vapor Hazards Ignition Study, A.D. Little, July 15, 1993, GAMA Task 2 Report.

Topping, R.F., Benedek, K.R., *Flammable Vapor Test Methodology Development for Gas-Fired Water Heaters*, Arthur D. Little, Inc., Cambridge, Massachusetts, April, 1996, GRI-96/0102.

Procedure to Estimate Infiltration Rate for Residential Structures

Paul Cabot
American Gas Association

Editor's Note: Users of the National Fuel Gas Code must determine whether a structure is of "unusually tight construction" (UTC) in order to decide whether combustion air can be obtained from the indoors or whether outdoor air openings must be installed. Indoor air can be used when the building has sufficient air infiltration. This supplement describes a method used by the American Society of Heating, Refrigerating, and Air Conditioning Engineers (ASHRAE) for calculating air changes per hour (ACH) and presents calculation examples.

NFPA is grateful to Paul Cabot of the American Gas Association for preparing this supplement. We also appreciate the permission from ASHRAE to reproduce tables from its Handbook of Fundamentals.

INTRODUCTION

Four criteria are described in the definition of *unusually tight construction* (UTC) that help users calculate air changes per hour. The definition of this term is structured such that if any one of the four criteria is not satisfied, the building is not considered unusually tight. The first three criteria are construction details that are the easiest to identify and, therefore, should be considered first. If any of these construction details is not part of the building, there is no need to evaluate the fourth criterion.

The fourth criterion, less than 0.35 air changes per hour (ACH), was added to the 1999 *National Fuel Gas Code* to provide guidance when a building is determined to contain the first three construction details but still may have enough air infiltration available to meet the appliance combustion air needs.

Air infiltration is also impacted by other building design and construction features. Some of these inherent features include the amount of window area, amount of exterior wall area, number of envelope penetrations, construction quality, age, and so forth.

ACH has traditionally been used in heat loss calculations for equipment sizing and in determining a building's overall energy efficiency. In this supplement, ACH is strictly a "gauge" used in a pass-fail evaluation.

ASHRAE METHOD

The American Society of Heating, Refrigerating, and Air Conditioning Engineers (ASHRAE) is the HVAC industry's principal professional organization. It develops and publishes state-of-the-art technical material for the benefit of design professionals and the general public well-being. This supplement provides a residential calculation method based on the ASHRAE equations and calculation methods described in the 1997 *ASHRAE Handbook of Fundamentals*, Chapter 25, "Ventilation and Infiltration." The mathematical formulas used to estimate infiltration from a building's effective air leakage area are calculated by using the following equation:

$$Q = A_L\sqrt{C_s\Delta t + C_w V^2}$$

where:

Q = airflow rate, cfm

A_L = effective air leakage area, in.^{2}

C_S = stack coefficient, cfm^2/(in.^{4} · °F)

Δt = average indoor-outdoor temperature difference for time interval of calculation, °F

C_W = wind coefficient, cfm^2/(in.^{4} · mph^2)

V = average wind speed measured at local weather station for time interval of calculation, mph

To convert Q in ft^3 per minute to air changes per hour, I,

$$I = \frac{Q \times 60 \text{ min/hr}}{V}$$

where:

I = air exchange rate, ft^3/hr

Q = airflow rate, cfm

V = interior volume of space, ft^3

FACTORS IN ESTIMATING AIR INFILTRATION

Two of the most important factors in estimating air infiltration are the determination of the effective air leakage area (A_L) and selection of the appropriate indoor-outdoor temperature difference for time interval of calculation (Δt). A brief discussion of these two factors follows:

- **Effective Air Leakage Area (A_L).** A_L can be estimated from a test value from a whole-building pressurization test or by utilizing tabulated data in conjunction with a building's blueprints or measurements.
  - **Whole-Building Pressurization Test.** Commonly known as a blower door test, the whole-building pressurization test can be used to estimate A_L for existing structures. This test must be conducted by a qualified person because it requires skill and experience to set up and interpret. For most code users, however, data in the ASHRAE handbook is a sufficient alternative.
  - **ASHRAE Data.** Since code users often must make combustion-air decisions before a building is constructed, the data in the *ASHRAE Handbook of Fundamentals* can be utilized. The examples in this supplement demonstrate use of this data.
- **Indoor-Outdoor Temperature Difference.** The 0.35 ACH criterion in the UTC definition is an average rate. Therefore, the user must select a time period within which to average the formula's indoor-outdoor temperature difference (Δt). For the purposes of the UTC evaluation, the time period chosen should be the coldest period of the season in which the building is likely to experience the highest Δt. This period is a baseline for the calculation and provides an indicator of a building's overall ability to supply combustion air. Its duration can vary and should be based on local weather conditions. Generally, a one-week period provides a conservative evaluation period.

CALCULATION EXAMPLES

As previously noted, the "less than 0.35 ACH" fourth criterion is not intended to replace the first three construction details in making the UTC determination. It is to be used when the construction details are present or when a question is raised regarding their applicability and the code user still wishes to demonstrate that the building has enough inherent leakage to supply combustion air. ACH is meant to be used as a measure in this determination.

Example 1: Two-Story House — Average Temperature, 20°F.

Estimate the average infiltration for a planned, two-story house that contains 1600 ft^2 of living space, with 8-ft ceilings. The house is located in a new subdivision. The space will be maintained at 70°F. The average outdoor temperature is 20°F during the coldest, one-week period. The house will be built with the first three construction details in the UTC definition.

Exhibit S5.1 *Effective air leakage areas (low-rise residential applications). (Courtesy of 1997 ASHRAE Handbook of Fundamentals, Chapter 25, Table 3.)*

	Units (see note)	Best Estimate	Mini-mum	Maxi-mum
Ceiling				
General	in^2/ft^2	0.026	0.011	0.04
Drop	in^2/ft^2	0.0027	0.00066	0.003
Ceiling penetrations				
Whole-house fans	in^2 ea	3.1	0.25	3.3
Recessed lights	in^2 ea	1.6	0.23	3.3
Ceiling/Flue vent	in^2 ea	4.8	4.3	4.8
Surface-mounted lights	in^2 ea	0.13		
Chimney	in^2 ea	4.5	3.3	5.6
Crawl space				
General (area for exposed wall)	in^2/ft^2	0.144	0.1	0.24
8 in. by 16 in. vents	in^2 ea	20		
Door frame				
General	in^2 ea	1.9	0.37	3.9
Masonry, not caulked	in^2/ft^2	0.07	0.024	0.07
Masonry, caulked	in^2/ft^2	0.014	0.004	0.014
Wood, not caulked	in^2/ft^2	0.024	0.009	0.024
Wood, caulked	in^2/ft^2	0.004	0.001	0.004
Trim	$in^2/lftc$	0.05		
Jamb	$in^2/lftc$	0.4	0.3	0.5
Threshold	$in^2/lftc$	0.1	0.06	1.1
Doors				
Attic/crawl space, not weatherstripped	in^2 ea	4.6	1.6	5.7
Attic/crawl space, weatherstripped	in^2 ea	2.8	1.2	2.9
Attic fold down, not weatherstripped	in^2 ea	6.8	3.6	13
Attic fold down, weatherstripped	in^2 ea	3.4	2.2	6.7
Attic fold down, with insulated box	in^2 ea	0.6		
Attic from unconditioned garage	in^2 ea	0	0	0
Double, not weatherstripped	in^2/ft^2	0.16	0.1	0.32
Double, weatherstripped	in^2/ft^2	0.12	0.04	0.33

	Units (see note)	Best Estimate	Mini-mum	Maxi-mum
Piping/Plumbing/Wiring penetrations				
Uncaulked	in^2 ea	0.9	0.31	3.7
Caulked	in^2 ea	0.3	0.16	0.3
Vents				
Bathroom with damper closed	in^2 ea	1.6	0.39	3.1
Bathroom with damper open	in^2 ea	3.1	0.95	3.4
Dryer with damper	in^2 ea	0.46	0.45	1.1
Dryer without damper	in^2 ea	2.3	1.9	5.3
Kitchen with damper open	in^2 ea	6.2	2.2	11
Kitchen with damper closed	in^2 ea	0.8	0.16	1.1
Kitchen with tight gasket	in^2 ea	0.16		
Walls (exterior)				
Cast-in-place concrete	in^2/ft^2	0.007	0.0007	0.026
Clay brick cavity wall, finished	in^2/ft^2	0.0098	0.0007	0.033
Precast concrete panel	in^2/ft^2	0.017	0.0004	0.024
Lightweight concrete block, unfinished	in^2/ft^2	0.05	0.019	0.058
Lightweight concrete block, painted or stucco	in^2/ft^2	0.016	0.0075	0.016
Heavyweight concrete block, unfinished	in^2/ft^2	0.0036		
Continuous air infiltration barrier	in^2/ft^2	0.0022	0.0008	0.003
Rigid sheathing	in^2/ft^2	0.005	0.0042	0.006
Window framing				
Masonry, uncaulked	in^2/ft^2	0.094	0.082	0.148
Masonry, caulked	in^2/ft^2	0.019	0.016	0.03
Wood, uncaulked	in^2/ft^2	0.025	0.022	0.039
Wood, caulked	in^2/ft^2	0.004	0.004	0.007
Windows				
Awning, not weatherstripped	in^2/ft^2	0.023	0.011	0.035
Awning, weatherstripped	in^2/ft^2	0.012	0.006	0.017
Casement, weatherstripped	$in^2/lftc$	0.011	0.005	0.14

Exhibit S5.1 *Continued.*

Item	Units				Item	Units			
Elevator (passenger)	in^2 ea	0.04	0.022	0.054	Casement, not weatherstripped	in^2/lftc	0.013		
General, average	in^2/lftc	0.015	0.011	0.021	Double horizontal slider, not weatherstripped	in^2/lftc	0.052	0.0009	0.16
Interior (pocket, on top floor)	in^2 ea	2.2			Double horizontal slider, wood, weatherstripped	in^2/lftc	0.026	0.0070	0.081
Interior (stairs)	in^2/lftc	0.04	0.012	0.070	Double horizontal slider, aluminum, weatherstripped	in^2/lftc	0.034	0.027	0.038
Mail slot	in^2/lftc	0.2			Double-hung, not weatherstripped	in^2/lftc	0.12	0.040	0.29
Sliding exterior glass patio	in^2 ea	3.4	0.46	9.3	Double-hung, weatherstripped	in^2/lftc	0.031	0.009	0.089
Sliding exterior glass patio	in^2/ft^2	0.079	0.009	0.22	Double-hung with storm, not weatherstripped	in^2/lftc	0.046	0.023	0.080
Storm (difference between with and without)	in^2 ea	0.9	0.46	0.96	Double-hung with storm, weatherstripped	in^2/lftc	0.037	0.021	0.05
Single, not weatherstripped	in^2 ea	3.3	1.9	8.2	Double-hung with pressurized track, weatherstripped	in^2/lftc	0.023	0.018	0.026
Single, weatherstripped	in^2 ea	1.9	0.6	4.2	Jalousie	in^2/louver	0.524		
Vestibule (subtract per each location)	in^2 ea	1.6			Lumped	in^2/lfts	0.022	0.00042	0.097
Electrical outlets/Switches					Single horizontal slider, weatherstripped	in^2/lfts	0.031	0.009	0.097
No gaskets	in^2 ea	0.38	0.08	0.96	Single horizontal slider, aluminum	in^2/lfts	0.04	0.013	0.097
With gaskets	in^2 ea	0.023	0.012	0.54	Single horizontal slider, wood	in^2/lfts	0.021	0.013	0.047
Furnace					Single horizontal slider, wood clad	in^2/lfts	0.030	0.025	0.038
Sealed (or no) combustion	in^2 ea	0	0	0	Single-hung, weatherstripped	in^2/lfts	0.041	0.029	0.058
Retention head or stack damper	in^2 ea	4.6	3.1	4.6	Sill	in^2/lftc	0.0099	0.0065	0.010
Retention head and stack damper	in^2 ea	3.7	2.8	4.6	Storm inside, heat shrink	in^2/lfts	0.00085	0.00042	0.00085
Floors over crawl spaces					Storm inside, rigid sheet with magnetic seal	in^2/lfts	0.0056	0.00085	0.011
General	in^2/ft^2	0.032	0.006	0.071	Storm inside, flexible sheet with mechanical seal	in^2/lfts	0.0072	0.00085	0.039
Without ductwork in crawl space	in^2/ft^2	0.0285			Storm inside, rigid sheet with mechanical seal	in^2/lfts	0.019	0.0021	0.039
With ductwork in crawl space	in^2/ft^2	0.0324			Storm outside, pressurized track	in^2/lftc	0.025		
Fireplace					Storm outside, 2-track	in^2/lftc	0.058		
With damper closed	in^2/ft^2	0.62	0.14	1.3	Storm outside, 3-track	in^2/lftc	0.116		
With damper open	in^2/ft^2	5.04	2.09	5.47					
With glass doors	in^2/ft^2	0.58	0.06	0.58					
With insert and damper closed	in^2/ft^2	0.52	0.37	0.66					
With insert and damper open	in^2/ft^2	0.94	0.58	1.3					
Gas water heater	in^2 ea	3.1	2.3	3.9					
Joints									
Ceiling-wall	in^2/lftc	0.070	0.0075	0.12					
Sole plate, floor/wall, uncaulked	in^2/lftc	0.2	0.018	0.26					
Sole plate, floor/wall, caulked	in^2/lftc	0.04	0.0035	0.056					
Top plate, band joist	in^2/lftc	0.005	0.0035	0.018					

Note: Air leakage areas are based on values found in the literature. The effective air leakage area (in square inches) is based on a pressure difference of 0.016 in. of water and $C_D = 1$.

Abbreviations: ft^2 = gross area in square feet   lftc = linear foot of crack
ea = each   lfts = linear foot of sash

Determining the Effective Air Leakage Area (A_L). Estimate the effective air leakage area (A_L) of a residential structure based on the figures shown in Exhibit S5.1. The table in this exhibit provides a list of extensive construction details, along with the effective air leakage areas for low-rise residential applications.

Using the house blueprints and Table S5.1, determine the construction component areas and features and select the appropriate leakage area units. For all components in this example, the best estimate number was selected from Exhibit S5.1. The calculated air leakage rates are shown in Table S5.1.

Table S5.1   Tabulation of Leakage Area Based on Component Leakage Areas

Component	Description	Size or Number	A_L per unit	A_L (in.^{2})
Ceiling	General	800 ft^2	0.026 in.^{2}/ft^2	20.8
Walls	Continuous air infiltration barrier	1857 ft^2	0.0022 in.^{2}/ft^2	4.1
Windows				
Exterior	Double hung	240 ft^2	0.031 in.^{2}/ft^2	7.4
Framing	Weather-stripped wood, caulked	240 ft^2	0.004 in.^{2}/ft^2	1.0
Doors				
Exterior	Double, weather-stripped	21 ft^2	0.12 in.^{2}/ft^2	2.5
Patio	Sliding exterior glass	42 ft^2	0.079 in.^{2}/ft^2	3.3
Framing	All doors, wood, caulked	63 ft^2	0.004 in.^{2}/ft^2	0.3
Attic	Fold-down, weather-stripped	1	3.4 in.^{2} ea.	3.4
Fireplace	With damper closed	9 ft^2	0.62 in.^{2}/ft^2	5.6
Exhaust vents				
Kitchen	Damper closed	1	0.8 in.^{2} ea.	0.8
Bathroom	Damper closed	2	1.6 in.^{2} ea.	3.2
Dryer	With damper	1	0.46 in.^{2} ea.	0.5
Penetrations				
Plumbing	Vent stacks, caulked	2	0.3 in.^{2} ea.	0.6
Vent	Furnace/water heater, caulked	1	4.8 in.^{2} ea.	4.8
Wiring	Caulked or gasketed	4	0.3 in.^{2} ea.	1.2
Electrical	Interior outlets located on outside walls, gasketed	20	0.023 in.^{2} ea.	0.5
Calculated total building effective air leakage area (A_L) =				59.9

Source: Adapted from *1997 ASHRAE Handbook of Fundamentals*, Chapter 25.

Determining ACH. During this period, the average indoor-outdoor temperature difference is 50°F, and the average wind speed is 7 mph. The house has a volume of 10,240 ft^3 (1600 ft^2 × 8 ft × 0.80) and an effective air leakage area (A_L) of 60 in.^{2}. Note that for this example, an 80 percent reduction factor is used in the volume calculation to account for interior walls, cabinets, and other structures that reduce the interior air volume available for combustion air. The house is located in a new subdivision, and there are adjacent structures or houses. The stack coefficient (C_s) for a two-story house is

0.0299 (see Table S5.2) and the local shielding class is #4 (see Table S5.3). The wind coefficient (C_W) is 0.0051 (see Table S5.4, two-story house, class #4).

The air flow rate is calculated as follows:

$$Q = A_L \sqrt{C_s \Delta t + C_w V^2}$$

$$Q = 60 \sqrt{(0.0299)(50) + (0.0051)(7)^2}$$

$$Q = 79 \text{ cfm}$$

Conversion to air changes per hour for the example building is as follows:

$$I = \frac{Q \times 60 \text{ min/hr}}{V}$$

$$I = \frac{(79)(60)}{(10,240)}$$

$$I = 0.46 \text{ ACH}$$

The building is not unusually tight, since 0.46 is more than 0.35 ACH.

Table S5.2 Stack Coefficient C_s

	House Height (Stories)		
	One	Two	Three
Stack coefficient	0.0150	0.0299	0.0449

Source: 1997 ASHRAE *Handbook of Fundamentals*, Chapter 25, Table 6.

Table S5.3 Local Shielding Classes

Class	Description
1	No obstructions or local shielding
2	Light local shielding; few obstructions, few trees, or small shed
3	Moderate local shielding; some obstructions within two house heights, thick hedge, solid fence, or one neighboring house
4	Heavy shielding; obstructions around most of perimeter, buildings or trees within 30 ft in most directions; typical suburban shielding
5	Very heavy shielding; large obstructions surrounding perimeter within two house heights; typical downtown shielding

Source: 1997 ASHRAE *Handbook of Fundamentals*, Chapter 25, Table 7.

Table S5.4 Wind Coefficient C_W

Shielding Class	House Height (Stories)		
	One	Two	Three
1	0.0119	0.0157	0.0184
2	0.0092	0.0121	0.0143
3	0.0065	0.0086	0.0101
4	0.0039	0.0051	0.0060
5	0.0012	0.0016	0.0018

Source: 1997 ASHRAE Handbook of Fundamentals, Chapter 25, Table 8.

Example 2: Two-Story House — Average Outdoor Temperature, 30°F.

Estimate the average infiltration for a two-story house during a period when the average outdoor temperature is 30°F. The house size is 50 ft × 30 ft, with 9-ft ceilings. The effective air leakage area is determined to be 110 in.^{2} (using Table S5.1). The space will be maintained at 70°F. The average wind speed over the period is 7 mph. The house contains the first three construction details in the UTC definition.

The stack coefficient (C_s) for a two-story house is 0.0299 (Table S5.2) and the local shielding class is #4 (Table S5.3). The wind coefficient (C_W) is 0.0051 (Table S5.4, two-story house, class #4).

The airflow rate due to infiltration is calculated as follows:

$$Q = 110\sqrt{(0.0299)(40) + (0.0051)(7)^2}$$
$$Q = 132 \text{ cfm}$$

Converting to air changes per hour for the example building,

$$I = \frac{(132)(60)}{(21,600)}$$
$$I = 0.37 \text{ ACH}$$

The building is not unusually tight, since 0.37 is more than 0.35 ACH.

Index

-A-

Aboveground outside piping, 3.2
Access
Duct furnaces, 6.10.3
Floor furnaces, 6.11.8
Gas utilization equipment, 5.2.1
On roofs, 5.4.3
Valves, 3.3.4, 3.10.2(a)
Wall heaters, 6.28.1(e)
Accessible
Definition, 1.7
Readily (definition), 1.7
Agency, qualified, 1.4
Definition, 1.7
Air
Circulating, *see* Circulating air
Combustion, *see* Combustion air
Dilution, 5.3
Definition, 1.7
Excess (definition), 1.7
Make-up, *see* Make-up air
Primary, *see* Primary air
Use under pressure, 5.1.5
Air conditioners, gas-fired, 6.2
Definition, 1.7
Air conditioning (definition), 1.7
Air ducts, *see* Ducts
Air shutters (definition), 1.7
Aircraft hangars
Duct furnaces in, 6.10.7
Heaters in, 5.1.11, 6.19.4, 6.27.5
Alternate materials, equipment, and procedures, 1.2
Altitude, high, 8.1.2
Ambient temperature (definition), 1.7; *see also* Weather conditions
Anchors, pipe, 3.3.6
Anodeless risers, 2.6.4
Definition, 1.7
Appliance categorized vent diameter/area (definition), 1.7

Appliances
Approval, 5.1.1, A.5.1.1
Automatically controlled (definition), 1.7
Category I venting systems, Chap. 10, App. G
Combination units
Chimneys, 7.5.5(c) to (d)
Venting, A.7.5.5(c)
Converted, 5.1.2 to 5-1.3
Counter (gas), *see* Food service equipment, gas
Decorative, *see* Decorative appliances
Definition, 1.7
Direct vent, *see* Direct vent appliances
Fan-assisted combustion
Definition, 1.7
Venting system, App. G
Gas input for typical, Table C.2
Household cooking, *see* Household cooking appliances
Illuminating, 6.16, Table 6.16.2(b)1
Installation of, Chap. 5
Low-heat
Definition, 1.7
Vent connectors, 7.10.2(d)
Outdoor cooking, *see* Outdoor cooking gas appliances
Placing in operation, 4.3.4
Pressure regulators, 2.8.4(b), 5.1.17 to 5.1.18
Purging, 4.3.4
Vented, *see* Vented appliances
Applicability of code, 1.1.1
Approved (definition), 1.7, A.1.7
Atmospheric pressure (definition), 1.7
Attics, appliances in, 7.10.2(b)

Authority having jurisdiction (definition), 1.7, A.1.7
Automatic damper regulators (definition), 1.7
Automatic firecheck, *see* Fire-checks, automatic
Automatic gas shutoff device (definition), 1.7
Automatic ignition, *see* Ignition, Automatic
Automatic vent damper devices, *see* Damper devices, automatic vent

-B-

Back pressure
Definition, 1.7
Protection, 2.10
Backfilling, 3.1.2(c)
Backfire preventers, *see* Safety blowouts
Baffle (definition), 1.7
Barometric draft regulators, 5.3.1(e), 7.12.4 to 7.12.6
Definition, 1.7
Bends
Pipe, 3.6, A.3.6.3
Vent connectors, 7.10.6
Bleeds
Diaphragm-type valves, 5.1.19
Direct makeup air heaters, 6.8.7
Industrial air heaters, 6.9.6
Blowers, mixing, *see* Mixing blowers
Boilers
Central heating, 6.3
Clearance, 6.3.1, Tables 6.2.3(a) to (b)
Cooling units used with, 6.3.8
Erection and mounting, 6.3.2
Low water cutoff, 6.3.4, 8.5

Boilers *(continued)*
    Steam safety and pressure
      relief valves, 6.3.5, A.6.3.5
    Temperature- or pressure-
      limiting devices, 6.3.3
    Hot water heating
      (definition), 1.7
    Hot water supply (definition), 1.7
    Low-pressure (definition), 1.7
    Steam heating (definition), 1.7
Bonding, electrical, 3.14
Branch lines, 3.3.2, 3.5.1, 3.9
    Definition, 1.7
Breeching, *see* Vent connectors
Broilers
    Definition, 1.7
    Open-top units, 6.20
      Commercial, 6.20 4
      Domestic, protection above,
        6.20.3
    Unit (definition), 1.7
Btu, A.2.4.2
    Definition, 1.7
Buildings
    Piping in, 3.3
      Concealed, 3.4
      Connection of gas utilization
        equipment to, 5.5, A.5.5.6
    Piping under, 3.1.6
    Structural members, 5.1.7
    Of unusually tight construction,
      5.3.2, A.5.3.2
**Built-in household cooking
    appliances,** 6.15.2
Burners
    Combination gas and oil burners,
      chimneys for, 7.5.5(d)
    Definitions, 1.7
    Gas conversion, *see* Conversion
      burners, gas
    Input adjustment, 8.1, Table
      8.1.1, A.8.1.1
    Primary air adjustment, 8.2,
      A.8.2
    In residential garages, 5.1.9(a)
Bypass valves, pool heaters, 6.22.4
**Bypasses, gas pressure
    regulators,** 2.8.5

-C-

Capping, of outlets, 3.8.2
Carbon steel (definition), 1.7

**Carpeting, installation of gas
    utilization equipment on,**
    5.2.3
**Casters, for floor-mounted food
    service equipment,** 6.12.6
Central furnaces, 6.3, 6.3.6(c),
    Tables 6.2.3(a) to (b)
    Definition, 1.7
    Direct vent central
      (definition), 1.7
Central premix system, 3.13,
    A.3.13
    Definition, 1.7
Chases, vertical, piping in, 3.5
**Chimneys (masonry, metal, and
    factory-built),** 7.5,
    A.7.5.1(c)
    Checking draft, 8.6, A.8.6
    Cleanouts, 7.5.4(c), 7.5.7
    Definitions, 1.7
    Exterior masonry, Tables 10.11
      to 10.13(a) and (b)
      Definition, 1.7
    Obstructions, 7.15, 10.1.1
    Vent connectors, 7.10.2(d) to (e),
      7.10.3(c), 7.10.12
    Venting system, 10.1.7, 10.1.9,
      10.2.17 to 10.2.18, App. G,
      Tables 10.8 to 10.9, Tables
      10.11 to 10.13
Circuits, electrical, 3.15, 5.6.3
Circulating air, 6.7.3, 6.10.5,
    6.11.3, 6.27.3, 6.28.3
    Definition, 1.7
    Ducts, 7.3.6
Clearances, 5.2.2
    Air-conditioning equipment,
      indoor installation, 6.2.3
    Boilers, central heating,
      Table 6.2.3(b), 6.3.1
    Clothes dryers, 6.4.1
    Domestic incinerators, 6.18.1
    Draft hoods, 7.12.8
    Food service equipment
      Counter appliances, 6.13.1 to
        6.13.3
      Floor-mounted, 6.12.1 to
        6.12.2
      Outdoor cooking appliances,
        6.21.2
    Furnaces
      Central, 6.3.1

    Duct, 6.10.1
    Floor, 6.11.7
    Gas-fired toilets, 6.26.1
    Heaters
      Direct makeup air, 6.8.2
      Industrial air, 6.9.4
      Infrared, 6.19.2
      Pool, 6.22.2
      Room, 6.24.3
      Unit, 6.27.2
      Water, 6.29.3
    Illuminating appliances, 6.16.1
      to 6.16.2, Table 6.16.2(b)1
    Refrigerators, gas, 6.23.1
    Single-wall metal pipe for vents,
      7.7.4(d)
    Vent connectors, 7.10.5
Clothes dryers
    Definition, 1.7
    Installation, 6.4
    Multiple family or public use,
      6.4.6
    Venting, 7.2.2(d)
Coal basket, *see* Decorative
    appliances, to install in
    vented fireplaces
Combustible material
    Clearances to, *see* Clearances
    Definition, 1.7
    Food service equipment mounted
      on/adjacent to, 6.12.3,
      6.12.5, 6.13.4, 6.14
    Roofs, metal pipe passing
      through, 7.7.4(e)
Combustion (definition), 1.7
Combustion air, 5-1.2(a), 5.3,
    A.5.3
    Ducts, 5.3.6
    Floor furnaces, 6.11.3
    Gas fireplaces, vented, 6.7.3
    Infrared heaters, 6.19.3
    Opening design, example of,
      App. L
    Unit heaters, 6.27.3
    Wall heaters, 6.28.3
**Combustion chamber
    (definition),** 1.7
**Combustion products
    (definition),** 1.7
**Compressed natural gas (CNG)
    vehicular fuel systems,**
    6.30

Concealed gas piping
In buildings, 3.4
Definition, 1.7
Condensate (condensation)
Definition, 1.7
Drain, 7.9
Confined spaces, *see* Spaces,
Confined
Connections
Air conditioners, 6.2.2
Chimney, 7.10.12
Electrical, 3.16, 5.6.1
Gas, 2.13.2
Branch, 3.5.1, 3.9
Concealed piping, 3.4.2
Plastic and metallic piping,
3.1.7(a) to (b)
Gas equipment, 5.5, A.5.5.6
Portable and mobile industrial
gas equipment, 5.5.3
Connectors
Gas hose, to gas utilization
equipment, 5.5.2
Vent, *see* Vent connectors
Construction
Chase, 3.5.2
Checklist, B.2
Coordination, B.1
Overpressure protection devices,
2.9.3
Single-wall metal pipe, 7.7.1
Unusually tight, 5.3.2, A.5.3.2
Definition, 1.7
Consumption (definition), 1.7
Control piping
Definition, 1.7
Overpressure protection
devices, 2.9.4
Controls
Definition, 1.7
Direct makeup air heaters,
6.8.5
Draft, 7.12
Duct furnaces, 6.10.4
As obstructions, 7.15, 10.1.1
Protective devices, 8.5
Safety shutoff devices, 8.3
Convenience outlets, gas, 3.8.1(6),
5.5.6
Definition, 1.7
Conversion burners, gas, 6.5
Definition, 1.7

Cooking appliances, *see* Food ser-
vice equipment, gas; House-
hold cooking appliances
Cooling units
Boilers, used with, 6.3.8
Furnaces, used with, 6.3.7
Corrosion, protection against,
2.6.6, 3.1.3, 3.4.5, A.3.1.3
Counter appliances, gas, *see* Food
service equipment, gas
Cubic feet (cu ft.) of gas, 2.4.2,
A.2.4.2
Definition, 1.7

-D-

Damper devices, automatic vent
Definition, 1.7
Electrically operated, App. I
Mechanically actuated, App. J
Thermally actuated, App. K
**Damper regulators, automatic
(definition),** 1.7
Dampers, 6.9.5(b)
Automatically operated vent,
7.14, 10.1.1
Chimney, free opening area of,
Table 6.6.2
Gravity, 6.9.7(b)
Manually operated, 7.13
**Decorative appliances, to install
in vented fireplaces,** 6.6,
A-6.6.1, Table 6.6.2
Definition, 1.7
Deep fat fryer, *see* Food service
equipment, gas
Defects, detection of, 2.6.5, 4.1.5
Definitions, 1.7, A.1.7
Design
Checklist, B.2
Coordination, B.1
**Design certification
(definition),** 1.7
Design pressure
Allowable pressure drop, 2.4.1,
2.4.4, A.2.4.1
Definition, 1.7
Maximum, 2.5.1, A.2.5.1(1)
Detectors, leak, *see* Leak detectors
Dilution air, 5.3
Definition, 1.7
**Direct gas-fired industrial air
heaters,** 6.9

Air supply, 6.9.5, A.6.9.5
Definition, 1.7
Prohibited installations, 6.9.2
**Direct gas-fired makeup air
heaters,** 6.8
Definition, 1.7
Input ratings, 6.8.6
Venting, 7.2.2(i)
Direct vent appliances
Definition, 1.7
Venting of, 7.2.5, 7.6.2(a), 7.8
**Distribution mains or gas mains
(definition),** 1.7
Diversity factor (definition), 1.7
Drafts, 5.7.2
Checking, 8.6, A.8.6
Definition, 1.7
Mechanical systems, 7.3.4,
7.6.2(a), 7.8(a) to (b),
7.10.4(b)
Requirements, 7.3.2
Draft controls, 7.12, 7.15
Draft hoods, 5.3.1(e), 6.10.4, 7.12,
8.6, A.7.12.4, A.8.6
Definition, 1.7
Domestic incinerators, 6.18.3
Vents and venting systems,
7.6.3(a), 7.7.5(a), 7.10.2(b)
to (c), 7.10.3, 10.1.2, 10.1.4,
10.1.9(3), 10.2.10 to
10.2.11, 10.2.15, 10.2.18(2),
10.2.19, Table 10.10, App. G
Draft regulators, 7.12.5, 7.15; *see
also* Barometric draft
regulators
Definition, 1.7
Drain, condensation, 7.9
Drip liquids, 1.6.2(a)
Drips, 3.3.2, 3.7
Definition, 1.7
Dry gas, 3.3.2 to 3.3.3
Definition, 1.7
Duct furnaces, 6.10
In commercial garages and air-
craft hangars, 6.10.7
Definition, 1.7
Use with refrigeration systems,
6.10.6
Ducts
Air, 5.3.6, 6.2.3(f), 6.3.6, 7.3.6
Exhaust, 6.4.4 to 6.4.5
Unit heaters, 6.27.4

-E-

Elbows, 3.6.4, 10.1.3, 10.2.6
Electrical systems, 3.14 to 3.16
  Air conditioning equipment,
    6.2.6
  Gas utilization equipment, 5.6
  Ignition and control devices,
    5.6.2
Engines, stationary gas, 6.25
Equipment, *see* Appliances; Gas
  utilization equipment
Excess air (definition), 1.7
Exhaust systems, mechanical, *See*
  Mechanical exhaust systems
Explosion heads (rupture discs),
  3.13.4, 3.13.6(d), A.3.13.4
  Definition, 1.7
Exterior masonry chimneys, *see*
  Chimneys (masonry, metal,
  and factory-built)

-F-

FAN Max, 10.1.6, 10.2.3(a),
  10.2.16, Tables 10.1 to 10.4,
  Tables 10.6 to 10.9
  Definition, 1.7
FAN Min, 10.1.1(2), 10.1.6,
  10.2.1(3), 10.2.3(b),
  10.2.16, Tables 10.1 to 10.4,
  Tables 10.6 to 10.9
  Definition, 1.7
Fan-assisted combustion system
  appliances, 10.2.19
  Definition, 1.7
  Venting system, 7.10.3(a),
    App. G
FAN+FAN, 10.2.1(3), Tables 10.6
  to 10.9
  Definition, 1.7
FAN+NAT, 10.2.1(2), (3),
  10.2.18(3), 10.2.18(4),
  Tables 10.6 to 10.9, Tables
  10.13(a) to (b)
  Definition, 1.7
Firechecks, automatic, 3.13.1(3),
  3.13.4, 3.13.6, A.3.13.4,
  A.3.13.6(a)
  Definition, 1.7
Fireplace insert, *see* Decorative
  appliances, to install in
  vented fireplaces
Fireplace screens, 6.6.3

Fireplaces, *see also* Decorative
  appliances, to install in
  vented fireplaces
  Definition, 1.7
  Gas, *see* Gas fireplaces
  Outlets, capping, 3.8.2(b)
  Vent connectors, 7.10.14
Fittings
  Concealed piping, 3.4.2
  Corrosion, protection against,
    2.6.6, 3.1.3, 3.4.5
  Gas pipe turns, 3.6, A.3.6.3
  Gas utilization equipment con-
    nections, 5.5.1(1) to (2)
  Metallic, 2.6.8, A.2.6.8(a),
    Table 2.6.7
  Overpressure relief device, 2.9.8
  Plastic, 2.6.4, 2.6.9
  Used, 2.6.1(b)
  Workmanship and defects, 2.6.5,
    4.1.5
Flame arresters, 3.13.2(2), 3.13.4,
  A.3.13.4
  Definition, 1.7
Flammable liquids, handling of, 1.6.2
Flammable vapors, gas appli-
  ances in area of, 5.1.8
Flange gaskets, 2.6.11
Flanges, 2.6.10
Floor furnaces, 6.11
  Definition, 1.7
  First floor installation, 6.11.12
  Upper floor installation, 6.11.11
Floor-mounted equipment
  Food service, 6.12, A.6.12.8
  Household cooking appliances,
    6.15.1
  Unit heaters, 6.27.2(b)
Floors, piping in, 3.4.5, 7.10.15
Flowmeters, 3.13.2(a)
Flues
  Appliance (definition), 1.7
  Chimney (definition), 1.7
Flue collars, 7.10.3(d), 10.1.2,
  10.1.4, 10.2.10 to 10.2.11,
  10.2.15, 10.2.19
  Definition, 1.7
  Multiple, on single appliance,
    7.10.3(b)
Flue gases
  Definition, 1.7
  Venting of, 5.1.13

Food service equipment, gas
  Combustible material adjacent to
    cooking top, 6.12.5
  Connections, 5.5.1(5)
  Counter appliances, 6.13
  Definitions, 1.7
  Floor-mounted, 6.12, A.6.12.8
  Venting, 7.2.2(a) to (c), (g)
Forced air furnaces, 6.3.7(a)
Foundations, piping through,
  3.1.5
Freezing, protection against,
  3.1.4, 3.3.2, A.3.1.4
Furnaces
  Central, *see* Central furnaces
  Downflow (definition), 1.7
  Duct, *see* Duct furnaces
  Enclosed (definition), 1.7
  Floor, *see* Floor furnaces
  Forced-air (definitions), 1.7
  Gravity (definitions), 1.7
  Horizontal (definition), 1.7
  Refrigeration coils and, 6.3.7
  Upflow (definition), 1.7
  Wall, *see* Wall furnaces

-G-

Garages
  Commercial
    Duct furnaces in, 6.10.7
    Gas utilization equipment in,
      5.1.10
    Heaters in, 5.1.10(b), 6.19.4,
      6.27.5
  Repair
    Definition, 1.7
    Gas utilization equipment in,
      5.1.10(b)
  Residential
    Definition, 1.7
    Gas utilization equipment in,
      5.1.9
Gas fireplaces, *see also* Decorative
  appliances, to install in
  vented fireplaces
  Direct vent (definition), 1.7
  Vented, 6.7, A.6.7.1
  Definition, 1.7
Gas log, *see* Decorative
  appliances, to install in
  vented fireplaces
Gas mains (definition), 1.7

Gas reliefs
Direct makeup air heaters, 6.8.7
Industrial air heaters, 6.9.6
Gas supplier regulations, 1.1.2
Gas utilization equipment
Accessibility, 5.2.1
Added or converted, 5.1.2 to 5.1.3
Air for combustion and ventilation, 5.3, A.5.3
Approval, 5.1.1, A.5.1.1
Bleed lines for diaphragm-type valves, 5.1.19
Carpeting, installation on, 5.2.3
Clearance to combustible materials, 5.2.2
Combination, 5.1.20
In confined spaces, 5.3.3, A.5.3.3
Connections to building piping, 5.5, A.5.5.6
Convenience outlets, 5.5.6
Coordination of design, construction, and maintenance, App. B
Definition, 1.7
Electrical systems, 5.6
Extra devices or attachments, 5.1.14
Installation, Chap. 5
Instructions
Installation, 5.1.21
Operating, 8.7
Louvers and grilles, 5.3.5
Mechanical exhausting and, A.5.3
Mobile, connections, 5.5.3
Operation procedures, Chap. 8
Outdoor, protection of, 5.1.22
Physical protection, 5.1.12, 5.5.1(6)
Piping
Capacity of, 5.1.15
Strain on, 5.1.16
Placing in operation, 4.3.4
Portable, connections, 5.5.3
Pressure regulators, 5.1.17 to 5.1.18
Protection from fumes and gases, 5.1.6, A.5.1.6
On roofs, 5.4
Specially engineered installations, 5.3.4
Type of gas(es), 5.1.3
In unconfined spaces, 5.3.2 to 5.3.3, A.5.3.2 to A.5.3.3
Venting, 5.1.13, Chap. 7
In well-ventilated spaces, 7.2.4
Gas vents, 7.6
Checking draft, 8.6, A.8.6
Definition, 1.7
Integral, equipment with, 7.2.6, 7.6.2(a)
Multistory design, Figs. G.13 to G.14
Serving equipment on more than one floor, 7.6.4
Size of, A.7.6.3(a)
Spaces surrounding, 7.5.8
Special, 7.4.3, 7.5.8
Wall heaters, 6.28.1(c) to (d)
Gas-air mixtures
Flammable, 3.13, A.3.13
Outside the flammable range, 3.12
Gases
Definition, 1.7
Dry, *see* Dry gas
Flow of, through fixed orifices, App. F
Flue
Definition, 1.7
Venting of, 5.1.13
LP-Gas systems, 2.5.2, 2.6.9(d), 5.1.4
Maximum demand, 2.4.1 to 2.4.2, A.2.4.1 to A.2.4.2
Purged, discharge of, 4.3.3
Used in gas utilization equipment, 5.1.3
Utility (definition), 1.7
Vent (definition), 1.7
Gaskets, flange, 2.6.11
Gas-mixing machines, 3.12, 3.13.1(1), 3.13.3
Definition, 1.7
Installation, 3.13.5, A.3.13.5
Governor, zero, 3.13.4
Definition, 1.7
Gravity, *see* Specific gravity
Grilles
Protecting openings, 5.3.5
Wall heaters, 6.28.1(e)
Grounding, electrical, 3.14

-H-
Hangars, *see* Aircraft hangars
Hangers, pipe, 3.3.6
Heat pumps, gas-fired, 6.2
Definition, 1.7
Heat reclaimers, 7.15
Heaters, *see also* Direct gas-fired industrial air heaters; Direct gas-fired makeup air heaters; Infrared heaters; Pool heaters; Room heaters; Unit heaters; Water heaters
In aircraft hangars, 5.1.11, 6.19.4, 6.27.5
In commercial garages, 5.1.10(b), 6.19.4, 6.27.5
Heating value (total)
Definition, 1.7
Hoods
Draft, *see* Draft hoods
Duct furnaces, 6.10.4
Ventilating, *see* Ventilating hoods
Hoop stress (definition), 1.7
Hot plates
Commercial, *see* Food service equipment, gas
Domestic, 6.14
Definition, 1.7
Venting, 7.2.2(c)
Hot taps, 1.5.3
Definition, 1.7
Household cooking appliances, 6.15; *see also* Outdoor cooking gas appliances
Built-in units, 6.15.2
Definition, 1.7
Floor-mounted, 6.15.1

-I-
Identification
Gas pressure regulators, 2.8.6
Meters, gas, 2.7.5
Ignition
Accidental, prevention of, 1.6
Automatic, 8.4
Definition, 1.7
Electrical, 5.6.2
Sources, 1.6.1
Definition, 1.7
Illuminating appliances, 6.16, Table 6.16.2(b)1

Incinerators
Commercial-industrial, 6.17
Domestic, 6.18
Definition, 1.7
Draft regulators, barometric,
7.12.5
Vent connectors, 7.10.2(e)
Venting, 7.3.4(a)
Indirect ovens (definition), 1.7
Industrial air heaters, *see* Direct
gas-fired industrial
air heaters
Infrared heaters, 6.19
Definition, 1.7
Inspections
Chimneys, 7.5.4
Draft, 8.6, A.8.6
Existing installation, 9.1, App. H
Gas piping, 4.1, A.4.1.1
Ignition, automatic, 8.4
Protective devices, 8.5
Safety, 9.1
Safety shutoff devices, 8.3
Vent connectors, 7.10.13
Installations
Chimneys, 7.5.1 to 7.5.2,
A.7.5.1(c)
Draft hoods and draft controls,
7.12.2 to 7.12.3
Electrical, 5.6
Equipment, Chap. 5
Gas piping, Chap. 3
Gas shutoff prior to, 1.5.2 to
1.5.3, A.1.5.2
Single-wall metal pipe for vents,
7.7.4
Specific equipment, Chap. 6
Instructions
Gas utilization equipment instal-
lation, 5.1.21
Manufacturers, 1.1.2
Operating, 8.7
Insulating millboard
(definition), 1.7
Interruption of service, 1.5,
A.1.5.2
Interruption of work, 1.5.4

-J-

Joining methods, 2.6, A.2.6
Joint compounds, thread, *see*
Thread joint compounds

Joints
Concealed piping, 3.4.2
Corrosion, protection against,
3.1.3, A.3.1.3
Definition, 1.7
Flared, 2.6.8(c)
Metallic, 2.6.8, A.2.6.8(a),
Table 2.6.7
Plastic, 2.6.9, 3.6.2(b)
Tubing, 2.6.8(b)
Vent connectors, 7.10.7

-K-

Kettle, gas-fired, *see* Food service
equipment, gas

-L-

Labeled (definition), 1.7
Laundry stoves, domestic, 6.14
Definition, 1.7
Venting, 7.2.2(c)
Leak check
Definition, 1.7
Suggested method, App. D
Leak detectors, 4.1.5
Definition, 1.7
Leakage
Emergency procedure for, App. E
Pressure test, 4.1.5
System and equipment test, 4.2,
A.4.2.3
Test, 1.5.2, A.1.5.2
Limit controls, 8.5, *see also* Temper-
ature limit controls/devices
Definition, 1.7
Listed (definition), 1.7, A.1.7
Loads, connected (definition), 1.7
Louvers
Heaters, 6.8.4
Industrial air heaters, 6.9.5(b),
6.9.7(b)
Protecting openings, 5.3.5
LP-Gas systems, 2.5.2, 2.6.9(d),
5.1.4

-M-

Main burners (definition), 1.7
Mains, gas or distribution
(definition), 1.7
Maintenance, B.1, B.3
Makeup air, 6.4.3; *see also* Direct
gas-fired makeup air heaters

Manifolds, gas (definition), 1.7
Manufactured home
(definition), 1.7
Manufacturer's instructions, 1.1.2
Marking
Gas vents, 7.6.6
Single-wall metallic pipe, 7.7.7
Masonry chimneys, *see* Chimneys
(masonry, metal, and
factory-built)
Maximum working pressure
Definition, 1.7
Overpressure relief device and,
2.9.5
Mechanical exhaust systems,
7.3.5, 7.6.2(a), A.7.3.5
Clothes dryers, 6.4.2, 6.4.4 to
6.4.5
Definition, 1.7
Metallic pipe, 2.6.2 to 2.6.3,
A.2.6.2(c), A.2.6.3(b)
Connection to plastic piping,
3.1.7(b)
Corrosion, protection against,
2.6.6, 3.4.5
Gas utilization equipment con-
nections, 5.5.1(1) to (2)
Joints and fittings, 2.6.8
Single-wall, for venting, 7.7,
10.2.20, A.7.7.5(a), Table
7.7.4(d), Table 10.10
Threads, 2.6.7
Turns, 3.6.1
Meters, gas, 2.7, A.2.7
Definition, 1.7
Millboard, insulating
(definition), 1.7
Mitered bends, 3.6.3, A.3.6.3
Mixing blowers, 3.13.3 to 3.13.4,
3.13.5(c), A.3.13.4
Definition, 1.7
Multistory installations, 10.2.12 to
10.2.14

-N-

NA, 10.1.1(2), 10.2.1(3)
Definition, 1.7
NAT Max, 10.1.1(1), 10.1.6,
10.2.1(1), 10.2.3(a), 10.2.16,
Tables 10.1 to 10.4, Tables
10.6 to 10.9
Definition, 1.7

NAT+NAT, 10.2.1(2), 10.2.18(3),
10.2.18(4), Tables 10.6 to
10.9, Tables 10.12(a) to (b)
Definition, 1.7
Noncombustible material, 6.2.4
Definition, 1.7
Thimbles, 7.7.4(e)
**Nondisplaceable valve member
(definition),** 1.7
**Notification of interrupted
service,** 1.5.1

-O-

Openings, *see also* Relief openings
Air opening design for combustion and ventilation, App. L
Chimneys, 7.5.5(b)
Confined spaces, 5.3.3, A.5.3.3
Pipe, size of, 2.9.8
**Operation of equipment,
procedures,** Chap. 8
Orifice cap (hood)
Definition, 1.7
Orifice spud (definition), 1.7
Orifices
Definition, 1.7
Fixed, flow of gas through, App.F
**Outdoor cooking gas
appliances,** 6.21
Definition, 1.7
Outlets
Convenience, *see* Convenience
outlets, gas
Piping, 3.8
Ovens
Baking and roasting, *see* Food
service equipment, gas
Indirect (definition), 1.7
**Overpressure protection
devices,** 2.9
Piping in vertical chases, 3.5.1
Setting, 2.9.5
Unauthorized operation, 2.9.6
Oxygen
As test medium, 4.1.2
Use under pressure, 5.1.5

-P-

Parking structures
Definition, 1.7
Gas utilization equipment in,
5.1.10(a)

Partitions
Piping in, 3.4.3
Tubing in, 3.4.4
Pilot (definition), 1.7
Pipe threads, metallic, 2.6.7, Table
2.6.7; *see also* Thread joint
compounds
Pipes and piping
Above-ceiling locations, 3.3.4
Aboveground outside, 3.2
Bends, 3.6, A.3.6.3
Branch, *see* Branch lines
Buildings, in, *see* Buildings,
Piping in
Bypass, 2.8.5
Clearances, underground piping,
3.1.1
Concealed, *see* Concealed gas
piping
Connections, *see* Connections;
Connectors
Control, *see* Control piping
Defects, 2.6.5, 4.1.5
Definition, 1.7
Drips, 3.7
Equivalent length (definition), 1.7
Floors, in, 3.4.5
Hangers and anchors, 3.3.6
Identification, multiple meter
installations, 2.7.5
Independent, 6.2.1
Inspection, testing, and purging,
Chap. 4
Installation, 2.1.1, Chap. 3, 5.5.8
Joining methods, 2.6, A.2.6
Materials, 2.6, A.2.6
Metallic, *see* Metallic pipe
Outlets, 3.8
Partitions, in, 3.4.3 to 3.4.4
Plastic, *see* Plastic pipe
Prohibited devices in, 3.11
Prohibited locations, 3.3.5
Protection, underground piping,
3.1.2
Protective coating, 2.6.6
Removal of, 3.3.7
Sediment traps, 3.7.3, 5.5.7
Sizing and capacities of, 2.4.1,
2.4.3, 5.1.15, Chap. 10,
A.2.4, A.2.4.1, App. C
Overpressure relief devices,
2.9.8

Sloped, 3.3.2 to 3.3.3
Strain on, 5.1.16
Supports, *see* Supports, Pipes
and piping
Turns, 3.6, A.3.6.3
Underground, 3.1, A.3.1.3
Used materials, 2.6.1(b)
Valves, 3.10
In vertical chases, 3.5
Workmanship, 2.6.5
Piping systems, gas, Chap. 2
Addition to existing, 2.1.2
Definition, 1.7
Design pressure
Allowable pressure drop,
2.4.1, 2.4.4, A.2.4.1
Maximum, 2.5.1
Flexibility, 2.13
Gas-air mixtures
Flammable, 3.13, A.3.13
Outside the flammable range,
3.12
Interconnections between, 2.3
Leakage tests, 4.1.5, 4.2
Local conditions, consideration
of, 2.13.2
Materials and joining methods,
2.6, A.2.6
Operating pressure limitations,
2.5, A.2.5.1(1)
Maximum, A.2.5.1(1)
Placing in operation,
purging for, 4.3.2
Plan, 2.1
Point of delivery, location of, 2.2
Pressure drop, allowable, 2.4.1,
2.4.4, A.2.4.1, A.2.4.4
Pressure testing and inspection,
4.1, A.4.1.1
Purging, 4.3, Table 4.3.1,
Table 4.3.2
Removal from service,
purging for, 4.3.1
Sizing of, 2.4, A.2.4
Thermal expansion, 2.13
Plastic pipe, 2.6.4, 2.6.9
Within chimney flue, 7.5.8
Connection of, 3.1.7(a) to (b)
Sizing, Tables 9.32 to 9.34
Tracer wire to locate, 3.1.7(c)
Turns, 3.6.2
As vent material, 7.4.2

Plenums, 3.3.4, 6.2.3(e) to (f),
6.3.6, 7.3.6
Definition, 1.7
Point of delivery, location of, 2.2
Pool heaters, 6.22
Definition, 1.7
Power, continuous, 5.6.4
Pressure
Atmospheric (definition), 1.7
Back
Definition, 1.7
Protection, 2.10
Definition, 1.7   ·
Design, *see* Design pressure
Limitations, A.2.5.1(1)
Low, protection, 2.11
Maximum working
Definition, 1.7
Overpressure relief device
and, 2.9.5
Operating, limitations on, 2.5
Pressure control (definition), 1.7
Pressure drop
Allowable, 2.4.1, 2.4.4, A.2.4.1,
A.2.4.4
Definition, 1.7
Pressure limiting devices, 6.3.3,
6.22.3, 6.29.4
Definition, 1.7
Pressure regulators, *see* Overpres-
sure protection devices;
Regulators
Pressure relief devices, 6.3.5, 8.5,
A.6.3.5; *see also* Overpres-
sure protection devices
Water heaters, 6.29.6
Pressure tests, 4.1, A.4.1.1
Definition, 1.7
Primary air, 8.2, A.8.2
Definition, 1.7
Protection
Back pressure, 2.10
Control piping, 2.9.4
Floor furnaces, 6.11.10
Gas pressure regulators, 2.8.3
Gas utilization equipment,
5.1.22
From fumes or gases, 5.1.6,
A.5.1.6
Physical, 5.1.12, 5.5.1(6)
Low pressure, 2.11
Meters, gas, 2.7.4

Open-top broiler units, 6.20.3
Piping, underground, 3.1.2
Protective devices, 8.5
Purge/purging, 4.3, 6.8.9, Table
4.3.1, Table 4.3.2, A.4.3
Definition, 1.7

-Q-

Qualified agency, 1.4
Definition, 1.7
Quick-disconnect devices,
3.8.1(6), 3.8.2(a), 5.5.5
Definition, 1.7

-R-

Radiant appliances, *see* Decora-
tive appliances, to install in
vented fireplaces
Ranges, *see* Food service equip-
ment, gas
Referenced publications,
Chap. 11, App. M
Refrigeration systems
Boilers used with, 6.3.8
Coils, 6.3.7
Duct furnaces used with, 6.10.6
Refrigerators, gas, 6.23
Definition, 1.7
Venting, 7.2.2(f)
Regulations, gas supplier, 1.1.2
Regulator vents (definition), 1.7
Regulators
Appliance pressure, 2.8.4(b)
Automatic damper
(definition), 1.7
Draft, *see* Draft regulators
Gas appliance (definition), 1.7
Gas appliance pressure, 5.1.17 to
5.1.18
Illuminating appliances, 6.16.5
Venting of, 5.1.18
Line gas, 2.8.4(a)
Definition, 1.7
Monitoring (definition), 1.7
Pressure, 2.8, 3.5.1, 5.1.17 to
5.1.18, A.2.8; *see also* Over-
pressure protection devices
Definition, 1.7
Venting of, 5.1.18
Protection of, 2.8.3
Series (definition), 1.7
Service (definition), 1.7

Relief openings
Definition, 1.7
Direct makeup air heaters, 6.8.8
Industrial air heaters, 6.9.7
Repairs, gas shutoff prior to, 1.5.2
to 1.5.3, A.1.5.2
Retroactivity of code, 1.3
Risers
Anodeless, *see* Anodeless risers
Corrosion, protection against,
3.4.5
Horizontal lines sloped to/from,
3.3.3
**Roofs, gas utilization equipment
on,** 5.4
Room heaters, 6.24
Institutions, installations in,
6.24.2
Prohibited installations, 6.24.1,
A.6.24.1
Unvented, 6.24.1, 7.2.2(h),
A.6.24.1
Definition, 1.7
Vented (definition), 1.7
Wall-type, 6.24.4
**Room large in comparison with
size of equipment
(definition),** 1.7

-S-

**Safety blowouts (backfire
preventers),** 3.13.1(4), 3.13.4,
3.13.6, A.3.13.4, A.3.13.6(a)
Definition, 1.7
**Safety inspection, existing appli-
ance installation,** App. H
Safety shutoff devices, 8.3
Definition, 1.7
Unlisted LP-gas equipment used
indoors, 5.1.4
Safety shutoff valves
Gas-mixing machines, 3.13.5(d),
A.3.13.5(d)
Manual, 3.13.5(d), A.3.13.5(d)
Scope of code, 1.1
Seepage pan, 6.11.9
**Separate users, interconnections
between,** 2.3.1
**Service head adapters
(definition),** 1.7
**Service meter assembly
(definition),** 1.7

Service regulators, *see* Regulators,
Pressure; Regulators, Service
Shall (definition), 1.7
**Shutoff devices, automatic gas
(definition),** 1.7
Shutoff procedure, 1.5.2 to 1.5.3,
A.1.5.2; , *see also* Safety
shutoff devices; Shutoff
valves
Shutoff valves, 2.12
Emergency, 3.10.3
Equipment, 1.5.2, 3.3.4, 3.8.2(b),
5.5.4, 7.5.5(b), A.1.5.2
Manual gas, 3.10, 5.5.4
For multiple systems, 3.10.2
Sources of ignition, *see* Ignition,
Sources
Spaces
Confined, 5.3.3, A.5.3.3
Definition, 1.7
Surrounding chimney lining or
vent, 7.5.8
Unconfined, 5.3.2 to 5.3.3,
A.5.3.2 to A.5.3.3
Definition, 1.7
Well-ventilated, 7.2.4
Specific gravity (definition), 1.7
**Standby fuels, interconnections
for,** 2.3.2
Steam cookers, *see* Food service
equipment, gas
Steam generators, *see* Food ser-
vice equipment, gas
Stress
Definition, 1.7
Hoop (definition), 1.7
**Structural members,
building,** 5.1.7
Supports
Chimneys, 7.5.6
Floor furnaces, 6.11.6
Gas utilization equipment, 5.1.7(c)
Gas vents, 7.6.5
Heaters
Infrared, 6.19.1
Unit, 6.22.1
Meters, gas, 2.7.3
Pipes and piping, 2.13.2, 3.3.6,
Table 3.3.6
Single-wall metallic pipe,
7.7.6
Vent connectors, 7.10.10

Suspended-type unit heaters,
6.27.2(a)
Switches, electrical supply line,
6.2.6

-T-
**Temperature, ambient
(definition),** 1.7; *see also*
Weather conditions
**Temperature limit controls/
devices,** 6.3.3
Floor furnaces, 6.11.2
Pool heaters, 6.22.3
Water heaters, 6.29.5
**Temperature relief devices, water
heaters,** 6.29.6
Tensile strength (definition), 1.7
Termination, venting systems,
see Venting systems
Testing, piping system
Defects, pressure test for, 4.1.5
Leakage
Pressure test for detection,
4.1.5
System and equipment test,
1.5.2, 4.2, A.1.5.2, A.4.2.3
Pressure, 4.1, A.4.1.1
Thermostats
Definitions, 1.7
Room temperature, 5.7
Thimbles, 7.7.4(e)
Thread joint compounds, 2.6.7(d)
Definition, 1.7
Toilets, gas-fired, 6.26
Traps, sediment, 3.7.3, 5.5.7
Trenches, 3.1.2(b)
Tubing
Connections, 3.4.2
Gas utilization equipment con-
nections, 5.5.1(2), (6)
Metallic, 2.6.3, 2.6.8(b),
A.2.6.3(b), A.2.6.8(a)
Partitions, in, 3.4.4
Plastic, 2.6.4, 2.6.9
Sizing and capacities of, Tables
9.13 to 9.18, Tables 9.27 to
9.28, Tables 9.33 to 9.34
Workmanship and defects, 2.6.5,
4.1.5
Type B gas vents, *see* Gas vents
Type B-W gas vents, *see* Gas vents
Type L vents, *see* Gas vents

-U-
Unconfined spaces, *see* Spaces,
Unconfined
Underground piping, 3.1, A.3.1.3
Unit broilers
Definition, 1.7
Open-top, 6.20
Unit heaters, 6.27
Definition, 1.7
**Unusually tight construction
(definition),** 1.7
Utility gases (definition), 1.7

-V-
**Vacuum relief devices, water
heaters,** 6.29.6
Valve members
Definition, 1.7
Nondisplaceable (definition), 1.7
Valves
Accessibility of, 3.3.4, 3.10.2(a)
Bypass, pool heaters, 6.22.4
Controlling multiple systems,
3.10.2
Definitions, 1.7
Diaphragm type, bleed lines for,
5.1.19
Mixing blowers, 3.13.4,
A.3.13.4
Pressure relief and steam safety,
boilers, 6.3.5, A.6.3.5
Shutoff, *see* Shutoff valves
Used, 2.6.1(b)
**Vapors, flammable, gas appli-
ances in area of,** 5.1.8
**Vehicular fuel systems, com-
pressed natural gas,** 6.30
Vents
Definition, 1.7
Gas, *see* Gas vents
Heater, 6.8.7, 6.9.6
Obstructions, 7.15, 10.1.1
Regulator (definition), 1.7
Sizes
Appliance categorized vent
diameter/area (definition),
1.7
Multiple appliance vents,
10.2, Tables 10.6 to 10.13
Single appliance vents, 10.1,
Tables 10.1 to 10.5
Toilets, gas-fired, 6.26.3

Vent connectors, 7.10 to 7.11,
    A.7.10.3, A.7.10.9(b)
  Bends, 7.10.6
  Chimneys, *see* Chimneys
  Clearance, 7.10.5
  Combination units, A.7.5.5(c)
  Dampers, 7.13 to 7.14
  Definition, 1.7
  Fireplaces, 7.10.14
  Inspection, 7.10.13
  Joints, 7.10.7
  Length, 7.10.9, A.7.10.9(b)
  Location, 7.10.11
  Materials, Table 7.4.1, 7.10.2,
    Tables 7.10.2(d) to (e)
  Mechanical draft systems,
    7.3.4(c)
  Obstructions, 7.15, 10.1.1
  Routing, 10.2.3
  Size of, 7.10.3, 10.1.10 to
    10.1.11, 10.2.4, 10.2.10,
    A.7.10.3, Figs. G.2 to G.4,
    G.6 to G.9, G.11
  Slope, 7.10.8
  Support, 7.10.10
  Through ceilings, floors, or
    walls, 7.10.15
  Toilets, gas-fired, 6.26.3
  Two or more appliances,
    7.10.3(c) to (d), 7.10.4
Vent damper devices, automatic,
  see Damper devices, auto-
  matic vent
Vent gases (definition), 1.7
Vented appliances
  Categories and definitions, 1.7,
    A.1.7
  Venting systems, Category I
    appliances, Chap. 10
Vented wall furnaces, 6.28.1(c)
    to (d)
  Definition, 1.7

Ventilating hoods, 6.20.3, 7.2.3,
    7.3.5, 7.6.2(a), A.7.3.5
Ventilation
  Air for, 5.3, A.5.3
    Opening design, example of,
      App. L
  Chase, 3.5.3, A.3.5.3
  Food service equipment, 6.12.8,
    A.6.12.8
  Gas utilization equipment,
    5.1.2(a)
  Industrial air heaters, 6.9.5,
    A.6.9.5
  Infrared heaters, 6.19.3
  Open-top broiler units, 6.20.3 to
    6.20.4
Venting, Chap. 7
  Definition, 1.7
  Equipment not requiring venting,
    7.2.2
  Flue gases, 5.1.13
  Overpressure protection devices,
    2.9.7
  Pressure regulators, 2.8.4, 3.5.1,
    5.1.18
Venting system
  Terminations
    Through the wall, A.7.8
  Two or more appliances, single
    vent, 7.10.3(c) to (d), 7.10.4,
    10.2.15
  Type of system to use, 7.4, Table
    7.4.1
  Ventilating hoods, A.7.2.3
Venting systems
  Category I, sizing of, Chap. 10
  Connection to, *see* Vent connec-
    tors
  Definition, 1.7, A.1.7
  Design and construction, 7.3
  Gas utilization equipment,
    5.1.2(c), 5.1.13

  Incinerators, 6.18.4
  Pool heaters, 6.22.5
  Refrigerators, 6.23.2
  Sizing of (appliances equipped
    with draft hoods or listed
    for use with Type B vent),
    App. G
  Specification for, 7.2
  Termination
    Chimneys, 7.5.2
    Gas vents, 7.6.2
    Mechanical draft systems,
      7.3.4(e)
    Single-wall metal pipe, 7.7.3
    Through the wall, 7.8

-W-

Wall furnaces, 6.28
  Direct vent (definition), 1.7
  Vented, *see* Vented wall furnaces
**Wall head adapter
  (definition),** 1.7
Wall-type room heaters, 6.24.4
Water heaters, 6.29
  Anti-siphon devices, 6.29.9,
    A.6.29.9
  Automatic instantaneous type,
    6.29.7
  Circulating tank type, 6.29.8
  Definitions, 1.7
  Prohibited installations, 6.29.1
  Venting, 7.2.2(e), 10.1.9(5),
    App. G
Weather conditions, 2.13.2, 3.3.2
  Floor furnaces and, 6.11.10
  Protection against, 3.1.4, A.3.1.4
  Single-wall metal pipe,
    use of, 7.7
Work interruptions, 1.5.4